# Metabolic Regulation and Metabolic Engineering for Biofuel and Biochemical Production

# Metabolic Regulation and Metabolic Engineering for Biofuel and Biochemical Production

**Kazuyuki Shimizu**
Institute for Advanced Biosciences
Keio University, Yamagata
Japan

**CRC Press**
Taylor & Francis Group
Boca Raton  London  New York

CRC Press is an imprint of the
Taylor & Francis Group, an **informa** business
A SCIENCE PUBLISHERS BOOK

CRC Press
Taylor & Francis Group
6000 Broken Sound Parkway NW, Suite 300
Boca Raton, FL 33487-2742

First issued in paperback 2021

© 2017 by Taylor & Francis Group, LLC
CRC Press is an imprint of Taylor & Francis Group, an Informa business

No claim to original U.S. Government works

ISBN-13: 978-0-367-78219-1 (pbk)
ISBN-13: 978-1-4987-6837-5 (hbk)

**Visit the Taylor & Francis Web site at**
**http://www.taylorandfrancis.com**

**and the CRC Press Web site at**
**http://www.crcpress.com**

# Preface

The increasing economic growth and the prosperity has accelerated worldwide with the increasing demand on the energy, mostly generated from fossil fuels since the beginning of the industrial revolution in the mid eighteenth century. This has brought about the rapidly increasing global warming, caused by the emission of the green-house gas such as $CO_2$, resulting in disastrous climate change, where this problem will become crucial even to the level of human survival by the end of this century. It is, therefore, the time to take action to protect further global warming by employing the alternative clean renewable sources of energy. Among the renewable energy sources such as nuclear, solar, wind energies, etc., bioenergy seems to be the most promising alternative source of renewable energy in the long range future.

This book intends to explain the current status and future perspectives for biofuel and biochemical production from biomass, and motivates the innovation for future energy problem. Moreover, attention is focused on the production of biofuels and biochemicals directly from atmospheric $CO_2$ by photosynthetic microorganisms. In particular, it is important to create new approaches that bring innovation or revolution for bio-based energy generation. The center for this is the metabolism of organisms. Although many papers have been published on the production of biofuels and biochemicals by metabolic engineering or synthetic biological approaches, less attention is focused on the metabolic regulation. Understanding the metabolism or metabolic regulation mechanisms in response to pathway modification or genetic modulation is critical for the innovative design of the cell metabolism for efficient biofuel production providing a solution to the future energy problems. Thus the present book intends to explain the metabolic regulation mechanisms including catabolic regulation prior to the metabolic engineering practices. The metabolic engineering practices are subdivided into two categories: one, using heterotrophic bacteria, yeast, and fungi, and second, the photoautotrophic autotrophic microorganisms such as algae and cyanobacteria. Finally, brief explanation is given for the systems biology approach and modeling.

# Contents

# Notations

(General)

| | |
|---|---|
| ABC transporter | ATP-binding cassette transporter |
| CBB cycle | Calvin-Benson-Bassham cycle |
| CBP | consolidated bioprocess |
| CCR | carbon catabolite repression |
| ETC | electron transport chain |
| FFA | free fatty acid |
| PP pathway | pentose phosphate pathway |
| TCA cycle | tri-carboxylic acid |
| PS | photo system |
| PTS | phosphotransferase system |
| ROS | reactive oxygen species |
| SHF | separate hydrolysis and fermentation |
| SSF | simultaneous saccharification and fermentation |

(Metabolite)

| | |
|---|---|
| AA | amino acid |
| AcAcCoA | acetoacetyl coenzyme A |
| ACC | 1-aminocyclopropane 1-carboxylic acid |
| AcCoA | acetyl coenzyme A |
| AcP | acetyl phosphate |
| ADP | adenosine diphosphate |
| ADPG | adenosine 5'-diphospho glucose |
| ALA | amino-levulinic acid |
| Alg | alginine |

| | |
|---|---|
| AMP | adenosine monophosphate |
| Asp | aspartic acid |
| ATP | adenosine triphosphate |
| BDO | butanediol |
| 1,3BPG | 1,3-bisphosphoglycerate |
| CHA | chorismic acid |
| CIT | citrate |
| DAHP | 3-deoxy-D-arabinoheptulosonate 7-phosphate |
| DHA | dihydroxyacetone |
| DHAP | dihydroxyacetone phosphate |
| DHQ | 3-dehydroqunic acid |
| DHS | 5-dehydro shikimic acid |
| DMAPP | demethylalyl diphosphate |
| E4P | erythrose 4-phosphate |
| EPSP | 5-enolpyruvate-shikimate-3-phosphate |
| F6P | fructose 6-phosphate |
| FBP | fructose 1,6-bisphosphate |
| Fru | fructose |
| FUM | fumarate |
| G1P | glucose 1-phosphate |
| G6P | glucose 6-phosphate |
| GABA | ɤ-amino butyric acid |
| GAP | glyceraldehyde 3-phosphate |
| GL3P | glycerol 3-phosphate |
| Glc | glucose |
| Glu | glutamic acid |
| Gln | glutamine |
| GOX | glyoxylate |
| GSH | glutathinon |
| GTP | guanosine triphosphate |
| HA | hydroxyalkanoate |

| | |
|---|---|
| HB | hydroxybutyric acid |
| IA | itaconic acid |
| ICI | isocitrate |
| KDPG | 2-keto-3-deoxy-6-phosphogluconate |
| αKG | α-ketoglutaric acid |
| KMBA | 2-keto-4-methyl-thiobutyric acid |
| Lys | lysine |
| MA | muconic acid |
| MAL | malate |
| 2MB | 2-methyl-1-butanol |
| MEP | 2-C-methyl-D-erythritol 4-phosphate |
| MG | methylglyoxal |
| MQ | menaquinone |
| MVA | mevalonic acid |
| OAA | oxaloacetate |
| OSP | oseltamivir phosphate |
| PDO | propanediol |
| 2PG | 2-phosphoglycerate |
| 3PG | 3-phosphoglycerate |
| PEP | phosphoenol pyruvate |
| PHA | polyhydroxy alkanoate |
| PHB | polyhydroxybutyrate |
| PLA | polylactic acid |
| PMA | plymalic acid |
| 6PG | 6-phosphogluconate |
| PGA | poly-β-1,6-N-acetyl-D-glucosamine |
| PPY | phenylpyruvate |
| PQ | plastoquinone |
| PQQ | pyrroloquinone |
| PUFA | polyunsaturated fatty acid |
| PYR | pyruvate |
| R5P | ribose 5-phosphate |
| R5PI | ribose phosphate isomerase |

| | |
|---|---|
| Ru5P | ribulose 5-phosphate |
| RuBP | ribulose 1,5-bisphosphate |
| SA | shikimic acid |
| S7P | sedoheptulose 7-phosphate |
| Ser | serine |
| SUC | succinate |
| TAG | triacylglycerol |
| Trp | tryptophane |
| Tyr | tyrosine |
| UQ | ubiquinone |
| UQOH | ubiquinol |
| X5P | xylulose 5-phosphate |
| XOS | xylo-oligosaccharide |
| | |
| (Protein) | |
| ACC | acyl-acyl carrier protein |
| Ack | acetate kinase |
| ACP | acyl carrier protein |
| Acs | acetyl coenzyme A synthetase |
| ADC (Adc) | acetoacetate decarboxylase |
| ADH | alcohol dehydrogenase |
| Adk | adenylate kinase |
| ADSL | adenylosuccinate lyase |
| AHL | acyl-homoserine lactone |
| AI | arabinose isomerase |
| Ald | aldolase |
| ALDC | 2-acetolactate decarboxylase |
| ALDH | aldehyde dehydrogenase |
| ALS | acetolactate synthase |
| AOR | aldehyde oxidoreductase |
| ASL | alginino-succinate lyase |
| AspP | adenosine diphosphate sugar pyrophosphatase |

| | |
|---|---|
| AtoB | acetyl transferase |
| BDH | butanediol dehydrogenase |
| CA | carboxylic anhydrase |
| cAMP | cyclic adenosine monophosphate |
| CcpA | catabolite control protein A |
| CimA | citramalate synthase |
| Cra | catabolite repressor/ acticvator |
| Crp | cAMP receptor protein |
| CS | citrate synthase |
| Csr | carbon storage regulator |
| Ctf | CoA transferase |
| Cya | adenylate cyclase |
| DXS | 1-deoxy-D-xylulose 5-phosphate synthase |
| EI | enzyme I |
| EII | enzyme II |
| EFE | ethylene forming enzyme |
| Eno | enolase |
| F1PK | D-fructose-1-phosphate kinase |
| FAA | fumarylacetoacetase |
| FAS | fatty acid synthase |
| FHL | formate hydrogenlyase |
| Fnr | fumarate nitrate reductase |
| Fum | fumarase |
| G6PDH | glucose 6-phosphate dehydrogenase |
| GADC | glutamate decarboxylase |
| GAPDH | glyceraldehydes 3-phosphate dehydrogenase |
| GDH | glutamate dehydrogenase |
| GlgA | glycogen synthase |
| GlgB | glycogen branching enzyme |
| GlgC | ADPG phosphorylase |
| GlgP | glycogen phosphorylase |

| | |
|---|---|
| Glk | glucokinase |
| Glx | glyoxalase |
| GlgX | glycogen debranching enzyme |
| GlpK | glycerol kinase |
| GlyDH | glycerol dehydrogenase |
| GOGAT | Glutamate synthase (glutamine oxoglutarate aminotransferase) |
| GS | glutamine synthetase |
| Hbd | 3-hydoxybutyryl-CoA dehydrogenase |
| HPr | histidine-phosphorylatable protein |
| HPrK | HPr kinase |
| Hxk | hexokinase |
| Hyd | hydrogenase |
| ICDH | isocitrate dehydrogenase |
| Icl | isocitrate lyase |
| IspS | isoprene synthase |
| KDC | ketoacid decarboxylase |
| Kivd | 2-keto-acid-decarboxylase |
| LADH | L-arabitol 4-dehydrogenase |
| LALDH | lactaldehyde dehydrogenase |
| LDH | lactate dehydrogenase |
| Lrp | leusine responsive regulatory protein |
| LXR | L-xylulose reductase |
| MarR | multiple antibiotic resistant regulator |
| MDH | malate dehydrogenase |
| Mez(Mae) | malic enzyme |
| Mgs | methylglyoxal synthase |
| MIOX | *myo*-inositol oxygenase |
| MS | malate synthase |
| Nac | nitrogen assimilation control protein |
| NR | nitrogen regulator |
| Pck | phosphoenolpyruvate caroxykinase |

| | |
|---|---|
| PDC | pyruvate decarboxylase |
| PDH | pyruvate dehydrogenase |
| PDOX | propanediol oxidoreductase |
| Pfk | phosphofructokinase |
| PGDH | 6-phosphogluconate dehydrogenase |
| Pgi | phosphoglucose isomerase |
| Pgm | phosphoglucomutase |
| Pgk | phosphoglycerate kinase |
| PhaA | β-ketothiolase |
| PhsB | acetoacetyl-CoA reductase |
| PhaC | polyhydroxy alkanoate (PHA) synthase |
| PI3K | phosphoinositide 3-kinase |
| Ppc | phosphoenolpyruvate carboxylase |
| Pps | phosphoenolpyruvate synthase |
| PRK | phosphoribulokinase |
| Pta | phosphotransacethylase |
| Pyc | pyruvate carboxylase |
| Pyk | pyruvate kinase |
| RhaD | rhamnose dehydrogenase |
| RK | ribulokinase |
| Rpe | ribulose 5-phosphate 4-epimerase |
| Rpi | ribulose 5-phosphate isomerase |
| RubisCo | ribulose 1,5-bisphosphate carboxylase |
| SADH | secondary alcohol dehydrogenase |
| SDH | succinate dehydrogenase |
| SOD | superoxide dismutase |
| Tal | transaldolase |
| TE | thioesterase |
| Ter | trans-2-enoyl-CoA reductase |
| Thl | thiolase |
| Tkt | transketolase |

| | |
|---|---|
| Udh | urinate dehydrogenase |
| XDH | xylitol dehydrogenase |
| XI | xylulose isomerase |
| XK | xylulokinase |
| XR | xylose reductase |

# 1

# Background

## ABSTRACT

The increasing economic growth and the prosperity has been accelerated worldwide with the increasing demand on the energy mostly generated from fossil fuels since the beginning of the industrial revolution in the mid eighteenth century. This has brought about rapid global warming caused by the emission of the greenhouse gases such as $CO_2$, resulting in the disastrous climate change, and this problem will become endangering even to the level of human survival by the end of this century. It is, therefore, the time to take action to prevent further global warming by employing the alternative clean renewable sources of energy. Among the renewable energy sources such as nuclear, solar, wind energies, etc., bioenergy seems to be the most promising alternative source of renewable energy in the long range future.

The so-called 1st generation biofuels have been produced from corn starch and sugarcane in USA and Brazil. However, this causes a problem of the so-called "food and energy issues" as the production scale increases. The 2nd generation biofuels production from lignocellulosic biomass has thus been paid more attention recently. However, it requires energy intensive pre-treatment for the degradation of lignocellulosic biomass. The 3rd generation biofuels production from photosynthetic organism such as cyanobacteria and algae has also come under attention, but the cell growth rate and thus the productivity of the fuels is significantly low. The typical processes for biofuel and biochemical production from biomass include pre-treatment of biomass, saccharification, fermentation, and separation of the dilute fermentation broth followed by purification.

### Keywords

Global warming, renewable energy, pretreatment, lignocellulosic biomass, biofuel, biochemical, 1st generation biofuel, 2nd generation biofuel, 3rd generation biofuel, low carbon society

## 1. Current Status of Global Warming and Action Plan

The increasing economic growth and the prosperity have been accelerated worldwide with the increasing demand on the energy generated from mostly fossil fuels since the beginning of the industrial revolution in the mid-eighteenth century. This has been promoted by the steam powered trains and ships, and then the internal combustion engines significantly changed the human's life style and production system in relation to transportation and industrial production. Nowadays, many people live comfortable lives spending much electricity for air conditioning and freezing of food, etc. Many people use cars and airplanes to move across the countries and the continents spending more and more fossil fuels as sources of energy. Unlike the beginning of the industrial revolution, where the population of the world was 700 million, the current population is 7 **billion** and is estimated to grow to 9 billion by 2,050, and even about 10 billion by 2,100 (Lee 2011). Unlike the developed countries, the population growth is more eminent in the developing countries like Asia and Africa, where the economic growth is accelerating year by year, caused by promotion of industrialization to catch up to the standard of posh and comfortable living of the developed countries. The **International Energy Agency (IEA)** has projected that the world's energy demand will increase from about 12 billion ton oil equivalents (t.o.e.) in 2,009 to either 18 billion t.o.e. or 17 billion t.o.e. by 2,035 under the current policies or new policies scenarios, respectively (International Energy Agency 2011). In association with this, carbon dioxide ($CO_2$) emissions are expected to increase from 29 giga tons (Gt) per year to 43 Gt/year or 36 Gt/year under the current policy and new policies, respectively. This creates significant climate-change risks in either policies, and we are now facing a risk to even human survival by the end of this century. Namely, we need another industrial revolution for the energy sources to be affordable, accessible, and **sustainable**. Energy efficiency and conservation, as well as decarbonizing the energy sources are essential to the revolution (Chu and Majumdar 2012).

**International Panel on Climate Change (IPCC)** keeps warning the global society on global warming caused by **greenhouse gases (GHGs)** such as $CO_2$ based on the accumulating data and the reliable prediction model. IPCC asks world societies to make decisions to take actions and to invest in the reduction of $CO_2$ emissions caused mostly by human activities. This may be also considered from the point of view of future cost caused by the severe climate change due to global warming. Namely, the global warming may cause serious local climate change as well as the rise in the sea level, which will severely damage the societies worldwide. In fact, every year we are experiencing disastrous climate change, and it seems to grow more and more severe. We have recently experienced disastrous phenomena as extreme heat waves, droughts due to segregated rain falls, wild fires in the broad forest, melting glaciers, etc. These phenomena may have been caused largely by the global warming most likely due to the accumulation of GHGs in the atmosphere, mainly $CO_2$ from the burning of fossil fuels. In 2013, a big news was broadcasted throughout the world warning that the $CO_2$ level in the earth's atmosphere had passed over 400 ppm (parts per million) for the first time in several million years. As shown in Fig. 1, the $CO_2$ level periodically changed in accordance with the repetition of warming and cooling of the earth probably due to the effect of black body irradiation from the surface of the sun since 400,000 years ago. However, the sharp increase in the $CO_2$ level caused by the human activities after industrial revolution becomes eminent, and this is becoming more and more severe.

The international society has recognized the importance of this problem since the late 20th century. The international political activities against climate change began at the Rio Earth summit in 1992, where Rio convention included the adoption of the **United Nations Framework Convention on Climate Change (UNFCCC)**. This convention attempted to take action for stabilizing atmospheric concentrations of **GHGs** to avoid "dangerous anthropogenic interference with climate system." The UNFCCC has currently a near universal membership of 195 parties including both developed and developing countries.

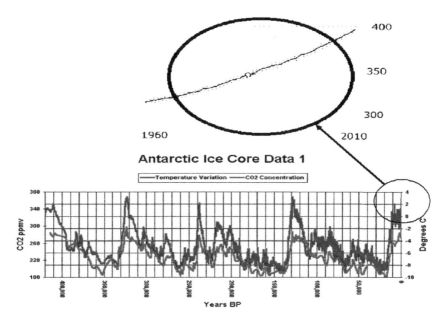

**Figure 1.** Global change in $CO_2$ concentration and the temperature on earth (https://stevengoddard.wordpress.com/2014/09/26/climate-is-not-difficult/, http://cdiac.ornl.gov/trends/co2/recent_mauna_loa_co2.html).

The annual **conference of parties (COP)** began in Berlin in 1995, followed by several conferences such as COP3 in Kyoto, where the **Kyoto Protocol** was adopted, COP11 in Montreal, where the action plan was determined, COP17 in Durban, where the Green Climate Fund was created. Unfortunately, some large $CO_2$ producing countries denied signing the agreement of Kyoto Protocol, and the efficient action against climate change has been unsuccessful. As mentioned above, however, the global warming steadily proceeds, where the glaciers and the iceberg in the north-pole are melting, causing sea level to rise, and causing drastic climatic change worldwide, and the atmosphere is becoming more and more humid due to vaporization of sea water at an unprecedented rate caused by the temperature rise. The international societies strongly recognize the current and future critical situations caused by $CO_2$ emission from fossil fuels.

In 2015, **COP21** (also known as the 2015 **Paris Climate Conference**) aimed to achieve a legally binding and universal agreement on climate, where the conference was attracted about 50,000 participants including 25,000 official delegates from government, intergovernmental organizations, united nation (UN) agencies, non-governmental organizations (NGOs) and civil societies. Then a historic agreement was signed to combat climate change and unleash actions and investment towards a **low carbon**, resilient and sustainable future. Each county must set individual target and must attain it, where the action result will be checked every 5 years, but this is rather effort target without legal constraint, even though this is a big one step towards preventing further global warming. The aim of the universal agreement is to keep the global temperature rise well below 2 degrees Celsius (2°C) during this century, and take strong action to limit the temperature rise even further to 1.5°C above pre-industrial level.

Now the developed countries face the problem of promoting industrialization without or much less emission of $CO_2$ with efficient transportation and production systems, or may have to change the life styles drastically, while the developing countries may have to keep industrialization without or less amount of emission of $CO_2$ with the aid of technology transfer and investment from the developed countries. Namely, green based innovation and revolution is necessary to attain such target and to realize a **low carbon society**. Now is the time to pay attention to innovations that provide an entirely new energy system including transportation and stationary systems together with energy generation systems. It is also of importance to investigate the ways of energy efficiency and the integration of energy sources with electricity transmission, distribution and storage (Chu and Majumdar 2012).

Currently, petroleum-oriented liquid fuels are the main source of energy in the transportation infrastructure throughout the world. The geographical distribution of petroleum resources changes as the new resources such as shale oil field are found and accessed by new technologies for discovery and production. The distribution of petroleum production does not coincide with the place of the demand. This means that fair amount of fuels must be transported spending energy for this purpose. For example, about 2.690 billion tons of oil were consumed, where 1.895 billion tons of crude oil and 0.791 billion tons of refined products crossed national borders (BP Statistical review of the world 2012). This also brings along with itself political problem of national-security for energy, where many countries are forced to import oils from the limited oil producing countries.

The overall loss caused by the increase in the number of heat waves, floods, wildfires, droughts, and storms may be estimated to be over $150 billion per year (Munich Re 2012), where these phenomena may possibly be caused by the climate change (Cambridge University Press 2012). The cost of renewable energy is rapidly becoming competitive with other sources of energy.

## 2. Attempts to Reduce Energy Consumption

Improvements in energy efficiency can contribute for the reduction of fuel usage. One idea is to use light-weight materials for vehicles. Some attempts have been made for ultra-high tensile strength steels, carbon-fiber reinforced composite materials, aluminium and magnesium alloys, and polymers (Gibbs et al. 2012). The potential of reducing the weight of vehicles such as cars and airplanes has already been shown without sacrificing safety, and it is expected to reduce more about 20–40% (Powers 2000). It is estimated that for every 10% weight reduction of the vehicle, 6–8% of improvement is expected in fuel consumption (Holmberg et al. 2012). Moreover, the reduction of the friction loss in vehicles may also contribute for the efficient energy usage, where several attempts have been made for the development of cost effective technologies such as tyres, braking and waste-heat energy recovery, etc. (Holmberg et al. 2012), where Rankin cycle may be utilized to convert waste heat to work by low-cost and high efficiency solid state thermoelectric systems (Yang and Caillat 2006).

Plug-in hybrid electric vehicles (PHEVs) or electric vehicles (EVs) have the opportunity to displace a significant amount of liquid fuel use in transportation, and reduce GHG emission. Such vehicles are becoming competitive, a mass-market car with the internal combustion engine (ICE) using liquid transportation fuel. The high efficiency of fuel-cell-powered electric vehicles may become a potential option in the near future, where there are inherent volumetric energy density issues for hydrogen-gas storage. The supply infrastructure and a low-carbon source of hydrogen are challenged in several countries. The technology advances in shale-gas production may give impact on the transportation system. In addition to the direct use of natural gas as a fuel, low-cost natural gas may give local reforming or hydrogen filling stations in the near future, where delivery to local filling stations may be made through high-pressure gaseous tube trailers in practice. Due to the low cost of natural gas as compared to petrol, natural-gas vehicles may become more widely used, where liquefied natural gas (LNG)-powered trucks and trains may be considered for the long distance transportation (Chu and Majumdar 2012).

## 3. Alternative to Petroleum-based Fuels

The time-scale and the mix of the current energy sources is shown in Fig. 2, where the ratios have been significantly changed after industrial revolution in mid 1800s.

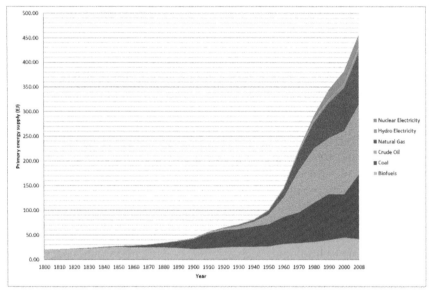

**Figure 2.** Change in energy usage after industrial revolution. World energy consumption by source, based on Vaclav Smil estimates from Energy Transitions: History, Requirements and Prospects together with BP Statistical Data for 1965 and subsequent. https://ourfiniteworld. com/2012/03/12/world-energy-consumption-since-1820-in-charts/.

### 3.1 Nuclear power energy

**Nuclear power** may play an important role in **decarbonizing** the energy production, where the percentage of nuclear energy usage is about 14% of the total electric power generated in 2009 (International Energy Agency

2010), although this percentage dropped to about 12% in 2011, mainly caused by the less usage in Japan due to Fukushima disaster and Germany (BP 2012, Ruhl 2011). After the Fukushima nuclear disaster in 2011, all the nuclear power plants were eventually shut down in Japan. This disaster affected the energy policies of various countries, where some countries keep running the nuclear power plants with careful caution; some have opted to reverse their decision on nuclear energy, or accelerating planned-phase-outs (Chu and Majumdar 2011). In some developing countries, however, the electricity shortage is critical, which prevents the comfortable life and industrialization, and thus the nuclear energy is among the top priority for such countries with the risk of safety. The most significant problem in using nuclear energy is treating the spent fuel before disposal, which is still contaminated with highly radioactive particles. Moreover, the numbers of old nuclear power plants (say more than about 40 years old) are increasing, and the cost of amortizing plants is not low. Hence the humans are yet to learn to control the nuclear energy.

### 3.2 Wind energy

Significant improvement has been made in turbines, blades and gearboxes, and the height of the wind towers to reduce the cost. As the height and size of the wind turbines increase, more and more mechanical stresses are imposed on the gearbox, blades and tower. Direct-drive wind turbines can convert the slow rotary motion directly into alternating current electrical power using electromagnetic generators, which consist of permanent magnets composed of rare-earth metals. The dependence of the materials on the rare-earth gives a problem of unforeseen future availability, since more than about 90% of the rare-earth metals are located in China (US Department of Energy: DOE 2012).

Offshore wind farms are attractive in the sense that they can be placed in near-proximity to cities, and yet far enough away to mitigate local opposition. However, it is not easy to construct a marine structure that can withstand the harsh marine environment for many years, and thus the cost remains high, which prevents its application in practice (Chu and Majumdar 2012).

As the scales and the number of wind turbines increase, the effect of such towers with fans on the aerodynamics, and the effect of high frequency sound on the human health must be carefully analyzed. Such problems could be avoided by installing the wind towers far from the population area, but another problem of energy loss during the long-distance transfer of energy may become eminent.

### 3.3 Solar energy

Although the installation cost for solar panels are becoming lower due to mass production, the apparent limitations are the locations and the areas to be used, and the dependence on the sunny weather. The costs may be further decreased by increasing the efficiency of the solar cells. The ultimate efficiency limit in the conversion of sunlight energy to electric energy is determined by a Carnot heat engine, where the limit is estimated to be about 94% based on the temperature of the sun's black body at 5,800 degree Kelvin (K) and the cell temperature of about 350 K. The Shockley-Queisser limit is a well-known efficiency limit of single-junction solar cells, which suffer from spectrum losses, recombination and black-body radiation, where the limit of the silicon is about 37%, but may actually be low at around 29% in practice (Swanson 2005). This efficiency may be more improved by considering the above theoretical upper bound.

The solar thin films are made of crystalline substances, and polycrystalline-silicon photovoltaics, direct-gap semi-conductors such as cadmium telluride and copper-indium gallium-selenide (CIGS). Cadmium-telluride thin films are used, but there is yet significant room for improvement. Improvement of solar-module efficiency may also come from multi-function cells that capture a large fraction of the solar spectrum, multi-exciton generation, multi-photon absorption or photon up-conversion and light concentration (Chu and Majumdar 2012).

Solar flat panels may be also replaced by the concentrated solar photovoltaics, where the main cost of the system is the mirrors, lenses, and their tracking system rather than solar cell at the focal point of the system (Angel and Olbert 2011).

In order to convert direct current to alternating current electricity and to integrate solar power into the grid, power electronics is crucial (US Department of Energy 2012).

### 3.4 Other sources of renewable energy

**Hydrolytic power energy** has long been employed in particular at mountainous regions, where water fall is necessary to catch the energy from the river. Although this may be important for the protection of flooding by the dam capacity during strong heavy rain, etc., this is limited to the mountainous locations.

**Geothermal energy** is another source of energy associated with volcanos, but the installation of the power plant may dry up the hot springs, and may disturb the geothermal environment.

Another investigated source of energy may be to utilize the **web** and **tidal powers**, but these may be limited to small scale in practice.

### 3.5 Grid energy storage and future perspectives of renewable energy sources

The information technologies may help the social system to change smart cities, and also most of the future infrastructure will be built by considering the opportunity of reducing mobility, thus significantly reducing fuel consumption (Chu and Majumdar 2012).

Grid energy storage and efficiency of the transmission and distribution system may be improved by **grid energy storage**. The mismatch between intermittent electricity supply by various energy sources and the demand in the cities occurs over time and energy scales, and this must be improved by sensing and controlling the grid.

In order to accelerate the use of economically viable renewable and clean-energy, government should be considered to stimulate invention, innovation, and marketing based on the increasing activities of science and engineering (Chu and Majumdar 2012).

Although the current percentage of biomass oriented energy may not be so high, this choice must be carefully analyzed as will be explained in the following chapters. Note that the petroleum is also the biomass oriented energy based on ancient plants.

## 4. Bio-based Energy Generation for the Reduced $CO_2$ Emission

The global carbon cycle has been perturbed by emissions from the combustion of fossil fuels and by changes in land use and land intensity. These perturbations have led to cumulative anthrogenic $CO_2$ emissions of $570 \pm 70$ peta gram carbon since 1750 to 2012 (Ganadell and Schulze 2014). 70% of these emissions originated from the combustion of fossil fuels (Ganadell and Schulze 2014).

According to the data of IEA (OECD 2011), total energy consumption in the world increased more than 78% over the last three decades. Major usage of fossil fuels causes serious environmental problems worldwide, and much attention has been focused on reducing their usage by alternative clean fuels. Namely, due to the global warming problem caused by the increased use of fossil fuels together with limited amount of fossil fuels and the fluctuating cost caused by unstable political disturbances, alternative renewable energy sources have recently been paid much attention (Schmidt and Dauenhauer 2007). In fact, at the present staggering rate of consumption, the world fossil oil reserves will be exhausted in less than 50 years (Rodolfi et al. 2009). Carbon neutral biofuels are needed to replace the petroleum oil which causes global warming caused by the emission of green house gases. Currently, the world consumes about 15 terawatts of energy per year, and only 7.8% of this is derived from renewable energy sources (Jones and Mayfield 2012). Moreover, in comparison with other forms of renewable energy such as wind, tidal, and solar energy, liquid biofuels allow solar energy to be stored, and also to be used directly in existing engines and transport infrastructure (Scott et al. 2010).

Annually, about $5,500 \times 10^{21}$ J of solar energy reaches the Earth's atmosphere (Smil 2005). Photosynthetic organisms including higher plants, microalgae, and cyanobacteria play the crucial roles of capturing solar energy and storing it as chemical energy (Larkum et al. 2012). The amount of solar energy currently captured by arable crops is limited by arable land area (about 3.9% of the Earth's surface area), fresh water (about 1% of global water), nutrient supply and solar energy-to-biomass conversion efficiency (Stephens 2010, Larkum 2010, Zhu et al. 2010). Terrestrial plants capture $121.7 \times 10^9$ metric tons of carbon from the atmosphere each year (Beer et al. 2010) using solar light and $CO_2$ as the energy and carbon sources.

Photosynthesized carbon is then chemically converted to a variety of chemical compounds, and it is attractive to use photosynthetic organisms as green factories for producing carbohydrates, liquid fuels, and pharmaceutical drugs as well as food and feed, thus contributing to the balancing of the atmospheric carbon (Yuan and Grotewold 2015).

The advantages of using photosynthetic microorganisms include the photosynthetic efficiency, location on non-arable land (about 25% of the Earth's surface), and the use of saline and wastewater source (Larkum et al. 2012), where less than 1% of the available solar energy flux is converted into chemical energy by photosynthesis (Overmann and Garcia-Pichel 2006), and much effort has been focused on the enhancement of photosynthetic carbon fixation.

The so-called 1st generation biofuels have been produced from corn starch and sugarcane. However, this causes the problem of the so-called "**food and energy issues**" as the production scale increases. The 2nd generation biofuels production from lignocellulosic biomass has thus been paid recent attention. However, it requires energy intensive pretreatment for the degradation of lignocellulosic biomass (Shimizu 2014). The 3rd generation biofuels production from photosynthetic organisms such as cyanobacteria and algae has also attracted some attention, but the cell growth rate is quite low, and thus the productivity of the metabolites is significantly low (Sheehan 2009).

In brief, the 1st generation biofuels are produced from feed stocks such as corn starch, sugarcane, and sugar beet, etc., where this brings serious problems of food versus fuel issues, and raises several environmental problems such as deforestation and ineffective land utilization. The 2nd generation biofuels are produced from lignocellulosic biomass such as wood, rice straw, corn stover, sugar cane bagasse, switch grass, etc. In this case, lignocellulosic biomass must be subjected to pretreatment to break down the complex structure of lignin and to decrease the fraction of crystalline cellulose by converting to amorphous cellulose. Most of the pretreatment process such as mechanical, heat explosion, and alkali and acid pretreatments are energy intensive with some environmental problem.

In contrast to these cases, the 3rd generation biofuels may be produced by photosynthetic organisms such as algae, where they do not contain lignin and hemicelluloses for supporting the cell structure since they are buoyant. However, most of the carbohydrates are entrapped within the cell wall and pretreatment such as extrusion and mechanical share is required to break down the cell wall. The significant innovation is necessary for the cell growth rate to improve the productivity in this case.

Although this must be kept in mind, the commercial scale production of biofuels will benefit to the societies worldwide. In particular, it is quite attractive in such countries as Southeast Asian, South American and African countries, where mass production of sugar can be made from sugarcane in such countries due to geological location of hot climate throughout the year. If this could be made on commercial scale without competing with food production, it would aid in overcoming the inherent energy shortage in such countries, and may help promoting the related industries as well as contributing to the improvement of the quality of people's life. As such, the technological innovation will make such countries promote industrialization, and assist economic growth without causing environmental problems.

## 5. Biofuel and Biochemical Production from Biomass

In order to avoid the competition with food production, the second generation biofuels produced from lignocellulosic biomass such as sugarcane bagasse, agricultural and forest residues, grasses, oilseed crops, and urban residues become more and more promising (Escamilla-Alvarado et al. 2012). The common feed stocks for biofuel production also include rapeseed, sunflower, switch grass, wheat, peanuts, sesame seeds, and soybean (Quintana et al. 2011). These renewable sources mostly consist of lignocellulose, where it is made-up of cellulose, hemicellulose, and lignin (Lynd et al. 2002). In Brazil, significant amount of sugarcane bagasse is produced in relation to bioethanol production (Goldenberg 2008, Amorim et al. 2011, Hofsetz and Silva 2012), where this bagasse is sometimes used to generate electricity by combustion or as animal feed stock. Yet, in certain cases, it remains stock piled in the field and factories, and its degradation could pose environmental problem (Cardonna et al. 2010). It is thus quite important to utilize lignocelluloses such as sugarcane bagasse as well as sugarcane waste juice for bioenergy production.

Since most microorganisms cannot directly assimilate lignocellulose, this must be first broken down into monosaccharides and fermented for biofuel production. However, the cost for such process is a major hurdle from the commercial application point of view (Stephanopoulos 2007). In particular, the breakdown of lignocellulose into monomeric sugars is the subject of recent intensive research (Galbe and Zacchi 2007, Merino and Cherry 2007, Mazzoli et al. 2012). The lignocellulosic biomass hydrolysates obtained after pretreatment and hydrolysis contain various hexoses and pentoses as well as lignin residues. In the case of barely straw, 34% of cellulose, 23% of hemicelluloses, and 13% of lignin (6% of moisture) are contained, where these can be converted to such monomeric sugars as glucose, xylose, arabinose, etc. by the pretreatment and enzymatic saccharification (Saha and Cotta 2010).

Although there is some variation in the sugar composition of hydrolysates from various feed stocks, glucose and xylose are the major sugars with small amounts of mannose, galactose, arabinose, galacturonic acid and rhamnose, etc. (Jojima et al. 2010). By contrast, hydrolysate from marine biomass such as microalgae mainly contains glucose and galactose. It may be also considered to use some fruits or fruits waste, where glucose and fructose may be the main sugars as well as some organic acids in such a case. Moreover, the recent considerable increase in biodiesel production resulted in an increased co-production of crude glycerol, where this can be also utilized to produce diols, ethanol, butanol, organic acids, polyols, and others (Almeida et al. 2012, Clomburg and Gonzalez 2013).

A variety of host organisms such as bacteria, fungi, and microalgae may be considered for the production of biofuels and biochemicals from $CO_2$ with sunlight. Although photosynthetic organisms offer the ability to produce biofuels and biochemicals directly from $CO_2$ and sunlight, significant innovation is inevitable for the process development in relation to large-scale cultivation, harvesting, and product separation, since the production rate is significantly low.

## 6. Brief Summary and the Outlook of the Book

The present book intends to explain the current status and future perspectives for biofuel and biochemical production from biomass, and motivates the innovation for future energy problem. The biofuels production from lignocellulosic biomass may comprise the following processes: (1) pretreatment or up-stream processes (hydrolysis of cellulose and hemi-cellulose), (2) fermentation processes (conversion of carbon sources obtained by the up-stream processes to biofuels and biochemicals), (3) down-stream processes (separation and purification of the target metabolite) (Fig. 3). Moreover, attention is focused on the biofuel and biochemical production directly from atmospheric $CO_2$ by photosynthetic microorganisms.

In particular, it is important to create new approaches that bring innovation or revolution for bio-based energy generation. The center for this is the metabolism of organisms. However, although many papers have been published for the production of biofuels and biochemical by metabolic engineering or synthetic biology approaches, less attention is focused on the metabolic regulation. Understanding on the metabolism or metabolic regulation mechanisms in response to pathway modification or genetic modulation is critical for the innovative design of the cell metabolism for the efficient biofuel production contributing to the future energy problems. Thus the present book intends to explain about the metabolic regulation mechanisms including catabolic regulation for co-consumption of multiple carbon sources prior to the metabolic engineering practices. The metabolic engineering practices are subdivided into two categories:

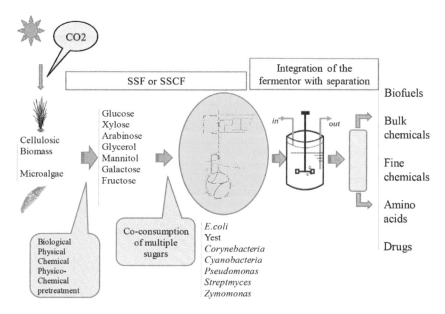

**Figure 3.** The overall process for biofuel and biochemical production from renewable Bioresources.

using heterotrophic bacteria, yeast, and fungi, and the photoautotrophic autotrophic microorganisms such as algae and cyanobacteria. Finally, brief explanation is given for the systems biology approach and modeling before closing.

## References

Almeida, J.R.M., L.C.L. Favaro and B.F. Quirino. 2012. Biodiesel biorefinery: opportunities and challenges for microbial production of fuels and chemicals from glycerol waste. Biotech for Biofuels. 5: 48.

Amorim, H.V., M.L. Lopes, J.V.C. Oliveira, M.S. Buckeridge and G.H. Goldman. 2011. Scientific challenges of ethanol production in Brazil. Appl Microbiol Biotechnol. 91: 1267–1275.

Angel, R. and B.H. Olbert. 2011. Method of manufacturing large dish reflectors for a solar concentrator apparatus. US Patent Application 20120125400.

Beer, C., M. Reichstein, E. Tomelleri, P. Ciais, M. Jung, N. Carvalhais and D. Papale. 2010. Terrestrial gross carbon dioxide uptake: global distribution and covariation with climate. Science. 329: 834–838.

BP. BP Statistical Review of World Energy. http://bp.com/statisticalreview2012. (BP, 2012).

Canadell, J.G. and E.D. Schulze. 2014. Global potential of biospheric carbon management for climate mitigation. Nature Commun. 5: 5282.

Cardonna, C.A., J.A. Quintero and I.C. Paz. 2010. Production of bioethanol from sugarcane bagasse: status and perspectives. Bioresor Technol. 101: 4754–4766.

Chu, S. and A. Majumdar. 2012. Opportunities and challenges for a sustainable energy future. Nature. 488: 294–303.

Clomburg, J.M. and R. Gonzalez. 2013. Anaerobic fermentation of glycerol: a platform for renewable fuels and chemicals. Trends in Biotechnol. 31(1): 20–28.

Escamilla-Alvarado, C., E. Rios-Leal, M.T. Ponce-Noyol and H.M. Poggi-Varaldo. 2012. Gas biofuels from solid substrate hydrogenogenic–methanogenic fermentation of the organic fraction of municipal solid waste. Process Biochem. 47: 1572–1587.

Galbe, M. and G. Zacchi. 2007. Pretreatment of lignocellulosic materials for efficient bioethanol production. Adv Biochem Eng Biotechnol. 108: 41–65.

Gibbs, J., A.A. Pesaran, P.S. Sklad and L.D. Marlino. 2012. pp. 426–444. In: D.S. Ginley and D. Cahen (eds.). Fundamentals of Materials for Energy and Environmental Sustainability. Cambridge Univ. Press.

Goldenberg, J. 2008. The Brazilian biofuels industry. Biotech for Biofules. 1: 1.

Hofsetz, K. and M.A. Silva. 2012. Brazilian sugarcane bagasse: Energy and non-energy consumption. Biomass and Bioeng. 6: 38.

Holmberg, K., P. Anderssona and A. Erdemirb. 2012. Global energy consumption due to friction in passenger cars. Tribol Int. 47: 221–234.

International Energy Agency. 2011. World Energy Outlook. 546–547.

Jojima, T., C.A. Omumasaba, M. Inui and H. Yukawa. 2010. Sugar transporters in efficient utilization of mixed sugar substrates: current knowledge and outlook. Appl Microbiol Biotechnol. 85: 471–480.

Jones, C.S. and S.P. Mayfield. 2012. Algae biofuels: versatility for the future of bioenergy. Curr Opin Biotechnol. 23(3): 346–351.

Larkum, A.W.D. 2010. Limitations and prospects of natural photosynthesis for bioenergy production. Curr Opin Biotechnol. 21: 271–276.

Larkum, A.W.D., I.L. Ross, O. Kruse and B. Hankamer. 2012. Selection, breeding and engineering of microalgae for bioenergy and biofuels production. Trends Biotechnol. 30(4): 198–205.

Lee, R. 2011. The outlook for population growth. Science. 333: 569–573.

Lynd, L.R., P.J. Weimer, W.H. van Zyl and I.S. Pretorius. 2002. Microbial cellulose utilization: fundamentals and biotechnology. Microbiol Mol Biol Rev. 66: 506–577.

Mazzoli, R., C. Lamberti and E. Pessione. 2012. Engineering new metabolic capabilities in bacteria: lessons from recombinant cellulolytic strategies. Trends Biotechnol. 30: 111–119.

Merino, S.T. and J. Cherry. 2007. Progress and challenges in enzyme development for biomass utilization. Adv Biochem Eng Biotechnol. 108: 95–120.

Overmann, J. and F. Garcia-Pichel. 2006. The photosynthetic way of life. pp. 32–85. *In*: M. Dworkin and S. Falkow (eds.). The Procaryotes: A Handbook on the Biology of Bacteria, Vol 2. Springer, New York.

Powers, W.F. 2000. Automotive materials in the 21st century. Adv Mater Process. 157: 38–41.

Quintana, N., F. Van der Kooy, M.D. Van de Rhee, G.P. Voshol and R. Verpoorte. 2011. Renewable energy from Cyanobacteria: energy production optimization by metabolic pathway engineering. Appl Microbiol Biotechnol. 91(3): 471–490.

Rodolfi, L., G. Chini Zittelli, N. Bassi, G. Padovani, N. Biondi, G. Bonini and M.R. Tredici. 2009. Microalgae for oil: Strain selection, induction of lipid synthesis and outdoor mass cultivation in a low-cost photobioreactor. Biotechnol Bioeng. 102(1): 100–112.

Ruhl, C. 2011. Energy in 2011—disruption and continuity BP Statistical Review of World Energy, http://www.bp.com/sectiongenericarticle800.do?categoryId=9037130&contentId=7068669 (BP, 2012).

Saha, B.C. and M.A. Cotta. 2010. Comparison of pretreatment strategies for enzymatic saccharification and fermentation of barley straw to ethanol. Nat Biotechnol. 27(1): 10–16.

Schmidt, L.D. and P.J. Dauenhauer. 2007. Chemical engineering: hybrid routes to biofuels. Nature. 447: 914–915.

Scott, S.A., M.P. Davey, J.S. Dennis, I. Horst, C.J. Howe, D.J. Lea-Smith and A.G. Smith. 2010. Biodiesel from algae: challenges and prospects. Curr Opin Biotechnol. 21(3): 277–286.

Sheehan, J. 2009. Engineering direct conversion of $CO_2$ to biofuel. Nat Biotechnol. 27: 1128–1129.

Shimizu, K. 2014. Biofuels and biochemical production by microbes. Nova Publ. Co.

Smil, V. 2005. Energy at the Crossroads: Global perspectives and uncertainties. MIT press.

Stephanopoulos, G. 2007. Challenges in engineering microbes for biofuels production. Science. 315: 801–804.

Stephens, A.I. 2010. Influence of nitrogen-limitation regime on the production by *Chlorella vulgaris* of lipids for biodiesel feedstocks. Biofuels. 1: 47–58.

Swanson, R.M. 2005. Proc. 31st IEEE Photovoltaic Specialists Conf. 889–894.

Yang, J. and T. Caillat. 2006. Thermoelectric materials for space and automotive power generation. MRS Bull. 31: 224–229.

Yuan, L. and E. Grotewold. 2015. Metabolic engineering to enhance the value of plants as green factories. Metab Eng. 27: 83–91.

Zhu, X.G., S.P. Long and D.R. Ort. 2010. Improving photosynthetic efficiency for greater yield. Annu Rev Plant Biol. 61: 235–261.

# 2

# Pretreatment of Biomass

## ABSTRACT

As compared to fossil fuels, lignocellulosic biomass is an abundant renewable resource geographically evenly available throughout the world, and can be used as the source of energy generation. The first generation biofuel production from corn starch or sugarcane has been industrialized, but this suffers from the so-called energy and food issue. The second generation biofuels production from lignocellulosic biomass has, therefore, been paid much attention. However, the appropriate pretreatment of lignocellulosic biomass is critical for the bio-energy production, where the process of hydrolyzing lignocelluloses to fermentable mono-saccharides is still problematic from the economic feasibility point of view. The degradation of cellulosic biomass into monomeric sugars can be made by biological, physical, chemical, and physiochemical pretreatments. The drawback of the pretreatment of cellulosic biomass is the slow degradation rate and energy intensive. The pretreatment and fermentation processes can be integrated into one process as simultaneous saccharification and fermentation (SSF), thus reducing the feedback inhibition by the sugars produced during saccharification, and the performance may be improved, where the appropriate culture conditions such as culture temperature and pH affect the performance due to different optimal conditions for saccharification and fermentation. The bio-transformations of lignocellulosic biomass to biofuels and biochemicals may be attained basically by three steps such as production of enzymes such as cellulase, etc., hydrolysis of biomass using such enzymes, and fermentation to produce the target metabolite. These three steps may be integrated in a single cell (consolidated biomass processing: CBP). One idea to attain this is to apply metabolic engineering to the cellulolytic organisms. Another approach is to display the enzymes such as amylase and cellulase on the cell surface by cell surface engineering technique, thus attaining SSF by a single cell.

**Keywords**

Lignocellulosic biomass, pretreatment of cellulosic biomass, simultaneous saccharification and fermentation (SSF), cell surface engineering, consolidated biomass processing (CBP)

## 1. Introduction

As mentioned in Chapter 1, the development of alternatives to fossil fuels is an urgent global priority, and in particular, among the renewable alternatives, biologically derived energy (bio-energy) or fuels (bio-fuels) relevant resources are the most promising in considering the long range future. As compared to fossil resources, the feedstock sources for lignocellulose-based biofuel and biochemical production are geographically more evenly distributed throughout the world, enabling the energy security in many countries. Moreover, the lignocellulose-based biofuels may be produced around the location of the feedstock resources, where the distribution of biofuel production may be more coincided with the place of demand, and much less transportation cost is required as compared to the case of fossil fuels. This also implies that the biofuel production from ligno-cellulosic biomass may help rural area to be industrialized and support employment. Biofuels are attractive for the energy content and portability, and also for their compatibility with the current existing petroleum-based production and transportation infrastructure (Hill et al. 2006).

Bioenergy is captured from solar energy via photosynthesis and stored typically in the form of cellulose, hemi-cellulose, and lignin mostly in the plant cell wall, where atmospheric carbon dioxide ($CO_2$) is fixed as a carbon source. Plants can be viewed as solar energy collectors and thermodynamical energy storage system, where this can be accessed via thermodynamical or enzymatic treatment and this is the main feature of the bioenergy generation as compared to other renewable energy production processes (Rubin 2008). Other lignocellulosic biomass such as grasses, wood, and crop residues may also give the potential as the source of bio-fuels, where it is an abundant renewable resource geographically evenly distributed throughout the world, and can be used as the source of energy generation. The energy stored in the lignocellulosic biomass may be taken out by various methods.

Although much attention has been paid to biofuels production from cellulosic biomass based on the optimistic prediction on the economic feasibility and environmental sustainability points of view, the actual production of cellulosic biofuels has been limited so far from the commercial scale application point of view. So far, biofuel industry primarily produces ethanol from corn starch in USA and from sugarcane in Brazil, and limited amount of biodiesel from vegetable oils and animal fats in various countries

(Weber et al. 2010). Although some commercial scale production of biofuels has been attained, these **first generation biofuels**, in particular, ethanol produced from corn starch or sugarcane causes some serious problems by competing with food production, which sometimes gives pressure on increasing the related food prices, the so-called **food and energy issue**. Therefore, a careful decision must be made on the use of the specific biomass. Although this must be kept in mind, the commercial scale production of biofuels will benefit to the societies worldwide. In particular, it is quite attractive in countries such as in African, Asian, and South American countries, where mass production of sugar from sugarcane is possible due to their geological location and hot climate throughout the year. If this could be made on a commercial scale, it would help overcome the inherent energy shortage in such countries, and may help promote the related industries as well as contributing to the improvement of the quality of people's life. As such, the technological innovation will help these countries promote industrialization, employment, and assist economic growth without causing environmental problems. In fact, this can also be considered even in small islands, where energy shortage and unemployment are the main problems.

In order to avoid competition with food production, the **second generation biofuels** produced from lignocellulosic biomass such as sugarcane bagasse, agricultural and forest residues, grasses, oilseed crops, and urban residues have attracted much attention (Escamilla-Alvarado et al. 2011). These renewable sources mostly consist of lignocellulose, where it is made-up of cellulose, hemicellulose, and lignin (Lynd et al. 2002). In Brazil, significant amount of sugarcane bagasse is produced in relation to bioethanol production from sugarcane (Goldemberg 2008, Amorim et al. 2011, Hofsetz and Silva 2012), where such bagasse is sometimes used to generate electricity by combustion or as animal feed stock. Yet, in certain cases, it remains stock piled in the field and factories, and its degradation may pose environmental problem (Cardona et al. 2010). It is thus desirable to convert lignocelluloses such as sugarcane bagasse as well as sugarcane waste juice to biofuels or biochemicals.

Since most microorganisms cannot directly assimilate lignocellulose, this must be first broken down into mono-saccharides for biofuels production by fermentation. However, the cost for such processing is a major hurdle from the commercial application point of view (Stephanopoulos 2007). In particular, the breakdown of lignocellulose into monomeric sugars is the subject of recent intensive research (Galbe and Zacchi 2007, Merino and Cherry 2007, Mazzoli et al. 2012). The lignocellulosic biomass hydrolysates obtained after pretreatment and hydrolysis contains various hexoses and pentoses as well as lignin residues (den Haan et al. 2013). Although there is some variation in the sugar composition of hydrolysates from various feed stocks, glucose and xylose are the major sugars with small amounts of mannose, galactose, arabinose, galacturonic acid and rhamnose, etc. (Jojima et al. 2010). By contrast, the hydrolysate from marine biomass such as microalgae mainly contains glucose and galactose. It may be also considered to use some fruits or fruits waste, where glucose and fructose may be the main sugars as well as some organic acids in such a case.

The typical components of lignocellulosic biomass are cellulose (40–50%), hemi-cellulose (25–30%), and lignin (15–20%) and other components (Knauf and Moniruzzaman 2004). Cellulose is the main structural component of plant cell walls, and is a linear long chain of glucose molecules linked by β-(1,4)-glycosidic bonds (Fig. 1) (van Wyk 2001). Hemicellulose is the second most abundant constituent of lignocellulosic biomass, and is a family of polysaccharides composed of 5- and 6-carbon monosaccharide units such as D-xylose, L-arabinose, D-mannose, D-glucose, D-galactose, and D-glucronic acid (Fig. 1). Lignin is a three dimensional polymer of phenyl-propanoid units such as p-coumaryl alcohol, coniferyl alcohol, and sinapyl alcohol, where this serves as the cellular glue to the plant tissues and fibers for the cell wall to be strong and stiff enough (DelRio et al. 2007), which may be also useful for resistance against insects and pathogens. Because of a complex hydrophobic, cross-linked aromatic polymer, the hydrolysis of lignin is quite difficult in practice.

The appropriate pretreatment of lignocelluloses is critical for the success of bioenergy production from cellulosic biomass. The process of hydrolyzing lignocelluloses to fermentable mono-saccharides is still technically problematic because the digestibility of cellulose is hindered by many factors, where pretreatment is an essential step from the economic feasibility point of view. The main aim of pretreatment is to break down the lignin structure and disrupt the crystalline structure of cellulose for enhancing enzyme accessibility to the cellulose during hydrolysis (Mosier et al. 2005a). In particular, it is strongly desired to remove lignin before or during pretreatment, since lignin preferentially binds to the elements (hydrophobic faces) of cellulose to which the cellulases also preferencially

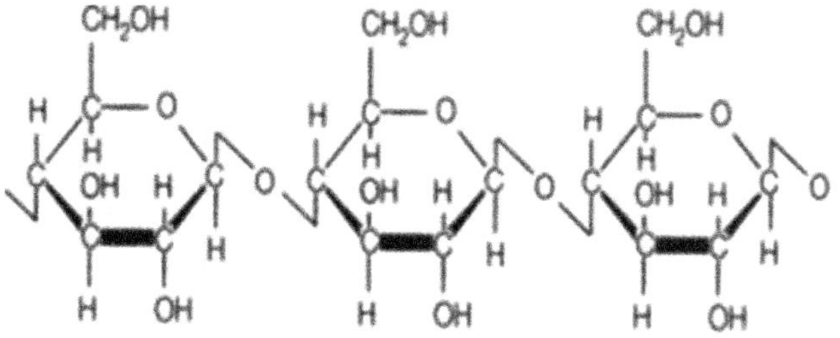

**Figure 1.** Chemical structure of cellulose and hemicelluloses (van Wyk et al. 2001).

binds and also to the specific residues on the cellulose-binding module of the cellulose that are critical for cellulose binding (Vermaas et al. 2015).

So far, a large number of pretreatment approaches have been investigated on a wide variety of feed stocks (Carvalheiro et al. 2008, Hendriks and Zeeman 2009, Taherzadeh and Karimi 2008, Yang and Wyman 2008, Alvira and Tomas-Pejo 2010, Menon and Rao 2012). Since different lignocellulosic biomass has different physiochemical characteristics, the most suitable pretreatment process should be selected among available technologies without causing environmental problem. Moreover, it must be kept in mind that the selected pretreatment process may affect the subsequent processes. In particular, toxic compounds must be minimized or removed before going into fermentation.

The current pretreatment technologies may be classified as biological, physical, chemical, and physiochemical pretreatments (Alvira and Tomas-Pejo 2010, Menon and Rao 2012), and the typical methods are briefly explained in the following sections, followed by the integration of saccharification and fermentation (SSF), and the related consolidated biomass processing (CBP).

## 2. Various Pretreatments

The cellulose chains are packed into microfibrils, where these are attached to each other by hemicelluloses and amorphous polymers of various sugars and pectin, which are covered by lignin layer. The microfibrils are often in the form of bundles or macrofibrils. The molecules of microfibrils in crystalline cellulose are packed so tightly that enzymes and even water molecule cannot enter deep inside. The cellulose is insoluble in water due to high molecular weight and ordered tertiary structure. Some parts of microfibrils are amorphous, where crystalline regions of cellulose are more resistant to degradation, while amorphous regions are a little easier. The isolation and derivatization/dissolutions of cellulose are crucial steps in determining the degree of polymerization of cellulose (Hallac and Ragauskas 2011). In general, plant cell wall may be subdivided into the outermost middle lamella, and the outer primary and inner secondary walls, where the secondary walls contain the major portion of cellulose, while the middle lamella is almost entirely composed of lignin (Fig. 2) (Pandy 2009).

The major obstacles for the economically feasible treatment are the degradation of cellulose and hemicellulose into sugars to be used in the fermentation by removing or modifying lignin. The brief outlook on the pretreatment processes is summarized in Fig. 3.

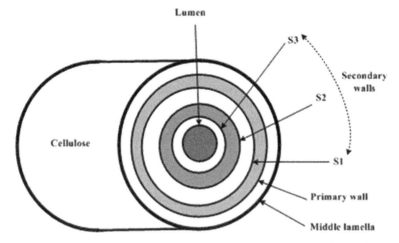

**Figure 2.** Structure of cellulose (Pereira et al., Polímeros vol. 25 no. 1 São Carlos Jan/Feb 2015).

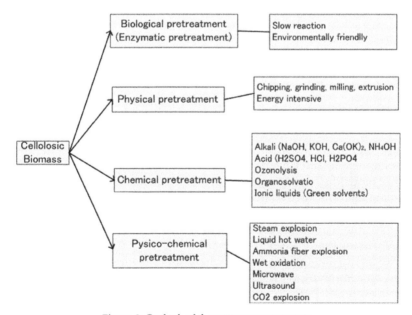

**Figure 3.** Outlook of the pretreatment process.

## 2.1 Physical pretreatment

Most of the ligno-cellulosic biomass resources may require some types of mechanical processing for its breakdown into small sizes. **Mechanical pretreatment** is effective, in particular by a combination of chipping, grinding or milling, irradiation using gamma rays, electron beam, microwave radiations, etc. However, the problem for such processing is relatively high power requirement with high energy requirement, which prevents the practical application (Hendriks and Zeeman 2009). Another **physical pretreatment** is to employ extrusion process, where the raw materials are subjected to heating, mixing and shearing, resulting in physical and chemical modifications during the passage through the extruder, and thus the lignocellulose's structure is disrupted causing defibrillation, fibrillation, and shortening of the fibers (Karunanithy et al. 2008). This pretreatment is effective for the degradation of cell wall of lignocelluloses as well as algae. Main drawback is the high energy requirement, and thus mechanical pretreatment may not be employed for the entire full-scale degradation, but may be employed just for size reduction before applying other pretreatment processes.

## 2.2 Biological pretreatment

As for **biological pretreatment**, fungal treatment has been extensively investigated, where the **enzymatic saccharification** is environmentally friendly and has long received much attention. Biological pretreatments employ such microorganisms as brown, white and soft-rot fungi which degrade lignin and hemicelluloses (Sanchez 2009). Lignin degradation is quite important, and this may be achieved by lignin-degrading enzymes such as peroxidase and laccase (Kumar and Wyman 2009). The biological pretreatment is attractive from such points of view as low-capital cost, low energy, no chemical requirement, and mild environmental condition. However, the main drawback is the low hydrolysis rate (Sun and Cheng 2002). One of the reasons for this is the product inhibition of enzyme activity by the sugars produced, where this may be improved to some extent by combining with fermentation as simultaneous saccharification and fermentation (SSF) as will be explained later.

Several types of cellulase activities are involved in cellulosic hydrolysis based on their structural properties such as endoglucanases or 1,4-β-D-glucan-4-glucanohydrolases (EC 3.2.1.4), exoglucanases including 1,4-β-D-glucan glucanohydrolases (or cellodextrinases) (EC 3.2.1.74), 1,4-β-D-glucan cellobiohydrolases (or cellobiohydrolases) (EC 3.2.1.91), and β-glucosidases or β-glucoside glucohydrolases (EC 3.2.1.21) (Menon and Rao 2012). Enzymes form the effective molecular structure, where cellulolytic systems can be associated into cellulosomes as multi-enzymatic complexes or unassociated as individual enzymes. The unassociated enzymes consist of a catalytic domain for the hydrolysis reaction and of a cellulose-binding domain. These two domains are linked by a sufficiently long and flexible peptide that allows enough orientation and operation. The cellulosomal enzymes are bound non-covalently to the cellulose integrating protein (Carrad et al. 2000).

Carbohydrate-binding modules must have the following properties: (i) a proximity effect, (ii) a target function, and (iii) disruptive function (Menon and Rao 2012). Carbohydrate binding module may be the limiting factor in the hydrolysis of cellulose. Although much knowledge on enzymatic reaction mechanisms has been accumulated, it is not still satisfactory for its application in practice due to (i) inherent complexity and heterogenity of natural cellulose, and (ii) insufficient understanding on the hydrolysis processes (Leschine 1995). Therefore, it is highly desirable to understand the molecular mechanisms for cellulose degradation and develop superior enzymes for the second generation bio-fuel production to become feasible.

It is important to separate lignin component from the celluloses and hemicelluloses for enzymatic hydrolysis. The lignin interferes with the hydrolysis by preventing the access of the cellulases to cellulose and by irreversibly binding to hydrolytic enzymes. Therefore, the removal of lignin is inevitable to improve the hydrolysis rate (McMillan 1994).

The most frequently investigated source of cellulases may be the fungus *Trichoderma reesei*, which produces extracellular cellulase. The problem for this is the low-glucosidase activity, and causes incomplete hydrolysis of cellobiose in the reaction mixture, resulting in serious inhibition of the enzymes (Holtzapple et al. 1992). Moreover, the enzyme inhibition by cellobiose and glucose slows down the hydrolysis rate, which may be overcome by SSF.

Hemicelluloses are multi-domain proteins that contain structurally discrete catalytic and carbohydrate binding domains targeting polysaccharide. Hemicelluloses are either glycoside hydrolases which hydrolyze glycosidic bonds, or carbohydrate esterases which hydrolyze ester linkage of acetate or ferulic acid groups. The two main glycosyl hydrolases depolymerizing the hemicelluloses backbone are end-1,4-β-D-xylanase and endo-1,4-β-D-mannase (Biely 1993). The former cleaves the glycosidic bonds in the xylan backbone. Since xylan is a complex component of hemicelluloses in wood, etc., its complete hydrolysis requires multiple enzymes such as β-xylanase, β-xylosidase, α-L-arabinofuranosidase, α-glucuronidase, acetyl xylan esterase, and hydroxycinnamic acid esterases that cleave side chain residues from the xylan backbone (Biely 1993). These enzymes play the roles for the specific hydrolysis, where for example, xylosidases play essential roles for the breakdown of xylan by hydrolyzing xylooligosaccharides to produce xylose (Deshpande et al. 1986).

D-mannose can be obtained from galactoglucomannans (EC 3.2.1.78) and exoacting β-mannosidases (EC 3.2.1.25) on cleaving the polymer backbone into single sugars. The side chain sugars attached at various points on mannans can be removed by such additional enzymes as β-glucosidases (EC 3.2.1.21), α-galactosidases (EC 3.2.1.22), and acetyl mannan esterases. The typical mannose producing microbes are *Bacillus* sp., *Streptomyces* sp., *Caldibacillus cellulovorans*, etc. (Menon and Rao 2012).

### 2.3 Chemical pretreatment

**Chemical pretreatments** have been extensively used in paper industry for de-lignifying cellular materials to increase the quality of paper products. Such pretreatment techniques may be modified for the effective and inexpensive pretreatment for the degradation of cellulose by removing lignin or hemicelluloses, and the degree of polymerization and crystallinity of the cellulose component (Menon and Rao 2012). As for chemical pretreatments, alkali pretreatments, acid pretreatment, ozonolysis, organosolution, ionic liquids pretreatment have been extensively investigated.

**Alkali pretreatments** increase cellulose digestibility, and in particular, effective for lignin solubilization (Carvalheiro et al. 2008). Sodium, potassium, calcium, and ammonium hydroxides are the suitable alkaline pretreatments. Sodium hydroxide (NaOH) causes swelling, increasing the internal surface of cellulose, and provokes lignin structure disruption (Taherzadeh and Karimi 2008, Kumar et al. 2009). Calcium hydroxide $(Ca(OH)_2)$ also known as lime, has been also extensively investigated, where lime pretreatment removes amorphous substances such as lignin. Lime has been shown to be effective at temperatures 85–150°C for 3–13 hrs with corn stover (Kim and Holtzapple 2006) or poplar wood (Chang et al. 2001). Pretreatment by lime has been shown to be lower cost and less safety requirements as compared to NaOH and potassium hydroxide (KOH) pretreatments, and lime can be easily recovered from hydrolysate by reaction with $CO_2$ (Mosier et al. 2005a). Addition of oxidant agent such as oxygen/$H_2O_2$ to alkaline treatment using NaOH or $Ca(OH)_2$ may contribute to lignin removal (Carvalheiro et al. 2008). Ethanol fermentation may be made by SSF and co-fermentation (SSCF) using *E. coli* from wheat straw pretreated with alkali peroxide (Saha and Cotta 2006). It is attractive, since no inhibitory furfural or HMF may not be formed in the hydrolysates obtained by alkaline peroxide pretreatment (Taherzadeh and Karimi 2008).

Although the hemicellulosic fraction of the biomass can be solvilized by **acid pretreatments**, the concentrated acid pretreatment may not be attractive due to the formation of inhibitory compounds and equipment corrosion problems, together with acid recovery, which require high operational and maintenance costs (Wyman 1996). On the other hand, diluted acid pretreatment may be considered in practice. This may be performed either at high temperature such as 180°C during a short period or at lower temperature such as 120°C for longer retention time of 30–90 min. By acid pretreatment, hemicelluloses, mainly xylan, can be converted to fermentable sugars, but some inhibitory compounds such as furfural, HMF and aromatic lignin degradation compounds may appear (Saha et al. 2005). Dilute sulphuric acid ($H_2SO_4$) may be the most popular acid for pretreatment, and has been commercially used for a variety of biomass such as switch grass (Digman et al. 2010, Li et al. 2010), corn stover (Du et al. 2010, Xu et al. 2009), spruce (softwood) (Shuai et al. 2010), and poplar (Wyman et al. 2009, Kumar et al. 2009). Other acids such as hydrochloric acid (HCl) (Wang et al. 2010), phosphoric acid ($H_3PO_4$) (Zhang et al. 2007, Marzialetti and Olarte 2008), and nitric acid ($HNO_3$) (Himmel et al. 1997) can be also considered (Moseier et al. 2005b). Organic acids such as fumaric acid or maleic acids may be also considered, where less furfural is formed as compared to the case of using $H_2SO_4$ (Kootstra et al. 2009). High throughput pretreatment and enzymatic hydrolysis has been extended to employ dilute acid for the saccharification of switch grass and poplar, etc. (Gao et al. 2012). Acid pretreatment removes hemicelluloses, while alkali pretreatment removes lignin, and thus relatively pure cellulose may be obtained by such pretreatments.

Ozone may be used to increase digestibility of lignocellulosic biomass such as sugarcane bagasse, where **ozonolysis** increased release of fermentable carbohydrates such as glucose and xylose during enzymatic hydrolysis (Travani et al. 2013). Although ozone may be considered for delignification with high efficiency, large amount of ozone is necessary, which is not economical in practice (Sun and Cheng 2002).

**Organosolvation** method may be of interest with the advantage of the recovery of relatively pure lignin as a by-product (Zhao et al. 2009). However, the solvents must be recycled by extraction and separation techniques such as evaporation and condensation, since the solvents are expensive for industrial applications (Sun and Cheng 2002).

The use of **ionic liquids** as solvents for pretreatment of cellulosic biomass has received recent attention (Bokinsky et al. 2011, Olivier-Bourbigou et al. 2010, Quijano et al. 2010, More-Pare et al. 2011, Mehmood et al. 2015), where the thermodynamics and kinetics of reactions carried out in ionic liquids are different from those in conventional molecular solvents. Ionic liquids consist of salts, typically composed of large organic cations and small inorganic anions (Hayes 2009). These factors tend to reduce the lattice energy of the crystalline form of the salt, and hence lower the melting point (Seddon 1998). The ionic liquids are non-volatile, non-flammable, and show high chemical and thermal stabilities (Olivier-Bourbigou et al. 2010, Quijano et al. 2010, Mora-Pere et al. 2011). Since neither toxic nor explosive gases are formed, ionic liquids are called "**green solvents**". Carbohydrates and lignin can be simultaneously dissolved in ionic liquids, where cellulose, hemicelluloses, and lignin are effectively disrupted with minimum formation of degradation product. Although the application has been investigated for such lignocellulosic feed stocks such as straw (Li et al. 2009) and wood (Lee et al. 2009), further investigation is needed for economic feasibility in practice (Alvira et al. 2010). Although ionic liquids are attractive as mentioned above, the following issues must be overcome before going into practical application: high cost of ionic liquids, regeneration requirement, lack of toxicological data and knowledge about basic physico-chemical characteristics, action mode on hemicelluloses and/or lignin contents of lignocellulosic materials and inhibitor issues together with solvent recovery (Menon and Rao 2012). In fact, it is difficult to completely remove ionic liquids after pretreatment, and give toxic or inhibitory effects on microorganisms, plants, and animal cells. For example, the ionic liquids such as 1-ethyl-3-methylimidazolium acetate and 1-ethyl-3-ethyl-3-methylimidazolium methylphosphonate inhibit the respiratory chain, and reduce the oxygen transfer rate (OTR) of *S. cerevisiae* at low concentration of less than 5% (Mehmood et al. 2015).

### *2.4 Physiochemical pretreatment*

The **physiochemical pretreatment** combines both physical and chemical processes, and the typical processes include steam explosion, liquid hot water, ammonia fiber explosion, wet oxidation, microwave pretreatment, ultrasound pretreatment, and $CO_2$ explosion (Alvira et al. 2010, Menon and Rao 2012, Mosier et al. 2005a, Brodeur et al. 2011).

**Steam explosion** may be the most widely employed physiochemical pretreatment for lignocellulosic biomass. It is a hydrothermal pretreatment in which the biomass is subjected to pressurized steam (pressure at 0.69–4.83 MPa and temperature at 160–260°C) for a period from several seconds to minutes, and then suddenly depressurized to atmospheric pressure (Alvira et al. 2010). The mixture of biomass and steam held under certain period may promote the hydrolysis of hemicelluloses due to its high temperature together with final explosive decomposition (Varga et al. 2004, Ruiz and Cara 2006, Kurabi et al. 2005). During steam pretreatment, part of lignocellulose hydrolyzes and forms acids, which may catalyze the hydrolysis of the soluble hemicellulose oligomers (Hendriks and Zeeman 2009). This pretreatment can be combined with mechanical forces and chemical effects due to the hydrolysis of acetyl groups present in hemicelluloses (Alvira et al. 2010). The steam explosion may be applied to sugarcane bagasse to degrade into cellulose, hemicelluloses, and lignin (Varma 2007). Steam explosion has been also shown to be effective for ethanol production from such raw materials as poplar (Oliva et al. 2003), olive residues (Cara et al. 2006), herbaceous residues as corn stover (Varga et al. 2004), and wheat straw (Ballesteros et al. 2006). The main drawbacks of steam explosion pretreatment are the partial hemicellulose degradation, and the generation of some toxic compounds such as furan derivatives like furfural, etc., weak acids, and phenolic compounds (Oliva et al. 2003, Palmqvist and Hahn-Hagerdal 2000).

**Liquid hot water** is another hydrothermal treatment which does not require rapid decomposition and does not employ any catalyst or chemicals. Pressure is applied to maintain water in liquid state at elevated temperature (160–240°C) and provoke alterations in the structure of the lignocelluloses (Alvira et al. 2010). Liquid hot water has been shown to remove up to 80% of the hemicelluloses and enhance the enzymatic digestibility of pretreated materials such as corn stover (Mosier et al. 2005b), and sugarcane bagasse (Perez et al. 2008). Although liquid hot water pretreatments may be attractive in that no catalyst requirement and low-cost reactor construction due to low corrosion potential, water demanding in the process and energetic requirement are higher for the practical application (Alvira et al. 2010).

**Ammonia fiber explosion (AFEX)** is the process in which lignocellulosic biomass is exposed to liquid ammonia (1–2 kg of ammonia/kg of dry biomass) at high temperature (90°C) with residence time of 30 min, and at high pressure, where the pressure is suddenly reduced. AFEX pretreatment can improve the fermentation rate of various herbaceous crops and grasses (Zhang et al. 2009). This pretreatment has been applied for such lignocellulosic biomass as alfalfa, wheat straw, and wheat chaff. The AFEX pretreatment gives de-crystallization of cellulose, partial de-polymerization of hemicelluloses, removal of acetyl groups predominantly on hemicelluloses, and cleavage of lignin, thus increase in accessible surface area (Gollapalli et al. 2002, Kumar et al. 2009). The ammonia used during AFEX process must be recovered and reused due to the cost of ammonia, and also environmental protection point of view (Holtzapple et al. 1992). Although over 90% of cellulose and hemicelluloses could be obtained, lignin removal is not effective, and thus this process may not be applied to such biomass as woods and nutshells, which contain relatively high lignin content (Taherzadeh and Karimi 2008). The success of this pretreatment depends on the cost of ammonia and its recovery after use.

Aqueous ammonia instead of liquid ammonia may be also used as **ammonia recycles percolation** (ARP), where aqueous ammonia (10–15 w%) passes through lignocellulosic biomass at high temperature (150–170°C) with a residence time of 14 min (Galbe and Zacchi 2007). Aqueous ammonia primarily reacts with lignin and causes de-polymerization of lignin cleavage of lignin-carbohydrate linkages. Since ammonia pretreatment does not produce inhibitory materials, a water wash is not necessary in the downstream processing (Jorgensen et al. 2007).

**Microwave pretreatments** are carried out by immersing the biomass in dilute chemical reagent and exposing the slurry to microwave radiation for about 5–20 min (Keshwani 2009). Several alkalis such as sodium hydroxide may be considered as effective reagent (Zhu et al. 2006). The effect of ultrasound on lignocellulosic biomass has also been investigated for extracting hemicellulose, cellulose, and lignin (Sun and Tomkinson 2002), where saccharification of cellulose has been shown to be enhanced efficiently by ultrasonic pretreatment (Yachmenev et al. 2009).

**Carbon dioxide explosion** is also used for lignocellulosic biomass pretreatment. The method is based on the utilization of $CO_2$ as supercritical fluid, which is in a gaseous form but is compressed at temperatures above its critical point to a liquid-like density, where supercritical $CO_2$ has been mostly used as an extraction solvent, but it is considered for non-extractive purpose due to its many advantages (Schacht et al. 2008).

## 3. Simultaneous Saccharification and Fermentation (SSF)

As mentioned above, the cost saving for pretreatment is critical for the feasibility of biofuels production from lignocellulosic biomass. This may be overcome to some extent by combining the pretreatment process with fermentation process by integration. The cellulosic biomass can be converted to biofuels or biochemicals by either **separate hydrolysis and fermentation (SHF)** or **simultaneous saccharification and fermentation (SSF)** of the pretreated biomass. In SHF, the lignocellulosic biomass is hydrolyzed to produce sugars using enzymes in the first step, and the reducing sugars produced in the first step are fermented to produce biofuels or biochemicals in the second step. The advantage of this process is that each step can be operated at its optimal conditions, while the disadvantage is the slow hydrolysis reaction due to the product inhibition of enzymes by the accumulated reducing sugars in the first step, and the investment cost for the two reactors for the first and second steps. In SSF, enzymatic hydrolysis and fermentation are integrated into one fermentor, thus reducing the number of reactors as compared to SHF. In SSF, the released sugars from enzymatic hydrolysis are simultaneously consumed by the microorganism, and therefore, the sugar concentration can be kept low, thus reducing the product inhibition of enzymes, where the product inhibition is one of the major problems in SHF (Galbe and Zacchi 2002, Oloffson et al. 2008a, Gong et al. 2013).

The schematic illustration for SSF is given in Fig. 4, where the major drawback is the difficulty in finding the favorable operating conditions, in particular for pH and temperature, where these must be determined based on the tradeoffs between the enzymatic hydrolysis and the cell growth rate/metabolite production rate. The optimal conditions for pH and temperature can be determined independently in the case of SHF, whereas a compromise must be made in SSF. In particular, it is preferred to operate at higher temperature and lower pH for enzymatic hydrolysis as compared to the

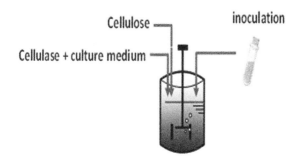

Cellulose

inoculation

Cellulase + culture medium

SSF at lower pH and higher temperature

**Figure 4.** Schematic illustration of SSF.

optimal conditions for the cell growth and metabolite formation. Thus the microbes to be used and the ways of engineering must be determined from the point of view of such characteristics in SSF. For example, the typical optimal temperature is about 30°C for *S. cerevisiae*, while it is about 37°C for *E. coli*, where the appropriate temperature for enzymatic hydrolysis may be more than about 50–60°C.

One of the advantages of using recombinant *E. coli* is the high growth rate, where *E. coli* can grow at low pH down to about 4 or even below to 2–3 and at higher temperature up to about 45°C. In the case of using *E. coli* for SSF, the effects of higher temperature (heat shock) and lower pH (acid shock) on the cell metabolism must be properly analyzed.

Another idea to overcome this problem is to use thermotolerant yeast strains such as *Fabospora fragilis*, *Saccharomyces uvarum*, *Candida brassicae*, *C. lusitaniae*, and *Kluyveromyces marxianus*, where the fermentation temperature can be increased closer to the optimal condition for enzyme reaction (Menon and Rao 2012). Other thermophiles may be also used, where co-culture of *Clostridium thermocellum* and *C. thermolacticum* was considered for ethanol fermentation at 57°C (Xu et al. 2011). The newly discovered *Caldicellulosiruptor* sp. was capable of producing large amount of ethanol from lignocelluloses in fermentation at higher temperature in a single step (Svetlitchnyi 2013).

The digestibility together with delignification of wood chips obtained from tripoid poplar enables SSF (Wang et al. 2013). The fed-batch type of SSF may be considered by appropriately adding enzymes (Hoyer et al. 2010), where a fed-batch separate enzymatic hydrolysis and fermentation gave ethanol titer of 49.5 g/l at a high solids loading (Lu et al. 2010). At high solid loadings (more than 15%), the conventional shaking and stirring is ineffective (Huang et al. 2011), where several gravitational tumbling in roller bottle reactors (Jorgensen et al. 2007, Roche et al. 2009), horizontal rotating shaft with paddlers (Zhang et al. 2010, Ferreira et al. 2010), and stirring with helical impellers (Zhang et al. 2010) have been used for SSF under high solids loadings. In the practical application of SSF to a variety of lignocellulosic biomass, multiple sugars produced during hydrolysis must be co-fermented. The co-consumption of multiple sugars can be attained in SSF in principle, since the sugar concentrations are kept low, and thus the catabolite repression is relaxed.

Simultaneous saccharification of cellulose and hemicelluloses and co-fermentation of multiple reducing sugars such as glucose and xylose (SSCF) has been shown to be effective. Since wild type *S. cerevisiae* cannot assimilate xylose, metabolic engineering must be made for the co-consumption of such multiple sugars to produce ethanol (Oloffson et al. 2010a), where the controlled feeding of cellulases could improve the conversion of xylose to ethanol by SSCF (Oloffson et al. 2010b). Two recombinant xylose-fermenting microorganisms such as *Zymomonas mobilis* and *S. cerevisiae* have been used for the SSCF of waste paper sludge for the production of ethanol with titer over 40 g/l and a yield of 0.39 g ethanol/carbohydrate on paper sludge at 37°C (Zhang and Lynd 2010). The SSCF of pretreated wheat straw using both enzyme and substrate feeding (Oloffson et al. 2008b), and SSCF of corncobs and fed-batch SSCF have been shown to be improved toward practical application of bioethanol production (Koppram et al. 2013).

Moreover, continuous SSCF was shown to be effective for the ethanol production from ammonia fiber expansion (AFEX) pretreated corn stover in a series of continuous stirred tank reactors (CSTRs) using enzyme and *S. cerevisiae* 424A (Jin et al. 2012).

## 4. Consolidated Biomass Processing (CBP)

The bio-transformations of ligno-cellulosic biomass to biofuels or biochemicals may be made in three steps such as production of cellulases and hemicellulases, hydrolysis of cellulose and hemicelluloses to monomeric sugars, and fermentation of such sugars to produce biofuels or biochemicals. These three processes may be integrated in a single cell, where it is called as **consolidated biomass processing (CBP)**. The integration of the three steps in a single cell may save the capital cost and utility cost for the cellulose production, but again the cultivation conditions such as pH and the culture temperature must be carefully determined based on the tradeoffs between hydrolysis rate and the cell growth/metabolite production rate.

One idea of constructing CBP organism is to utilize the organisms that naturally digest lignocelluloses, but may limit the range of metabolite production. It is thus necessary to design the metabolism of the cellulolytic organisms for biofuel and biochemical production by metabolic engineering approach. The CBP organism such as *Clostridium phytofermentans* has been investigated for ethanol production from AFEX-treated corn stover (Jin et al. 2011). The incorporation of heterologous genes for *n*-butanol synthesis into *Clostridium cellulolyticum* enables the *n*-butanol production from crystalline cellulose, where 120 mg/L of *n*-butanol could be produced from cellobiose and crystalline cellulose in 20 days (Gaida et al. 2016).

Noting that it is not easy to maintain the enzyme activities for hydrolysis of cellulosic biomass for a long period in the conventional system in which enzymes are secreted into the medium or immobilized on a solid support, an idea was realized by displaying amylolytic enzymes on the surface of yeast cells (Kondo et al. 2002), indicating that the enzymes are generally self-immobilized on the cell surface so that the enzyme activities could be retained as long as the yeast continues growing (Tanaka and Ueda 2000) (Fig. 5). This so-called **cell surface engineering** has been extensively applied to a wide variety of applications for the simultaneous improvement of saccharification and ethanol production from lignocellulosic biomass (Khaw et al. 2006, Hasunuma et al. 2012, Matano et al. 2012). The direct ethanol production from cellulosic biomass could be also made by co-expressing foreign endoglucanase and β-glucosidase genes on *Zymobacter palmae* (Kojima et al. 2013). The direct ethanol production from cellulosic biomass can be also made by co-expressing foreign endoglucanase and β-glucosidase genes on *Zymobacter palmae* (Kojima et al. 2013). The limiting step is the degradation of lignocelluloses to produce sugars on the cell surface, where the sugars produced are quickly consumed by the cell, and this technique may be combined with the conventional enzymatic treatment, thus reducing the amount of enzymes to be used, since the enzyme is expensive in practice. It may be also considered that the glucose produced on the cell surface is taken up quickly by the cell, and thus the glucose concentration around the cell surface is quite low, indicating that the catabolite repression is relaxed, which enables co-consumption of multiple sugars.

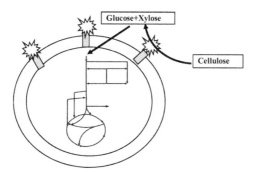

**Figure 5.** Display of cellulase or amylase on the cell surface for consolidated biomass process.

## 5. Concluding Remarks

Lignocellulose based biofuel production involves cultivation and collection of biomass, break down of cell wall polymers into sugars (pretreatment and saccharification), conversion of sugars to biofuels (fermentation), and purification (down-stream processing), where the enzyme hydrolysis step remains as a major techno-economic bottleneck for biofuel production from lignocelluloses.

In either case of SSF or cell surface engineering, the temperature and pH affect the hydrolysis of lignocellulosic biomass, where the higher temperature and lower pH is preferred for this, but the higher temperature and lower pH affect negatively for the cell growth and product formation, and thus it must be careful to determine the microbes to be used and the operating condition as mentioned above. One idea to overcome the constraint on the operating condition in SSF may be the periodic operation, where the optimal condition for enzymatic hydrolysis and that for the cell growth/metabolite production may be alternatively changed with respect to time using on-line computer control. The potential for such kind of dynamic operation has not been fully investigated so far.

## References

Alvira, P.E. and M. Tomas-Pejo. 2010. Pretreatment technologies for an efficient bioethanol production process based on enzymatic hydrolysis: A review. Bioresource Technol. 101: 4851–4861.

Amorim, H.V., M.L. Lopes, J.V.C. Oliveira, M.S. Buckeridge and G.H. Goldman. 2011. Scientific challenges of ethanol production in Brazil. Appl Microbiol Biotechnol. 91: 1267–1275.

Ballesteros, I., M. Negro, J.M. Oliva, A. Cabanas, P. Manzanares and M. Ballesteros. 2006. Ethanol production from steam-explosion pretreated wheat straw. Appl Biochem Biotechnol. 130: 496–508.

Biely, P. 1993. Biochemical aspects of the production of microbial hemicelluloses. *In*: Hemicellulose and Hemicellulase. Cambridge Portland Press.

Bokinsky, G., P.P.P. Peralta-Yayya, A. George, B.M. Holmes, E.J. Steen, J. Dietrich, T.S. Lee, D. Tullman-Ercek, C.A. Voight, B.A. Simmons and J.D. Keasling. 2011. Synthesis of three advanced biofuels from ionic liquid-pretreated switchgrass using engineered *Escherichia coli* PNAS USA. 108(50): 19949–19954.

Brodeur, G., E. Yau, K. Badel, J. Collier, K.B. Ramachandran and S. Ramachandran. 2011. Chemical and physicochemical pretreatment of lignocellulosic biomass. A review. Enzym Res 2011, Article ID 787532, 1–17.

Cara, C., E. Ruiz, I. Ballesteros, M.J. Negro and E. Castro. 2006. Enhanced enzymatic hydrolysis of olive tree wood by steam explosion and alkaline peroxide delignification. Process Biochem. 41: 423–429.

Cardona, C.A., J.A. Quintero and I.C. Paz. 2010. Production of bioethanol from sugarcane bagasse: status and perspectives. Bioresour Technol. 101: 4754–4766.

Carrad, G., A. Koivula, H. Soderlund and P. Beguin. 2000. Cellulose-binding domains promote hydrolysis of different steps on crystalline cellulose. PNAS USA. 97: 10342–10347.

Carvalheiro, F., L.C. Duarte and F.M. Girio. 2008. Hemicellulose biorefineries: a review on biomass pretreatments. J Sci Ind Res. 67: 849–864.

Chang, V.S., M. Nagwani, C.H. Kim and M.T. Holtzapple. 2001. Oxidative lime pretreatment of high-lignin biomass: poplar wood and newspaper. Appl Biochem Biotechnol. 94: 1–28.

DelRio, J.C., G. Marques, J. Rencoret, A.T. Martinez and A. Gutierrez. 2007. Occurrence of naturally acetylated lignin units. J Food Chem. 55: 5461–5468.

den Haan, R., H. Kroukamp, M. Mert, M. Bloom, J.F. Görgens and W.H. van Zyl. 2013. Engineering *Saccharomyces cerevisiae* for next generation ethanol production. J Chem Technol Biotechnol. 88(6): 983–991.

Deshpande, V., A. Lachke, C. Mishra, S. Keskar and M. Rao. 1986. Mode of action and properties of xylanase and L-xylosidase from Neurospora crassa. Biotechnol Bioeng. 26: 1832–1837.

Digman, M.F., K.J. Shinners, M.D. Casler, B.S. Dien, R.D. Hatfield, H.-J.G. Jung et al. 2010. Optimizing on-farm pretreatment of perennial grasses for fuel ethanol production. Bioresour Technol. 101: 5305–5314.

Du, B., L.N. Sharma, C. Becker, S.-F. Chen, R.A. Mowery, G.P. Walsum et al. 2010. Effect of varying feedstock-pretreatment chemistry combinations on the formation and accumulation of potentially inhibitory degradation products in biomass hydrolysates. Biotechnol Bioeng. 107: 430–440.

Escamilla-Alvarado, C., J.A. Vazquez-Barragán, M.T. Ponce-Noyola, H.M. Poggi-Varaldo. 2011. A novel biorefinery approach for biofuels and holocelullolytic enzymes production from organic wastes. In: G.B. Wickramanayake and H. Rectanus (eds.). Bioremediation and Sustainable Environmental Technologies-2011. Book in CD-ROM. ISBN 978-0-9819730-4-3

Ferreira, F.J.T.E., M.V. Cistelecan and A.T. de Almeida. 2010. Voltage Unbalance Impact on the Performance of Line-Start Permanent-Magnet Synchronous Motors. pp. 123–136, Proc. of the 6th Int Conf. eemods '09: Energy efficiency in motor driven systems, Nantes, France, Sept. 14–17.

Gaida, S.M., A. Liedtke, A.H.W. Jentges, D. Engels and S. Jennewein. 2016. Metabolic engineering of *Clostridium cellulolyticum* for the production of *n*-butanol fron crystalline cellulose. Microb Cell Fact. 15: 6.

Galbe, M. and G. Zacchi. 2002. A review of the production of ethanol from softwood. Appl Microbiol Biotechnol. 59: 618–628.

Galbe, M. and G. Zacchi. 2007. Pretreatment of lignocellulosic materials for efficient bioethanol production. Adv Biochem Eng Biotechnol. 108: 41–65.

Gao, X., R. Kumar, J.D. DeMartini, H. Li and C.E. Wyman. 2012. Application of high throughput pretreatment and co-hydrolysis system to thermochemical pretreatment. Part 1: Dilute acid. Biotechnol Bioeng. 110(3): 754–762.

Goldenberg, J. 2008. The Brazilian biofuels industry. Biotech for Biofules. 1: 1.

Gollapalli, L.E., B.E. Dale and D.M. Rivers. 2002. Predicting digestibility of ammonia fibre explosion (AFEX) treated rice straw. Appl Biochem Biotechnol. 100: 23–35.

Gong, Z., H. Shen, Q. Wang, X. Yang, H. Xie and Z.K. Zhao. 2013. Efficient conversion of biomass into lipids by using the simultaneous saccharification and enhanced lipid production process. Biotechnol Biofuels. 6: 36.

Hallac, B.B. and A.J. Ragauskas. 2011. Analyzing cellulose degree of pomerization and its relevancy to cullulosic ethanol. Biofuels Bioprodu Bioref. 5: 215–225.

Hasunuma, T., F. Okazaki, N. Okai, K.Y. Hara, J. Ishii and A. Kondo. 2012. A review of enzymes and microbes for lignocellulosic biorefinery and the possibility of their application to consolidated bioprocessing technology. Bioresour Technol. in press.

Hayes, D.J. 2009. An examination of biorefining processes, catalysts and challenges. Catal Today. 145: 138–151.

Hendriks, A.T.W.M. and G. Zeeman. 2009. Pretreatments to enhance the digestibility of lignocellulosic biomass. Bioresour Technol. 100: 10–18.

Hill, J., E. Nelson, D. Tilman, S. Polasky and D. Tiffany. 2006. Environmental, economic, and energy costs and benefits of biodiesel and ethanol biofuels. PNAS USA. 103: 11206–11210.

Himmel, M.E., W.S. Adney, J.O. Baker, R. Elander, J.D. McMillan, R.A. Nieves et al. 1997. Advanced bioethanol production technologies: a perspective. Fuels Chem Biomass. 666: 2–45.

Hofsetz, K. and M.A. Silva. 2012. Brazilian sugarcane bagasse: Energy and non-energy consumption. Biomass Bioenergy. 46: 564–573.

Holtzapple, M.T., J.E. Lundeen and R. Sturgis. 1992. Pretreatment of lignocellulosic municipal solid waste by ammonia fiber expansion (AFEX). Appl Biochem Biotechnol. 34: 5–21.

Hoyer, K., M. Galbe and G. Zacchi. 2010. Effects of enzyme feeding strategy on ethanol yield in fed-batch simultaneous saccharification and fermentation of spruce at high dry matter. Biotechnol Biofuels. 3: 14–25.

Huang, R., R. Su, W. Qi and Z. He. 2011. Bioconversion of lignocelluloses into bioethanol: process intensification and mechanism research. Bioenerg Res. 4(4): 225–245.

Jin, M., C. Gunawan, V. Balan, X. Yu and B.E. Dale. 2012. Continuous SSCF of AFEX™ pretreated corn stover for enhanced ethanol productivity using commercial enzymes and *Saccharomyces cerevisiae* 424A (LNH-ST). Biotechnol Bioeng. 110(5): 1302–1311.

Jin, M., V. Balan, C. Gunawan and B.E. Dale. 2011. Consolidated bioprocessing (CBP) performance of *Clostridium phytofermentans* on AFEX-treated corn stover for ethanol production. Biotechnol Bioeng. 108: 1290–1297.

Jojima, T., C.A. Omumasaba, M. Inui and H. Yukawa. 2010. Sugar transporters in efficient utilization of mixed sugar substrates: current knowledge and outlook. Appl Microbiol Biotechnol. 85: 471–480.

Jorgensen, H., J. Vibe-Pedersen, J. Larsen and C. Felby. 2007. Liquifaction of lignocelluloses at high solid concentrations. Biotechnol Bioeng. 96: 862–870.

Jorgensen, H., J.B. Kristensen and C. Felby. 2007. Enzymatic conversion of lignocelluloses into fermentable sugars: challenges and opportunities. Biofuels Bioprod Bioref. 1: 119–134.

Karunanithy, C., K. Muthukumarappan and J.L. Julson. 2008. Influence of high shear bioreactor parameters on carbohydrate release from different biomasses. American Soc. of Agricultural and Biological Engineers Annual International Meeting, ASABE 084114. ASABE, St. Joseph, Mich.

Keshwani, D.R. 2009. Microwave Pretreatment of Switchgrass for Bioethanol Production. Thesis Dissertation. North Carolina State University.

Khaw, T.S., Y. Katakura, J. Koh, A. Kondo, M. Ueda and S. Shioya. 2006. Evaluation of performance of different surface-engineering yeast strains for direct ethanol from raw starch. Appl Microbiol Biotechnol. 70: 573–579.

Kim, S. and M.T. Holtzapple. 2006. Delignification kinetics of corn stover in lime pretreatment. Bioresour Technol. 97: 778–785.

Knauf, M. and M. Moniruzzaman. 2004. Lignocellulosic biomass processing: a perspective. Int Sugar J. 106: 147–150.

Kojima, M., K. Okamoto and H. Yanase. 2013. Direct ethanol production from cellulosic materials by *Zymobacter palmae* carrying *Cellulomonas* endoglucanase and *Ruminococcus* β-glucosidase genes. Appl Microbiol Biotechnol. 97(11): 5137–5147.

Kondo, A., H. Shigeuchi, M. Abe, K. Uyama, T. Matsumoto, S. Takahashi, M. Ueda, A. Tanaka, M. Kishimoto and H. Fukuda. 2002. High-level ethanol production from starch by a flocculent *Saccharomyces cerevisiae* strain displaying cell-surface glucoamylase. Appl Microbiol Biotechnol. 58: 291–296.

Kootstra, A.M.J., H.H. Beeftink, E.L. Scott and J.P.M. Sanders. 2009. Comparison of dilute mineral and organic acid pretreatment for enzymatic hydrolysis of wheat straw. Biochem Eng J. 46: 126–131.

Koppram, R., F. Nielsen, E. Albers, A. Lambert, S. Wannstrom, L. Welin, G. Zacchi and L. Olsson. 2013. Simultaneous saccharification and co-fermentation for bioethanol production using corncobs at lab, PDU and demo scale. Biotechnol for Biofuels. 6: 2.

Kumar, P., D.M. Barrett, M.J. Delwiche and P. Stroeve. 2009. Methods for pretreatment of lignocellulosic biomass for efficient hydrolysis and biofuel production. Ind Eng Chem Res. 48: 37–39.

Kumar, R. and C.E. Wyman. 2009. Effects of cellulase and xylanase enzymes on the deconstruction of solids from pretreatment of poplar by leading technologies. Biotechnol Prog. 25: 302–314.

Kumar, R. and C.E. Wyman. 2009. Access of cellulose to cellulose and lignin for popular solids produced by leading pretreatment technologies. Biotechnol Prog. 25: 807–819.

Kurabi, A., A. Berlin, N. Gilkes, D. Kilburn, R. Bura, J. Robinson et al. 2005. Enzymatic hydrolysis of steam-exploded and ethanol organosolv-pretreated DoughlaspFirby novel and commercial fungal cellulases. Appl Biochem Biotechnol. 121: 219–230.

Lee, S.H., T.V. Doherty, R.J. Linhardt and J.S. Dordick. 2009. Ionic liquid-mediated selective extraction of lignin from wood leading to enhanced enzymatic cellulose hydrolysis. Biotechnol Bioeng. 102: 1368–1376.

Leschine, S.B. 1995. Cellulose degradation in anaerobic environment. Ann Rev Microbiol. 49: 399–426.

Li, C., B. Knierim, C. Manisseri, R. Arora, H.V. Schller, M. Auer et al. 2010. Comparison of dilute acid and ionic liquid pretreatment of switchgrass: biomass recalcitrance, delignification, and enzymatic saccharification. Bioresour Technol. 101: 4900–4906.

Li, Q., Y.C. He, M. Xian, G. Jun, X. Xu, J.M. Yang and L.Z. Li. 2009. Improving enzymatic hydrolysis of wheat straw using ionic liquid 1-ethyl-3-methyl imidazolium diethyl phosphate pretreatment. Bioresour Technol. 100: 3570–3575.

Lu, Y.F., Y.H. Wang, G.Q. Xu, J. Chu, Y.P. Zhuang and S.L. Zhang. 2010. Influence of high solid concentrations on enzymatic hydrolysis and fermentation of steam-exploded corn stover biomass. Appl Biochem Biotechnol. 160: 360–369.

Lynd, L.R., P.J. Weimer, W.H. van Zyl and I.S. Pretorius. 2002. Microbial cellulose utilization: fundamentals and biotechnology. Microbiol Mol Biol Rev. 66: 506–577.

Marzialetti, T., M.B.V. Olarte, C. Sievers, T.J.C. Hoskins, P.K. Agrawal and C.W. Jones. 2008. Dilute acid hydrolysis of loblolly pine: a comprehensive approach. I & EC Res. 47: 7131–7140.

Matano, Y., T. Hasunuma and A. Kondo. 2012. Simultaneous improvement of saccharification and ethanol production from crystalline cellulose by alleviation of irreversible adsorption of cellulose with a cell surface-engineered yeast strain. Appl Microbiol Biotechnol. 97: 2231–2237.

Mazzoli, R., C. Lamberti and E. Pessione. 2012. Engineering new metabolic capabilities in bacteria: lessons from recombinant cellulolytic strategies. Trends Biotechnol. 30: 111–119.

McMillan, J.D. 1994. Enzymatic conversion of biomass for fuels production. Washington DC: American Chem Society. ACS Symposium Series, Vol. 566, Chapter 15, pp. 292–324.

Mehmood, N., E. Husson, C. Jacquard, S. Wewetzer, J. Büchs, C. Sarazin and I. Gosselin. 2015. Impact of two ionic liquids, 1-ethyl-3-methylimidazolium acetate and 1-ethyl-3-methylimidazolium methylphosphonate, on *Saccharomyces cerevisiae*: metabolic, physiologic, and morphological investigations. Biotechnol for Biofuels. 8: 17.

Menon, V. and M. Rao. 2012. Trends in bioconversion of lignocellulose: biofuels, platform chemicals & bio-refinery concept. Progress in Energy and Combustion Science. 38: 522–550.

Merino, S.T. and J. Cherry. 2007. Progress and challenges in enzyme development for biomass utilization. Adv Biochem Eng Biotechnol. 108: 95–120.

Mora-Pale, M., L. Meli, T.V. Doherty, R.J. Linhardt and J.S. Dordick. 2011. Room temperature ionic liquids as emerging solvents for the pretreatment of lignocellulosic biomass. Biotechnol Bioeng. 108: 1229.

Mosier, N., C.E. Wyman, B.E. Dale, R.T. Elander, Y.Y. Lee, M. Holtzapple and C.M. Ladisch. 2005a. Features of promising technologies for pretreatment of lignocellulosic biomass. Bioresour Technol. 96: 673–686.

Mosier, N., R. Hendrickson, N. Ho, M. Sedlak and M.R. Ladisch. 2005b. Optimization of pH controlled liquid hot water pretreatment of corn stover. Bioresour Technol. 96: 1986–1993.

Oliva, J.M., F. Saez, I. Ballesteros, A. Gonzalez, M.J. Negro, P. Manzanares and M. Ballesteros. 2003. Effect of lignocellulosic degradation compounds from steam explosion pretreatment on ethanol fermentation by thermotolerant yeast Kluyveromyces marxianus. Appl Microbiol Biotechnol. 105: 141–154.

Olivier-Bourbigou, H., L. Magna and D. Morvan. 2010. Ionic liquids and catalysis: recent progress from knowledge to applications. Appl Catal A. 373: 1.

Olofsson, K., M. Bertilsson and G. Liden. 2008a. A short review on SSF-an interesting process option for ethanol production from lignocellulosic feedstocks. Biotechnol for Biofuels. 1: 7.

Oloffson, K., A. Rudolf and G. Liden. 2008b. Designing simultaneous saccharification and fermentation for improved xylose conversion by a recombinant strain of *Saccharomyces cerevisiae*. J Biotechnol. 134: 112–120.

Oloffson, K., B. Palmqvist and G. Liden. 2010a. Improving simultaneous saccharification and co-fermentation of pretreated wheat straw using both enzyme and substrate feeding. Biotechnol Biofuels. 3: 17.

Oloffson, K., M. Wiman and G. Liden. 2010b. Controlled feeding of cellulases improves conversion of xylose in simultaneous saccharification and co-fermentation for bioethanol production. J Biotechnol. 145: 186–175.

Palmqvist, E. and B. Hahn-Hagerdal. 2000. Fermentation of lignocellulosic hydrolysates II: inhibitors and mechanism of inhibition. Bioresour Technol. 74: 25–33.

Pandy, A. 2009. Handbook of Plant-based Biofuels. CRC Press. Florida.

Perez, J.A., I. Ballesteros, M. Ballesteros, F. Saez, M.J. Negro and P. Manzanares. 2008. Optimizing liquid hot water pretreatment conditions to enhance sugar recovery from wheat straw for fuel-ethanol production. Fuel. 87: 3640–3647.

Quijano, G., A. Couvert and A. Amrane. 2010. Ionic liquids: applications and future trends in bioreactor technology. Bioresour Technol. 101: 8923.

Roche, C.M., C.J. Dibble and J.J. Stickel. 2009. Laboratory scale method for enzyme saccharification of lignocellulosic biomass at high solids loadings. Biotechnol Biofuels. 2: 8.

Rubin, E.M. 2008. Genetics of cellulosic biofuels. Nature. 454: 841–845.

Ruiz, Z.E., C. Cara, M. Ballesteros, P. Manzanares, I. Ballesteros and E. Castro. 2006. Ethanol production from pretreated Olive tree wood and sunflower stalks by an SSF process. Appl Microbiol Biotechnol. 129: 631–643.

Saha, B.C., L.B. Iten, M.A. Cotta and Y.V. Wu. 2005. Dilute acid pretreatment, enzymatic saccharification and fermentation of wheat straw to ethanol. Process Biochem. 40: 3693–3700.

Saha, B.C. and M.A. Cotta. 2006. Ethanol production from alkaline peroxide pretreated enzymatically saccharified wheat straw. Biotechnol Prog. 22: 449–453.

Sanchez, C. 2009. Lignocellulosic residues: biodegradation and bioconversion by fungi. Biotechnol Adv. 27: 185–194.

Schacht, C., C. Zetzl and G. Brunner. 2008. From plant materials to ethanol by means of supercritical fluid technology. J Supercrit Fluids. 46: 299–321.

Seddon, K.R. 1998. In Molten Salt Forum *In*: H. Wendt (ed.). Proc. of 5th International Conference on Molten Salt Chemistry and Technology. 5-6: 53–62.

Shuai, L., Q. Yang, J.Y. Zhu, F.C. Lu, R.J. Weimer, J. Ralph et al. 2010. Comparative study of SPORL and dilute-acid pretreatments of spruce of cellulosic ethanol production. Bioresour Technol. 101: 3106–3114.

Stephanopoulos, G. 2007. Challenges in engineering microbes for biofuels production. Science. 315: 801–804.

Sun, R.C. and R.C. Tomkinson. 2002. Characterization of hemicelluloses obtained by classical and ultrasonically assisted extractions from wheat straw. Carbohydr Polym. 50(3): 263–271.

Sun, Y. and J. Cheng. 2002. Hydrolysis of lignocellulosic materials for ethanol production: a review. Bioresour Technol. 83: 1–11.

Svetlitchnyi, V.A., O. Kensch, D.A. Falkenhan, S.G. Korseska, N. Lippert, M. Prinz, J. Sassi, A. Schickor and S. Curvers. 2013. Single-step ethanol production from lignocelluloses using novel extremely thermophilic bacteria. Biotechnol for Biofuels. 6: 31.

Taherzadeh, M.J. and K. Karimi. 2008. Pretreatment of lignocellulosic wastes to improve ethanol and biogas production: a review. Int J Mol Sci. 1621–1651.

Tanaka, K. and M. Ueda. 2000. Cell surface engineering of yeast: construction of arming yeast with biocatalyst. J Biosci Bioeng. 90: 125–136.

Travaini, R., M.D.M. Otero, M. Coca, R. Da-Silva and S. Bolado. 2013. Sugarcane bagasse ozonolysis pretreatment: effect on enzymatic digestibility and inhibitory compound formation, in press.

van Wyk, J.P. 2001. Biotechnology and the utilization of bio-waste as a resource for bio-product development. Trends Biotechnol. 19: 172–177.

Varga, E., K. Reczey and G. Zacchi. 2004. Optimization of steam pretreatment of corn stover to enhance enzymatic digestibility. Appl Biochem Biotechnol. 113-116: 509–523.

Varma, A.J. 2007. Indian patent application. 1893/DEL/2007. Dated 27.08.07.

Vermaas, J.V., L. Petridis, X. Qi, R. Schulz, B. Lindner and J.C. Smith. 2015. Mechanism of lignin inhibition of enzymatic biomass deconstruction. Biotechnol for Biofuels. 8: 217.

Wang, K., H. Yang, W. Wang and R. Sun. 2013. Structural evaluation and bioethanol production by simultaneous saccharification and fermentation with biodegraded tripoid poplar. Biotechnol Biofuels. 6: 42.

Wang, H., J. Wang, Z. Fang, X. Wang and H. Bu. 2010. Enhanced bio-hydrogen production by anaerobic fermentation of apple pomace with enzyme hydrolysis. Int J Hydrogen Energy. 35: 8303–8309.

Weber, C., A. Farwick, F. Benish, D. Brat, H. Dietz, T. Subtil and E. Boles. 2010. Trends and challenges in the microbial production of lignocellulosic bioalcohol fuels. Appl Microbiol Biotechnol. 87: 1303–1315.

Wyman, C.E. 1996. Handbook on Bioethanol: Production and Utilization. Taylor Francis, Washington, p. 417.

Wyman, C.E., B.E. Dale, R.T. Elander, M. Holtzapple, M.R. Ladisch, Y.Y. Lee et al. 2009. Comparative sugar recovery and fermentation data following pretreatment of poplar wood by leading technologies. Biotechnol Prog. 25: 333–339.

Xu, L. and U. Tschirner. 2011. Improved ethanol production from various carbohydrates through anaerobic thermophilic co-culture. Bioresour Technol. 102: 10065–10071.

Xu, J., M.H. Thomsen and A.B. Thomsen. 2010. Pretreatment on corn stover with low concentration of formic acid. J Microbiol Biotechnol. 19: 845–850.

Yachmenev, V., B. Condon, T. Klasson and A. Lambert. 2009. Acceleration of the enzymatic hydrolysis of corn stover and sugar cane bagasse celluloses by low intensity uniform ultrasound. J Biobased Mater Bioenergy. 3: 25–31.

Yang, B. and C.E. Wyman. 2008. Pretreatment: the key to unlocking low-cost cellulosic ethanol. Biofuels Bioprod Bior. 2: 26–40.

Zhang, J. and L.R. Lynd. 2010. Ethanol production from paper sludge by simultaneous saccharification and co-fermentation using recombinant xylose-fermenting microorganisms. Biotechnol Bioeng. 107: 235–244.

Zhang, Y., Z. Pan and R. Zhang. 2009. Overview of biomass pretreatment for cellulosic ethanol production. Int J Agric Biol Emg. 2: 51–68.

Zhang, Y.H.P., S.Y. Ding, J.R. Mielenz, J.-B. Cui, R.T. Elander, M. Laser et al. 2007. Fractionating recalcitrant lignocelluloses at modest reaction conditions. Biotechnol Bioeng. 97: 214–223.

Zhao, X., K. Cheng and D. Liu. 2009. Organosolv pretreatment of lignocellulosic biomass for enzymatic hydrolysis. Appl Microbiol Biotechnol. 82: 815–827.

Zhu, S., Y. Wu, Z. Yu, C. Wang, F. Yu, S. Jin, Y. Ding, R. Chi, J. Liao and Y. Zhang. 2006. Comparison of three microwave/chemical pretreatment processes for enzymaric hydrolyisis of rice straw. Biosyst Eng. 93: 279–283.

3

# Transport of Nutrients and Carbon Catabolite Repression for the Selective Carbon Sources

## ABSTRACT

It is important to understand the metabolism and its regulation for the proper design of the cell systems for the efficient production of biofuels and biochemicals. Here, the brief explanation is given for transport of carbohydrates and carbon catabolite repression (CCR). Various carbohydrates are transported through outer membrane porin proteins, where the two-component signal transduction system composed of sensor protein and the response regulator plays important roles. The typical porins are OmpC and OmpF in *Escherichia coli*, and the porin genes are typically controlled by such two-component system as EnvZ-OmpR. The carbohydrates transported into the periplasm are internalized into the cytosol by various transport mechanisms, where the typical group translocation is made by the phosphotransferase system (PTS) in bacteria and some archaea. The PTS is related to the catabolite repression for the selective assimilation of carbohydrates.

**Keywords**

Nutrient uptake, porin, two-component system, carbohydrate uptake, transporter, phosphotransferase system, PTS, EnvZ-OmpR, carbon catabolite repression, CCR

## 1. Introduction

Although vast amount of metabolic engineering practices have been attempted to improve the specific target metabolite production by gene manipulation, the resultant phenotypes are often suboptimal giving marginal improvement, and sometimes unsatisfactory due to distant effects of gene modifications or unknown regulatory interactions. It is, therefore, important to understand and get deep insight into the metabolic regulation mechanisms to clarify the reason why the expected outcomes cannot be attained. Namely, before going into or in parallel with the metabolic design of the cell system for the efficient biofuel and biochemical production, it is inevitable to understand the metabolism and its regulation mechanisms of the organisms of concern.

In living organisms or cells, thousands of biochemical reactions as well as transport processes are linked together to break down various nutrients to generate energy and to synthesize cellular constituents, where the enzyme reactions are not static, but change dynamically in response to the change in the growth environment. Moreover, among possible topological networks, only a subset is active at any given point in time and growth condition. Complex signaling networks interconvert signals or stimuli for the efficient uptake of nutrients and their breakdown for energy generation and biomass synthesis. The living organism must maintain the cell system by the effective sensing of the external and/or internal state to survive in response to the variety of environmental changes (Janga et al. 2007). The enzymes which form the metabolic pathways are subject to multiple levels of regulation, where enzyme level and transcriptional regulations play important roles for metabolic regulation (Harbison et al. 2004, Luscombe et al. 2004, Balazsi et al. 2005, Moxley et al. 2009). It is thus important to understand the regulatory processes that govern the cellular metabolism.

Living organisms sense the changes in the environmental condition by detecting extracellular signals such as the available amount of nutrients, and the growth condition such as pH, oxygen level, temperature, etc. These signals eventually feed into the regulatory systems, and affect the physiological and morphological changes that enable the cell system to adapt effectively for survival (Seshasayee et al. 2006). In particular, **transcription factors** (TFs) play important roles for this, where the typical TFs contain two-headed molecules which constitutes of a DNA-binding domain and an allosteric site to which metabolites bind non-covalently or to which enzymes covalently modify in order to modulate the regulatory activity of TF. In particular, the **two-component signal transduction system** is important for efficiently detecting extracellular signals and transducing the signals into cytosol for metabolic regulation. These typically involve a phospho-relay from a transmembrane histidine protein kinase sensor to the target response regulator (Fig. 1), where the two components are usually

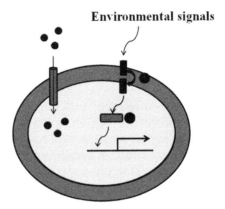

**Figure 1.** Transcription factors and their function.

located within the same operon, enabling their coordinated expression, while some of the kinases and regulators are not adjacent on the chromosome, and it is not straightforward to link the partner in such a case. There might exist a cross-talk between non-cognate sensors and regulators (Yamamoto et al. 2007). As typically seen in Lac repressor in *Escherichia coli*, another method for sensing exogenous signals is by TF binding of transported small molecules, which regulate the enzymatic pathways that process these molecules (Seshasayee et al. 2006). In addition to exogenous signals, cell can recognize the cell's state by detecting the intracellular metabolites. There exists, yet hybrid type of TFs, where they sense the metabolites that are transported from the culture environment or synthesized endogenously (Fig. 1). This can be typically seen in the regulation of amino acid synthetic pathways, possibly because it is preferable for the cell to import essential metabolites when they are freely available rather than expend energy on their synthesis (Seshasayee et al. 2006).

## 2. Variety of Regulation Mechanisms

Biological systems are known to be robust and adaptable to the culture environment. Such robustness is inherent in the biochemical and genetic networks. Living organisms have complex but efficient mechanisms to respond to the change in culture environment, where this is mainly achieved by the so-called global regulators. The global transcriptional regulators are themselves regulated by posttranscriptional regulators. Thus, the global regulation forms a cascade of regulations (Timmermans and Melderen 2010). In relation to global regulators, sigma factors play also important roles, where they allow RNA polymerase to be recruited at the specific DNA sequences in the promoter regions from which they initiate transcription

(Fig. 2). In *E. coli*, seven sigma factors have been found so far, and those play the specific roles depending on the environmental stimuli (Table 1). In *Bacillus subtilis*, multiple sigma factors control sporulation (Kroos 2007). H-NS (histone-like nucleotide structuring protein) is another type of global regulator, found in enterobacteria. It is a small DNA-associated protein that binds preferentially to a curved AT-rich DNA without showing sequence preferences (Timmermans and Melderen 2010). H-NS regulates a variety of physiological aspects such as metabolism, fimbriae expression, virulence flagella synthesis, and proper function.

Other types of global regulators are the signaling molecules such as cyclic-AMP (cAMP) and cyclic-di-GMP (Bruckner and Titgemeyer 2002, Gutierrez-Rios et al. 2007). The cAMP binds to Crp (cAMP receptor protein), also known as CAP (catabolite activation protein), and cAMP-Crp complex becomes an active TF in relation to catabolite regulation as also explained

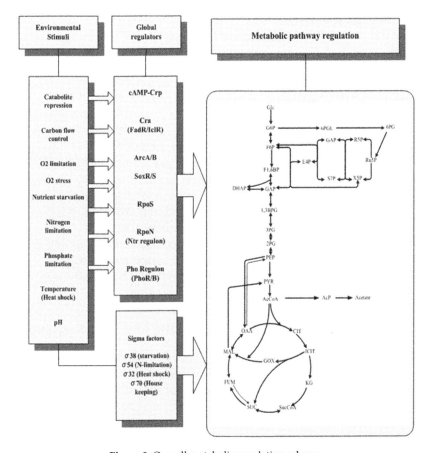

**Figure 2.** Overall metabolic regulation scheme.

Table 1. Seven sigma factors in *E. coli.*

| | |
|---|---|
| $\sigma^{19}$ | Ion transport |
| $\sigma^{24}$ | Extreme temperature |
| $\sigma^{28}$ | Flagella genes |
| $\sigma^{32}$ | Heat shock |
| $\sigma^{38}$ | Stationary phase or carbon starvation, and so forth |
| $\sigma^{54}$ | Nitrogen regulation |
| $\sigma^{70}$ | House keeping |

later. The cAMP regulates not only catabolite regulation, but also flagellum synthesis, biofilm formation, quorum sensing, and nitrogen regulation, etc.

As another level of regulation, small noncoding RNAs (sRNAs) play important roles in the posttranscriptional regulation (Gottesman 2005). The sRNAs are mainly involved in the stress response regulation, pathogenesis, and virulence. A single sRNA can affect multiple targets, where sRNAs modify the translation or stability of the targets and chaperone. One such example is SgrS in *E. coli*, which binds to the mRNA of *ptsG* gene, which encodes EIIBC$^{Glc}$ for glucose uptake (Vanderpool and Gottesman 2004). SgrS encodes a small protein SgrT, where SgrT is also involved in the inhibition of glucose uptake, and thus regulate *ptsG* activity (Wadler and Vanderpool 2007). Another examples are DsrA (sRNA in *E. coli*) which regulates $\sigma^{38}$ expression (Majdalani et al. 2005), and CsrB in *E. coli* (Babitzke and Romeo 2007), of which sRNAs regulate the activity of CsrA global regulator for carbon storage regulation as also explained later.

Another level of posttranscriptional regulation is the control of protein stability and folding carried out by ATP dependent proteases and chaperones (Timmermans and van Melderen 2010). Such examples are Lon ATP-dependent proteases that regulate flagella expression by degrading $\sigma^{38}$ as well as acid shock tolerance regulon by regulating the amount of GadE, where *gadE* is under $\sigma^{38}$ transcriptional control in *E. coli*. The posttranslational control is mediated by the C1PXP ATP-dependent protease, which degrades $\sigma^{38}$.

## 3. Porin Proteins in the Outer Membrane and their Regulation

The gram-negative bacteria such as *E. coli* have two concentric membranes surrounding the cytoplasm, where the space between these two membranes is called as **periplasm**. The **outer membrane** and **inner** or **cytoplasmic membrane** constitute a hydrophobic barrier against polar compounds. The outer membrane contains channel proteins, where the specific molecules can only move across these channels. In the outer membrane of *E. coli*, 108 channels are formed by the porin proteins (Nikaido and Nakae 1980). **Porins** are the outer membrane proteins that produce large, open but regulated water-filled pores that form substrate-specific, ion-selective, or nonspecific channels that allow the influx of small hydrophilic nutrient molecules and the efflux of waste products (de la Cruz et al. 2007). They also exclude many antibiotics and inhibitors that are large and lipophilic (Nikaido 2003). Porins including **OmpC** and **OmpF** of *E. coli* form stable trimers with a slight preference for cations over anions (Saier et al. 2006, Saier et al. 2009). The OmpC and OmpF are the most abundant porins present under typical growth condition representing up to 2% of the total cellular protein (Nikaido 1996). OmpF seems to have slightly larger channel than OmpC, where these are the constitutive porins. Their relative abundance changes depending on such factors as osmolarity, temperature, and growth phase (Hall and Silhavy 1981, Lugtenberg et al. 1976, Pratt and Silhavy 1996). These porins serve for glucose to enter into the periplasm when glucose is present at higher concentration than about 0.2 mM (Nikaido and Vaara 1985, Death and Ferenci 1994). The diffusion rate of glucose molecule is about two-fold higher through OmpF than through OmpC (Nikaido and Rosenberg 1983). Under glucose limitation, the outer membrane glycoporin LamB is induced (Death and Ferenci 1994), where this protein permeates several carbohydrates such as maltose, maltodextrins, and glucose (von Meyenburg and Nikaido 1977), where about 70% of the total glucose import is contributed by LamB (Death and Ferenci 1994). Glucose transport by diffusion through porins of the outer membrane is a passive process (Gosset 2005).

The porin genes are under control of two-component system such as EnvZ-OmpR system, where **EnvZ** is an inner membrane sensor kinase and **OmpR** is the cytoplasmic response regulator (Fig. 3). In response to the environmental signals such as osmolarity, pH, temperature, nutrients, and toxins, EnvZ phosphorylates OmpR to form phosphorylated OmpR (OmpR-P), where OmpR-P increases its binding affinity for the promoter regions of porin genes such as *ompC* and *ompF* (de la Cruz et al. 2007). Acetyl phosphate (AcP) can function as a phosphate donor for OmpR under certain condition. OmpR controls cellular processes such as chemotaxis and virulence as well (Brzostek et al. 2007). In terms of virulence, abolition of porin synthesis diminishes pathogenesis (de la Cruz et al. 2007).

There are several other porins such as OmpU and OmpT (*Vibrio cholerae*), OmpH and OmpL (*Photobacterium*), OmpD (*Salmonella typhimurium*), OmpS1 and OmpS2 (*S. enterica*), and OmpW (*S. enterica, E. coli, and V. cholerae*). Porin genes are also under control of other regulators such as CpxR (under extra-cytoplasmic stress) (Brzostek et al. 2007), PhoB (under phosphate limitation) (von Kruger et al. 2006), Lrp (under starvation) (Ferrario et al. 1995), Rob (for cationic peptides), MarA (under weak acids), SoxS (under oxidative stress) (Delihas and Forst 2001), CadC (at low pH) (Lee et al. 2007),

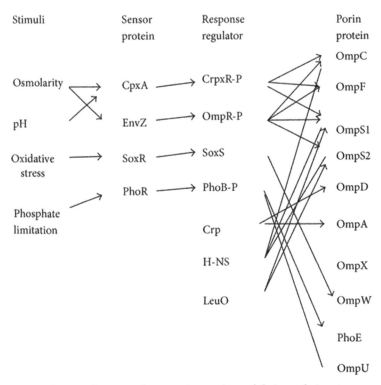

**Figure 3.** Outer membrane porin proteins and their regulations.

Crp (under catabolite repression), Fnr (for anaerobiosis) (Santiviago et al. 2003), ToxR (for virulence) (Miller and Mekalanos 1988), H-NS, StpA, Ihf, Hu (for nucleotide proteins) (Deighan et al. 2000), and LeuO (for stringent response) (de la Cruz et al. 2007, Fernandez-Mora et al. 2004).

CpxA and CpxR form a two-component system, where CpxA is the transmembrane sensor kinase, while CpxR is the response regulator. CpxA can be induced by a variety of stimuli such as higher pH (alkali), misfolded proteins, and alterations in the membrane composition (Dorel et al. 2006, Raivio 2005). Upon activation of the kinase activity of CpxA, the phosphorylated CpxR (CpxR-P) controls the expression of *ompC* and *ompF* genes (de la Cruz et al. 2007).

**PhoR** and **PhoB** also form a two-component system, where phoR is the sensor kinase and detects a low concentration of phosphate or phosphate starvation and phosphorylate PhoB (Wanner 1996), where the phosphorylated PhoB (PhoB-P) activates the transcription of *phoE* gene, where PhoE porin is induced under phosphate limitation (de la Cruz et al. 2007). Moreover, PhoB negatively regulates OmpT, OmpU, and OmpA porins in *V. cholera* (von Kruger et al. 2006). More detailed explanation on the roles of PhoR-PhoB is given later.

**Lrp** (leucine responsive regulator protein) is a global regulator which regulates mainly amino acid metabolism, where its activity is stimulated in minimal medium (which means low nutrient availability), while it is repressed in rich medium such as LB medium (free amino acids available in the medium) (Calvo and Metth 1994). In minimal medium, Lrp represses *ompC* gene expression, while activates *ompF* gene expression.

MarA, SoxS, and Rob are the members of the AraC/XylS family of transcriptional regulators (Gallegos et al. 1997). These three regulators repress *ompF* expression. SoxR and SoxS form the two-component system, where SoxR is a cytoplasmic sensor protein that detects the oxidative stress and activates the SoxS regulator, where the detailed regulation mechanisms are explained later. MarA responds to weak acids like salicylic acid, etc. (Balague and Vescovi 2001) and certain antibiotics (Hachler et al. 1991). Rob may be a general regulator and might be stimulated by cationic peptides (Oh et al. 2000).

CadC is an inner membrane transcriptional activator that acts both as a signal sensor and as a transcriptional regulator, where it activates OmpC and OmpF at low pH (Kuper and Jung 2006, Rhee et al. 2005).

Crp plays an essential role for catabolite regulation, where Crp regulates *ompR-envZ* operon by binding directly to the promoter region (Huang et al. 1992). The *ompA* gene in *E. coli* is activated by Crp (Gilbert and Barbe 1990), while *ompX* is repressed by Crp by means of CyaR, a small RNA (sRNA) (Papenfort et al. 2008). In *S. typhimurium*, *ompD* porin gene is activated by cAMP-Crp (Santiviago et al. 2003). The roles of cAMP-Crp on the catabolite repression is explained later, where the increase in cAMP-Crp level implies the nutrient limitation.

Fnr is a DNA-binding protein that senses oxygen level and regulates the metabolism under anaerobic condition together with ArcA/B system, where Fnr activates *ompD* gene expression under anaerobic condition by the posttranscriptional regulation (Santiviago et al. 2003). The detailed mechanisms for redox regulation by Fnr and ArcA/B system are explained later.

ToxR is a transmembrane DNA-binding protein, and it is an important regulator of virulence gene expression in *V. cholera*. ToxR activates *ompT* porin gene expression. The increased osmolarity enhances OmpT formation and diminishes OmpU formation, similar to that of OmpR on *ompC* and *ompF* in *E. coli*. Moreover, TorX represses *ompW* gene expression at high osmolarity in *V. cholerae* in the presence of glucose (Nandi et al. 2005).

Bacteria possess small nucleotide proteins such as H-NS, StpA, Ihf, and Hu with functional similarity to eukaryotic histones, which affect several porin genes (de la Cruz et al. 2007). H-NS is a master global regulator, which controls the expression of several porin genes such as *ompC, ompF, ompS1*, and *ompS2*. H-NS represses *ompC* gene expression and represses the activity of OmpF. StpA is a paralogue of H-NS and is an RNA chaperone. H-NS and StpA repress *ompS1* and *ompS2* gene expression in *E. coli* and *S. typhimurium* (Hommais et al. 2001, Navarre et al. 2006). On the other hand, H-NS and StpA stimulate the synthesis of the outer membrane maltoporin LamB through posttranscriptional control of the maltose regulon activator MalT (Johansson et al. 1998).

**Ihf** (integration host factor) is one of the most abundant sequence-specific DNA binding proteins and is a global regulator. The Ihf protein represses *ompC* expression, and it is necessary for the negative osmo-regulation of *ompF*. Ihf affects *ompC* and *ompF* in two distinct ways, directly by binding upstream to the promoter regions and indirectly by influencing the expression of EnvZ-OmpR (Huang et al. 1990).

**LeuO** (regulator of leucine biosynthesis) is a LysR-type regulator that controls the expression of several genes in response to stress, virulence, and biofilm accumulation. The OmpS1 and OmpS2 quiescent porins are silenced by H-NS (Navarre et al. 2006, Flores-Valdez et al. 2003), while LeuO acts as an antagonist of H-NS, thereby derepressing *ompS1* and *ompS2* gene expression (de la Cruz et al. 2007, Fernandez-Mora 2004).

Small untranslated regulatory RNAs, often referred to as non-coding RNA, also affect porin regulation. MicF is one such example. In general, they inhibit translation of the transcripts by direct RNA-RNA interaction (de la Cruz et al. 2007). The sRNAs have been found to play diverse physiological roles in response to stress, metabolic regulation, control of bacterial envelope composition, and bacterial virulence (Najdalani et al. 2005, Romby et al. 2006, Storz et al. 2005, Vogel and Papenfort 2006). Enterobacteria use many sRNAs such as MicC, MicA, InvR, RybB, CyaR, IpeX, and RseX to fine-tune the outer membrane composition at the posttranscriptional level (Vogel and Papenfort 2006).

## 4. Transport of Carbohydrates and PTS

The first step in the metabolism of carbohydrates is the transport of these molecules into the cell (Fig. 4). In bacteria, various carbohydrates can be taken up by several mechanisms. Primary transport of sugars is driven by ATP, while secondary transport is driven by the electrochemical gradients of the translocated molecules across the membrane (Poolman and Konings 1993), where the secondary transport systems contain the symporters which co-transport two or more molecules, **uniporters** that transport single molecule, and **antiporters** that counter transport two or more molecules. Sugar symporters usually couple the uphill movement of the sugar to the downhill movement of proton (or sodium ion). Namely, the electrochemical proton (or sodium ion) gradient drives the sugar transport (Gunnewijk et

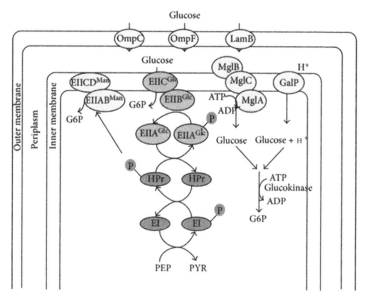

**Figure 4.** Outer and inner membrane and periplasm and glucose transport by PTS and non-PTS.

al. 2001). Sugar uptake by group translocation is unique to bacteria and is involved in the **phosphotransferase system (PTS)** (Fig. 4). PTS is a transport system that catalyzes the uptake of a variety of carbohydrates and their conversion into their respective phosphoesters during transport (Deutscher et al. 2006, 2014, Kotrba et al. 2001, Gosset 2005, Luo et al. 2014).

Once sugar was transported inside the periplasm, it is internalized into the cytoplasm by the PTS. The sugar concentration in the periplasm may be low due to active transport systems in the cytoplasmic membrane (Gosset 2005). Once inside the periplasm, sugar can be transported into the cytosol by PTS, where PTS is widespread in bacteria, in some archaea, and absent in eukaryotic organisms (Postma et al. 1996). Although archaea have been considered to be devoid of any PTS (Barabote and Saier 2005), recent investigation indicates that among about 230 archaea, more than about 50 contain the PTS components of EI and HPr, where most of them belong to the genus of *Haloferax*, and the PTS is specific to the transport and phosphorylation of fructose (Pickl et al. 2012, Cai et al. 2014) or phosphorylation of dihydroxyacetone (DHA) (Bachler et al. 2005). Gram-positive bacteria usually contain only EI and HPr, while *Enterobacteriaceae* contains several EI and HPr homologues or paralogues, where nitrogen related EI[Ntr] and fructose specific FPr are such examples. Many sugars and sugar derivatives such as sugar alcohols, amino-sugars, glycuronic acids, disaccharides, and many other carbohydrates can be transported via PTS (Deutscher et al. 2014).

The PTS is typically composed of one membrane-spanning protein and four cytosolic and soluble proteins, where the phosphate of PEP is transferred in turn via these proteins up to the carbohydrates to be transported into the cytosol. The first two proteins are the soluble and non-sugar-specific components Enzyme I (EI) encoded by *ptsI* and the phosphor-histidine carrier protein (HPr) encoded by *ptsH*, which are involved in the uptake of all PTS carbohydrates in most organisms (Fig. 4). These EI and HPr transfer phosphoryl group from PEP to the sugar-specific enzymes IIA, IIB, and IIC (IID in some cases such as mannose specific PTSs), where IIC is an integral membrane protein permease that recognizes and transports the sugar molecules, where it is phosphorylated by EIIB (Fig. 4). The carbohydrate specificity of EII is designated as the superscript such as EII$^{Glc}$, indicating the glucose-specific EIIA component. The genes encoding the PTS components are normally form an operon to be co-transcribed simultaneously.

Various carbohydrates are taken up and transported and phosphorylated by the carbohydrate-specific PTS (Table 2), where the phosphorylation of carbohydrates is made by the phosphate originated from PEP. The phosphorylation cascade begins with EI, where it auto-phosphorylates at the specific position of a conserved histidyl residue at the expense of PEP (Alpert et al. 1985). This phosphorylated EI (P~His-EI) transfers the phosphoryl group to the specific position of His-15 in HPr (Gassner et al. 1977), and P~His-EI passes the phosphoryl group on to the sugar-specific E II As. EIIAs are also phosphorylated at the specific position of a histidyl

**Table 2** Various Enzyme II complexes of PTS

---

(1) • **Glucose-fructose-lactose superfamily**
  Glucose-glucoside (Glc) family
    *E. coli* EIIA$^{Glc}$/EIICB$^{Glc}$, *B. subtilis* EIICBA$^{Glc}$
  • Fructose-mannitol (Fru) family
    *E. coli* EIICBA$^{Mtl}$
  • Lactose-*N*,*N*9-diacetylchitobiose-$\beta$-glucoside (Lac) family
    *L. casei* EII$^{Lac}$/EIICB$^{Lac}$
  • Glucitol (Gut) family
  • Galactitol (Gat) family
  • Mannose-fructose-sorbose (Man) family

(2) • **Ascorbate-galactitol superfamily**
    *E. coli* SgaA/SgaB/SgaT
  • Galactitol family
    *E. coli* EIIA$^{Gat}$/EIIB$^{Gat}$/EIIC$^{Gat}$

(3) **Mannose family**
    *E. coli* EIIAB$^{Man}$/EIIC$^{Man}$/EIID$^{Man}$
    *B. subtilis* EIIA$^{Lev}$/EIIB$^{Lev}$/EIIC$^{Lev}$/EIID$^{Lev}$

(4) **Dihydoxyacetone family**
    *E. coli* EIIA-HPr-EI$^{Dha}$, EIIA$^{Pha}$ in filmicutes

---

residue (Deutscher et al. 1982, Dorschug et al. 1984). P~E II As phosphorylate their cognate E II B at a cysteine residue (Pas and Robillard 1988), while E II Bs of mannose PTS family are phosphorylated at the specific position of a conserved histidine (Charrier et al. 1997). In the last step of phosphate transfer, P~E II B transfers its phosphoryl group to a carbohydrate molecule bound to the cognate E II C.

There exist 21 different enzyme II complexes in *E. coli* that are involved in the transport of about 20 different carbohydrates (Tchieu et al. 2001) (Table 2). In *E. coli*, EII$^{Glc}$ and EII$^{Man}$ are involved in the transport of glucose. The EII$^{Glc}$ is composed of the soluble EIIA$^{Glc}$ encoded by *crr* (Crr stands for catabolite repression resistance) and of the integral membrane permease EIICB$^{Glc}$ encoded by *ptsG*. The EII$^{Man}$ complex is composed of the EIIAB$^{Man}$ homodimer enzyme and the integral membrane permease EIICD$^{Man}$ (Fig. 4), where these proteins are encoded in the *manXYZ* operon (Gosset 2005). In addition to mannose, these proteins can also transport glucose, fructose, N-acetylglucosamines, and glucosamine with similar efficiency (Curtiz and Epstein 1975). In a wild-type strain growing on glucose, *ptsG* is induced, while *manXYZ* operon is repressed.

In *ptsG* mutant, the glucose can be transported by EII$^{Man}$ complex, and the cell can grow with less growth rate than the wild-type strain (Chou et al. 1994). When the extracellular glucose concentration is less than about 1 $\mu$M, or it is more than about 2 g/L for *pts* mutants, this can be also utilized (Flores et al. 2005). The induction of these genes is caused by the intracellular galactose that functions as an auto-inducer of the system (Death and Ferenci 1994). One of the genes induced under glucose limitation is *galP* that codes for the low-affinity galactose: H$^+$ symporter GalP (Fig. 4).

The genes in the *mglABC* operon encode an ATP-binding protein, a galactose/glucose periplasmic binding protein, and an integral membrane transporter protein, respectively, forming Mgl system for the galactose/glucose (methyl galactoside) import (Gosset 2005). This high-affinity porter belongs to the **ATP-binding cassette (ABC) superfamily** of the primary active class of transporters. When extracellular glucose concentration is very low, the Mgl system together with LamB attains high-affinity glucose transport (Gosset 2005). The glucose molecule transported either by GalP or Mgl systems must be phosphorylated by glucokinase (Glk) encoded by *glk* from ATP to become G6P (Fig. 4) (Lunin et al. 2004). PTS seems to be quite efficient as it consumes one mole of PEP for each internalized and phosphorylated glucose, where one mole of PEP is equivalent to one mole of ATP, since the conversion of PEP to PYR by Pyk would yield one mole of ATP by substrate-level phosphorylation. The high-affinity Mgl-glucokinase system is energetically the most expensive, as it consumes two moles of ATP per glucose. The GalP-glucokinase system requires one mol of H$^+$ that is internalized into the cytoplasm and one mol of ATP (Fig. 4).

## 5. Carbohydrate Uptake by Various PTSs and without PTS

*Escherichia coli* contains different EII complexes, where EII complexes are formed either by distinct proteins that contain EI and/or HPr domains exist (Deutscher et al. 2006). A prominent example for the latter is **FPr**, which consists of HPr and EIIA domain and mediates the phosphotransfer in the uptake of fructose by *E. coli*. As shown in Fig. 5, fructose can be transported and phosphorylated by the fructose PTS (EIIBC$^{Fru}$) or ATP dependent manno-fruct kinase Mak (Aulkemeyer et al. 1991). The EIIBC$^{Fru}$ encoded by *fruA* phosphorylates fructose and concomitantly transports it to become D-fructose 1-phosphate (F1P), which is further converted to fructose 1,6-bisphosphate (FBP) in the upper glycolysis (Kornberg 1990).

There are three pathways for the utilization of fructose as shown in Fig. 6 (Kornberg 2001). In the primary pathway, fructose (Fru) is transported via the membrane-spanning protein FruA and concomitantly phosphorylated by a PEP: D-fructose 1-phosphotransferase (fructose PTS) system (ATP:

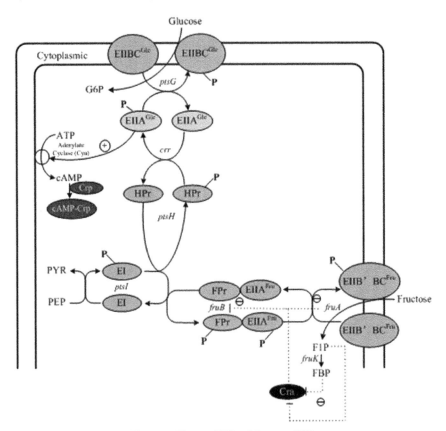

**Figure 5.** Glucose PTS and fructose PTS.

D-fructose 1-phosphotransferase, EC 2.7.1.3), which is induced by D-fructose and enter the cell as F1P, where this process is affected by the transfer of a phosphoryl moiety from PEP to the hexose by the concerted action of two cytoplasmic proteins: EI of PTS and a membrane-associated diphosphoryl transfer protein (DTP). F1P is then converted to FBP by ATP and by the inducible enzyme D-fructose-1-phosphate kinase (F1PK) (ATP: D-fructose-1-1phosphate 6-phosphotransferase).

In the second pathway, fructose enters the cell via membrane-spanning proteins that have the ability to recognize sugars possessing the 3,4,5-D-arabino-hexose configuration which include the permeases for mannose (ManXYZ), glucitol (SrlA), and mannitol (MtlA) (Kornberg 2001). D-fructose is converted to F6P by a specific sucrose-induced D-fructokinase (ATP: D-fructose 6-phosphotransferase, EC 2.7.1.4), and then converted to FBP by Pfk of the upper glycolysis.

In the 3rd pathway, fructose enters the cell by diffusion, using an isoform of the glucose transporter PtsG. Since this mode of entry does not involve the PTS, the free fructose has to be phosphorylated by ATP to become F6P.

There are two entry points such as FBP and F6P in the upper glycolysis to which fructose is converted, where this affects the flux toward the oxidative pentose phosphate (PP) pathway.

D-xylose is converted to D-xylulose by xylose isomerase (D-xylose ketoisomerase, EC 5.3.1.5) (Fig. 7). D-Xylulose is subsequently phosphorylated by xylulokinase (ATP: D-xylulose 5-phosphotransferase, EC 2.7.1.17) to form D-xylulose 5-phosphate (X5P). Under anaerobic condition, xylulose reductase (XR) is induced, and xylitol and xylitol 5-phosphate are produced, where they may inhibit the cell growth.

Glycerol is oxidized to dihydroxyacetone (DHA) by a glycerol dehydrogenase (glycerol: NAD oxidoreductase, EC 1.1.1.6).

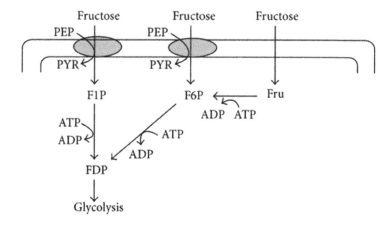

**Figure 6.** Fructose uptake pathways.

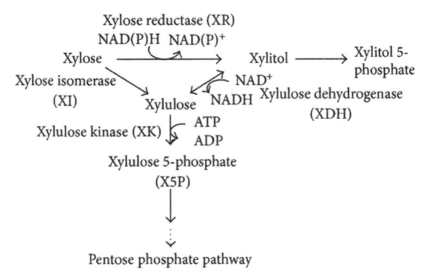

**Figure 7.** Xylose uptake pathways.

Dihydroxyacetone is then phosphorylated by a kinase using ATP (Fig. 8). Another pathway for glycerol utilization is that glycerol is phosphorylated by glycerol kinase (ATP: glycerol phosphotransferase, EC 2.7.1.30) to form L-glycerol 3-phosphate (GL3P), which then is converted to glyceraldehyde 3-phosphate (GAP) in the glycolysis.

## 6. Nitrogen PTS

In addition to carbohydrate PTS, most proteobacteria possess a paralogous system such as nitrogen phosphotransferase system PTS$^{Ntr}$, where it consists of EI$^{Ntr}$ encoded by *ptsP*, NPr encoded by *ptsO*, and E II A$^{Ntr}$ encoded by *ptsN*, which are paralogues to the carbohydrate PTS components such as EI, HPr, E II A, respectively (Powell et al. 1995, Peterkofski et al. 2006, Pfluger-Grau and Gorke 2010). *E. coli* PTS$^{Ntr}$ plays a role in relation to K$^+$ uptake, where dephosphorylated E II A$^{Ntr}$ binds to and regulates the low affinity K$^+$ transporter TrkA (Lee et al. 2007) and the K$^+$-dependent sensor kinase KdpD (Pfluger-Grau and Gorke 2010, Luttmann et al. 2009). K$^+$ regulates global gene expression involving both σ$^{70}$- and σ$^{38}$-dependent promoters (Lee et al. 2010). Moreover, dephosphorylated E II A$^{Ntr}$ modulates the phosphate starvation response through interaction with sensor kinase PhoR (Luttmann et al. 2012). Dephosphorylated form of PTS$^{Ntr}$ interacts with and inhibits LpxD, which catalyzes biosynthesis of lipidA of the lipopolysaccharide (LPS) layer (Kim et al. 2011).

   Although the physiological role of PTS$^{Ntr}$ has not been well known, glutamine and αKG, reciprocally regulate the phosphorylation state of the

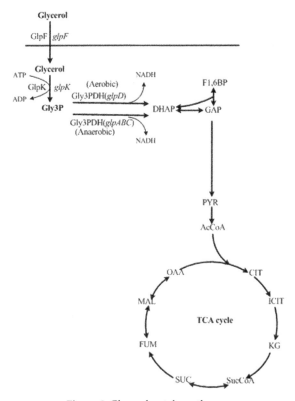

**Figure 8.** Glycerol uptake pathways.

PTS$^{Ntr}$ by direct effects on EI$^{Ntr}$ auto-phosphorylation. This implies that PTS$^{Ntr}$ senses nitrogen availability (Lee et al. 2013).

## 7. Carbon Catabolite Repression for the Selective Carbon Source Uptake

Among the culture environment, carbon sources are by far important for the cell from the point of view of energy generation and biosynthesis. Most living organisms use various compounds as carbon sources, where these can be either co-metabolized or selectively consumed with preference for the specific carbon sources among available carbon sources. One typical example of selective carbon-source usage is the **diauxie phenomenon** when a mixture of glucose and other carbon sources such as lactose was used as a carbon source, where this phenomenon was first observed by Monod (1942). Subsequent investigation on this phenomenon has revealed that selective carbon source utilization is common and that glucose is the preferred carbon source in many organisms. Moreover, the presence of glucose often prevents the use of other carbon sources. This preference of glucose over other carbon sources has been named as **glucose repression**, or more generally **carbon catabolite repression (CCR)** (Magasanik 1961). CCR is observed in most heterotrophic bacteria which include facultatively autotrophic bacteria that repress the genes for $CO_2$ fixation in the presence of organic carbon source (Bowien and Kusian 2002). CCR is important for the cells to compete with other organisms in nature, where it is crucial to select a preferred carbon source in order to promote the cell growth, which then results in survival as compared to other competing organisms. Moreover, CCR has a crucial role in the expression of virulence genes, which often enable the organism to access new sources of nutrients. The ability to select the appropriate carbon source that allows fastest growth may be the driving force for the evolution of CCR (Gorke and Stulke 2008).

The *E. coli lac* operon is only expressed if allolactose (a lactose isomer formed by *β*-galactosidase) binds and inactivates the *lac* repressor. Lactose cannot be transported into the cell in the presence of glucose, because the lactose permease, LacY is inactive in the presence of glucose (Winler and Wilson 1967). As shown in Fig. 9, phosphorylated EIIA$^{Glc}$ is dominant when glucose is absent and does not interact with LacY, whereas unphosphorylated EIIA$^{Glc}$ can bind and inactivates LacY when glucose is present (Hogema et al. 1999, Nelson et al. 1983). This only occurs if lactose is present (Smirnova et al. 2007). The same mechanism may be seen for the transport of other secondary carbon sources such as maltose, melibiose, raffinose, and galactose (Titgemeyer et al. 1994, Misko et al. 1987).

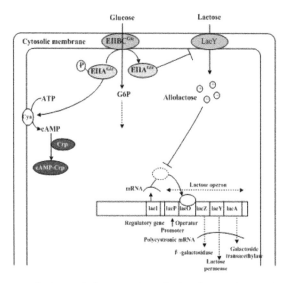

**Figure 9.** Inducer exclusion and the activation of adenylate cyclase in the glucose-lactose system.

The central players in CCR in *E. coli* are the transcriptional activator Crp, cAMP, adenylate cyclase (Cya), and the PTSs, where these systems are involved in both transport and phosphorylation of carbohydrates.

The glucose-specific PTS in *E. coli* consists of the cytoplasmic protein EIIA$^{Glc}$ encoded by *crr* and the membrane-bound protein EIICB$^{Glc}$ encoded by *ptsG*, which transport and concomitantly phosphorylate glucose as explained before. The phosphoryl groups are transferred from PEP via successive phosphor-relay reactions in turn by EI, HPr, EIIA$^{Glc}$ and EIICB$^{Glc}$ to glucose. Unphosphorylated EIIA$^{Glc}$ inhibits the uptake of other non-PTS carbohydrates by the so-called **inducer exclusion** (Boris 2008), while phosphorylated EIIA$^{Glc}$ (EIIA$^{Glc}$-P) activates Cya, which generates cAMP from ATP and leads to an increase in the intracellular cAMP level (Park et al. 2006) (Fig. 9). The cAMP-Crp complex and the repressor Mlc are involved in the regulation of *ptsG* gene and *pts* operon expression. Unphosphorylated EIICB$^{Glc}$ can relieve the expression of *ptsG* gene expression by sequestering Mlc from its binding sites through a direct protein-protein interaction in response to glucose concentration. In contrast to Mlc, where it represses the expression of *ptsG*, *ptsHI,* and *crr* (Plumbridge 1998), cAMP-Crp complex activates *ptsG* gene expression (de Reuse and Danchin 1988) (Fig. 10). Since intracellular cAMP levels are low during growth on glucose, these two antagonistic regulatory mechanisms guarantee a precise adjustments of *ptsG* expression level under various conditions (Bettenbrock et al. 2006) (Fig. 10). In the absence of glucose, Mlc binds to the upstream of *ptsG* gene and prevents its transcription. If glucose is present in the medium, the amount of unphosphorylated EIICB$^{Glc}$ increases due to the phosphate transfer to glucose. In this situation, Mlc binds to EIICB$^{Glc}$,

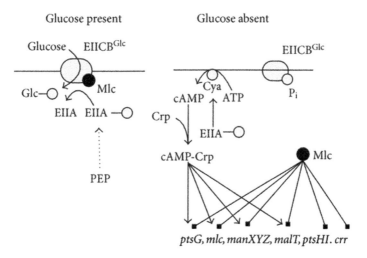

**Figure 10.** The multiple regulations by Mlc and cAMP-Crp.

and thus it does not bind to the operator region of *pts* genes (Bettenbrock et al. 2006, Tanaka et al. 2000, Lee et al. 2000).

If the concentration ratio between PEP and PYR (PEP/PYR) is high, EIIA$^{Glc}$ is predominantly phosphorylated, whereas if this ratio is low, then EIIA$^{Glc}$ is predominantly dephosphorylated (Bettenbrock et al. 2007, Hogema et al. 1998). EIIA$^{Glc}$ is preferentially dephosphorylated when *E. coli* cells grow rapidly with glucose as a carbon source. The cAMP levels are low during growth with non-PTS carbohydrates such as lactose, where PEP/PYR ratio is the key factor that controls phosphorylation of EIIA$^{Glc}$, which explains dephosphorylation of EIIA$^{Glc}$, resulting in low cAMP pool (Bettenbrock et al. 2007, Hogema et al. 1998). As stated above, inducer exclusion is the dominant factor for the glucose-lactose diauxie (Hogema et al. 1999, Inada et al. 1996). The roles of cAMP-Crp is then to express *lac* operon, and it is involved in CCR by activating the expression of *ptsG* and EIICB domain of the glucose-specific PTS, and therefore, the transport of glucose (Kimata et al. 1997).

In the case of using glucose and other carbon sources, glucose is preferentially consumed first before other carbon sources are assimilated by the "rigid" CCR system probably by inducer exclusion as well as the lower level of cAMP-Crp, while the other non-glucose carbon sources may be ranked in a hierarchy, where the higher the cell growth rate, the higher the sugar in the hierarchy in *E. coli* (Aiderberg et al. 2014). This hierarchy is "soft" in the sense that lower ranked sugars are not completely repressed, where such hierarchy can be explained by the different promoter activity by cAMP-Crp, showing the ranking for the non-PTS sugars from higher to lower as lactose > arabinose > xylose > sorbitol > rhamnose > ribose in *E. coli* (Aidelberg et al. 2014). Another mechanism for the hierarchy may be caused by the cross regulation, where a lower-ranking sugar system is repressed by the regulator of the higher ranking sugar. For example, Arabinose specific AraC represses the xylose utilization promoters (Desai and Rao 2010), while XylR represses *srl, rha,* and SrlR represses *rha,* etc. (Fig. 11) (Aidelberg et al. 2014, Perez-Alfaro et al. 2014). Another CCR hierarchy may be seen for the preferential consumption of sugars over a short-chain fatty acid such as propionate, where the expression of *prpBCDE* is repressed due to down regulation of cAMP-Crp in *E. coli* (Park et al. 2012).

CCR depends on the growth rate, where slowly glowing *E. coli* cells (at less than about 0.1 h$^{-1}$) do not show CCR, and co-consume multiple carbon sources (Zhou et al. 2013). As the cell growth rate increases from 0.1 h$^{-1}$, CCR becomes eminent eventually showing overflow metabolism by forming acetate in *E. coli,* and the cell volume increases to match the faster biomass accumulation, and the buoyant density also increases (Zhou et al. 2013). The extensive CCR may be seen above 0.4 h$^{-1}$ of the cell growth rate.

(a)

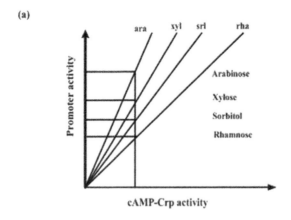

(b)

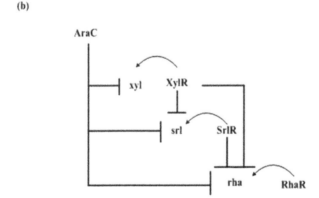

**Figure 11.** Two possible regulatory mechanisms that can implement a hierarchal decision in sugar utilization. (a) Hierarchy can be obtained if CRP shows differential regulation for the different sugar systems so that the induction curves of each system as a function of CRP-cAMP activity are separated. (b) Hierarchy can also be obtained by cross regulation so that systems lower in the hierarchy are directly repressed, for example by the sugar-specific transcription factors of the better sugar systems (Aidelberg et al. 2014).

In the cultivation of *E. coli* using a mixture of glucose, maltose, L-lactate, galactose, and glycerol, three different phases may be seen, where glucose, which provides the highest growth rate, is first preferentially consumed, followed by the second phase of simultaneous consumption of maltose, L-lactate, and galactose, and then glycerol and (produced and secreted) acetate are consumed in the final phase (Beg et al. 2007).

The slow glycerol uptake may be due to allosteric inhibition of glycerol kinase (GlpK) by FBP, where this is believed to avoid the accumulation of dihydroxyacetone phosphate (DHAP), which may direct toward toxic methylglyoxal producing pathway upon DHAP accumulation (Freedberg et al. 1971). The GlpK is also inhibited by unphosphorylated E II $A^{Glc}$ (Novotny et al. 1985), where the inhibition of GlpK by FBP is dominant in the glucose control of glycerol utilization (Holtman et al. 2001). The glycerol assimilation pathway enzymes are encoded by *glpFKX* operon, which is activated by cAMP-Crp, while it is repressed by GlpR in the absence of intracellular glycerol, and thus *glpFKX* is under glucose repression. The inhibition of GlpK by FBP can be relaxed by evolutional mutation, and the glycerol uptake rate can be increased, where cAMP-Crp decreases showing glycerol induced catabolite repression (Applebee et al. 2011). Although the glycerol uptake rate can be improved by such mutation, the overflow metabolism cannot be avoided due to lower level of cAMP-Crp (Cheng et al. 2014), where the phosphate of E II $A^{Glc}$ transferred from PEP may be utilized for the reaction of GlpK instead of E II $A^{Glc}$.

## 8. CCR in other Bacteria than *E. coli*

The key players in CCR in *Bacillus subtilis* are the pleiotropic transcription factor **CcpA (catabolite control protein A)**, the HPr protein of the PTS, the bifunctional HPr kinase/phosphorylase (HPrK) and the glycolytic intermediates such as FBP and G6P (Titgemeyer and Hillen 2002, Warner and Lolkema 2003, Henkin et al. 1991) (Fig. 12). Unlike *E. coli*, HPr phosphorylation plays an important role, where phosphorylated HPr serves as the effector for the dimeric CcpA, which controls the expressions of CCR genes (Gorke and Stulke 2008). The phosphorylation of HPr is catalyzed by HPrK, that binds ATP, and its activity is triggered by the availability of FBP as an indicator of high glycolytic activity (Galinier et al. 1988, Reizer et al. 1998, Jault et al. 2000). By contrast, phosphorylase activity prevails under nutrient limitation, and the activation is stimulated by the inorganic phosphate in the cell (Jault et al. 2000, Mijakovic et al. 2002). Under nutrient rich condition, HPrK acts as a kinase and phosphorylates HPr, and the cofactor for CcpA is formed. The interaction between CcpA and the phosphorylated HPr is enhanced by FBP and G6P (Seidel et al. 2005, Schumacher et al. 2007).

The inducer exclusion has also been reported for Gram positive bacteria, and HPr is the major player in these organisms. In *Lactobacillus brevis*, HPr (Ser-P) is formed when glucose is present and binds and inactivate permease (Djordjecic et al. 2001). By contrast, the lactose permease of *Streptococcus thermophilus* is controlled by HPr-(His-P-)dependent phosphorylation. In the absence of glucose, HPr (His-P) can phosphorylate PTS-like domain, thereby activating the permease for lactose transport (Poolman et al. 1995). When glucose is present, HPr becomes phosphorylated on Ser46 and can no longer activate the lactose permease (Gunnewijk and Poolman 2000).

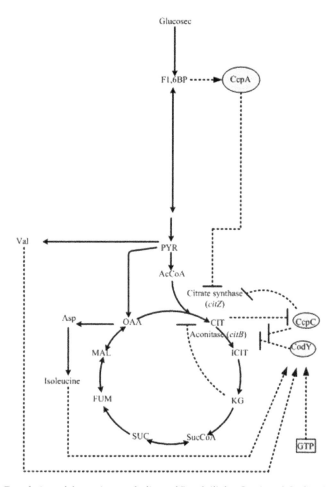

**Figure 12.** Regulation of the main metabolism of *B. subtilis* by CcpA and CodY (Gorke 2008).

With the exception of the mycoplasmas, a Firmicutes also use HPr, HPrK, and CcpA for CCR (Titgemeyer and Hillen 2002). CcpA in lactic acid bacteria such as *Lactococcus lactis* represses not only genes of carbon metabolism, but also controls metabolic pathway genes such as glycolysis and lactic acid formation pathway genes (Gorke and Stulke 2008).

Another phenomenon can be seen in *Corynebacterium glutamicum*, where co-assimilation of glucose and other carbon sources is made, but it is highly regulated (Frunzke et al. 2006, Wendisch et al. 2000). *C. glutamicum* is important in the industrial production of amino acids, where it prefers to use multiple carbon sources simultaneously. Diauxic growth is observed for the case of using a mixture of glutamate or ethanol and glucose, where the repressor protein RamB is activated when glucose is present and binds to the promoter regions of the genes involved in acetate and ethanol catabolism (Arndt and Eikmanns 2007, Gerstmeir et al. 2004). The *ramB* expression is regulated by the feedback of RamB and RamA, where RamA is activated when acetate is present.

It is of interest that for some bacteria such as *Streptococcus thermophilus*, *Bifidobacterium longum*, and *Pseudomonas aeruginosa*, glucose is not a primary carbon source, and the genes for glucose utilization are repressed when preferred carbon sources are available, where this phenomenon is called as **reverse CCR** (van den Bogaard et al. 2000, Parche et al. 2006, Collier 1996). CCR is one of the most important regulatory phenomena in many bacteria (Liu et al. 2005, Blencke et al. 2003, Moreno et al. 2001).

*Pseudomonas putida* can assimilate various aromatic and aliphatic hydrocarbons, where the use of hydrocarbons is repressed by succinate, and this seems to be a general feature of CCR in this organism (Collier et al. 1996, Muller et al. 1996). Under CCR, the translation of operon-specific regulators is inhibited by the binding of an RNA-binding protein Crc to mRNAs of the regulator transcript, and thus CCR seems to be governed by an RNA-binding protein at the level of posttranscriptional control rather than by a DNA-binding transcriptional regulator (Moreno et al. 2007, Moreno and Rojo 2008).

In *Streptomyces coelicolor* and the related species, glucose kinase is the key player of CCR, where it is independent of the PTS (van Wezel et al. 2006).

CCR is crucial for the expression of virulence genes and for pathogenicity in many pathogenic bacteria. Note that the primary aim of pathogenic bacteria is to gain access to nutrients rather than to cause damage to the host and that the expression of virulent genes is linked to the nutrient supply of the bacteria (Gorke and Stulke 2008). Some pathogenic bacteria such as *Chlamydia trachomatis* and *Mycoplasma pneumonia* seem to lack CCR, where these are adapted to nutrient-rich host environments (Nicholson et al. 2004, Halbedel et al. 2007).

In many Firmicutes, the mutants devoid of the HPr kinase grow significantly slower than wild-type cells. It is, therefore, suggested that HPr kinase, which generates the cofactor for CcpA, might be a suitable drug target, where the compound that inhibits the kinase activity of HPr has been identified, and this compound inhibits the growth of *B. subtilis* but not of *E. coli*, where *E. coli* does not contain HPr kinase (Gorke and Stulke 2008). Crp and cAMP are essential for the expression of virulence genes in enteric bacteria, and therefore, the corresponding *crp* and *cya* mutant strains of *Salmonella enterica* and *Yersinia enterocolitica* can be used as live vaccines in mice and pigs (Petersen and Young 2002, Ramstrom et al. 2004, Curtiss and Kelly 1987).

## 9. Concluding Remarks

Here, the brief explanation on the transport of carbohydrates and carbon catabolite repression is given for a variety of organisms. These are tightly regulated, and it is important to understand the overall regulation mechanisms before going into the metabolic design for biofuel and biochemical production.

Here, the catabolite repression for the selective carbon source uptake is explained, where its regulation systems differ depending on the organism and the culture environment. More in-depth explanation for catabolite regulation and metabolic regulation will be explained in the following chapters.

## References

Aidelberg, G., B.D. Towbin, D. Rothschild, E. Dekel, A. Bren and U. Alon. 2014. Hierarchy of non-glucose sugars in *Escherichia coli*. BMC Syst Biol. 8: 133.

Alpert, C.-A., R. Frank, K. Stüber, J. Deutscher and W. Hengstenberg. 1985. Phosphoenolpyruvate-dependent protein kinase enzyme I of *Streptococcus faecalis*. Purification and properties of the enzyme and characterization of its active center. Biochemistry. 24: 959–964.

Applebee, M.K., A.R. Joyce, T.M. Conrad, D.W. Pettigrew and B.O. Palsson. 2011. Functional and metabolic effects of adaptive glycerol kinase (GLPK) mutants in *Escherichia coli*. J Biol Chem. 286(26): 23150–23159.

Arndt, A. and B.J. Eikmanns. 2007. The alcohol dehydrogenase gene *adhA* in *Corynebacterium glutamicum* is subject to carbon catabolite repression. J Bacteriol. 189(20): 7408–7416.

Aulkemeyer, P., R. Ebner, G. Heilenmann et al. 1991. Molecular analysis of two fructokinases involved in sucrose metabolism of enteric bacteria. Mol Microbiol. 5(12): 2913–2922.

Babitzke, P. and T. Romeo. 2007. CsrB sRNA family: sequestration of RNA-binding regulatory proteins. Curr Opin Microbiol. 10(2): 156–1637.

Bächler, C., K. Flükiger-Brühwiler, P. Schneider, P. Bähler and B. Erni. 2005. From ATP as substrate to ADP as coenzyme: functional evolution of the nucleotide binding subunit of dihydroxyacetone kinase. J Biol Chem. 280: 18321–18325.

Balague, C. and E.G. V´escovi. 2001. Activation of multiple antibiotic resistance in uropathogenic *Escherichia coli* strains by aryloxoalcanoic acid compounds. Antimicrob Agents and Chemotherapy. 45(6): 1815–1822.

Balazsi, G., A.L. Barab´asi and Z.N. Oltvai. 2005. Topological units of environmental signal processing in the transcriptional regulatory network of *Escherichia coli*. PNAS USA. 102(22): 7841–7846.

Barabote, R.D. and M.H. Saier Jr. 2005. Comparative genomic analyses of the bacterial phosphotransferase system. Microbiol Mol Biol Rev. 69: 608–634.

Beg, O.K., A. Vazquez, J. Ernst, M.A. de Menezes, Z. Gar-Joseph, A.-L. Barabasi and Z.N. Oltvai. 2007. Intracellular crowding defines the model and sequence of substrate uptake by *Escherichia coli* and constraints its metabolic activity. PNAS USA. 104(31): 12663–12668.

Bettenbrock, K., S. Fischer, A. Kremling, K. Jahreis, T. Sauter and E.D. Gilles. 2006. A quantitative approach to catabolite repression in *Escherichia coli*. J Biol Chem. 281(5): 2578–2584.

Bettenbrock, K., T. Sauter, K. Jahreis, A. Kremling, J.W. Lengeler and E.D. Gilles. 2007. Correlation between growth rates, EIIA$^{Crr}$ phosphorylation, and intracellular cyclic AMP levels in *Escherichia coli* K-12. J Bacteriol. 189(19): 6891–6900.

Blencke, H.M., G. Homuth, H. Ludwig, U. Mader, M. Hecker and J. Stulke. 2003. Transcriptional profiling of gene expression in response to glucose in *Bacillus subtilis*: regulation of the central metabolic pathways. Metab Eng. 5(2): 133–149.

Boris, G. 2008. Carbon catabolite repression in bacteria: many ways to make the most out of nutrients. Nat Rev Microbiol. 6(8): 613–624.

Bowien, B. and B. Kusian. 2002. Genetics and control of $CO_2$ assimilation in the chemoautotroph *Ralstonia eutropha*. Arch Microbiol. 178(2): 85–93.

Bruckner, R. and F. Titgemeyer. 2002. Carbon catabolite repression in bacteria: choice of the carbon source and autoregulatory limitation of sugar utilization. FEMS Microbiol Lett. 209(2): 141–148.

Brzostek, K., M. Brzostkowska, I. Bukowska, E. Karwicka and A. Raczkowska. 2007. OmpR negatively regulates expression of invasion in *Yersinia enterocolitica*. Microbiol. 153(8): 2416–2425.

Cai, L., S. Cai, D. Zhao, J. Wu, L. Wang, X. Liu, M. Li, J. Hou, J. Zhou, J. Liu, J. Han and H. Xiang. 2014. Analysis of the transcriptional regulator GlpR, promoter elements, and posttranscriptional processing involved in fructose-induced activation of the phosphoenolpyruvate-dependent sugar phosphotransferase system in *Haloferax mediterranei*. Appl Environ Microbiol. 80: 1430–1440.

Calvo, J.M. and R.G. Matthews. 1994. The leucine-responsive regulatory protein, a global regulator of metabolism in *Escherichia coli*. Microbiol Rev. 58(3): 466–490.

Charrier, V., J. Deutscher, A. Galinier and I. Martin-Verstraete. 1997. Protein phosphorylation chain of a *Bacillus subtilis* fructose-specific phosphotransferase system and its participation in regulation of the expression of the *lev* operon. Biochemistry. 36: 1163–1172.

Cheng, K.K., B.-S. Lee, T. Masuda, T. Ito, K. Ikeda, A. Hirayama, L. Deng, J. Dong, K. Shimizu, T. Soga, M. Tomita, B.O. Palsson and M. Robert. 2014. Global metabolic network reorganization by adaptive mutations allows fast growth of *Escherichia coli* on glycerol. Nature Commun. 5: 3233.

Chou, C.H., G.N. Bennett and K.Y. San. 1994. Effect of modulated glucose uptake on high-level recombinant protein production in a dense *Escherichia coli* culture. Biotechnol Prog. 10(6): 644–647.

Collier, D.N., P.W. Hager and P.V. Phibbs Jr. 1996. Catabolite repression control in the *Pseudomonads*. Res Microbiol. 147(6-7): 551–561.

Curtis, S.J. and W. Epstein. 1975. Phosphorylation of D glucose in *Escherichia coli* mutants defective in glucose phosphotransferase, mannose phosphotransferase, and glucokinase. J Bacteriol. 122(3): 1189–1199.

Curtiss, R. and S.M. Kelly. 1987. *Salmonella typhimurium* deletion mutants lacking adenylate cyclase and cyclicAMP receptor protein are avirulent and immunogenic. Infect Immunity. 55(12): 3035–3043.

de la Cruz, M.A., M. Fern´andez-Mora, C. Guadarrama et al. 2007. LeuO antagonizes H-NS and StpA-dependent repression in *Salmonella enterica mpS1*. Mol Microbiol. 66(3): 727–743.

de Reuse, H. and A. Danchin. 1988. The *ptsH*, *ptsI*, and *crr* genes of the *Escherichia coli* phosphoenolpyruvate-dependent phosphotransferase system: a complex operon with several modes of transcription. J Bacteriol. 170(9): 3827–3837.

Death, A. and T. Ferenci. 1994. Between feast and famine: endogenous inducer synthesis in the adaptation of *Escherichia coli* to growth with limiting carbohydrates. J Bacteriol. 176(16): 5101–5107.

Deighan, P., A. Free and C.J. Dorman. 2000. A role for the *Escherichia coli* H-NS-like protein StpA in OmpF porin expression through modulation of *micF* RNA stability. Mol Microbiol. 38(1): 126–139.

Delihas, N. and S. Forst. 2001. MicF: an antisense RNA gene involved in response of *Escherichia coli* to global stress factors. J Mol Biol. 313(1): 1–12.

Desai, T.A. and C.V. Rao. 2010. Regulation of arabinose and xylose metabolism in *Escherichia coli*. Appl Environ Microbiol. 76: 1524–1532.

Deutscher, J., K. Beyreuther, M.H. Sobek, K. Stüber and W. Hengstenberg. 1982. Phosphoenolpyruvate-dependent phosphotransferase system of *Staphylococcus aureus*: factor III[Lac], a trimeric phospho-carrier protein that also acts as a phase transfer catalyst. Biochemistry. 21: 4867–4873.

Deutscher, J., C. Francke and P.W. Postma. 2006. How phosphotransferase system-related protein phosphorylation regulates carbohydrate metabolism in bacteria. Microbiol Mol Biol Rev. 70(4): 939–1031.

Deutscher, J., F. Moussan, D. Ake, M. Derkaoui, A.C. Zebre, T.N. Cao, H. Bouraoui, T. Kentache, A. Mokhtari, E. Milohanic and P. Joyet. 2014. The bacterial phosphoenolpyru vate:carbohydrate phosphotransferase system: regulation by protein phosphorylation and phosphorylation-dependent protein-protein interactions. Microbiol Mol Biol Rev. 78(2): 231–256.

Djordjevic, G.M., J.H. Tchieu and M.H. Saier Jr. 2001. Genes involved in control of galactose uptake in *Lactobacillus brevis* and reconstitution of the regulatory system in *Bacillus subtilis*. J Bacteriol. 183(10): 3224–3236.

Dorel, C., P. Lejeune and A. Rodrigue. 2006. The Cpx system of *Escherichia coli*, a strategic signaling pathway for confronting adverse conditions and for settling biofilm communities? Res Microbiol. 157(4): 306–314.

Dörschug, M., R. Frank, H.R. Kalbitzer, W. Hengstenberg and J. Deutscher. 1984. Phosphoenolpyruvate-dependent phosphorylation site in enzyme IIIglc of the *Escherichia coli* phosphotransferase system. Eur J Biochem. 144: 113–119.

Erni, B. 2013. The bacterial phosphoenolpyruvate:sugar phosphotransferase system (PTS): an interface between energy and signal transduction. J Iran Chem Soc. 10: 593–630.

Fernandez-Mora, M., J.L. Puente and E. Calva. 2004. OmpR and LeuO positively regulate the *Salmonella enterica* serovar typhi *ompS2* porin gene. J Bacteriol. 186(10): 2909–2920.

Ferrario, M., R. Ernsting, D.W. Borst, D.E. Wiese II, R.M. Blumenthal and R.G. Matthews. 1995. The leucine-responsive regulatory protein of *Escherichia coli* negatively regulates transcription of *ompC* and *micF* and positively regulates translation of *ompF*. J Bacteriol. 177(1): 103–13.

Flores, N., S. Flores, A. Escalante et al. 2005. Adaptation for fast growth on glucose by differential expression of central carbon metabolism and gal regulon genes in an *Escherichia coli* strain lacking the phosphoenol pyruvate: carbohydrate phosphotransferase system. Metab Eng. 7(2): 70–87.

Flores-Valdez, M.A., J.L. Puente and E. Calva. 2003. Negative osmoregulation of the *Salmonella ompS1* porin gene independently of OmpR in an hns background. J Bacteriol. 185(22): 6497–6506.

Freedberg, W.B., W.S. Kistler and E.C.C. Lin. 1971. Lethal synthesis of methylglyoxal by *Escherichia coli* during unregulated glycerol metabolism. J Bacteriol. 108(1): 137–144.

Frunzke, J., V. Engels, S. Hasenbein, C. Gätgens and M. Bott. 2008. Co-ordinated regulation of gluconate catabolism and glucoseuptake in *Corynebacterium glutamicum* by two functionally equivalent transcriptional regulators, GntR1 and GntR2. Mol Microbiol. 67(2): 305–322.

Galinier, A., M. Kravanja, R. Engelmann et al. 1998. New protein kinase and protein phosphatase families mediate signal transduction in bacterial catabolite repression. PNAS USA. 95(4): 1823–1828.

Gallegos, M.T., R. Schleif, A. Bairoch, K. Hofmann and J.L. Ramos. 1997. AraC/XylS family of transcriptional regulators. Microbiol Mol Biol Rev. 61(4): 393–410.

Gassner, M., D. Stehlik, O. Schrecker, W. Hengstenberg, W. Maurer and H. Rüterjans. 1977. The phosphoenolpyruvate-dependent phosphotransferase system of *Staphylococcus aureus*. 2. 1H and 31P nuclear magnetic-resonance studies on the phosphocarrier protein HPr, phosphohistidines and phosphorylated HPr. Eur J Biochem. 75: 287–296.

Gerstmeir, R., A. Cramer, P. Dangel, S. Schaffer and B.J. Eikmanns. 2004. RamB, a novel transcriptional regulator of genes involved in acetate metabolism of *Corynebacterium glutamicum*. J Bacteriol. 186(9): 2798–2809.

Gibert, I. and J. Barbe. 1990. Cyclic AMP stimulates transcription of the structural gene of the outer-membrane protein ompA of *Escherichia coli*. FEMS Microbiol Lett. 68(3): 307–311.

Gorke, B. and J. Stulke. 2008. Carbon catabolite repression in bacteria: many ways to make the most out of nutrients. Nat Rev Microbiol. 6(8): 613–624.

Gosset, G. 2005. Improvement of *Escherichia coli* production strains by modification of the phosphoenolpyruvate: sugar phosphotransferase system. Microb Cell Fact. 4: 14.

Gottesman, S. 2005. Micros for microbes: non-coding regulatory RNAs in bacteria. Trends in Genetics. 21(7): 399–404.

Gunnewijk, M.G.W. and B. Poolman. 2000. Phosphorylation state of Hpr determines the level of expression and the extent of phosphorylation of the lactose transport protein of *Streptococcus thermophilus*. J Biol Chem. 275(44): 34073–34079.

Gunnewijk, M.G.W., P.T.C. van den Bogaard, L.M. Veenhoff et al. 2001. Hierarchical control versus autoregulation of carbohydrate utilization in bacteria. J Mol Microbiol Biotechnol. 3(3): 401–413.

Gutierrez-Ríos, R.M., J.A. Freyre-Gonzalez, O. Resendis, J. Collado-Vides, M. Saier and G. Gosset. 2007. Identification of regulatory network topological units coordinating the genome-wide transcriptional response to glucose in *Escherichia coli*. BMC Microbiol. 7: 53.

Hachler, H., S.P. Cohen and S.B. Levy. 1991. *marA*, a regulated locus which controls expression of chromosomal multiple antibiotic resistance in *Escherichia coli*. J Bacteriol. 173(17): 5532–5538.

Halbedel, S., H. Eilers, B. Jonas et al. 2007. Transcription in *Mycoplasma pneumoniae*: analysis of the promoters of the *ackA* and *ldh* genes. J Mol Biol. 371(3): 596–607.

Hall, M.N. and T.J. Silhavy. 1981. The *ompB* locus and the regulation of the major outer membrane porin proteins of *Escherichia coli* K12. J Mol Biol. 146(1): 23–43.

Harbison, C.T., D.B. Gordon, T.I. Lee et al. 2004. Transcriptional regulatory code of a eukaryotic genome. Nature. 430: 99–104.

Henkin, T.M., F.J. Grundy, W.L. Nicholson and G.H. Chambliss. 1991. Catabolite repression of *α*-amylase gene expression in *Bacillus subtilis* involves a trans-acting gene product homologous to the *Escherichia coli* lacI and galR repressors. Mol Microbiol. 5(3): 575–584.

Hogema, B.M., J.C. Arents, R. Bader et al. 1998. Inducer exclusion in *Escherichia coli* by non-PTS substrates: the role of the PEP to pyruvate ratio in determining the phosphorylation state of enzyme IIA$^{Glc}$. Mol Microbiol. 30(3): 487–498.

Hogema, B.M., J.C. Arents, R. Bader and P.W. Postma. 1999. Autoregulation of lactose uptake through the LacY permease by enzyme IIA$^{Glc}$ of the PTS in *Escherichia coli* K-12. Mol Microbiol. 31(6): 1825–1833.

Holtman, C.K., A.C. Pawlyk, N.D. Meadow and D.W. Pettigrew. 2001. Reverse genetics of *Escherichia coli* glycerol kinase allosteric regulation and glucose control of glycerol utilization *in vivo*. J Bacteriol. 183(11): 3336–3344.

Hommais, F., E. Krin, C. Laurent-Winter et al. 2001. Large-scale monitoring of pleiotropic regulation of gene expression by the prokaryotic nucleoid-associated protein, H-NS. Mol Microbiol. 40(1): 20–36.

Huang, L., P. Tsui and M. Freundlich. 1990. Integration host factor is a negative effector of *in vivo* and *in vitro* expression of *ompC* in *Escherichia coli*. J Bacteriol. 172(9): 5293–5298.

Huang, L., P. Tsui and M. Freundlich. 1992. Positive and negative control of *ompB* transcription in *Escherichia coli* by cyclic AMP and the cyclic AMP receptor protein. J Bacteriol. 174(3): 664–670.

Inada, T., K. Kimata and H. Aiba. 1996. Mechanism responsible for glucose-lactose diauxie in *Escherichia coli*: challenge to the cAMP model. Genes to Cells. 1(3): 293–301.

Janga, S.C., H. Salgado, A. Mart´nez-Antonio and J. Collado-Vides. 2007. Coordination logic of the sensing machinery in the transcriptional regulatory network of *Escherichia coli*. Nucleic Acids Res. 35(20): 6963–6972.

Jault, J.M., S. Fieulaine, S. Nessler et al. 2000. The HPr kinase from *Bacillus subtilis* is a homo-oligomeric enzyme which exhibits strong positive cooperativity for nucleotide and fructose 1,6-bisphosphate binding. J Biol Chem. 275(3): 1773–1780.

Johansson, J., B. Dagberg, E. Richet and B.E. Uhlin. 1998. H-NS and StpA proteins stimulate expression of the maltose regulon in *Escherichia coli*. J Bacteriol. 180(23): 6117–6125.

Kim, H.-J., C.-R. Lee, M. Kim, A. Peterkofsky and Y.-J. Seok. 2011. Dephosphorylated NPr of the nitrogen PTS regulates lipid A biosynthesis by direct interaction with LpxD. Biochem Biophys Res Commun. 409(3): 556–561.

Kimata, K., H. Takahashi, T. Inada, P. Postma and H. Aiba. 1997. cAMP receptor protein-cAMP plays a crucial role in glucoselactose diauxie by activating the major glucose transporter gene in *Escherichia coli*. PNAS USA. 94(24): 12914–12919.

Kornberg, H.L. 2001. Routes for fructose utilization by *Escherichia coli*. J Mol Microbiol Biotechnol. 3(3): 355–359.

Kotrba, P., M. Inui and H. Yukawa. 2001. Bacterial phosphotransferase system (PTS) in carbohydrate uptake and control of carbon metabolism. J Biosci Bioeng. 92(6): 502–517.

Kroos, L. 2007. The Bacillus and Myxococcus developmental networks and their transcriptional regulators. Ann Rev Genetics. 41: 13–39.

Kuper, C. and K. Jung. 2006. CadC-mediated activation of the *cadBA* promoter in *Escherichia coli*. J Mol Microbiol Biotechnol. 10(1): 26–39.

Lee, C.R., S.H. Cho, M.J. Yoon, A. Peterkofsky and Y.J. Seok. 2007. *Escherichia coli* enzyme IIA[Ntr] regulates the K[+] transporter TrkA. PNAS USA. 104: 4124–4129.

Lee, C.-R., Y.-H. Park, M. Kim, Y.-R. Kim, S. Park, A. Peterkofsky and Y.-J. Seok. 2013. Reciprocal regulation of the autophosphorylation of enzyme I[Ntr] by glutamine and α-ketoglutarate in *Escherichia coli*. Mol Microbiol. 88(3): 473–485.

Lee, S.J., W. Boos, J.P. Bouche and J. Plumbridge. 2000. Signal transduction between a membrane-bound transporter, PtsG, and a soluble transcription factor, Mlc of *Escherichia coli*. EMBO J. 19(20): 5353–5361.

Lee, Y.H., B.H. Kim, J.H. Kim, W.S. Yoon, S.H. Bang and Y.K. Park. 2007. CadC has a global translational effect during acid adaptation in Salmonella enterica serovar *Typhimurium*. J Bacteriol. 189: 2417–2425.

Liu, M., T. Durfee, J.E. Cabrera, K. Zhao, D.J. Jin and F.R. Blattner. 2005. Global transcriptional programs reveal a carbon source foraging strategy by *Escherichia coli*. J Biol Chem. 280(16): 15921–15927.

Lugtenberg, B., R. Peters, H. Bernheimer and W. Berendsen. 1976. Influence of cultural conditions and mutations on the composition of the outer membrane proteins of *Escherichia coli*. Mol General Genetics. 147(3): 251–262.

Kornberg, H.L. 1990. Fructose transport by *Escherichia coli*. Philosoph Transact Royal Soc B: Biol Sci. 326(1236): 505–513.

Lunin, V.V., Y. Li, J.D. Schrag, P. Iannuzzi, M. Cygler and A. Matte. 2004. Crystal structures of *Escherichia coli* ATP-dependent glucokinase and its complex with glucose. J Bacteriol. 186(20): 6915–6927.

Luo, Y., T. Zhang and H. Wu. 2014. The transport and mediation mechanisms of the common sugars in *Escherichia coli*. Biotechnol Adv. 32: 905–919.

Luscombe, N.M., M.M. Babu, H. Yu, M. Snyder, S.A. Teichmann and M. Gerstein. 2004. Genomic analysis of regulatory network dynamics reveals large topological changes. Nature. 431: 308–312.

Lüttmann, D., R. Heermann, B. Zimmer, A. Hillmann, I.S. Rampp, K. Jung and B. Görke. 2009. Stimulation of the potassium sensor KdpD kinase activity by interaction with the phosphotransferase protein IIANtr in *Escherichia coli*. Mol Micobiol. 72(4): 978–994.

Lüttmann, D., Y. Göpel and B. Görke. 2012. The phosphotransferase protein EIIANtr modulates the phosphate starvation response through interaction with histidine kinase PhoR in *Escherichia coli*. Mol Microbiol. 86(1): 96–110.

Magasanik, B. 1961. Catabolite repression, cold spring harbor symposia on quantitative biology. 26: 249–256.

Majdalani, N., C.K. Vanderpool and S. Gottesman. 2005. Bacterial small RNA regulators. Critical Rev Biochem Mol Biol. 40(2): 93–113.

Mijakovic, I., S. Poncet, A. Galinier et al. 2002. Pyrophosphate producing protein dephosphorylation by HPr kinase/phosphorylase: a relic of early life? PNAS USA. 99(21): 13442–13447.

Miller, V.L. and J.J. Mekalanos. 1988. A novel suicide vector and its use in construction of insertion mutations: osmoregulation of outer membrane proteins and virulence determinants in *Vibrio cholerae* requires toxR. J Bacteriol. 170(6): 2575–2583.

Misko, T.P., W.J. Mitchell, N.D. Meadow and S. Roseman. 1987. Sugar transport by the bacterial phosphotransferase system. Reconstitution of inducer exclusion in *Salmonella typhimurium* membrane vesicles. J Biol Chem. 262(33): 16261–16266.

Monod, J. 1942. Recherches sur la Croissance de cultures Bacteriennes [thesis]. Hermann et Cie, Paris, France.

Moreno, M.S., B.L. Schneider, R.R. Maile, W. Weyler and M.H. Saier Jr. 2001. Catabolite repression mediated by the CcpA protein in *Bacillus subtilis*: novel modes of regulation revealed by whole genome analyses. Mol Microbiol. 39(5): 1366–1381.

Moreno, R., A. Ruiz-Manzano, L. Yuste and F. Rojo. 2007. The *Pseudomonas putida* Crc global regulator is an RNA binding protein that inhibits translation of the AlkS transcriptional regulator. Mol Microbiol. 64(3): 665–675.

Moreno, R. and F. Rojo. 2008. The target for the *Pseudomonas putida* Crc global regulator in the benzoate degradation pathway is the BenR transcriptional regulator. J Bacteriol. 190(5): 1539–1545.

Moxley, J.F., M.C. Jewett, M.R. Antoniewicz et al. 2009. Linking high-resolution metabolic flux phenotypes and transcriptional regulation in yeast modulated by the global regulator Gcn4p. PNAS USA. 106(16): 6477–6482.

Muller, C., L. Petruschka, H. Cuypers, G. Burchhardt and H. Herrmann. 1996. Carbon catabolite repression of phenol degradation in *Pseudomonas putida* is mediated by the inhibition of the activator protein PhlR. J Bacteriol. 178(7): 2030–2036.

Nandi, B., R.K. Nandy, A. Sarkar and A.C. Ghose. 2005. Structural features, properties and regulation of the outer-membraneprotein W (OmpW) of *Vibrio cholera*. Microbiol. 151(9): 2975–2986.

Navarre, W.W., S. Porwollik, Y. Wang et al. 2006. Selective silencing of foreign DNA with low GC content by the H-NS protein in *Salmonella*. Science. 313(5784): 236–238.

Nelson, S.O., J.K. Wright and P.W. Postma. 1983. The mechanism of inducer exclusion. Direct interaction between purified $III^{Glc}$ of the phosphoenolpyruvate: sugar phosphotransferase system and the lactose carrier of *Escherichia coli*. EMBO J. 2: 715–720.

Nicholson, T.L., K. Chiu and R.S. Stephens. 2004. *Chlamydia trachomatis* lacks an adaptive response to changes in carbon source availability. Infect Immun. 72(7): 4286–4289.

Nikaido, H. 1996. Outer Membrane, ASM Press, Washington DC, USA. edited by F.C. Neidhardt.

Nikaido, H. 2003. Molecular basis of bacterial outer membrane permeability revisited. Microbiol Mol Biol Rev. 67(4): 593–656.

Nikaido, H. and T. Nakae. 1980. The outer membrane of gram-negative bacteria. Adv Microb Physiol. 20: 163–250.

Nikaido, H. and E.Y. Rosenberg. 1983. Porin channels in *Escherichia coli*: studies with liposomes reconstituted from purified proteins. J Bacteriol. 153(1): 241–252.

Nikaido, H. and M. Vaara. 1985. Molecular basis of bacterial outer membrane permeability. Microbiol Rev. 49(1): 1–32.

Novotny, M., W.L. Frederickson, E.B. Waygood and M.H. Saier Jr. 1985. Allosteric regulation of glycerol kinase by enzyme $III^{glc}$ of the phosphotransferase system in *Escherichia coli* and *Salmonella typhimurium*. J Bacteriol. 162(2): 810–816.

Oh, J.T., Y. Cajal, E.M. Skowronska et al. 2000. Cationic peptide antimicrobials induce selective transcription of *micF* and *osmY* in *Escherichia coli*. Biochim Biophys Acta. 1463(1): 43–54.

Papenfort, K., V. Pfeiffer, S. Lucchini, A. Sonawane, J.C.D. Hinton and J. Vogel. 2008. Systematic deletion of *Salmonella* small RNA genes identifies CyaR, a conserved CRP-dependent riboregulator of OmpX synthesis. Mol Microbiol. 68(4): 890–906.

Parche, S., M. Beleut, E. Rezzonico et al. 2006. Lactose-over-glucose preference in Bifidobacterium longum NCC2705: *glcP*, encoding a glucose transporter, is subject to lactose repression. J Bacteriol. 188: 1260–1265.

Park, J.M., P. Vinuselvi and S.K. Lee. 2012. The mechanism of sugar-mediated catabolite repression of the propionate catabolic genes in *Escherichia coli*. Gene. 504: 116–121.

Park, Y.H., B.R. Lee, Y.J. Seok and A. Peterkofsky. 2006. *In vitro* reconstitution of catabolite repression in *Escherichia coli*. J Biol Chem. 281(10): 6448–6454.

Pas, H.H. and G.T. Robillard. 1988. *S*-Phosphocysteine and phosphohistidine are intermediates in the phosphoenolpyruvate-dependent mannitol transport catalyzed by *Escherichia coli* EII$^{Mtl}$. Biochemistry. 27: 5835–5839.

Pérez-Alfaro1, R.S., M. Santillán, E. Galán-Vásquez and A. Martínez-Antonio. 2014. Regulatory switches for hierarchical use of carbon sources in *E. coli*. Network Biol. 4(3): 95–108.

Peterkofski, A., G. Wang and Y.-J. Seok. 2006. Parallel PTS systems. Arch Biochem Biophys. 453(1): 101–107.

Petersen, S. and G.M. Young. 2002. Essential role for cyclic AMP and its receptor protein in *Yersinia enterocolitica* virulence. Infect Immun. 70(7): 3665–3672.

Pflüger-Grau, K. and B. Görke. 2010. Regulatory roles of the bacterial nitrogen-related phosphotransferase system. Trend Microbiol. 18(5): 205–214.

Pickl, A., U. Johnsen and P. Schönheit. 2012. Fructose degradation in the haloarchaeon *Haloferax volcanii* involves a bacterial type phosphoenolpyruvate-dependent phosphotransferase system, fructose-1-phosphate kinase, and class II fructose-1,6-bisphosphate aldolase. J Bacteriol. 194: 3088–3097.

Plumbridge, J. 1998. Expression of *ptsG*, the gene for the major glucose PTS transporter in *Escherichia coli*, is repressed by Mlc and induced by growth on glucose. Mol Microbiol. 29(4): 1053–1063.

Poolman, B. and W.N. Konings. 1993. Secondary solute transport in bacteria. Biochim Biophys Acta. 1183(1): 5–39.

Poolman, B., J. Knol, B. Mollet, B. Nieuwenhuis and G. Sulter. 1995. Regulation of bacterial sugar-H+ symport by phosphoenolpyruvate-dependent enzyme I/HPr-mediated phosphorylation. PNAS USA. 92(3): 778–782.

Poolman, B., J. Knol, B. Mollet, B. Nieuwenhuis and G. Sulter. 1995. Regulation of bacterial sugar-H$^+$ symport by phosphoenolpyruvate-dependent enzyme I/HPr-mediated phosphorylation. PNAS USA. 92(3): 778–782.

Postma, P.W., J.W. Lengeler and G.R. Jacobson. 1996. Phosphoenolpyruvate: carbohydrate phosphotransferase systems. pp. 1149–1174. *In*: F.C. Neidhardt (ed.). *Escherichia coli* and Salmonella: Cellular and Molecular Biology. ASM Press, Washington DC, USA.

Powell, B.S., D.L. Court, T. Inada, Y. Nakamura, V. Michotey, X. Cui et al. 1995. Novel proteins of the phosphotransferase system encoded within the *rpoN* operon of *Escherichia coli*. Enzyme IIA$^{Ntr}$ affects growth on organic nitrogen and the conditional lethality of an erats mutant. J Biol Chem. 270: 4822–4839.

Pratt, L.A. and T.J. Silhavy. 1996. The response regulator SprE controls the stability of RpoS. PNAS USA. 93(6): 2488–2492.

Raivio, T.L. 2005. Envelope stress responses and gram-negative bacterial pathogenesis. Mol Microbiol. 56(5): 1119–1128.

Ramstr¨om, H., M. Bourotte, C. Philippe, M. Schmitt, J. Haiech and J.J. Bourguignon. 2004. Heterocyclic bis-cations as starting hits for design of inhibitors of the bifunctional enzyme histidine containing protein kinase/phosphatase from *Bacillus subtilis*. J Medicinal Chem. 47(9): 2264–2275.

Reizer, J., C. Hoischen, F. Titgemeyer et al. 1998. A novel protein kinase that controls carbon catabolite repression in bacteria. Mol Microbiol. 27(6): 1157–1169.

Rhee, J.E., K.S. Kim and S.H. Choi. 2005. CadC activates pH dependent expression of the *Vibrio vulnificus* cadBA operon at a distance through direct binding to an upstream region. J Bacteriol. 187(22): 7870–7875.

Romby, P., F. Vandenesch and E.G.H. Wagner. 2006. The role of RNAs in the regulation of virulence-gene expression. Curr Opin Microbiol. 9(2): 229–236.

Saier, M.H., C.V. Tran and R.D. Barabote. 2006. TCDB: the transporter classification database for membrane transport protein analyses and information. Nucleic Acids Res. 34: D181–D186.

Saier, M.H., M.R. Yen, K. Noto, D.G. Tamang and C. Elkan. 2009. The transporter classification database: recent advances. Nucleic Acids Res. 37(1): D274–D278.

Santiviago, C.A., C.S. Toro, A.A. Hidalgo, P. Youderian and G.C. Mora. 2003. Global regulation of the *Salmonella enterica* serovar *Typhimurium* major porin, OmpD. J Bacteriol. 185(19): 5901–5905.

Schumacher, M.A., G. Seidel, W. Hillen and R.G. Brennan. 2007. Structural mechanism for the fine-tuning of CcpA function by the small molecule effectors glucose 6-phosphate and fructose 1,6-bisphosphate. J Mol Biol. 368(4): 1042–1050.

Seidel, G., M. Diel, N. Fuchsbauer and W. Hillen. 2005. Quantitative interdependence of coeffectors, CcpA and *cre* in carbon catabolite regulation of *Bacillus subtilis*. FEBS J. 272(10): 2566–2577.

Seshasayee, A.S., P. Bertone, G.M. Fraser and N.M. Luscombe. 2006. Transcriptional regulatory networks in bacteria: from input signals to output responses. Curr Opinion Microbiol. 9(5): 511–519.

Smirnova, I., V. Kasho, J.Y. Choe, C. Altenbach, W.L. Hubbell and H.R. Kaback. 2007. Sugar binding induces an outward facing conformation of LacY. PNAS USA. 104(42): 16504–16509.

Storz, G., S. Altuvia and K.M. Wassarman. 2005. An abundance of RNA regulators. Annual Rev Biochem. 74: 199–217.

Tanaka, Y., K. Kimata and H. Aiba. 2000. A novel regulatory role of glucose transporter of *Escherichia coli*: membrane sequestration of a global repressor Mlc. EMBO J. 19(20): 5344–5352.

Tchieu, J.H., V. Norris, J.S. Edwards and M.H. Saier Jr. 2001. The complete phosphotransferase system in *Escherichia coli*. J Mol Microbiol Biotechnol. 3(3): 329–346.

Timmermans, J. and L. van Melderen. 2010. Post-transcriptional global regulation by CsrA in bacteria. Cellular and Mol Life Sci. 67(17): 2897–2908.

Titgemeyer, F., R.E. Mason and M.H. Saier Jr. 1994. Regulation of the raffinose permease of *Escherichia coli* by the glucose-specific enzyme IIA of the phosphoenolpyruvate: sugar phosphotransferase system. J Bacteriol. 176(2): 543–546.

Titgemeyer, F. and W. Hillen. 2002. Global control of sugar metabolism: a gram-positive solution. Antonie van Leeuwenhoek. 82(1-4): 59–71.

Van den Bogaard, P.T.C., M. Kleerebezem, O.P. Kuipers and W.M. Vos. 2000. Control of lactose transport, $\beta$-galactosidase activity, and glycolysis by CcpA in Streptococcus thermophilus: evidence for carbon catabolite repression by a non- phosphoenolpyryvate-dependent phosphotransferase system sugar. J Bacteriol. 182: 5982–5989.

Vanderpool, C.K. and S. Gottesman. 2004. Involvement of a novel transcriptional activator and small RNA in post-transcriptional regulation of the glucose phosphoenolpyruvate phosphotransferase system. Mol Microbiol. 54(4): 1076–1089.

Vogel, J. and K. Papenfort. 2006. Small non-coding RNAs and the bacterial outer membrane. Curr Opin Microbiol. 9(6): 605–611.

Von Kruger, W.M., L.M. Lery, M.R. Soares et al. 2006. The phosphate-starvation response in *Vibrio cholerae* O1 and *phoB* mutant under proteomic analysis: disclosing functions involved in adaptation, survival and virulence. Proteomics. 6: 1495–1511.

von Meyenburg, K. and H. Nikaido. 1977. Outer membrane of gram negative bacteria. XVII. Specificity of transport process catalyzed by the $\lambda$-receptor protein in *Escherichia coli*. Biochem Biophys Res Commun. 78(3): 1100–1107.

Wadler, C.S. and C.K. Vanderpool. 2007. A dual function for a bacterial small RNA: SgrS performs base pairing-dependent regulation and encodes a functional polypeptide. PNAS USA. 104(51): 20454–20459.

Wanner, B.L. 1996. Phosphorus assimilation and control of the phosphate regulon. pp. 1357–1381. *In*: F.C. Neidhardt, I.I.I. Curtiss, R.J.L. Ingraham et al. (eds.). *Escherichia coli* and Salmonella: Cellular and Molecular Biology. ASM Press, Washington, DC, USA.

Warner, J.B. and J.S. Lolkema. 2003. A Crh-specific function in carbon catabolite repression in *Bacillus subtilis*. FEMS Microbiol Lett. 220(2): 277–280.

Wendisch, V.F., A.A. de Graaf, H. Sahm and B.J. Eikmanns. 2000. Quantitative determination of metabolic fluxes during cultivation of two carbon sources: comparative analyses with *Corynebacterium glutamicum* during growth on acetate and/or glucose. J Bacteriol. 182(11): 3088–3096.

Winkler, H.H. and T.H. Wilson. 1967. Inhibition of $\beta$-galactoside transport by substrates of the glucose transport system in *Escherichia coli*. Biochim Biophys Acta. 135(5): 1030–1051.

Yamamoto, K., K. Hirao, T. Oshima, H. Aiba, R. Utsumi and A. Ishihama. 2005. Functional characterization *in vitro* of all two component signal transduction systems from *Escherichia coli*. J Biol Chem. 280(2): 1448–1456.

Zhou, H.X., G. Rivas and A.P. Minton. 2008. Macromolecular crowding and confinement: biochemical, biophysical, and potential physiological consequences. Ann Rev Biophys. 37: 375–397.

4

# Catabolite Regulation of the Main Metabolism

## ABSTRACT

Living organisms have sophisticated but well organized regulation system. It is important to understand the metabolic regulation mechanisms in relation to growth environment for the efficient design of cell factories for biofuel and biochemical production. The metabolism is controlled by both enzyme level regulation and transcriptional regulation via transcription factors such as cAMP-Crp and Cra in *Escherichia coli*. Moreover, multiple regulations are coordinated by the intracellular metabolites, where fructose 1,6-bisphosphate (FBP), phosphoenol pyruvate (PEP), and pyruvate (PYR) play important roles for enzyme level regulation as well as transcriptional control. Effect of Pyk mutation on the metabolic changes is explained for *E. coli*, *Corynebacterium glutamicum*, *Bacillus subtilis*, and cyanobacteria. The overflow metabolism is also explained in terms of enzyme level and transcriptional regulations.

**Keywords**
Catabolite regulation, glycolytic flux, overflow metabolism, enzyme level regulation, transcriptional regulation, FBP, Pyk isozymes, Pyk mutation

## 1. Introduction

The living organisms on earth survive by manipulating the cell system in response to the change in growth environment by sensing signals of both external and internal states of the cell. The complex signaling networks interconvert signals or stimuli for the cell to function properly. The transfer of information in signal transduction pathways and cascades is designed to respond to the variety of growth environment. Metabolism is the core for energy generation **(catabolism)** and cell synthesis **(anabolism)**. Metabolic network, defined as the set and topology of metabolic biochemical reactions within a cell, plays an essential role for the cell to survive, where it is under organized control. The set of enzymes changes dynamically in accordance with the change in growth environment and the cell's state. The enzymes which form the metabolic pathways are subject to multiple levels of regulation, and it is important to deeply understand the overall regulation mechanism. Although huge amount of information is embedded in the genome, only a subset of the pathways among possible topological networks is active at certain point in time under certain growth condition.

Recent investigation on the metabolism is widespread, ranging from bacteria to human (Jone and Thompson 2009, Hsu and Sabatini 2008, Kroemer and Pouyssegur 2008, Folger et al. 2011). The metabolic capabilities allow various organisms to grow in various limiting conditions and environmental niches in the ecological biosphere (McInerney et al. 2009, Dolfing et al. 2008, Lin et al. 2006). Many efforts have been focused on the emerging challenges in sustainable energy, green society, as well as pharmaceuticals for human health by modifying the metabolic pathways (Peralta-Yahya et al. 2012, Atsumi et al. 2008, Yim et al. 2011, Dellomonaco et al. 2011, Bond-Watts et al. 2011, Shen et al. 2011, Keasling 2010, Shimizu 2014).

Deep understanding on the metabolic regulation mechanism is essential for all these efforts for manipulating and redesigning the metabolism, and it is critical to understand the basic principles that govern the cell metabolism (Bar-Even et al. 2012, Flamholz et al. 2013, Shimizu 2013, 2014, 2016, Chubukov et al. 2014). Such principles may be in common to various organisms, or some set of organisms, while some are the specific to the organisms of concern.

Biochemical logic of metabolic pathways may be determined based on key biochemical constraints such as thermodynamic favorability, availability of enzymatic mechanisms, and physicochemical properties of pathway intermediates (Bar-Even et al. 2012). More specifically, there might be a connection between an organism's growth environment and thermodynamic and biochemical properties for the determination of pathways (Flamholz et al. 2013). How do organisms select the pathways among available pathways? For example, there are several glycolysis pathways such as Embden-Meyerhoff-Parnas (EMP) pathway and Entner-Doudoroff (ED) pathway, etc., but how is the pathway selected among them? The glucose metabolism may reflect a trade off between a pathway's energy (ATP) yield and the amount of enzymatic protein required to catalyze the pathway flux. From this point of view, some microorganisms such as *Zymomonas mobilis* and *Pseudomonas* sp. utilize ED pathway instead of most popular energy intensive EMP pathway due to less requirement of enzymatic protein together with thermodynamic preference (Flamholz et al. 2013).

In fact, the decision may be made not only by the above consideration, but it is also made by transcriptional regulation together with global regulators or transcription factors. Moreover, some specific metabolites are also involved in the coordinated regulation of the metabolism. Here, somewhat in-depth explanation is given for the catabolic regulation of microbes with special interest on the coordination of regulation systems.

## 2. Regulation of the Glycolytic Flux

Among the main metabolic pathways, **glycolysis** is common to many organisms and plays an essential role for both catabolism and anabolism, where the glycolysis signifies the breakdown of carbon sources by **Embden-Meyerhoff-Parnas (EMP) pathway**. In particular, the glycolysis provides most of the ATP by **substrate level phosphorylation** when terminal electron acceptors are not available (catabolism), and some of its intermediates are the precursors for the cell synthesis (anabolism). Moreover, the terminal metabolites such as phosphoenolpyruvate (PEP) and pyruvate (PYR) are the starting metabolites for the rest of the main metabolic pathways. Thus, the modulation of the glycolytic flux is by far important for the cells to survive under varying environmental conditions and for the metabolic engineering toward the useful metabolite production.

Much effort has thus been focused on increasing the glycolyic flux. However, the problem of controlling the glycolytic flux is somewhat as-yet-uncharacterized issue in cellular metabolism. Namely, the enhancement or modulation of enzyme activities of the glycolysis may not necessarily improve the glycolytic flux in *Escherichia coli* (Ruyter et al. 1991, Babul et al. 1993, Emmeling et al. 1999, Haverkorn van Rijsewijk et al. 2011), in yeast (Muller et al. 1997, Schaaff et al. 1989, Hauft et al. 2000, Smits et al. 2000), in *Lactococcus lactis* (Solem et al. 2003, Ramos et al. 2004, Koebman et al. 2006, Solem et al. 2010), in *Bacillus* sp. (Chubukov et al. 2013), in *Corynebacterium glutamicum* (Dominguez et al. 1998), and in *Zymomonas mobilis* (Snoep et al. 1995). All of these results imply that the glycolytic flux may not be much controlled by the glycolytic enzymes, but the controlling factor may reside outside of the glycolysis itself.

On the other hand, the specific glucose consumption rate or the glycolytic flux increases under certain culture conditions such as under nitrogen limitation in *E. coli* (Hua et al. 2003) and in *Saccharomyces cerevisiae* (Larsson et al. 1997), under oxygen limitation in *E. coli* (Emmerling et al. 2000), in *S. cerevisiae* (van der Brink et al. 2008), and in *L. lactis* (Neves et al. 2002), under acidic condition in *L. lactis* subsp. *cremoris* (Even et al. 2003), and at higher temperature in *S. cerevisiae* (Postmus et al. 2008, 2012). The cofactor such as NADH may also affect the metabolism and the glycolytic flux. Namely, the glycolytic flux may be increased by the overexpression of water-forming NADH oxidase (NOX) and the soluble $F_1$-ATPase, where NADH/NAD⁺ ratio becomes lower for the former, while acetate is more formed and ATP/ADP ratio becomes lower for the latter (Holm et al. 2010). In the case of *L. lactis*, the overexpression of NOX affects the glucose consumption rate (Neves et al. 2002), where ATP is less produced. The NADH/NAD⁺ ratio also decreases by overexpression of the *ndh* gene which encodes NADH dehydrogenase (NDH), and then the glucose uptake rate increased in *C. glutamicum* under oxygen deprivation (Tsuge et al. 2015). Moreover, the specific glucose uptake rate increases by the reduction of proton motive force (PMF) in the respiratory chain in *E. coli* (Kihira et al. 2012), in *C. glutamicum* (Sekine et al. 2001), and in multivitamin-auxotrophic yeast, *Torulopsis glabrata* (Liu et al. 2006a,b). In all the above cases, ATP production is less than the control. The negative correlation between the glycolytic flux and ATP/ADP ratio has thus been illustrated for *E. coli* (Koebmann et al. 2002) and for *S. cerevisiae* (Larsson et al. 1997), where as the ATP/ADP ratio decreases by modulating ATPase, the glycolytic flux increases. This indicates that the majority of the control of the cell growth rate resides in the anabolism, and the cell's state is carbon or energy limitation. Namely, the control of the glycolytic flux may reside in the ATP demand (Fig. 1) (Koebmann et al. 2002). In fact, it was shown that the cell is able to sense the energetic state, and regulates the glycolysis (Aledo et al. 2008), and that the glycolytic flux of *S. cerevisiae* may be regulated by ATP by changing the ambient temperature and altering NADH (Chen et al. 2014). Thus, various strategies have been attempted to manipulate the ATP production for the useful metabolite production (Zhou et al. 2009). However, ATP is usually produced in excess in the cell in the typical growth condition (Yang et al. 2004), and it is not obvious that ATP level directly controls the glycolytic flux.

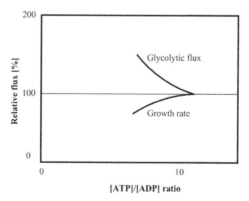

**Figure 1.** Effect of ATP/ADP ratio on the specific glucose consumption rate and the cell growth rate (Koebman et al. 2002).

## 3. Enzyme Level Regulation of the Glycolysis

The metabolic regulation is made essentially by fast enzyme level regulation and relatively slow transcriptional regulation (Ralser et al. 2009). As for enzyme level regulation, the intermediate of the glycolysis such as fructose 1,6-bisphoaphate (FBP) plays an essential role for the allosteric regulation of the glycolytic flux, where FBP senses the flux of the upper glycolysis and regulates the lower glycolytic flux by the allosteric activation of pyruvate kinase (Pyk) (Fig. 2) (Waygood et al. 1976). The relationship between FBP and the flux of the lower glycolysis may be considered as **"flux sensing"**, where the metabolic fluxes may be sensed by molecular systems, and in turn this may transmit the sensed signals (such as FBP concentration) to the regulatory machinery in *E. coli* (Kotte et al. 2010) (Fig. 3). Namely, FBP pool size affects the enzyme activity of Pyk in the lower glycolysis to match the upper glycolytic flux, making the linear relationship between the glycolytic flux and FBP pool size (Fig. 2) (Kochanowski et al. 2013). This mechanism is important for the cell survival. Namely, the ATP (and also PEP in the case of phosphotransferase system: PTS) is required in the upper glycolysis, while the ATP is gained in the lower part of the glycolysis. Thus if the upper glycolytic flux surpasses over the lower glycolytic flux during the transient metabolism, the cell encounters the problem of energy shortage, and endangers the survival (van Heerden et al. 2014). This risk can be reduced with the incorporation of such fast feed-forward regulation (Fig. 2).

The similar mechanism may be seen in other organisms such as *S. cerevisiae*, where the glycolytic flux is also sensed by FBP, and FBP

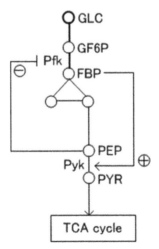

**Figure 2.** Allosteric regulation of FBP on Pyk and linear relationship between FBP pool size and the glycolytic flux.

regulates the respiratory activity (Diaz-Ruiz et al. 2008), resulting in the linear relationship between the glycolytic flux and the ethanol production rate (Christen and Sauer 2011, Huberts et al. 2013).

In the case of lactic acid bacteria such as *L. lactis*, FBP allosterically activates not only Pyk but also lactate dehydrogenase (LDH), which allows the cell to consume glucose quickly, and instantly produce pyruvate via PTS, and then convert it to lactate when glucose is available (Fig. 4), thereby acidifying the surrounding environment, preventing other organisms or competitors to make access to glucose and other nutrients (Voit et al. 2006). This may be effective for lactic acid bacteria in such environment as gastrointestinal tract where the nutrients are infrequently available, and many microorganisms compete for them.

This feedforward activation plays also an important role when glucose was depleted. Namely, when glucose was depleted, FBP pool size decreases and deactivates Pyk or shuts off the flux from PEP to PYR, causing accumulation of PEP. In the case of *E. coli*, phosphoenol pyruvate carboxylase (Ppc) is also allosterically activated by FBP (Xu et al. 2012a),

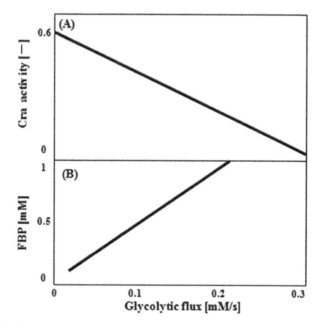

**Figure 3. (A)** Cra activity as a function of glycolytic flux, **(B)** FBP concentration as a function of glycolytic flux (Kochanowski et al. 2013).

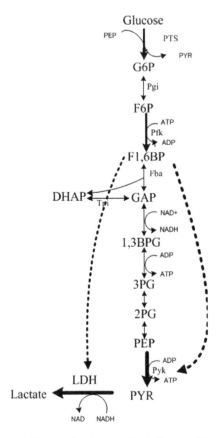

**Figure 4.** Enzyme level regulation in Lactobacillus sp.

and thus the consumption of PEP through Pyk and Ppc is stopped upon glucose depletion (Xu et al. 2012a). In fact, Ppc requires another co-activator, acetyl CoA (AcCoA) (Izui et al. 1981, Yang et al. 2003, Xu et al. 2012a), where AcCoA pool size decreases in accordance with the shut-off of Pyk, and thus Ppc is tightly deactivated by the decrease in both FBP and AcCoA (Fig.5). In the case of *E. coli*, PEP accumulation is backed up by the transcriptional regulation as well, i.e., the decreased FBP upon glucose depletion activates the transcription factor (TF) such as catabolite repressor/activator (Cra), which represses *pykF* and *ppc* gene expression, while activating *pckA* and *ppsA* gene expression (Shimada et al. 2011a), enhancing the accumulation of PEP (Fig. 6). Note that part of PEP is converted to PYR through PTS without Pyk pathway, but it becomes low upon glucose depletion. The accumulated PEP can be used as a substrate for the phosphorylation of the newly internalized glucose through PTS when glucose becomes available

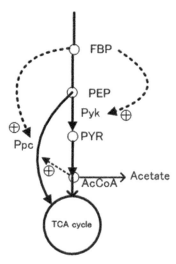

**Figure 5.** Enzyme level regulation of the main metabolism in *E. coli*.

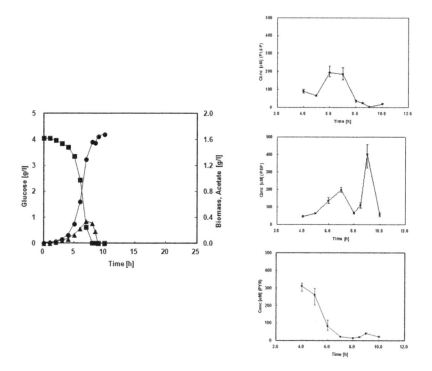

**Figure 6.** Time profile of the batch culture of wild type *E. coli* (left), and the corresponding change in the intracellular metabolite concentrations of FBP, PEP, and PYR, which indicate the accumulation of PEP upon depletion of glucose (Toya et al. 2010).

again, and the glucose can be consumed quickly as shown for *L. lactis* (Voit et al. 2006) and for *E. coli* (Xu et al. 2012a).

The ultrasensitive FBP regulation on Pyk can be also seen in *S. cerevisiae* (Xu et al. 2012b), where this organism does not furnish PTS, so that the glucose phosphorylation does not rely on PEP but ATP, and thus PEP accumulation mechanism might be a way to store ATP equivalents, and in fact the quick glucose consumption was observed upon glucose addition after glucose depletion (Xu et al. 2012b).

This **ultrasensitive allostery** becomes critical for the practical applications such as DO-stat and pH-stat operations, where the repeated fed-batch is made based on the signals of carbon source starvation obtained by the dissolved oxygen (DO) sensor and pH sensor. Without such ultrasensitive allostery, the fermentation will be prolonged and also give eventual damage to the cell due to frequent and longer period of starvation. Moreover, the nutrients are infrequently available to the living organisms depending on their environment or habitat in nature, where the cell's state may be one of feast or famine, or rather in-between hunger with carbon-scavenging system (Koch 1971, Ferenci 2001). In the mammalian gastrointestinal tract, the availability of the preferred nutrients is limited and variable, which causes bacterial cells to be hungry or occasionally starving. It is, therefore, important for the cells to furnish such regulation mechanism for survival.

This feedforward activation of Pyk by FBP is conserved in many organisms such as bacteria as *E. coli*, *L. lactis*, *Streptococcus lactis* (Collins and Thomas 1974), yeast, and higher organisms such as rat liver (Bailey et al. 1968), mammarian (Jurica et al. 1998), human liver (Carminatti et al. 1968), human erythrocyte (Mattevi et al. 1996), stem cells (Ochocki and Simon 2013), and cancer cell (Vander Heiden et al. 2009).

On the other hand, this mechanism may not be seen in Gram-positive bacteria such as *B. subtilis* (Diesterhaft and Freese 1972), *B. stearothermophilus* (Sakai et al. 1986, Walker et al. 1992, Lowell et al. 1998), *C. glutamicum* (Jetten et al. 1994), pathogens such as *Mycobacterium smegmatis* (Kapoor and Venkitasubramanian 1983), and *Z. moblis* (Pawluk et al. 1986). In the case of *B. subtilis*, the FBP pool size correlates with the glycolytic flux (Chubukov et al. 2013), where FBP may modulate the glycolysis not by Pyk, but by the transcriptional regulation via such transcription factors as CcpA and CggR (Deutscher et al. 1995, Doan et al. 2003). The Pyk of *B. subtilis* is activated by adenosine monophosphate (AMP) and ribose 5-phosphate (R5P) (Dieterhaft and Frees 1972), while that of *Z. mobilis* lacks any allosteric regulation (Pawluk et al. 1986). The reason for the latter may be that the Entner-Doudoroff (ED) pathway is dominant instead of EMP pathway in *Z. mobilis*. The Pyk activity of photosynthestic organism such as cyanobacteria is also activated by R5P and AMP as well as glycerol 3-phosphate (GL3P), while inhibited by FBP, inorganic phosphate ($P_i$), and ATP, etc. (Knowles et al. 2001).

## 4. Regulation of Pyruvate Kinase

### *4.1 Regulation of Pyk expression by isozymes*

Although the feedforward activation of Pyk by FBP is important as stated above, the activity of Pyk depends on the isozymes to be expressed. In fact, there are several functionally different types of Pyk (Munoz and Ponce 2003), and the isozymes are transcriptionally regulated depending on the culture condition such as the types of carbon sources available. There are two types of Pyk isozymes such as PykI (PykF) and Pyk II (PykA) in *E. coli* (Ponce et al. 1995) and in *Salmonella typhimurium* (Garcia-Olalla and Garrido-Pertierra 1987), where PykI is activated by FBP, while PykII is activated by AMP (Waygood et al. 1976, Garcia-Olalla and Garrido-Pertierra 1987) and also by R5P (Malcovati and Valentini 1982). These two isozymes are under transcriptional regulation, where PykF is dominant for the case of using glucose as a carbon source, while PykA is activated when glycerol is used (Peng et al. 2003). *S. cerevisiae* also has two isozymes such as Pyk1 activated by FBP, and Pyk2 that is insensitive to FBP (Murcott et al. 1991, Barwell et al. 1971, Jurica et al. 1998, Boles et al. 1997). In some organisms such as *Trypansoma brucei*, Pyk is activated by fructose 2,6-bisphosphate (Van Schaftingen et al. 1985). In the case of mammarian tissues, there are four types of isozymes, where $M_1$ is found in skeleton muscle, $M_2$ in kidney, adipose tissue, and lungs, L in liver, and R in red blood cells. Kinetic properties of isozymes are different depending on the specific metabolic requirements of the tissues. While $M_1$ isozyme lacks allosteric regulation, $M_2$, L, and R isozymes are all allosterically regulated (Muirhead et al. 1986, Demina et al. 1999). In particular, $M_2$ isozyme is under allosteric regulation by FBP, while L-isozyme gene transcription is positively regulated by glucose and insulin, and negatively regulated by glucagon and cAMP (Vaulont et al. 1986, Kahn et al. 1997).

### *4.2 Effects of Pyk mutation on the metabolism*

As mentioned above, Pyk is important for enzyme level regulation in many organisms. If Pyk gene was disrupted, PEP accumulates, and in turn the intermediates of the glycolysis above PEP up to G6P also accumulate as typically see in *E. coli* (Toya et al. 2010). The accumulated PEP causes an increase in the flux from PEP to oxaloacetate (OAA) via Ppc, where Ppc is also allosterically activated by the increased FBP. The increased OAA and in turn malate (MAL) causes an increase in the flux from MAL to PYR via malic enzyme (Mez), and thus the decreased PYR caused by Pyk mutation is backed up by this alternative pathway (Fig. 7a), and the cell growth is little affected by this mutation in *E. coli* (Siddiquee et al. 2004a,b, Toya et al. 2010). Note again that PTS allows part of PEP to be converted to PYR without using

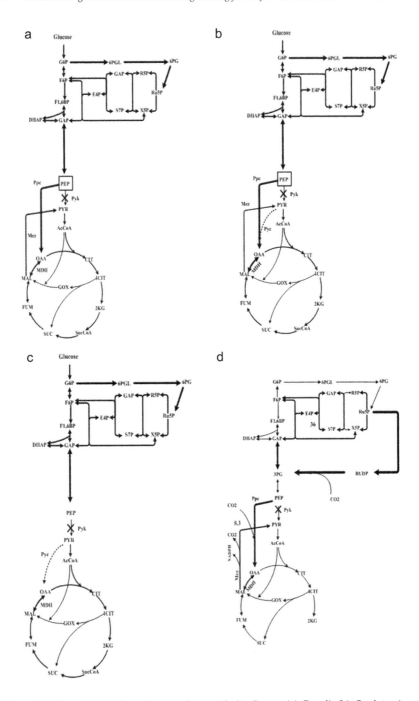

**Figure 7.** Effect of Pyk disruption on the metabolic fluxes: (a) *E. coli*, (b) *C. glutamicum*, (c) *B. subtilis*, (d) cyanobacteria.

Pyk pathway. Moreover, the increase in PEP pool size allosterically inhibits Pfk activity, and in turn F6P and G6P tend to accumulate (Blangy et al. 1968), and the accumulated G6P flows into the oxidative PP pathway (Siddiquee et al. 2004a,b, Toya et al. 2010). Both increases in PEP and E4P caused by the activated PP pathway flux may activate shikimic acid formation pathway as well as aromatic amino acid formation pathway in the practical application of the mutant. It is notable that NADPH is overproduced by the increased flux of the oxidative PP pathway and Mez (Toya et al. 2010), where this may be utilized for the fatty acid or poly-hydroxy butyrate (PHB) production in the practical application or this may contribute to the defense against oxidative stress.

Although the reduced amount of pyruvate is backed up by the alternative pathways in Pyk mutant, its amount may be reduced, and causes the reduction of acetate formation in *E. coli* (Ponce et al. 1999, Zhu et al. 2001). Even with reduced amount, pyruvate is backed up via the alternative pathway, and thus some amount of AcCoA may still be present (Siddiquee et al. 2004a,b), keeping Ppc to be active to some extent even if FBP pool size was reduced by the depletion of glucose, and thus the ultrasensitive allostery may not be retained in Pyk mutant (Toya et al. 2010). Upon Pyk gene knockout, cAMP level increases (Cunningham et al. 2009) due to increase in PEP/PYR ratio (Siddiquee et al. 2004a,b), but marginal for the activation of cAMP-Crp to modulate the metabolism.

*C. glutamicum* is a predominantly aerobic, nonpathogenic, biotin-auxotrophic Gram-positive soil bacterium belonging to the class of Actinobacteria, which also includes *Mycobacterium* and *Streptomyces* (Gao and Gupta 2012). *C. glutamicum* is widely employed for the industrial fermentation of amino acids such as lysine and glutamate, etc. (Leuchtenberger et al. 2005). *C. glutamicum* has only one Pyk (Jetten et al. 1994), and pyruvate carboxylase (Pyc) is dominant for the anaplerotic pathway in the typical growth condition (Becker et al. 2008). Pyk mutation causes the change in the anaplerotic pathway from Pyc to Ppc, and pyruvate can be backed up by the alternative pathway via Ppc-Mae (Fig. 7b), while Pck activity decreases (Sawada et al. 2010), where overflow metabolism may be seen by the accumulation of dihydroxyacetone (DHA) and glycerol (Becker et al. 2008). Moreover, TCA cycle is activated, and aspartate and glutamate production rates may be increased, and NADPH production at ICDH is increased, while its production at PP pathway and Mae decreases (Becker et al. 2008). Pyk is essential for high-level lysine production in *C. lactofermentum*, and thus lysine production significantly decreases in Pyk mutants as compared to its parent strain (Gubler et al. 1994).

In the case of low GC Gram-positive, spore-forming, soil bacterium, *B. subtilis*, the Pyk mutation causes significant reduction in the growth rate and acid production, since this organism does not have Ppc but has only Pyc (Fig. 7c), where PEP significantly accumulates, which inhibits Pfk activity, resulting in the accumulation of G6P, which flows into the oxidative PP pathway, where this may be useful for the production of aromatic amino acids, etc., but the cell growth rate is depressed (Fry et al. 2000, Zhu et al. 2001).

In Gram-negative pathogenic bacterium such as *Helicobacter pylori*, Pyk is missing, while gluconeogenic Pps is present, where Pyk is not resistant at lower pH as compared to Pps (Podesta and Plaxton 1992). *H. pylori* may adapt to the strong acidic condition in the stomach by using Pps without Pyk (Huynen et al. 1998). This implies that gluconeogenic pathways may be more active at acidic condition as compared to glycolysis in general. Another pathway such as pyruvate dikinase may be also used without using Pyk in such organisms as *Mycobacterium tuberculosis* and *Treponema pallidium* (Dandekar et al. 1999).

In the case of photosynthetic organism such as cyanobacteria, Pyk activity is inhibited under light condition (Young et al. 2011), where this pathway is bypassed via the alternative pathways of Ppc and Mez in the similar way as mentioned above for Pyk mutant *E. coli* to back up pyruvate (Fig. 7d). In the photosynthetic organisms such as cyanobacteria and C4 plants, Ppc is important for fixing $CO_2$ (Owittrim and Colman 1988) to produce C4 acids in the TCA cycle under light conditions. Moreover, one of the C4 acids such as malate is decarboxylated to produce $CO_2$ and NADPH through Mez pathway (Bricker et al. 2004), and backup pyruvate, where NADPH instead of NADH becomes the donor of the electron transport chain to gain ATP in the photosynthetic organisms. This alternate pathway flux together with Pps is dominant, since cyanobacteria do not have Pck pathway for gluconeogenesis. In an attempt to produce lactate by cyanobacteria, heterologous Pyk expression together with knockdown of Ppc may increase pyruvate and then lactate production, but the cell growth is depressed due to decreased $CO_2$ fixation at Ppc and less production of NADPH at Mez (Angermayr et al. 2014).

In most proliferating cells, embryonic tissue, and many tumors, the glycolytic flux increases, and lactate is produced, while the TCA cycle activity and the oxidative phosphorylation in the respiration is reduced even in the presence of excess oxygen, where this phenomenon is called as "**aerobic glycolysis**" or the **Warburg effect** (Koppeol et al. 2011). Since proliferating mammalian cells are usually exposed to a continuous supply of glucose and nutrients in circulating blood, high ATP/ADP ratio can be retained by the high glycolytic flux or by converting two ADPs to one ATP and AMP by signaling pathways with the activation of adenylate kinases (ADKs) with the help of AMP-activated protein kinase (AMPK), otherwise cells undergo apoptosis (Izyumov et al. 2004, Vander Heiden et al. 2009). As mentioned before, there are four Pyk isoenzymes, where $PykM_1$ and $PykM_2$ are encoded by the same but differentially spliced gene (Noguchi et al. 1986). PykM$_1$ is primarily expressed in normal tissues such as brain and muscle with active oxidative phosphorylation (Marie et al. 1976, Gugen-Guillouzo et al. 1977), while PykM$_2$ is expressed in proliferating tissues such as embryonic cells, adult stem cells, and all tumor cells as well as cancer cell (Max-Audit et al. 1984, Christofk et al. 2008). PykM$_1$ exists only as a tetramer with high affinity to PEP, while PykM$_2$ forms both tetramers with high affinity to PEP and dimers with low affinity to PEP (van Berkel et al. 1974). PykM$_2$ requires FBP to form the active tetramers, while PykM$_1$ does not (Muroya et al. 1976, Eigenbrodt and Schoner 1977). Activation of growth factor receptors leads to both tyrosine kinase and phosphoinositide 3-kinase (PI3K) signaling pathways, where PI3K signaling pathway is linked to both growth control and glucose metabolism, where glucose transporter to capture glucose molecule such as hexokinase (Hxk), and Pfk are activated. The phosphorylation of PykM$_2$ by tyrosine kinase makes the conformational change from tetramer to less active dimeric form (Ye et al. 2012). This molecular switch is made by the growth factor stimulation, and the aerobic glycolysis is dominant in consistent with proliferating cell metabolism, where this regulation via phosphorylation of PykM$_2$ may change the metabolism from catabolic oriented aerobic respiration to more biosynthetic pathways for amino acids and nucleotide formation together with increased flux toward PP pathway for overproduction of NADPH (Vander Heiden et al. 2009, Christofk et al. 2008). The metabolic switch to aerobic glycolysis also changes the metabolism from TCA cycle and respiration to lactate formation by LDH in response to cell proliferation (Fantin et al. 2006), where lactate is excreted as waste. The *TIGAR* gene is induced by p53, and leads to inhibition of Pfk, thus rerouting the glycolytic flux toward the oxidative PP pathway, resulting in overproduction of NADPH.

In addition to FBP, serine (Ser) also activates PykM$_2$ (Eigenbrodt et al. 1983), where serine is the precursor for the synthesis of lipids such as phosphatidylserine and sphingolipids (Mazurek 2011). The inactivation of PykM$_2$ by tyrosine kinase may cause the accumulation of the glycolytic intermediates including 3-phosphogrycerate (3PG) from which serine is formed. In this way, accumulated serine may activate PykM$_2$ and activates glycolysis (Ye et al. 2012). The switch of PykM$_2$ activity thus contributes to endogenous serine synthesis and to maintain mammalian target of rapamycine complex 1 (mTORC1) activity (Ye et al. 2012), where this protein kinase complex has the ability to sense a variety of nutrients in response to the fluctuation of the nutrient availability as well as growth factor and stresses, and modulate the metabolism by adjusting catabolic and anabolic processes (Howell and Manning 2011, Sengupta et al. 2010, Duvel et al. 2010).

The growth of Human tumor cells is driven by *MYC* oncogene, and sensitive to glutamin (Gln) metabolism or glutaminolysis (Yuneva et al.

2007), where glutamin transported by glutamine transporter is converted to glutamate (Glu) by glutaminase, and in turn glutamate is converted to $\alpha$KG in the TCA cycle. In this reaction, NADPH and $NH_3$ are produced, where $NH_3$ is excreted as waste. Lipid synthesis is made from citrate in the proliferating cell, where OAA is limiting for citrate formation, and thus OAA must be replenished by anaplerosis. Although Pyc may play a role for this, it may be minor. Instead, the glutamine-dependent anaplerosis plays an important role. As mentioned above, glutamine can be converted to $\alpha$KG in the cytosol as well as in the mitochondria, and $\alpha$KG in the cytosol can enter into the mitochondria. The mitochondrial glutamate can be converted to $\alpha$KG by glutamate dehydrogenase (GDH). This $\alpha$KG is converted to replenish OAA via part of the TCA cycle. Glutamine carbon is converted to lactate by glutaminolysis (deBerardinis et al. 2008). In the case where glutamine is limited, the anaplerosis by the above pathway is limited, and the anaplerosis is mainly made by Pyc from glucose (Cheng et al. 2011).

## 5. Transcriptional Regulation of the Glycolysis

As the upper glycolytic flux increases, FBP concentration increases, and the increased FBP allosterically activates Pyk, and in turn PEP concentration decreases, and PEP/PYR ratio becomes lower in *E. coli* (Bettenbrock et al. 2007, Kremling et al. 2008). The decreased PEP concentration may activate Pfk activity because Pfk is allosterically inhibited by PEP, and accelerate the upper glycolytic flux, but this is marginal under typical growth condition. This feedforward with feed-back regulation mechanism may be considered as "axel" operation (Fig. 9). This feed-forward regulation is enzyme level and allosteric regulation, and its response is fast on the order of seconds to minutes, where it is crucial to quickly uptake the nutrients as soon as those are available in the ever changing environment. The feed-forward regulation with axel operation must be adjusted by "brake" action. This may be accomplished by the transcriptional regulation, where this response is relatively slow on the order of minutes to hours.

In the case of *E. coli*, two transcription factors such as cyclic adenocine 5'-phosphate (cAMP)-catabolite repressor protein (Crp) complex (cAMP-Crp) and Cra play important roles for catabolic regulation. FBP inhibits the activity of Cra, which regulates the carbon flow by activating the expression of gluconeogenic genes while repressing the expression of the glycolysis genes by the transcriptional regulation. Therefore, the glycolytic flux is negatively correlated with Cra activity irrespective of carbon sources used in *E. coli* (Fig. 3) (Kochanowski et al. 2013). Cra activates the TCA cycle and glyoxylate pathway genes such as *icdA*, *aceA*, etc. (Shimada et al. 2011a). As the glycolytic flux increases, FBP pool size increases, and Cra activity is inhibited, and in turn the glycolytic flux further increases. This indicates that Cra accelerates the glycolytic flux in addition to enzyme level feed-forward activation of Pyk by FBP. In this case, however, TCA cycle is repressed by the down regulation of *icdA* and *aceA* by the lower activity of Cra.

The brake action for the glucose uptake rate is made by cAMP-Crp in relation to PTS, where the phosphate of PEP is transferred via EI, HPr, and carbohydrate-specific E II A$^{Glc}$, to membrane-bound E II BC$^{Glc}$ encoded by *ptsG* as explained in Chapter 3. In the case of excess glucose, the phosphate of phosphorylated E II A (E II A-P) is transferred to E II BC, and thus un-phosphorylated E II A becomes dominant, and this inhibits the uptake of other carbon sources by the so-called inducer exclusion (Hogema et al. 1998) as also shown by the structural analysis of E II A$^{Glc}$ complex with an ATP-binding cassette (ABC) transporters (Chen et al. 2013). Moreover, the PEP/PYR ratio decreases as the glycolytic flux increases (Bettenbrock et al. 2007), and this causes dephosphorylation of E II A, and thus cAMP-Crp level decreases irrespective of the type of sugars, since E II A-P activates the adenylate cyclase (Cya), which in turn increases cAMP, where cAMP level is relatively higher for non-PTS carbohydrates as compared to PTS carbohydrates at the same growth rate (Kremling et al. 2007, 2008). The

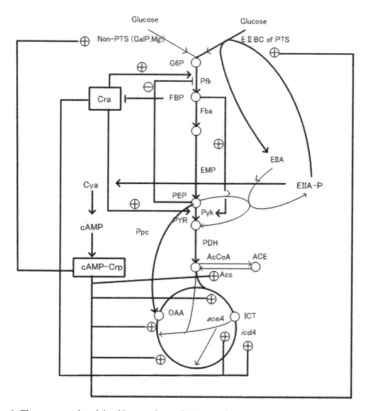

**Figure 8.** The enzyme level feedforward regulation and transcriptional feedback regulation.

cAMP acts as a transcription factor in combination with pleiotropic regulator Crp, where Crp activates the expression of *pps* and *pckA* genes (Shimada et al. 2011b), and thus the increase in Crp activity causes PEP concentration to be increased, causing PEP/PYR ratio to be increased, and in turn increases E II A-P and cAMP level, and thus cAMP and Crp changes in coordination together. Moreover, Crp activates the expression of *ptsG* and *ptsH* genes (Shimada et al. 2011b). Therefore, the cAMP-Crp level becomes lower as the glucose consumption rate increases, and in turn the expression of *ptsG* gene as well as *ptsH* gene decreases (Shimada et al. 2011b), resulting in the brake action to the glucose consumption rate (Fig. 8). This is the feed-back regulation in response to the increase in the glucose uptake rate, and this is important for the robustness against perturbations (Kremling et al. 2008). Moreover, other PTS genes such as *fruBKA* for fructose transport, *manXYZ* for mannose transport, *mltAD* for mannitol transport, as well as some ABC-family transporter genes such as *ugpG* for glycerol 3-phosphate transport and *xylF* for xylose transport, etc. are also activated by Crp (Shimada et al. 2011b), and thus the expression of such genes are down-regulated in the presence of glucose for the selective carbon source (glucose) utilization.

## 6. Overflow Metabolism and the Oxidative Stress Regulation

Since cAMP-Crp activates the TCA cycle gene expression (Appendix A) (Shimada et al. 2011b), the decrease in cAMP-Crp level together with lower activity of Cra at higher catabolic rates causes the down-regulation of the TCA cycle activity in *E. coli*. Although the detailed transcriptional regulation mechanisms may be different depending on the organisms, the phenomenon of the repression of the TCA cycle activity in response to the increase in the glycolytic flux may be common to many organisms probably to avoid the oxidative stress at the higher catabolic rate as will be also explained later. Therefore, there might be a tradeoff between energy generation and the protection against oxidative stress for the cell survival. Note that the TCA cycle is rather activated when glycolytic flux is relatively low, while TCA cycle activity is eventually repressed as the glycolytic flux increases beyond some threshold value (Valgepea et al. 2010, Yao et al. 2011).

The increase in the glycolytic flux and the repression of the TCA cycle or the reduction in the respiratory capacity cause the formation of by-products such as acetate in *E. coli* and ethanol in yeast, known as "**overflow metabolism**" (Wolfe 2005) or "**Crabtree effect**" (Crabtree 1929). In the case of yeast, the so-called Crabtree-positive species such as *S. cerevisiae*, *S. bayanus*, *S. exiguous*, *Kluyveromyces thermotolerance* show such phenomenon, while Crabtree-negative species such as *Yarrowia lipolytica*, *Pichia angusta* and *Candida rugosa* respire without overflow metabolism, where the glucose uptake rate for the latter species are all lower as compared to the former species (Christen and Sauer 2011). One of the short term events for Crabtree effect in yeast is an overflow through pyruvate decarboxylase (PDC), where the increased FBP in the upper glycolysis due to higher glucose consumption rate represses the mitochondrial oxidative phosphorylation, causing lower respiratory flux, where this phenomenon is also observed in the rat liver mitochondria (Diaz-Ruiz et al. 2008).

In *B. subtilis* and other *Firmicutes*, the major players for overflow metabolism are two regulatory proteins such as CcpA and CodY, where the former is activated by FBP, while the latter is activated by GTP and branched chain amino acids such as isoleusine and valine (Sonenshein 2007). CcpA is a member of LacI family, and a global regulator of catabolic pathways in many Gram-positive bacteria (Warner and Lolkema 2003). In response to the increase in the glycolytic flux, such global regulators repress the TCA cycle activity, and activate the expression of acetate, lactate, and acetoin synthetic pathway genes, while repressing the expression of the assimilating pathway genes (Sonenshein 2007). When cells grow rapidly on glucose or other rapidly metabolizing carbon sources, HPr in PTS, and the closely related protein Crh are phosphorylated on a serine residue by HPr kinase. These phosphorylated proteins independently bind to CcpA, and increases its affinity for **catabolite responsive element** (*cre*) site (Seidel et al. 2005). The HPr kinase is activated by ATP and FBP (Nessler et al. 2003), which links the activation of CcpA by FBP and G6P in complex with phosphorylated HPr (Deutscher et al. 2006, Seidel et al. 2005, Schumacher et al. 2007). CodY helps cells to adapt to poor nutritional availability, and represses the development of genetic competence and sporulation for adaptation to nutrient limitation through the signals of GTP and branched chain amino acids (Sonenshein 2005, Bergara et al. 2003). The TCA cycle enzymes such as citrate synthase (CS), aconitase (Acn), and isocitrate dehydrogenase (ICDH) are encoded by *citZ*, *citB*, and *citC*, respectively, where *citZ* and *citB* form an operon together with *citH* encoding malate dehydrogenase (MDH) (Jin and Sonenshein 1994, Jin et al. 1996). The *citZCH* operon is repressed by CcpA in response to the increase in FBP pool size (Blencke et al. 2003, Kim et al. 2002). The expression of *citZ* and *citB* is repressed by the regulatory protein CcpC, a member of LysR family, where CcpC activity is repressed by citrate (CIT) (Jourlin-Castelli et al. 2000, Kim et al. 2002, Blencke et al. 2006) (Fig. 9). The overflow metabolism is seen by the repression of the TCA cycle activity during the cell growth on glucose, where FBP pool size increases and activates CcpA, which in turn represses *citZCH* operon. Moreover, the overflow metabolism is also seen for the cells grown on a rich medium that contains a mixture of amino acids, where CodY is activated by valine and isoleucine, and in turn *citB* is repressed by CodY. Moreover, CS is allosterically inhibited by $\alpha$KG (Sonenshein 2007).

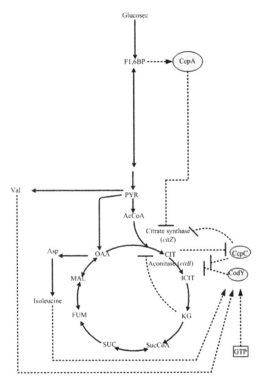

**Figure 9.** Transcriptional regulation of the glycolysis and TCA cycle in *Bacillus subtilis* (Gorke 2008).

One of the fast enzyme level repressions of TCA cycle is made by NADH, where the activities of CS and ICDH are repressed by the higher level of NADH caused by the activation of the TCA cycle in response to the increased glycolytic flux. Therefore, overexpression of NOX may contribute

---

**Box 1. Pasteur, Crabtree, and Warburg effects.**

Pasteur, Crabtree, and Warburg effects are the three aspects of the similar phenomenon of the overflow metabolism, while glucose and oxygen may be the two sides of the same coin (Valdaconda et al. 2013). The Pasteur Effect appears under oxygen limitation, and implies that "fermentation is an alternate form of life and that fermentation is suppressed by respiration" (Pasteur 1857). The Warburg effect or "aerobic glycolysis" implies that the repressed respiration or TCA cycle promotes fermentation of glucose even in the presence of excess oxygen; typically appear in cancer cell metabolism (Warburg 1956). The Crabtree effect implies that the higher growth uses aerobic glycolysis with the use of fermentation in the presence of oxygen at the high glucose consumption rate (Crabtree 1929, Pfeiffer and Morley 2014).

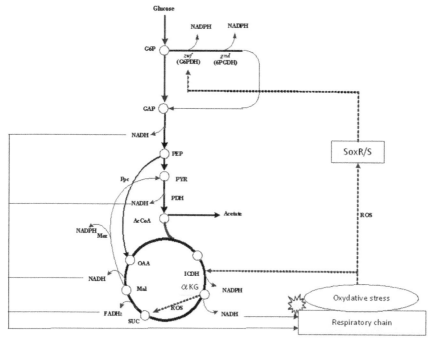

**Figure 10.** Oxidative stress regulation.

to the activation of the TCA cycle activity by reducing the amount of NADH as shown for *E. coli* (Holms et al. 2010, Vemuri et al. 2006) and for *S. cerevisiae* (Vemuri et al. 2007) with the reduction of ATP production in the respiratory chain causing the lower cell growth rate. On the other hand, the overflow metabolism such as acetate excretion may be enhanced (and even lactate and succinate may be also produced) by increasing NADH in *E. coli* even with ample oxygen (Berrios-Rivera et al. 2002). The overflow metabolism may be enhanced also when ATP production was lowered by the mutation of $F_1$ part of $F_0F_1$-ATP synthase in *E. coli* (Koebman et al. 2002).

The **overflow metabolism** occurs due to the repression of the TCA cycle in response to the increased glycolytic flux, where the TCA cycle or the respiration is constrained by some threshold value. This may be reasoned by the protection against oxidative stress caused by the **reactive oxygen species (ROSs)** generated in the respiratory chain reaction. Most oxygen-dependent organisms furnish deliberate defense mechanisms against oxidative stress caused by ROSs (Fig. 10). The TCA cycle generates NADH, and this may generate ROSs during respiration, but also scavenges them by the production of NADPH at ICDH in some organisms such as *E. coli* and mammals, while some bacteria such as *Pseudomonas fluorescens*, etc.

have both NADH- and NADPH-producing ICDH (Mailloux et al. 2007). The typical enzymes for detoxification of ROS are catalase, superoxide dismutase, and glutathione peroxidase, where the glutathione plays also an important role for this (Urso and Claekson 2003). The oxidative stress is transcriptionally regulated by such transcription factors as SoxR/S, OxyR, etc., and depending on NADPH availability (Nordberg and Arner 2001). RpoS plays important roles for general stress response including oxidative stress, where the expression of *rpoS* which encodes the alternative sigma factor $\sigma^S$ is repressed by Crp (Barth et al. 2009, Basak and Jiang 2012). As the cell growth rate increases, cAMP-Crp level decreases, and thus RpoS is activated to cope with the oxidative stress.

## 7. Constraint on ATP Production by Respiration

As the cell growth rate increases, cell size and ribosomal content increases, and the metabolism changes from the energy efficient pathways at low glucose consumption rates to energetically inefficient mode causing **energy spilling** at higher glucose consumption rates (Russel 2007, Valgepea et al. 2010, 2011). This metabolic shift from the state of full oxidization of glucose to the overflow metabolism is typically seen in *E. coli* (Vemuri et al. 2006), in *S. cerevisiae* (Vemuri et al. 2007), in lactic acid bacteria (Thomas et al. 1979, Teusink et al. 2006), and in *B. subtilis* (Sonenshein 2007). One possible reason of inefficient overflow metabolism may be due to the protection against the oxidative stress as mentioned above, while another reason may be the capacity constraint imposed by the laws of physics and chemistry (Molenaar et al. 2009). Such constraints may be a minimal lipid-to-transporter protein required for membrane integrity and the total concentration of intracellular protein to be less than the certain maximal value.

Since ATP production rate is proportional to the cell growth rate (Yao et al. 2011, Toya et al. 2010), the primal objective for the cell survival is the efficient ATP production, where ATP synthesis is made by both substrate level and oxidative phosphorylation, and the latter produces more ATP as compared to the former. Therefore, ATP production by respiration is more efficient. As seen before, however, many facultative aerobes employ fermentative pathways together with respiration at higher catabolic rate even with the presence of ample oxygen, and causes overflow metabolism.

It has long been considered that this phenomenon is due to a hypothetical limitation on the respiratory capacity (Majewski and Domach 1990), where the selection of the specific acetate formation pathway may be due to additional ATP production in *E. coli* (Majewski and Domach 1990, Varma and Palsson 1994, Pfeiffer et al. 2001), whereas this may not be the case for other organisms. The switch from total respiration at lower catabolic rate to both respiration and fermentation at higher catabolic rate may be due to the constraint on the respiration caused by the expensive synthesis costs of respiratory enzymes (Pfeiffer and Bobhoeffer 2004, Molenaar et al. 2009). The tradeoff between catabolic rate and the ATP yield may be determined by the fraction of intracellular volume occupied by the glycolytic and respiratory enzymes (Vazquez et al. 2008). This constraint by volume exclusion by the presence of macromolecules is called as "**molecular crowding**", and can explain acetate production in *E. coli* to some extent. However, this may not necessarily be able to explain other experimentally observed phenomena for the cell growth rate and yield, etc. (Zhuang et al. 2011). Experimental evidence indicates that the efficiency of the respiratory chain reaction itself may be adjusted by less-efficient dehydrogenases and cytochromes, where there exists a thermodynamic tradeoff between the turnover and the energetic efficiency of such enzymes (Meyer and Jones 1973, Waddell et al. 1997, Pfeiffer and Bonhoeffer 2002).

## 8. Respiratory Pathways and the Competition with Catabolic Transport

The fraction of ribosomal proteins increases almost proportionally with the cell growth rate, and the optimal cell shape, or volume-to-surface ratio increases with the increase in the glucose uptake rate to accommodate the increased proportion of the transporters (Molenaar et al. 2009). Growth rate may be bounded by the diffusion rates of enzyme and substrate (Beg et al. 2007), and by other physical and biochemical laws. These principles are the universal limits to the cell growth and give the basic reason why the cell growth rate is upper-bounded (Ehrenber and Kurland 1984).

The lipid membrane allows various transmembrane proteins such as substrate or metabolite transporters and respiratory enzymes to express, but constrains their amount or expression (protein-to-lipid) within certain level to maintain membrane integrity (Molenaar et al. 2009). At higher catabolic rates, the prokaryotic cytoplasmic membrane may become saturated, and therefore, various transmembrane proteins must compete for the limited membrane area. In particular, at the higher catabolic rates, the substrate transporters compete with the allowable membrane space with the respiratory enzymes. The similar situation may be expected in the mitochondrial membrane inside eukaryotic cells such as yeast, where pyruvate transporter may compete with respiratory enzymes.

Since the membrane requirement of an enzyme is inversely related to its turnover, and the membrane cost is inversely related to oxygen affinity to the enzyme, the faster but inefficient respiratory enzymes such as CydI and Cyd II might be preferred over the slower and efficient enzyme such as Cyo in *E. coli* at higher catabolic rates, thus leading to the change in the respiratory stoichiometry (Zhuang et al. 2011). One of the reasons for the higher glucose uptake rate under anaerobic condition as compared to aerobic condition may be the release of the transmembrane space by the respiratory enzymes,

**Box 2. Respiratory pathways.**

In the **respiratory chain reaction**, NADH is oxidized by either NDH1 encoded by *nuoABCDEFGHIJKLM* or NDH2 encoded by *ndh*, producing the proton and translocated to periplasm as 2 H$^+$/e$^-$ for the former, and retain the proton in the cytosol for the latter such as 0 H$^+$/e$^-$, where the resulting electrons are stored as quinol pool in *E. coli*. Although there may exist other dehydrogenase such as WrbA, YhdH, and QOR, little is known about these, and these may not be electrogenic (Bekker et al. 2009). This quinol is oxidized to quinon by the quinol oxidases or cytochrome oxidases such as Cyo and Cyd, where the former is encoded by *cyoABCDE* giving 2 H$^+$/e$^-$, while the latter is subdivided into CydI and Cyd II, where Cyd I is encoded by *cydAB*, giving 1 H$^+$/e$^-$, while Cyd II is encoded by *appCB* (also known as *cyxAB* or *cbdAB*) giving some controvertial values of 0–0.94 H$^+$/e$^-$ (Fig. 11) (Borisov et al. 2011, Bekker et al. 2009).

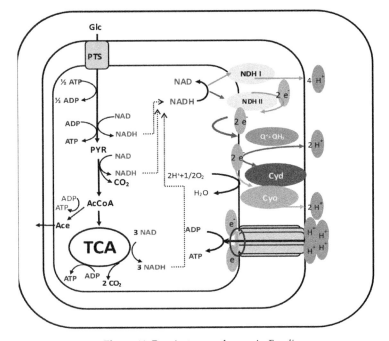

**Figure 11.** Respiratory pathways in *E. coli*.

and predominantly occupied by the glucose transporters (Zhuang et al. 2011). The three cytochromes such as Cyo, CydI, and Cyd II have different turnover rates and energetic efficiency defined in terms of the proton translocation stoichiometry and the associated ATP yield (Bekker et al. 2009). Since the relative membrane cost of an enzyme is inversely related to its turnover rate, the fast and inefficient Cyd II has a much lower cost than the slow and highly efficient Cyo (Zhuang et al. 2011). The relative cost of the moderately efficient CydI is similar to that of Cyo under fully aerobic condition, while the cost of CydI becomes low under micro-aerobic conditions due to high affinity to oxygen (Bekker et al. 2009). In the end, at the lower glucose uptake rate (less than about 3.2 mmol/gDCW.h), Cyo is used under aerobic condition, while some of the cytochrome oxidases are replaced by Cyd II at the medium glucose uptake rate (more than about 3.2

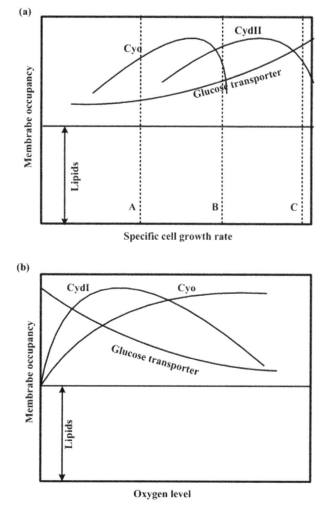

**Figure 12.** The *E. coli* cytoplasmic membrane occupancy of Cyo, Cyd - I, and Cyd - II, and the glucose transporter with respect to the cell growth rate under aerobic condition (a), and with respect to oxygen level under microaerobic conditions (b).

mmol/gDCW.h and less than about 8 mmol/gDCW.h). At higher glucose uptake rate (more than about 8 mmol/gDCW.h), the membrane is occupied predominantly by Cyd II for the respiration (Zhuang et al. 2011). Further increase in the glucose uptake rate makes Cyd II be replace by the glucose transporters, decreasing the rate of respiration (Fig. 12) (Zhuang et al. 2011). The maximum growth rate may be, therefore, determined by the maximal ATP production by adjusting the increase in the glucose uptake rate and the decrease in the respiration.

The transcriptional regulation of *cyoABCDE* and *cydAB* is made by ArcA, where *cyo* is repressed under aerobic condition, while *cydAB* is activated under micro-aerobic conditions by ArcA in *E. coli*. Since CydI is less costly than Cyo under micro-aerobic conditions, *arcA* mutant shows the decreased glucose uptake rate due to the decrease in the glucose transporters by the activated Cyo (Nikel et al. 2009).

## 9. Coordination of the Metabolism by cAMP-Crp at Higher Catabolic Rate

As shown in Fig. 13a, the cAMP level or the phosphorylated E II A (E II A–P) decreases linearly with respect to the specific cell growth rate irrespective of the type of carbon sources in *E. coli* (You et al. 2013, Valgepea et al. 2011, Bettenbrock et al. 2007, Kremling et al. 2008). As the cell growth rate increases, the overflow metabolism appears and acetate is formed at the cell growth rate of more than about 0.27 $h^{-1}$ (Valgepea et al. 2011). The specific $CO_2$ production rate also increases in accordance with the increase in the cell growth rate with the activation of the TCA cycle, and reached to the maximal value at the cell growth rate of about 0.46 $h^{-1}$. As mentioned before for the expression of cytochromes, Cyo is predominantly expressed under typical aerobic conditions, but it is eventually replaced by the lower cost of Cyd II as the catabolic rate increases, and all cytochromes are occupied by Cyd II, where the specific $O_2$ consumption rate becomes the highest among the cytochromes (Bekker et al. 2009). Further increase in the catabolic rates push the cell to increase the substrate transporters by replacing part of the area for respiration (Zhuang et al. 2011), resulting in the lower $O_2$ consumption rate (Fig. 12).

The cAMP-Crp may play an important role for such regulation. Namely, as the catabolic rate increases, cAMP-Crp level decreases as mentioned before, where Crp activates the expression of the respiratory chain genes such as *nuoABCEFGHIJKLMN*, *cyoABCD*, and *atpIBEFHAGDC* as well as TCA cycle genes (Appendix A) (Shimada et al. 2011b), and thus the expression of these genes will be down-regulated in accordance with the decrease in the cAMP-Crp level. Note that Cra may also affect the expression of the respiratory pathway genes such as *cyoA* and *ndh* (Shimada et al.

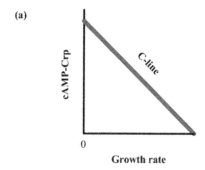

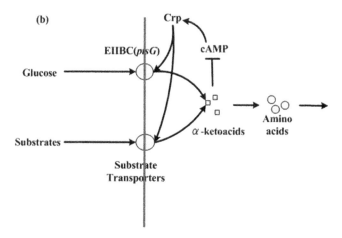

**Figure 13.** The cAMP level and the cell growth rate (a), Coordination between catabolism and anabolism by α-keto acids (b) (You et al. 2015).

2011a), and the expression of these genes may be activated by the lower level of Cra at higher catabolic rate. In fact, Cra controls *crp* and thus those are coordinately control the catabolism and respiration (Shimada et al. 2011a). As mentioned before, Cra activity decreases with the increase in FBP, where Pyk is activated, and PEP decreases, and thus PEP/PYR ratio decreases, and in turn E II A–P decreases, and then cAMP level decreases. Thus the change in the Cra activity and cAMP-Crp level are coordinated in such situation.

Moreover, limitation of other nutrients such as nitrogen, sulfur, phosphate, etc. also reduces cAMP level (You et al. 2013, Mandelstam et al. 1962, FcFall and Magasanik 1962, Clark et al. 1964), where this may be made for the coordination between catabolism and anabolism (Fig. 13b).

Namely, αketo acids such as αketo glutarate (αKG) may accumulate under nitrogen limitation, and inhibits EI of PTS, thus reducing the glucose uptake rate in coordination with the reduced demand for anabolism (Doucette et al. 2010). This coordination is made not only by αKG but also by other αketo acids such as oxaloacetate (OAA) and pyruvate as well, where such αketo acids inhibit Cya, causing cAMP level to be decreased (You et al. 2013), and the substrate uptake rate may be adjusted to match the requirement of anabolic demand. Namely, the lower level of cAMP-Crp represses the glucose uptake rate by reducing the expression of *ptsG* gene in accordance with the reduced anabolic demand caused by the nitrogen limitation (Doucette et al. 2010). Note, however, that the specific substrate consumption rate or the glycolytic flux becomes rather increased under nutrient limitation in response to lower energy level (ATP/ADP, etc.) as mentioned before. Namely, under nitrogen limitation, the glycolytic flux is rather increased by the lower energetic state. Some of the experimental data are given for the chemostat culture, and the comparison is made at the same growth rate (dilution rate) (Hua et al. 2003, 2004, Kumar and Shimizu 2010), where the anabolic demand is similar. Thus the difference comes from the increased substrate transporters made by the available spaces on the cytosolic membrane by the reduced expression of Cyo (and possibly activated Cyd II with low cost) in accordance with the lower level of cAMP-Crp (Fig. 12b).

In the case of oxygen limitation under micro-aerobic conditions, CydI is predominantly expressed by the activation of ArcA, where the cost of producing this protein is lower as compared to Cyo, and more substrate transporters can express in the cytosolic membrane as compared to the case under aerobic condition. Moreover, Crp activity decreases as the aerobiosis decreases (Rolfe et al. 2012, Ederer 2014), which may cause the down-regulation of *nuoA-N* and *cyoABCD*. Under anaerobic condition, Fnr represses the expression of both *cyo* and *cyd* genes, and thus the substrate transporters can expand the area on the cytosolic membrane, resulting in the increased catabolic rate.

In the case where NOX was overexpressed, NADH is reduced, and thus the respiratory enzymes may be less active, and then the glycolytic flux increased due to increased transporters as compared to the strain without NOX.

## 10. Carbon Catabolite Repression

When multiple carbon sources are available, most organisms either selectively assimilate one of them first among others or co-assimilate them simultaneously. They maintain strict control for nutrient assimilation to ensure the efficient usage for survival.

## 10.1 Carbon catabolite repression in E. coli

The typical example of selective carbon source assimilation may be the *lac* operon in *E. coli* (Jacob and Monod 1961), where the growth on glucose lowers the activity of certain enzymes associated with the transport and metabolism of other carbon sources, known as **glucose effect** or **glucose repression** (Roseman and Meadow 1990), resulting in biphasic growth pattern, called as **diauxie phenomenon** (Monod 1942). The mechanism behind this phenomenon is the inhibitory effect of glucose uptake on cAMP synthesis via PTS (Deutscher et al. 2006), where cAMP is required for the expression of most of the substrate specific transporters through its activation of pleoiotropic regulator Crp (Shimada et al. 2011b).

There might be a specific hierarchy for the utilization of carbon sources with glucose being on top of it, although it depends on the organisms due to their habitat and evolution. The phenomenon that the preferred carbon sources repress the synthesis of the enzymes of transport and metabolism of less favorable carbon sources is known as **carbon catabolite repression (CCR)** (Gorke and Stulke 2008), being more general as compared to glucose repression. Central to CCR is PTS, where it is widespread in bacteria and absent in Archeae, and eukaryotic and higher organisms (Techieu et al. 2001). The cAMP is known to mediate CCR (Magasanik 1961), where this is rather ubiquitous phenomenon observed in many organisms. The inhibitory effect of glucose uptake on cAMP synthesis is not restricted to PTS sugars, but also non-PTS sugars (Hogema et al. 1997, Bettenbrock et al. 2007).

CCR significantly contributes to the survival and proliferation of the cells in the varying environment, and therefore the mutation of *pts* genes causes the slower glucose uptake in *E. coli* (Flores et al. 2007, Yao et al. 2011, Nichols et al. 2001, Fuentes et al. 2013). The similar phenomenon may be seen for *pgi* and *pfk* gene mutants (Siedler et al. 2012, Toya et al. 2010). The *ptsG* mutant cells show lower buoyant density but larger cell volume than the wild type strain, implying that physiological cell density and volume regulation is intertwined with CCR (Zhou et al. 2013). It is interesting that *E. coli* cells grow faster in the mixed substrates than in the individual substrates (Hermsen et al. 2015, Zhou et al. 2013). Those data suggest that substrate and the cell growth rate relates to "macromolecular crowding", and that CCR enables the optimal cell growth (Zhou et al. 2013), where macromolecular crowding has various effects on metabolic reactions (Morelli et al. 2011, Zhou et al. 2008).

Glucose repression is mainly made by the lower level of cAMP-Crp, and thus this may be relaxed by increasing *crp* (*crp⁺*) (Gosset et al. 2004), or enhancing *crp* gene (*crp\**) (Khankal et al. 2009), where co-consumption of multiple carbon sources may be attained by such mutation, where the expression of the TCA cycle genes and glyoxylate pathway genes as well as some of the gluconeogenic genes such as *ppsA* is upregulated. The increased

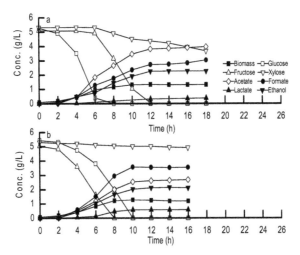

**Figure 14.** Batch culture of *E. coli* under anaerobic condition using a mixture of glucose, fructose, and xylose as a carbon source: Wild type (upper figure), and *cra* mutant (lower figure).

activity of Pps causes PEP/PYR ratio to be increased, which increases the phosphorylation of E II A, and in turn cause cAMP level to be increased (Khankal et al. 2009). The TCA cycle will be activated by enhancing *crp* gene expression, causing less acetate overflow metabolism, while glucose uptake rate decreased (Yao et al. 2011). The decrease in the glucose uptake rate in *crp* overexpression may be the activation of *mlc* gene, where Mlc represses *ptsG* gene expression, where the overexpression of Mlc by mutation of *mlc* gene (*mlc\**) represses *ptsG* gene expression, and shows similar co-consumption phenomenon as *ptsG* mutant (Nakashima and Tamura 2012). Another reason may be the activation of *cyo*, *nuo*, and *atpA* genes, and the glucose transporters may be reduced by the activated cAMP-Crp, where Cra as well as cAMP-Crp may also affect CCR, and thus *crp-cra* double mutant may be free from glucose repression (Gosset et al. 2004). Even though *pts* mutant shows slow glucose consumption rate, co-consumption of multiple carbon sources such as glucose and glycerol makes this strain metabolize more carbon source per unit time, where PEP tends to accumulates, and thus aromatic compounds may be more produced with additional mutation on *tktA* and *aroG* (Martinez et al. 2008). Glucose consumption rate may be increased in PTS⁻ mutant by enhancing *galP*, etc. (glc⁺) to recover the cell growth rate, where the enhancement of *crp* (*crp⁺*) in *mgsA, pgi*, an *ptsG* mutants allows co-consumption of multiple carbon sources with less production of acetate for useful metabolite production (Escalante et al. 2012).

In the case of *crp* gene knockout mutant (Δ*crp*), TCA cycle gene expression decreases, causing more acetate overflow metabolism (Yao et al. 2011). Note that Δ*crp* strain causes RpoS to be activated, and contribute for the oxidative stress responses.

Noting that Cra, as originally called as FruR, represses fructose consuming pathway genes such as *fruBKA*, the *cra* gene knockout mutant may activate the fructose assimilating pathways. The *fruBKA* is also under control of cAMP-Crp, and thus catabolite repression occurs when a mixture of glucose and fructose is used as a carbon source, and fructose consumption is repressed by glucose. In Cra mutant (Δ*cra*), this catabolite repression is relaxed, resulting in co-consumption of both sugars, where fructose is rather consumed faster as compared to glucose (Fig. 14) (Yao et al. 2013).

### 10.2 Catabolite regulation in Corynebacteria

The carbon source uptake system in *C. glutamicum* is different from others (Ikeda 2012), and co-consumes glucose and other carbon sources such as sucrose, fructose, ribose, acetate, and pyruvate exhibiting monophasic growth pattern (Dominguez et al. 1997, Wendisch et al. 2000). When a mixture of glucose and gluconate was used, GntR1 and GntR2 coordinately regulate gluconate metabolism, and repress *ptsG* of glucose PTS, resulting in the reduced glucose consumption rate (Frunzke et al. 2008), where gluconate catabolic genes such as *gntP* and *gntK* are also under catabolite repression by PTS sugars (Letek et al. 2006). This organism also co-consumes glucose and acetate, where the glucose uptake rate is also reduced by about half as compared to the case of using only glucose without acetate as a carbon source (Wendisch et al. 2000). The transcriptional regulator SugR represses the expression of the PTS genes during growth on gluconeogenic carbon sources such as acetate and pyruvate, etc. during growth on sugars (Teramoto et al. 2010, 2011).

The transcriptional regulator FruR represses fructose PTS such as *ptsI*, *ptsH*, and *ptsF* during growth on fructose, where *fruR* encoding FruR is repressed by SugR in *C. glutamicum* (Tanaka et al. 2008). NADPH is less formed when fructose is used as compared to glucose, where fructose is mainly assimilated by the fructose PTS (small amount can be taken up by glucose PTS), forming fructose 1-phosphate (F1P) and then converted to FBP in the glycolysis, thus resulting in the reduction in the flux toward the oxidative PP pathway (Kiefer et al. 2004). The similar situation may occur when sucrose was used as a carbon source, where sucrose can be taken up by sucrose PTS, becoming sucrose 6-phosphate (S6P), and in turn it becomes fructose and G6P. The fructose unit thus generated may diffuse out of the cytosol and taken up again by fructose PTS, resulting in the formation of F1P and then FBP (Moon et al. 2005), and thus the flux toward PP pathway is reduced. Note that the fructokinase of *Cl. acetobutylicum* converts intracellular fructose to F6P instead of F1P (Moon et al. 2005), and thus the flux toward PP pathway may be retained by the heterologous expression of such gene.

When a mixture of glutamate and PTS sugars is used, the glutamate uptake is repressed by the PTS sugars, exhibiting diauxic growth with sequential utilization of such carbon sources in *C. glutamicum* (Kramer and Lambert 1990, Kramer et al. 1990). This CCR and also acetate metabolism may be modulated by cAMP-Crp in terms of GlxR (Park et al. 2010). This regulatory system may be different from *E. coli*, where cAMP increases during growth on glucose in *C. glutamicum* (Kim et al. 2004), while GlxR plays as a transcriptional repressor in response to the increase in cAMP in *C. glutamicum* (Toyoda et al. 2011).

### 10.3 Catabolite regulation in Baccili

In *Bacillus subtilis*, the **catabolite control protein (CcpA)** is the essential regulator, which is a member of the LacI-GalR family, and appears to be widespread among low-G+C Gram-positive bacteria (Sonenshein 2007). CcpA binds to *cre* site to enhance CCR. In *B. subtilis*, the increased FBP pool size triggers an ATP-dependent protein kinase that phosphorylates HPr at residue Ser-46. The phosphorylated HPr subsequently enhances the binding of CcpA to *cre* site, and controls glycolytic activity by CCR (Deutscher et al. 1995, Fujita et al. 1995, Jones et al. 1997). In *B. subtilis*, the expression of *alsS* encoding α-acetolactate synthase, and *ackA* genes is activated in the presence of glucose, etc. This mechanism may be also seen in *L. lactis*, where CcpA activates *las* operon, and activates glycolysis by activating Pfk, Pyk and LDH (Luesink et al. 1998).

### 10.4 Catabolite regulation in Clostridia

*Cl. acetobutylicum* is an industrially important anaerobe for acetone-butanol-ethanol (ABE) fermentation, and is a low-GC content Gram-positive spore-forming bacterium. Like *Bacillus* sp., CCR is mediated by a pleiotropic regulator CcpA, which regulates its target genes by binding to *cre* sites, where CcpA is activated by the phosphorylated HPr (Schumacher et al. 2004), and enhanced by G6P and FBP (Gorke and Stulke 2008). One of the drawbacks of *Cl. acetobutylicum* may be the poor pentose assimilation for ABE fermentation from lignocellulosic biomass due to CCR as stated above in practice. This may be overcome to some extent by modulating CcpA and enhancing solvent producing genes such as *ctfA*, *ctfB*, and *adhE1* (Wu et al. 2015). The pentose sugars such as xylose and arabinose are taken up via symporters and ATP transporters, while the disaccharides such as sucrose, lactose, maltose, and cellobiose, and hexoses such as glucose, mannose, galactose, and fructose are taken up via PTS and gluconate: $H^+$ (GntP) transporter (Servinsky et al. 2010).

Although glucose and fructose have their own PTS, and these are translocated differently in *Cl. acetobutylicum* (Hutkins and Kashker 1986), their metabolism is similar (Servinsky et al. 2010), where the cell growth rates are similar with slightly higher growth on glucose as compared to fructose (Voigt et al. 2014). This may be due to the fact that the cell growth rate is low, and NADPH requirement at PP pathway becomes less under anaerobic condition.

## 11. Heteroginity of the Cell Population and CCR

We often encounter a longer lag phase without apparent cell growth and substrate consumption prior to the exponential growth when inoculation was made into the synthetic medium. This lag phase is considered to be attributed to the period of physiological adaptation to the new nutrient condition, where it is much less when rich media such as LB media were used. During this initial lag phase, the glycolytic flux is significantly low, and thus FBP concentration is low, and in turn cAMP-Crp level may be high. Unlike the case of late log phase, where gluconeogenesis is becoming dominant, the glycogen may be degraded to be used as a startup of the glycolysis (Yamamotoya et al. 2012).

It has long been believed that the lag phase was due to the time required for transcriptional reprogramming (Monod 1942). Recent investigation pays attention to the heterogeneity of the cell population, where a clonal population uses random or stochastic sensing of variable and unpredictable environments, and switch phenotypically different states. Such so-called "**bet-hedging**" strategies can generate phenotypic diversity independent of the current environmental conditions (New et al. 2014). This diversity is important for the evolution ensuring some portion of the population is always able to adapt to unforeseen future nutrient conditions. This strategy can reduce the duration of the lag phase (Kussell and Leibler 2005, Levy et al. 2012, Acar et al. 2005, Donaldson-Matasci et al. 2008, Thattai and van Oudenaarden 2004). Dominant portion of the population may employ "specialist" strategy that gives high cell growth rate with more stringent catabolic repression, but slower transcriptional reprogramming, while the other portion may employ "generalist" strategy that gives lower fitness in glucose, but allows faster transcriptional reprogramming, giving shorter lag phases during the transition from the preferred carbon source such as glucose to the less favorable carbon source(s) (New et al. 2014). The tradeoffs associated with different levels of catabolite repression depend upon the frequency and duration of the past environmental change as a memory, where this was shown by the cultivation of yeast in glucose-maltose mixture system in *S. cerevisiae* (New et al. 2014).

Clonal populations of *S. cerevisiae* show broad distributions of the cell growth rates, and the slow growth predicts resistance to heat killing in a probabalistic manner, where a trehalose-synthesis regulator Tsl1 plays an

---

**Box 3. Diauxie phenomenon.**

When multiple carbon sources are present, a short lag phase without growth appears after the preferred carbon source such as glucose was consumed before assimilating the next favorable carbon source(s), known as the diauxie lag phase (Monod 1949), where it may be considered for the cells to prepare the necessary enzymatic adaptation to switch from one substrate to another (Stanier 1951). Namely, when preferred carbon source was depleted, the enzymes for the transport and metabolism of the less favorable carbon source(s) must be synthesized, and this may take some time causing lag phase before the cell is ready to grow again on the less favorable carbon source(s). The degree of diauxie or the lag phase period depend on the inoculum condition, where glucose in the inoculum promotes the maximum diauxic effect, while maltose does not in the glucose-maltose system, which implies that bacteria may have the past memory on the growth condition (Roseman and Meadow 1990).

In the glucose-lactose system in *E. coli*, diauxie involves more than induction of *lac* operon. Namely, after the preferred glucose was depleted, and before the *lac* operon is induced, the general stress and stringent responses are induced, and CCR is relaxed, where guanosine 3',5'-bispyrophosphate (ppGpp) accumulates by RelA (ppGpp synthase) (and SpoT) binds RNA polymerase (RNAP) to down-regulate or inhibit ribosome production and increase transcription of amino acid biosynthesis (Cashel and Gallant 1969). Induction is also made for general stress response and carbon scavenging genes ensuring the survival during growth arrest and switching to alternative carbon sources, where these genes are under control of RpoS and Crp in *E. coli*. Stringent control and general stress responses are connected where ppGpp is required for RpoS accumulation (Gentry et al. 1993). Likewise, carbon scavenging and the general stress response is connected by RpoS and Crp (King et al. 2004). During diauxie, Crp and RpoS as well as RelA-dependent ppGpp play important roles for global regulation during carbon starvation, where diauxie was delayed in the *relA* mutant (Traxler et al. 2006).

---

important role for such resistance (Levy et al. 2012). This implies that yeast bet hedging results from multiple epigenetic growth states determined by a combination of stochastic and deterministic factors (Levy et al. 2012).

In the cultivation of *Lactococcus lactis* using a mixture of glucose and cellobiose, two stable cell types with alternative metabolic strategies emerge and coexist, where the fraction of each metabolic phenotype depends on the level of catabolite repression and induction of stringent response caused by the ratio of the two carbon sources in the pre-culture, as well as on epigenetic cues (Solopova et al. 2014). The production of alternative metabolic phenotypes potentially entails a bet-hedging strategy, where it gives a selective advantage (Solopova et al. 2014).

In the typical batch culture of bacteria such as *E. coli* using glucose as a carbon source, there is a significant metabolic transition from glycolysis to gluconeogenesis in accordance with glucose exhaustion (Enjalbert et al. 2013). Although metabolism continuously changes and adapts to unpredictable environmental changes, and certain substrate uptake pathways exhibit phenotypic bi-stability (Ozbudak et al. 2004, Acar et al. 2005), the main metabolism has been considered as a whole to be operated deterministically. A clonal cell population splits into two stochastically generated phenotypic subpopulations after glucose-gluconeogenic substrate shifts, and only a stochastically generated subpopulation of cells adapts to growth under the new substrate conditions, where the central metabolism does not ensure the gluconeogenic growth of individual cells, but uses a population-level adaptation resulting in responsive diversification upon nutrient changes (Kotte et al. 2014).

During the cell growth phase, the rod-shaped *E. coli* grows primarily by elongation (Beg and Danachie 1985), while during energy starvation, *E. coli* becomes spherical and much smaller, allowing the cell surface to increase and to scavenge for multiple nutrient sources simultaneously (Zheng et al. 2011, Volkmer and Heinemann 2011).

## 12. Carbon Storage Regulation

Most organisms accumulate carbon and energy reserves under certain conditions as energy maintenance in preparation for the future starvation of the available nutrients for the continuous survival (Wilson et al. 2010, Preiss 1984, Goh and Klaenhammer 2014). Glycogen formation is the typical strategy for such storage in bacteria, yeast and mammals. Glycogen is a soluble multi-branched glucose homo-polysaccharide, consisting of glucose units in a branched structure having $\alpha$-1,4 glycosyl linkages and less number of $\alpha$-1,6 linkages, and its metabolic pathway may serve as a carbon capacitor that regulates the downstream carbon metabolism (Goh and Klaenhammer 2014).

The glycogen synthesis and degradation is highly conserved in many bacteria (Ballicora et al. 2003, Preiss 2009), where glucose 1-phosphate (G1P) is formed from G6P by phosphogluco mutase (Pgm). G1P is then converted to ADP-glucose (ADPG) by means of ADPG phosphorylase (GlgC). The glycogen is then formed from ADPG by glycogen synthase (GlgA), where glycogen branching enzyme (GlgB) catalyzes the formation of branched oligosaccharide chains having $\alpha$-1,6-glucoside linkage. During the stationary phase after depletion of substrate, glucose units are removed from the nonreducing ends of the glycogen by glycogen phosphorylase (GlgP) and debranching enzyme (GlgX) (Dauvillee et al. 2005, Alonso-Casajus et al. 2006). GlgP activity increases when GlgP binds to HPr of PTS, and this allows the accumulation of glycogen at the late growth phase or the onset of the stationary phase where substrate is still present (Deutscher et al. 2006). In order to prevent ADPG excess accumulation, where it may

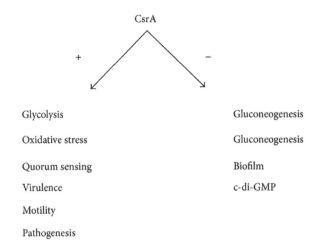

**Figure 15.** Regulation by carbon storage regulator (Csr).

divert the carbon flux to other metabolic pathways, adenosine diphosphate sugar pyrophosphatase (AspP) catalyzes the breakdown of ADPG (Moran-Zorzano et al. 2008), where its activity is allosterically activated by FBP (Moran-Zorzano et al. 2007).

### 12.1 Carbon storage regulation in E. coli

In *E. coli* and *Salmonella*, *glgBX* and *glgCAP* are two tandemly arranged operons, where *glgC* and *glgA* expression may be activated by cAMP-Crp and guanosine 5'-(tri)diphosphate 3'-diphosphate ((p)ppGpp) (Romeo and Preiss 1989). In *E. coli*, glycogen metabolism is interconnected with a variety of cellular processes and is tightly regulated in response to nutritional and energetic status (Eydallin et al. 2007, 2009).

The carbon storage regulator (Csr) influences a variety of physiological processes such as central carbon metabolism, biofilm formation, motility, peptide uptake, virulence and pathogenesis, quorum sensing, and oxidative stress response (Timmermans and van Melderen 2010, Romeo 1998, Jonas et al. 2010, Yakhnin et al. 2007) (Fig. 15). Csr is controlled by the RNA-binding protein CsrA, a posttranscriptional global regulator that regulates mRNA stability and translation (Romeo and Gong 1993, Romeo et al. 1993, Romeo 1996). CsrA binds to the 5' untranslated region of its target mRNAs, often in the region spanning the Shine-Dalgarno (SD) site (Baker et al. 2007). CsrA is regulated by two sRNAs such as CsrB and CsrC in *E. coli* (Dubey et al. 2005, Suzuki et al. 2002, Weilbacher et al. 2003). These sRNAs are composed of multiple CsrA-binding sites that bind and sequester CsrA (Babitzke and Romeo 2007). CsrA is a global regulator and regulates a variety of

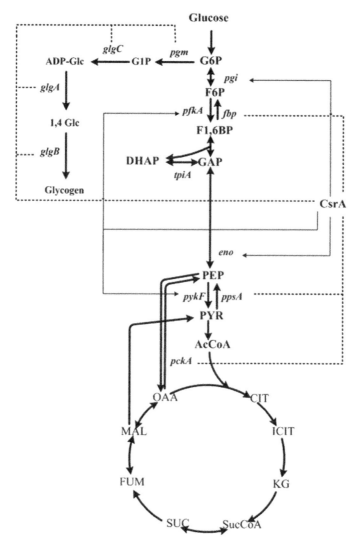

**Figure 16.** Regulation of Csr on the main metabolic pathways.

pathways, where CsrA represses glycogen accumulation by regulating the expression of *glgCAP* operon and *glgB* of *glgBX* operon (Timmermans and van Melderen 2010, Baker et al. 2002). As illustrated in Fig. 16, CsrA represses the expression of the glycogen synthesis pathway genes such as *pgm*, *glgC*, *glgA*, and *glgB*, as well as gluconeogenic genes such as *fbp*, *ppsA*, and *pckA* genes, while it activates the glycolysis genes such as *pgi*, *pfkA* (but not *pfkB*), *tpiA*, *eno,* and *pykF* (Timmermans and van Melderen 2010, Mckee et al. 2012).

Phenylalanine production could be enhanced by manipulation of Csr (Yakandawara et al. 2008). More recently, biofuel production could be enhanced by manipulating (enhancing) CsrB in *E. coli* (Mckee et al. 2012).

## 12.2 Carbon storage regulation in a variety of organisms

In nature, the nutrients are infrequently available to the living organisms, depending on their environment or habitat, where the cell's state may be one of feast or famine, or rather in-between hunger with carbon-scavenging system (Koch 1971, Ferenci 2001). In the mammalian gastrointestinal tract, the availability of the preferred nutrients is limited and variable, which causes the bacterial cells to be hungry or occasionally starving. The intracellular carbon and energy storage may play important roles for competition and persistence in the intestine. Glycogen is synthesized when carbon sources are abundant but other nutrients are limiting (Preiss 1984). Glycogen breakdown yields maltose and maltodextrins, where maltose consists of two glucose molecules connected by $\alpha$-1,4 linkage, and maltodextrins are longer glucose polymers. MalQ uses maltodextrin as maltose acceptors by releasing glucose, where it enters into glycolysis via glucokinase (Glk) (Boos and Shuman 1998). MalP cleaves glycosyl residues from maltodextrins, giving G1P, which enters into glycolysis after conversion to G6P via Pgm (Boos and Shuman 1998). Exogenous maltose gives a competitive advantage *in vivo* to pathogenic *E. coli* O157:H7 and commensal *E. coli* K12, whereas maltodextrin does not (Jones et al. 2008).

In contrast to Gram-negative bacteria, where the outer membrane is composed of phospholipids and lipopolysaccharides, the hydrophobic outer lipid layer of the Gram-positive bacteria such as *Corynebacteria* and the related mycobacteria, nocardia, rhodococci, etc. is constituted of the covalently cell wall linked mycolic acid esters and non-covalently bound glycolipids (Puech et al. 2001). *C. glutamicum* accumulates significant amounts of glycogen under sugar excess condition during the cell growth phase. The glycogen synthesis might be involved in osmotically regulated trehalose synthesis (Tzvetkov et al. 2003, Wolfe et al. 2003, Padilla et al. 2004b). Trehalose synthesis is made by such pathways as OtsA-OtsB pathway from UDPG and G6P, TreY-TreZ pathway from malto-oligosaccharides ($\alpha$-1,4-glucans), or TreS pathway from maltose (Tzvetkov et al. 2003). Unlike in *E. coli*, the glycogen is formed at the early growth phase, and is degraded at the late growth phase before depletion of sugar substrates in *C. glutamicum* (Seibold et al. 2007), where *glgX* gene product is essential for glycogen degradation, and glycogen is constantly recycled serving as a carbon storage for trehalose synthesis by TreY-TreZ pathway in response to hyperosmotic stress (Seibold and Eikmanns 2007).

In Gram-positive bacteria such as *B. subtilis* and *B. stearothermophillus*, the glycogen synthesis genes form a *glg* operon in such order as *glgBCDAP*

(Kiel et al. 1994, Tanaka et al. 1997), and glycogen synthesis may play a role in the sporulation in *B. subtilis* (Kiel et al. 1994), and in differentiation in *Streptomyces coelicolor* (Martin et al. 1997).

*Streptococcus mutans* is the oral pathogen, and a facultative colonizer of the human dental plaque that covers the surface of the teeth (Busuioc et al. 2009). Various sugars can be utilized, and metabolized through glycolysis, and organic acids such as lactate, etc. are produced (Carlsson 1997). Sucrose is used to produce extracellular polysaccharides to form the biofilm matrix, becoming part of dental plaque, and the acids produced via glycolysis dissolve tooth enamel (Loesche 1986). If excess sugars are available, intracellular polysaccharide such as glycogen is formed, and this can be used as a carbon source during sugar starvation between meals, and lowering pH of the resting plaque. The inactivation of *glgA* gene encoding a glycogen synthase prevents the accumulation of intracellular polysaccharide (Busuioc et al. 2009).

Lactobacilli are widely used for yogurt production, probiotic supplement, and ideal vehicle for mucosal-targeted delivery of vaccines and bio-therapeutics (Goh et al. 2014). The glycogen metabolism is important for the survival and probiotic functionalities of lactobacilli in the gastrointestinal tract. This metabolic pathway is encoded by *glgBCDAP-amy-pgm* genes, where all genes are co-transcribed as a polycistronic mRNA and *glg* operon (Goh et al. 2014). Glycogen biosynthesis in *L. acidophilus* and others is highly dependent on the type of available sugar substrates (Goh and Klaenhammer 2013). Among the sugar substrates, raffinose induced the highest expression of *glg* operon and intracellular glycogen accumulation, followed by the disaccharides such as trehalose and lactose, whereas glucose represses *glg* expression and glycogen biosynthesis (Goh and Klaenhammer 2013, Barrangou et al. 2006). The glycogen metabolism is under catabolic regulation with a *cre* site located upstream of the *glg* operon in *L. acidophilus*, whereas glycogen accumulation and gene expression were induced by glucose in *C. glutamicum*, *S. mutans*, and *Salmonella enteritidis*, etc. (Barrangou et al. 2006). In *L. acidophilus*, glucose may be a signal for nutrient abundance and gives the priority of carbon flow towards glycolysis as well as other biosynthesis pathways (Goh and Klaenhammer 2013). On raffinose or trehalose, the glycogen was synthesized and accumulated during the early growth phase (Goh and Kaenhammer 2014).

*Propionibacterium freudenreichii* is of food and probiotic interest in practice, where it grows in cheese during ripening at warm temperature from 20°C to 40°C. During the storage of cheese at lower temperature such as 4°C, the cells undergo cold stress, and reroute their carbon metabolism toward trehalose and glycogen synthesis for the long-term survival (Dalmasso et al. 2012).

*Mycobacterium tuberculosis* causes a serious health problem with about one-third of the world population infected by such pathogen (Sambou et al. 2008). Proper understanding of the molecular and metabolic regulation mechanisms that enables such pathogen to multiply in macrophase and survival in the host is important for the design of powerful strategies to combat the diseases (Sambou et al. 2008). In particular, the cell envelope plays important roles in host-pathogen interactions and resistance for the pathogens to chemotherapeutic treatments (Jarlier and Nikaido 1994, Daffe and Draper 1998). The outermost compartment of the cell envelope of pathgens consists of a loosely bound structure called capsule, where this envelope makes it difficult for the host to degrade (Daffe and Draper 1998). *M. tuberculosis* and other pathogenic mycobacteria produce large amounts of a glycogen-like $\alpha$-glucan which forms their outermost capsular layer, where glucan and glycogen are synthesized by $\alpha$-1,4-glycosyl transferase Rv3032 and GlgACB.

In *Mycobacterium smegmatis*, glycogen is continuously synthesized and degraded by GlgE during exponential growth phase, where this constant recycling of glycogen controls the downstream availability of carbon and energy. Thus, in this organism, glycogen may serve as a carbon capacitor for the glycolysis during the exponential growth phase in addition to its conventional storage role (Belanger and Hatfull 1999).

Cholera is an acute intestinal infection caused by the Gram-negative bacterium *Vibrio cholerae*, where it is still a major global burden as outbreaks frequently occur in developing countries (Sack et al. 2006). *V. cholerae* is acquired by ingestion of contaminated food or water, and once acquired it colonizes and multiplies in the human small intestine. The induction of cholera toxin after colonization of the small intestine causes the production of a secretory diarrhea, or rice-water stool (Lee et al. 1999, Reidl and Klose 2002). *V. cholerae* survives as it passes between the human small intestine and aquatic environment, where it faces dramatic change in its living environment with much less available nutrient. Therefore, *V. cholerae* may prepare for such harsh condition prior to exiting the host at the late stage of infection (Schild et al. 2007). The surviving strategies may be to form biofilms (Watnick and Kolter 2000) to produce large amounts of polyphosphate (Ogawa et al. 2000, Jahid et al. 2006), and to accumulate glycogen during human infection (Bourassa and Camilli 2009).

In *S. cerevisiae*, the specific ethanol production rate or the specific glucose consumption rate increases by nitrogen limitation, while glycogen and trehalose accumulation is enhanced (Larsson et al. 1997). In yeast, glycogen is synthesized from UDP-glucose (UDPG) by a self-glycosylating initiator protein, glycogenin (Farkas et al. 1991, Cheng et al. 1995), and debranching enzyme (Rowen et al. 1992). UDPG is synthesized from UTP and G1P by UDPG pyrophosphorylase encoded by *UGP1* (Daran et al. 1995). UDPG is also utilized for galactose utilization as a carbon source, trehalose production, and cell wall $\beta$-glucan, etc. (Daran et al. 1997). UDPG is used for the cell wall during the cell growth phase, while it is used for glycogen synthesis during stationary phase in response to both nutritional status and cell integrity signaling (Grose et al. 2007, Smith and Rutter 2007).

Eukaryotic glycogen synthases cannot initiate glycogen synthesis *de novo*, where they function to elongate a pre-existing oligosaccharide primer attached to glycogenin. Glycogenin participates in the initiation of glycogen synthesis in yeast and mammals, whereas it may not exist in bacteria. The expression of glycogen synthase, glycogen phosphorylase, branching enzyme, debranching enzyme, and glycogenin is tightly regulated, where the Ras/Protein kinase A (PKA) pathway controls transcription of the related genes (Wilson et al. 2010).

In photosynthetic organism such as cyanobacteria, polysaccharides mostly glycogen are synthesized and accumulated from $CO_2$ via ADP-glucose pyrophosphorylase (AGPase), glycogen synthase, and branching enzyme during unfavorable growth condition (Nakamura et al. 2005). Such oligosaccharides play an important role in tolerance to salt or oxidative stress (Suzuki et al. 2010).

## 13. Concluding Remarks

All the cell constituents are synthesized from the precursor metabolites generated in the central metabolism, and this restricts the common use of the main metabolism, where lower glycolysis is retained in almost all organisms, while the upper EMP pathway may be replaced by the ED pathway, where this pathway is rather ancient due to thermodynamic and protein costs points of view.

The metabolic regulation is made by both the offensive fast enzyme level regulation and defensive slow transcriptional regulation.

The overflow metabolism is a realization of the transcriptional regulation of the TCA cycle and the enzyme level regulation of the glycolytic flux. The mechanism behind this phenomenon is the transcriptional repression of the TCA cycle activity to protect the cell against the oxidative stress caused by ROSs generated during the respiration, where there is still some debate on the mitochondrial function, ROSs, and the regulation of the oxidative phosphorylation (Murphy 2009, Kadenbach 2012). Moreover, the glycolytic flux may be determined by the adenylate phosphate esters such as ATP, ADP, and AMPs. Note that as ATP production increases, ATP consuming anabolic processes such as ribosome and protein synthesis might increasingly constrain growth (Scott et al. 2010).

It is important to comprehensively understand the diversity of the metabolic strategies present in a variety of organisms living in various environments and to derive the common principles which govern the cell metabolism. The deep understanding on the mechanisms behind the organisms will contribute to the understanding on the life or its origin, and the metabolic engineering for the next generation biofuel and chemical production.

## References

Acar, M., A. Becskei and A. van Oudenaarden. 2005. Enhancement of cellular memory by reducing stochastic transitions. Nature. 435: 228–232.

Aidelberg, G., B.D. Towbin, D. Rothschild, E. Dekel, A. Bren and U. Alon. 2014. Hierarchy of non-glucose sugars in *Escherichia coli*. BMC Syst Biol. 8: 133.

Aledo, J.C., S. Jiménez-Rivérez, A. Cuesta-Munoz and J.M. Romero. 2008. The role of metabolic memory in the ATP paradox and energy homeostasis. FEBS J. 275(21): 5332–5342.

Alonso-Casajús, N., A.M. Viale, F.J. Muñoz, E. Baroja-Fernández, M.T. Morán-Zorzano and J. Pozueta-Romero. 2006. Glycogen phosphorylase, the product of the *glgP* gene, catalyzes glycogen breakdown by removing glucose units from the non-reducing ends in *Escherichia coli*. J Bacteriol. 188: 5266–5272.

Andreas Angermayr, A.A., A.D. van der Woude, D. Correddu, A. Vreugdenhil, V. Verrone and K.J. Hellingwerf. 2014. Exploring metabolic engineering design principles for the photosynthetic production of lactic acid by *Synechocystis* sp. PCC6803. Biotechnol Biofuels. 7: 99.

Atsumi, S., T. Hanai and J.C. Liao. 2008. Non-fermentative pathways for synthesis of branched-chain higher alcohols as biofuels. Nature. 451: 86–90.

Babitzke, P. and T. Romeo. 2007. CsrB sRNA family: sequestration of RNA-binding regulatory proteins. Curr Opinion Microbiol. 10(2): 156–163.

Babul, J., D. Clifton, M. Kretschmer and D.G. Fraenkel. 1993. Glucose metabolism in *Escherichia coli* and the effect of increased amount of aldolase. Biochem. 32: 4685–4692.

Bailey, E., F. Stirpe and C.B. Taylor. 1968. Regulation of rat liver pyruvate kinase. Biochem J. 108: 427–436.

Baker, C.S., I. Morozov, K. Suzuki, T. Romeo and P. Babitzke. 2002. CsrA regulates glycogen biosynthesis by preventing translation of *glgC* in *Escherichia coli*. Mol Microbiol. 44(6): 1599–1610.

Baker, C.S., L.A. Eöry, H. Yakhnin, J. Mercante, T. Romeo and P. Babitzke. 2007. CsrA inhibits translation initiation of *Escherichia coli hfq* by binding to a single site overlapping the Shine-Dalgarno sequence. J Bacteriol. 189(15): 5472–5481.

Ballicora, M.A., A.A. Iglesias and J. Preiss. 2003. ADP-glucose pyrophosphorylase, a regulatory enzyme for bacterial glycogen synthesis. Microbiol Mol Biol Rev. 67: 213–225.

Bar-Even, A., A. Flamholz, E. Noor and R. Mil. 2012. Rethinking of glycolysis: on the biochemical logic of metabolic pathways. Nature Chem Biol. 8: 509–517.

Barrangou, R., M.A. Azcarate-Peril, T. Duong, S.B. Conners, R.M. Kelly and T.R. Klaenhammer. 2006. Global analysis of carbohydrate utilization by *Lactobacillus acidophilus* using cDNA microarrays. PNAS USA. 103: 3816–3821.

Barth, E., K.V. Gora, K.M. Gebendorfer, F. Settele, U. Jakob and J. Winter. 2009. Interplay of cellular cAMP levels, {sigma}S activity and oxidative stress resistance in *Escherichia coli*. Microbiol. 155(Pt 5): 1680–1689.

Barwell, C.J., B. Woodward and R.V. Brunt. 1971. Regulation of pyruvate kinase by fructose 1, 6-diphosphate in *Saccharomyces cerevisiae*. Eur J Biochem. 18: 69–64.

Basak, S. and R. Jiang. 2012. Enhancing *E. coli* tolerance towards oxidative stress via engineering its global regulator cAMP receptor protein (CRP). PLoS ONE. 7(12): e51179.

Becker, J., C. Klopprogge and C. Wittmann. 2008. Metabolic responses to pyruvate kinase deletion in lysine producing *Corynebacterium glutamicum*. Microb Cell Fact. 7: 8.

Beg, Q.K., A. Vazquez, J. Ernst, M.A. de Menezes, Z. Bar-Joseph, A.-L. Barabási and Z.N. Oltvai. 2007. Intracellular crowding defines the mode and sequence of substrate uptake by *Escherichia coli* and constrains its metabolic activity. PNAS USA. 104(31): 12663–12668.

Begg, K.J. and W.D. Donachie. 1985. Cell shape and division in *Escherichia coli*: experiments with shape and division mutants. J Bacteriol. 163: 615–622.

Bekker, M., S. de Vries, A. Ter Beek, K.J. Hellingwerf and M.J. Teixeira de Mattos. 2009. Respiration of *Escherichia coli* can be fully uncoupled via the nonelectrogenic terminal cytochrome *bd*-II oxidase. J Bacteriol. 191(17): 5510–5517.

Belanger, A.E. and G.F. Hatful. 1999. Exponential-phase glycogen recycling is essential for growth of *Mycobacterium smegmatis*. J Bacteriol. 181: 6670–6678.

Bergara, F. et al. 2003. CodY is a nutritional repressor of flagellar gene expression in *Bacillus subtilis*. J Bacteriol. 185: 3118–3126.

Berrios-Rivera, S.J., G.N. Bennet and K.Y. San. 2002. Metabolic engineering of *Escherichia coli*: increase of NADH availability by overexpressing an NAD$^+$-dependent formate dehydrogenase. Metabolic Eng. 4: 217–229.

Bettenbrock, K., T. Sauter, K. Jahreis, A. Klemling, J.W. Lengeler and E.D. Gilles. 2007. Correlation between growth rates, EIIACrr phosphorylation, and intra-cellular cyclic AMP levels in *Escherichia coli* K-12. J Bacteriol. 189: 6891–6900.

Blangy, D., H. Buc and J. Monod. 1968. Kinetics of the allosteric interactions of phosphofructokinase from *Escherichia coli*. J Mol Biol. 31: 13–35.

Blencke, H.M. et al. 2003. Transcriptional profiling of gene expression in response to glucose in *Bacillus subtilis*: regulation of the central metabolic pathways. Metab Eng. 5: 133–149.

Blencke, H.M. et al. 2006. Regulation of *citB* expression in *Bacillus subtilis*: integration of multiple metabolic signals in the citrate pool and by the general nitrogen regulatory system. Arch Microbiol. 185: 136–146.

Boles, E., F. Schulte, T. Miosga, K. Freidel, E. Schluter, F.K. Zimmermann. et al. 1997. Characterization of a glucose-repressed pyruvate kinase (Pyk2p) in *Saccharomyces cerevisiae* that is catalytically insensitive to fructose 1,6-bisphosphate. J Bacteriol. 179: 2987–2993.

Bond-Watts, B.B., R.J. Bellerose and M.C. Chang. 2011. Enzyme mechanism as a kinetic control element for designing synthetic biofuel pathways. Nat Chem Biol. 7: 222–227.

Boos, W. and H. Shuman. 1998. Maltose/maltodextrin system of *Escherichia coli*: transport, metabolism, and regulation. Microbiol Mol Biol Rev. 62: 204–229.

Borisov, V.B., R. Murali, M.L. Verkhovskaya, D.A. Bloch, H. Han, R.B. Gennis and M.I. Verkhovsky. 2011. Aerobic respiratory chain of *Escherichia coli* is not allowed to work in fully uncoupled mode. PNAS USA. 108(42): 17320–17324.

Bourassa, L. and A. Camilli. 2009. Glycogen contributes to the environmental persistence and transmission of *Viblio cholarea*. Mol Microbiol. 72(1): 124–138.

Bricker, T.M., S. Zhang, S.M. Laborde, P.R. Mayer II, L.K. Frankel and J.V. Moroney. 2004. The malic enzyme is required for optimal photoautotrophic growth of Synechocystis sp. strain PCC6803 under continuous light but not under a diurnal light regimen. J Bacteriol. 186(23): 8144–8148.

Busuioc, M., K. Mackiewicz, B.A. Buttaro and P.J. Piggot. 2009. Role of intracellular polysaccharide in persistence of *Streptococcus mutans*. J Bacteriol. 191(23): 7315–7322.

Carlsson, J. 1997. Bacterial metabolism in dental biofilms. Adv Dent Res. 11: 75–80.

Carminatti, H., L.J. Asua, E. de Recondo, S. Passeron and E. Rozengurt. 1968. Some kinetic properties of liver pyruvate kinase (Type I). J Biol Chem. 243: 3051–3056.

Cashel, M. and J. Gallant. 1969. Two compounds implicated in the function of the *RC* gene of *Escherichia coli*. Nature. 221: 838–841.

Chen, S., M.L. Oldham, A.L. Davidson et al. 2013. Carbon catabolite repression of the maltose transporter revealed by X-ray crystallography. Nature. 499: 364–368.

Chen, Y., Q. Liu, X. Chen, J. Wu, J. Xie, T. Guo, C. Zhu and H. Ying. 2014. Control of glycolytic flux in directed biosynthesis of uridine-phosphoryl compounds through the manipulation of ATP availability. Appl Microbiol Biotechnol. 98(15): 6621–6632.

Cheng, C., J. Mu, I. Farkas, D. Huang, M.G. Goebl and P.J. Roach. 1995. Requirement of the self-glucosylating initiator proteins Glg1p and Glg2p for glycogen accumulation in *Saccharomyces cerevisiae*. Mol Cell Biol. 15: 6632–6640.

Cheng, T., J. Sudderth, C. Yang, A.R. Mullen, E.S. Jin, J.M. Mates and R.J. DeBerardinis. 2011. Pyruvate carboxylase is required for glutamine-independent growth of tumor cells. PNAS USA. 108(21): 8674–8679.

Christen, S. and U. Sauer. 2011. Intracellular characterization of aerobic glucose metabolism in seven yeast species by 13C flux analysis and metabolomics. FEMS Yeast Res. 11: 263–272.

Christofk, H.R. et al. 2008. The M2 splice isoform of pyruvate kinase is important for cancer metabolism and tumor growth. Nature. 452: 230–233.

Chubukov, V., L. Gerosa, K. Kochanowski and U. Sauer. 2014. Coodination of microbial metabolism. Nat Rev. 12: 327–340.

Chubukov, V., M. Uhr, L. Le Chat, R.J. Kleijn, M. Jules, H. Link, S. Aymerich, J. Stelling and U. Sauer. 2013. Transcriptional regulation is inefficient to explain substrate-induced flux changes in *Bacillus subtilis*. Mol Syste Biol. 9: 709.

Clark, D.J. and A.G. Marr. 1964. Studies on the repression of beta-galactosidase in *Escherichia coli*. Biochim Biophys Acta. 92: 85–94.

Collins, L. and T.D. Thomas. 1974. Pyruvate kinase of *Streptococcus lactis*. J Bacteriol. 120(1): 52–58.

Crabtree, H.G. 1929. Observations on the carbohydrate metabolism of tumours. Biochem J. 23: 536–45.

Cunningham, D.S., Z. Liu, N. Domagalski, R.R. Koepsel, M.M. Ataai and M.M. Domach. 2009. Pyruvate kinase-deficient *Escherichia coli* exhibits increased plasmid copy number and cAMP levels. J Bacteriol. 191(9): 3041–3049.

Daffe, M. and P. Draper. 1998. The envelope layers of mycobacteria with reference to their pathogenicity. Adv Micobiol Physiol. 39: 131–203.

Dalmasso, M., J. Aubert, S. Even, H. Falentin, M.-B. Meillard, S. Parayre, V. Loux, J. Tanskanen and A. Thierry. 2012. Accumulation of intracellular glycogen and trehalose by *Propionibacterium freudenreichii* under conditions mimicking cheese ripening in the cold. Appl Environ Microbiol. 78(17): 6357–6364.

Dandekar, T., S. Schuster, B. Snel, M. Huynen and P. Bork. 1999. Pathway alignment: application to the comparative analysis of glycolytic enzymes. Biochem J. 343: 115–124.

Daran, J.M., N. Dallies, D. Thines-Sempoux, V. Paquet and J. Francois. 1995. Genetic and biochemical characterization of the UGP1 gene encoding the UDPG pyrophosphorylase from *Saccharomyces cerevisiae*. Eur J Biochem. 233: 520–30.

Daran, J.M., W. Bell and J. Francois. 1997. Physiological and morphological effects of genetic alterations leading to a reduced synthesis of UDPG in *Saccharomyces cerevisiae*. FEMS Microbiol Lett. 153: 89–96.

Dauvillée, D., I.S. Kinderf, Z. Li, B. Kosar-Hashemi, M.S. Samuel, L. Rampling, S. Ball and M.K. Morell. 2005. Role of the *Escherichia coli glgX* gene in glycogen metabolism. J Bacteriol. 187: 1465–1473.

DeBerardinis, R.J., J.J. Lum, G. Hatzivassiliou and C.B. Thompson. 2008. The biology of cancer: Metabolic reprogramming fuels cell growth and proliferation. Cell Metab. 7: 11–20.

Dellomonaco, C., J.M. Clomburg, E.N. Miller and R. Gonzalez. 2011. Engineered reversal of the β-oxidation cycle for the synthesis of fuels and chemicals. Nature. 476: 355–359.

Demina, A., K.I. Varughese, J. Barbot, L. Forman and E. Beutler. 1999. Six previously undescribed pyruvate kinase mutations causing enzyme deficiency. Blood. 92: 647–652.

Desai, T.A. and C.V. Rao. 2010. Regulation of arabinose and xylose metabolism in *Escherichia coli*. Appl Environ Microbiol. 76: 1524–1532.

Deutscher, J., C. Francke and P.W. Postma. 2006. How phosphotransferase system-related protein phosphorylation regulates carbohydrate metabolism in bacteria. Microbiol Mol Biol Rev. 70: 939–1031.

Deutscher, J., E. Kuster, U. Bergstedt, V. Charrier and W. Hillen. 1995. Protein kinase-dependent HPr/CcpA interaction links glycolytic activity to carbon catabolie repression in Gram-positive bacteria. Mol Microbiol. 15: 1049–1053.

Diaz-Ruiz, R., N. Averet, D. Araiza, B. Pinson, S. Uribe-Carvajal, A. Devin and M. Rigoulet. 2008. Mitochondrial oxidative phosphorylation is regulated by fructose 1, 6-bisphosphate. J Biol Chem. 283: 26948–26955.

Diesterhaft, M. and E. Freese. 1972. Pyruvate kinase of *Bacillus subtilis*. Biochim Biophys Acta. 268(2): 373–80.

Doan, T. and S. Aymerich. 2003. Regulation of the central glycplytic genes in *Bacillus subtilis*: binding of the repressor CggR to its single DNA target sequence is modulated by fructose-1, 6-bisphosphate. Mol Microbiol. 47: 1709–1721.

Dolfing, J., B. Jiang, A.M. Henstra, A.J. Stams and C.M. Plugge. 2008. Syntrophic growth on formate: a new microbial niche in anoxic environments. Appl Environ Microbiol. 74: 6126–6131.

Dominguez, H. et al. 1998. Carbon-flux distribution in the central metabolic pathways of *Corynebacterium glutamicum* during growth on fructose. Eur J Biochem (FEBS). 254: 96–102.

Dominguez, H., M. Cocaign-Bousquet and N.D. Lindley. 1997. Simultaneous consumption of glucose and fructose from sugar mixtures during batch growth of *Corynebacterium glutamicum*. Appl Microbiol Biotechnol. 47: 600–603.

Donaldson-Matasci, M.C., M. Lachmann and C.T. Bergstrom. 2008. Phenotypic diversity as an adaptation to environmental uncertainty. Evolutionary Ecology Res. 10: 493–515.

Doucette, C.D., D.J. Schwab, N.S. Wingreen and J.D. Rabinowitz. 2011. α-Ketoglutarate coordinates carbon and nitrogen utilization via enzyme I inhibition. Nat Chem Biol. 7: 894–901.

Dubey, A.K., C.S. Baker, T. Romeo and P. Babitzke. 2005. RNA sequence and secondary structure participate in high-affinity CsrA-RNA interaction. RNA. 11(10): 1579–1587.

Duvel, K., J.L. Yecies, S. Menon, P. Raman, A. Lipovsky, A.L. Souza et al. 2010. Activation of a metabolic gene regulatory network downstream of mTOR complex 1. Mol Cell. 39: 171–183.

Ederer, M., S. Steinsiek, S. Stagge, M.D. Rolfe, A. Ter Beek, D. Knies, M.J.T. de Mattos, T. Sauter, J. Green, R.K. Poole, K. Bettenbrock and O. Sawodny. 2014. A mathematical model of metabolism and regulation provides a systems-level view of how *Escherichia coli* responds to oxygen. Front. Microbiol. 5: 124.

Ehrenberg, M. and C.G. Kurland. 1984. Costs of accuracy determined by a maximal growth rate constraint. Q Rev Biophys. 17(1): 45–82.

Eigenbrodt, E., S. Leib, W. Kromer, R.R. Friis and W. Schoner. 1983. Structural and kinetic difference between the M2 type pyruvate kinase from lung and various tumors. Biomed Biochim Acta. 42: 5278–5282.

Eigenbrodt, E. and W. Schoner. 1977. Modification of interconversion of pyruvate kinase type M2 from chiken liver by fructose 1, 6-bisphosphate and l-alanine. Hoppe Seylers Z Physiol Chem. 358: 1057–1067.

Emmerling, M., J.E. Bailey and U. Sauer. 1999. Glucose catabolism of *Escherichia coli* strains with increased activity and altered regulation of key glycolytic enzymes. Metab Eng. 1: 117–127.

Emmerling, M., J.E. Bailey and U. Sauer. 2000. Altered regulation of pyruvate kinase or co-overexpression of phosphofructokinase increases glycolytic fluxes in resting *Escherichia coli*. Biotechnol Bioeng. 67(5): 623–627.

Enjalbert, B., F. Letisse and J.-C. Portais. 2013. Physiological and molecular timing of the glucose to acetate transition in *Escherichia coli*. Metabolites. 3: 820–837.

Escalante, A., A.S. Cervantes, G. Gosset and F. Bolivar. 2012. Current knowledge of the *Escherichia coli* phosphoenolpyruvate-carbohydrate phosphotransferase system: peculiarities of regulation and impact on growth and product formation. Appl Microbiol Biotechnol. 94: 1483–1494.

Even, S., N.D. Lindley and M. Cocaign-Bousquet. 2003. Transcriptional, translational and metabolic regulation of glycolysis in *Lactococcus lactis* subsp. *cremoris* MG1363 grown in continuous acidic cultures. Microbiol. 149: 1935–1944.

Eydallin, G., A.M. Viale, M.T. Morán-Zorzano, F.J. Muñoz, M. Montero, E. Baroja-Fernandez and J. Pozueta-Romero. 2007. Genome-wide screening of genes affecting glycogen metabolism in *Escherichia coli* K-12. FEBS Lett. 581: 2947–2953.

Eydallin, G., M. Montero, G. Almagro, M.T. Sesma, A.M. Viale, F.J. Muñoz, M. Rahimpour, E. Baroja-Fernández and J. Pozueta-Romero. 2009. Genome-wide screening of genes whose enhanced expression affects glycogen accumulation in *Escherichia coli*. DNA Res. doi:10.1093/dnares/dsp028.

Fantin, V.R., J. St-Pierre and P. Leder. 2006. Attenuation of LDH-A expression uncovers a link between glycolysis, mitochondrial physiology, and tumor maintenance. Cancer Cell. 9: 425–434.

Farkas, I., T.A. Hardy, M.G. Goebl and P.J. Roach. 1991. Two glycogen synthase isoforms in *Saccharomyces cerevisiae* are coded by distinct genes that are differentially controlled. J Biol Chem. 266: 15602–15607.

Ferenci, T. 2001. Hungry bacteria: definition and properties of a nutritional state. Environ Microbiol. 3: 605–611.

Flamholz, A., E. Noor, A. Bar-Even, W. Liebermeister and R. Milo. 2013. Glycolytic strategy as a tradeoff between energy yield and protein cost. PNAS USA. 110(24): 10039–10044.

Flores, N., L. Leal, J.C. Sigala, R. Anda, A. Escalante, A. Martinez, O.T. Ramirez, G. Gosset and F. Bolivar. 2007. Growth recovery on glucose under aerobic conditions of an *Escherichia coli* strain carrying aphosphenolpyruvate: carbohydrate phosphotransferase system deletion by inactivating *arcA* and overexpressing the genes coding for glucokinase and galactose permease. J Mol Microbiol Biotechnol. 13(1-3): 105–116.

Folger, O. et al. 2011. Predicting selective drug targets in cancer through metabolic networks. Mol Syst Biol. 7: 501.

Frunzke, J., V. Engels, S. Hasenbein, C. Gatgens and M. Bott. 2008. Co-ordinated regulation of gluconate catabolism and glucose uptake in *Corynebacterium glutamicum* by two functionally equivalent transcriptional regulators, GntR1 and GntR2. Mol Microbiol. 67(2): 305–322.

Fry, B., T. Zhu, M.M. Domach, R.R. Koepsel, C. Phalakornkule and M.M. Ataai. 2000. Characterization of growth and acid formation in a *Bacillus subtilis* pyruvate kinase mutant. Appl Environ Microbiol. 66: 4045–4049.

Fuentes, L.G., A.R. Lara, L.M. Martinez, O.T. Ramirez, A. Martinez, F. Bolivar and G. Gosset. 2013. Modification of glucose import capacity in *Escherichia coli*: physiolosic consequences and utility for improving DNA vaccine production. Microb Cell Fact. 12: 42.

Fujita, Y., Y. Miwa, A. Galinier and J. Deutscher. 1995. Specific recognition of the *Bacillus subtilis gnt cis*-acting catabolite-responsive element by a protein complex formed between CcpA and seryl-phosphorylated HPr. Mol Microbiol. 17: 953–960.

Gao, B. and R.S. Gupta. 2012. Phylogenetic framework and molecular signature for the main clades of the phylum Actinobacteria. Microbiol Mol Biol Rev. 76: 66–112.

Garcia-Olalla, C. and A. Garrido-Pertierra. 1987. Purification and kinetic properties of pyruvate kinase isoenzymes of *Salmonella typhimurium*. Biochem J. 241: 573–581.

Gentry, D.R., V.J. Hernandez, L.H. Nguyen, D.B. Jensen and M. Cashel. 1993. J Bacteriol. 175: 7982–7989.

Goh, Y.J. and T.R. Klaenhammer. 2013. A functional glycogen biosynthesis pathway in *Lactobacillus acidophilus*: expression and analysis of *glg* operon. Mol Microbiol. 89: 1187–1200.

Goh, Y.J. and T.R. Klaenhammer. 2014. Insights into glycogen metabolism in *Lactobacillus acidophilus*: impact on carbohydrate metabolism, stress tolerance and gut retention. Microb Cell Fact. 13: 94.

Gorke, B. and J. Stulke. 2008. Carbon catabolite repression in bacteria: many ways to make the most out of nutrients. Nat Rev Microbiol. 6(8): 613–624.

Gosset, G., Z. Zhang, S. Nayyar, W.A. Cuevas and M.H. Saier Jr. 2004. Transcriptome analysis of *crp*-dependent catabolite control of gene expression in *Escherichia coli*. J Bacteriol. 186(11): 3516–3524.

Grose, J.H., T.L. Smith, H. Sabic and J. Rutter. 2007. Yeast PAS kinase coordinates glucose partitioning in response to metabolic and cell integrity signaling. EMBO J. 26: 4824–4830.

Gubler, M., M. Jetten, S.H. Lee and A.J. Sinskey. 1994. Cloning of the pyruvate kinase gene (*pyk*) of *Corynebacterium glutamicum* and site-specific inactivation of *pyk* in a lysine-producing *Corynebacterium lactofermentum* strain. Appl Environ Microbiol. 60(7): 2494–2500.

Guguen-Guillouzo, C., M.F. Szajnert, J. Marie, D. Delain and F. Schapira. 1977. Differentiation *in vivo* and *in vitro* of pyruvate kinase isomers in rat muscles. Biochimie. 59: 65–71.

Hauf, J., F.K. Zimmermann and S. Muller. 2000. Simultaneous genomic overexpression of seven glycolytic enzymes in the yeast *Saccharomyces cerevisiae*. Enzyme Microbiol Micobiol. 26: 688–698.

Haverkorn van Rijsewijk, B.R.B., A. Nanchen, S. Nallet, R.J. Kleijn and U. Sauer. 2011. Large-scale 13C-flux analysis reveals distinct transcriptional control of respiratory and fermentative metabolism in *Escherichia coli*. Mol Syste Biol. 7: 477.

Hermsen, R., H. Okano, C. You, N. Werner and T. Hwa. 2015. A growth-rate composition formula for the growth of *E. coli* on co-utilized carbon substrates. Mol Syst Biol. 11(4): 801.

Hogema, B.M. et al. 1997. Catabolite repression by glucose 6-phosphate, gluconate and lactose in *Escherichia coli*. Mol Microbiol. 24: 857–867.

Hogema, B.M., J.C. Arents, R. Bader, K. Eijkemans et al. 1998. Inducer exclusion in *Escherichia coli* by non-PTS substrates: The role of the PEP to pyruvate ratio in determining the phosphorylation state of enzyme IIA$^{Glc}$. Mol Microbiol. 30: 487–498.

Holm, A.K., L.M. Blank, M. Oldiges et al. 2010. Metabolic and transcriptional response to cofactor perturbations in *Escherichia coli*. J Biol Chem. 285(23): 17498–17506.

Howell, J.J. and B.D. Manning. 2011. mTOR couples cellular nutrient sensing to organismal metabolic homeostasis. Trends in Endocrinol Metabol. 22(3): 94–102.

Hsu, P.P. and D.M. Sabatini. 2008. Cancer cell metabolism: Warburg and beyond. Cell. 134: 703–707.

Hua, Q., C. Yang, T. Baba, H. Mori and K. Shimizu. 2003. Responses of the central metabolism in *Escherichia coli* to *pgi* and glucose-6-phosphate dehydrogenase knockouts. J of Bacteriol. 185: 7053–7067.

Hua, Q., Y. Chen, T. Oshima, H. Mori and K. Shimizu. 2014. Analysis of Gene Expression in *Escherichia coli* in response to changes of growth-limiting nutrient in chemostat cultures. Appl Environ Microbiol. 70: 2354–2366.

Huberts, D.H.E.W., B. Niebel and M. Heinemann. 2011. A flux-sensing mechanism could regulate the switch between repiration and fermentation. FEMS Yeast Res. 12: 118–128.

Hutkins, R.E. and E.R. Kashket. 1986. Phosphotransferase activity in *Clostridium acetobutylicum* from acidogenic and solventogenic phases of growth. Appl Environ Microbiol. 51: 1121–1123.

Huynen, M.A., T. Dandekar and P. Bork. 1998. Differential genome analysis applied to the species-specific features of *Helicobacter pylori*. FEBS Lett. 426: 1–5.

Ikeda. 2012. Sugar transport systems in *Corynebacterium glutamicum*: features and applications to strain development. Appl Microbiol Biotechnol. 96: 1191–1200.

Izui, K., M. Taguchi, M. Morikawa and H. Katsuki. 1981. Regulation of *Escherichia coli* phosphoenol pyruvate carboxylase by multiple effectors *in vivo*. II. Kinetic studies with a reaction system containing physiological concentrations of ligands. J Biochem. 90: 1321–1331.

Izyumov, D.S. et al. 2004. Wages of fear: transient threefold decrease in intracellular ATP level imposes apoptosis. Biochim Biophys Acta. 1658: 141–147.

Jacob, F. and J. Monod. 1961. Genetic regulatory mechanisms in the synthesis of proteins. J Mol Biol. 3: 318–356.

Jahid, I.K., A.J. Silva and J.A. Benitez. 2006. Polyphosphate stores enhance the ability of *Viblio cholerae* to overcome environmental stresses in a low-phosphate environment. Appl Environ Microbiol. 72: 7043–7049.

Jarlier, V. and H. Nikaido. 1994. Mycobacterial cell wall: structure and role in natural resistance to antibiotics. FEMS Microbiol Lett. 123(1-2): 11–8.

Jetten, M.S.M., M.E. Gubler, S.H. Lee and A.J. Sinskey. 1994. Structural and functional analysis of pyruvate kinase from *Corynebacterium glutamicum*. Appl Environ Microbiol. 60(7): 2501–2507.

Jin, S. and A.L. Sonenshein. 1994. Identification of two distinct *Bacillus subtilis* malate dehydrogenase gene. J Bacteriol. 176: 4669–4679.

Jin, S., M. De Jesus-Perrios and A.L. Sonenshein. 1996. A *Bacillus subtilis* malate dehydrogenase gene. J Bacteriol. 178: 560–563.

Jonas, K., A.N. Edwards, I. Ahmad, T. Romeo, U. Romling and O. Melefors. 2010. Complex regulatory network encompassing the Csr, c-di-GMP and motility systems of *Salmonella typhimurium*. Environ Microbiol. 12(2): 524–540.

Jone, R.G. and C.B. Thompson. 2009. Tumor suppressors and cell metabolism: a recipe for cancer growth. Genes Dev. 23: 537–548.

Jones, B.E., V. Dossonnet, E. Kuster, W. Hillen, J. Deutscher and R.E. Klevit. 1997. Binding of the catabolite repressor protein CcpA to its DNA target is regulated by phosphorylation of its corepressor HRp. J Biol Chem. 272: 26530–26535.

Jones, S.A., M. Jogensen, F.Z. Chowdhury, R. Rodgers, J. Hartline, M.P. Leatham, C. Struve, K.A. Krogfelt, P.S. Cohen and T. Conway. 2008. Glycogen and maltose utilization by *Escherichia coli* O157:H7 in the mouse intestine. Infect Immun. 76(6): 2531–2540.

Jurica, M.S. et al. 1998. The allosteric regulation of pyruvate kinase by fructose-1, 6-bisphosphate. Structure. 6: 195–210.

Jurica, M.S., A. Mesecar, P.J. Heath, W. Shi, T. Nowak and B.L. Stoddard. 1998. The allosteric regulation of pyruvate kinase by fructose-1, 6-bisphosphate. Structure. 15; 6(2): 195–210.

Jourlin-Castelli, C., N. Mani, M.M. Nakano and A.L. Sonenshein. 2000. CcpC, a novel regulator of the *citB* gene in *Bacillus subtilis*. J Mol Biol. 295: 865–878.

Kadenbach, B. 2012. Introduction to mitochondrial oxidative phosphorylation. Adv Exp Med Biol.

Kahn, A. 1997. Transcriptional regulation by glucose in the liver. Biochimie. 79: 113–118.

Kapoor, R. and T.A. Venkitasubramanian. 1983. Purification and properties of pyruvate kinase from *Mycobacterium smegmatis*. Arch Biochem Biophys. 225(1): 320–330.

Keasling, J.D. 2010. Manufacturing molecules through metabolic engineering. Science. 330: 1355–1358.

Khankal, R., J.W. Chin, D. Ghosh and P.C. Cirino. 2009. Transcriptional effects of CRP* expression in *Escherichia coli*. J Biol Eng. 3: 13.

Kiefer, P., E. Heinzle, O. Zelder and C. Wittmann. 2004. Comparative metabolic flux analysis of lysine-producing *Corynebacterium glutamicum* cultured on glucose or fructose. Appl Environ Microbiol. 70: 229–239.

Kiel, J.A., J.M. Boels, G. Beldman and G. Venema. 1994. Glycogen in *Bacillus subtilis*: molecular characterization of an operon encoding enzymes involved in glycogen biosynthesis and degradation. Mol Microbiol. 11: 203–218.

Kihira, C., Y. Hayashi, N. Azuma, S. Noda, S. Maeda, S. Fukiya, M. Wada, K. Matsushita and A. Yokota. 2012. Alterations of glucose metabolism in *Escherichia coli* mutants defective in respiratory-chain enzymes. J Biotechnol. 158: 215–223.

Kim, et al. 2002, Jourlin-Castelli et al. 2000, Blencke et al. 2006, Holms et al. 2010, Vemuri et al. 2006, Vemuri et al. 2007, Berrios-Rivera et al. 2002, Koebman et al. 2002, Vadlakorda et al. 2013, Pasteur 1857. Warburg 1956, Crabtree 1929

Kim, H.J., A. Roux and A.L. Sonenshein. 2002. Direct and indirect roles of CcpA in regulation of *Bacillus subtilis* Krebs cycle genes. Mol Microbiol. 48: 179–190.

Kim, H.J., T.H. Kim, Y. Kim and H.S. Lee. 2004. Identification and characterization of *glxR*, a gene involved in regulation of glyoxylate bypass in *Corynebacterium glutamicum*. J Bacetriol. 186: 3453–3460.

King, T., A. Ishihama, A. Kori and T. Ferenci. 2004. J Bacteriol. 186: 5614–5620.

Knowles, V.L., C.S. Smith, C.R. Smith and W.C. Plaxton. 2001. Structural and regulatory properties of pyruvate kinase from the Cyanobacterium Synechococcus PCC 6301. J Biol Chem. 276(24): 20966–20972.

Koch, A.L. 1971. Theadaptive responses of *Escherichia coli* to a feast and famine existence. Adv Microbiol Physiol. 6: 147–217.

Kochanowski, K., B. Volkmer, L. Gerosa, H.B.R. van Rijsewijk, A. Schmidt and M. Heinemann. 2013. Functioning of a metabolic flux sensor in *Escherichia coli*. PNAS USA. 110: 1130–1135.

Koebmann, B.J., H.V. Westerhoff, J.L. Snoep, D. Nilsson and P.R. Jensen. 2002. The glycolytic flux in *Escherichia coli* is controlled by the demand for ATP. J Bacteriol. 184(14): 3909–3916.

Koebmann, B., C. Solem and P.R. Jensen. 2006. Control analysis of the importance of phosphogrycerate enolase for metabolic fluxes in *Lactococcus lactis* subsp. *lactis* IL 1403. Syst Biol. 153: 346–349.

Koppenol, W.H., P.L. Bounds and C.V. Dang. 2011. Otto Warburg's contributions to current concepts of cancer metabolism. Nat Rev Cancer. 11: 325–337.

Kotte, O., B. Volkmer, J.L. Radzikowski and M. Heinemann. 2014. Phenotypic bistability in *Escherichia coli*'s central carbon metabolism. Mol Syst Biol. 10: 736.

Kotte, O., J.B. Zaugg and M. Heinemann. 2010. Bacterial adaptation through distributed sensing of metabolic fluxes. Mol Syst Biol. 6: 355.

Kramer, R. and C. Lambert Ebbighausen. 1990. Uptake of glutamate in *Corynebacterium glutamicum*. 1. Kinetic properties and regulation by internal pH and potassium. Eur J Biochem. 194: 929–935.

Kramer, R. and C. Lambert. 1990. Uptake of glutamate in *Corynebacterium glutamicum*. 2. Evidence for a primary active transport system. Eur J Biochem. 194: 037–944.

Kremling, A., K. Bettenbrock and E.D. Gilles. 2007. Analysis of global control of *Escherichia coli* carbohydrate uptake. BMC Syst Biol. 1: 42.

Kremling, A., K. Bettenbrock and E.D. Gilles. 2008. A feed-forward loop guarantees robust behavior in *Escherichia coli* carbohydrate uptake. Bioinformatics. 24: 704–710.

Kroemer, G. and J. Pouyssegur. 2008. Tumor cell metabolism: cancer's Achilles' heel. Cancer Cell. 13: 472–482.

Kumar, R. and K. Shimizu. 2010. Metabolic regulation of *Escherichia coli* and its *gdhA, glnL, gltB, D* mutants under different carbon and nitrogen limitations in the continuous culture. Microbial Cell Fact. 9: 8.

Kussell, E. and S. Leibler. 2005. Phenotypic diversity, population growth, and information in fluctuating environments. Science. 309: 2075–2078.

Larsson, C., A. Nilsson, A. Blomberg and L. Gustafsson. 1997. Glycolytic flux is conditionally correlated with ATP concentration in *Saccharomyces cerevisiae*: a chemostat study under carbon- or nitrogen-limiting conditions. J Bacteriol. 179(23): 7243–7250.

Lee, S.H., D.L. Hava, M.K. Waldor and A. Camilli. 1999. Regulation and temporal expression patterns of *Viblio cholerae* vilulence genes during infection. Cell. 99: 625–634.

Letek, M., N. Valbuena, A. Ramos, E. Ordonez, J.A. Gil and L.M. Mateos. 2006. Characterization and use of catabolite-repressed promoters from gluconate genes in *Corynebacterium glutamicum*. J Bacteriol. 188: 409–423.

Leuchtenberger, W., K. Huthmacher and K. Drauz. 2005. Biotechnological production of amino acids and derivatives: current status and perspects. Appl Microbiol Biotechnol. 69: 1–8.

Levy, S.F., N. Ziv and M.L. Siegal. 2012. Bet hedging in yeast by heterogenous, age-correlated expression of a stress protectant. PLoS Biol. 10: e1001325.

Lin, L.H. et al. 2006. Long-term sustainability of a high-energy, low-diversity crustal biome. Science. 314: 479–482.

Liu, L., Y. Li, H. Li and J. Chen. 2006a. Significant increase of glycolytic flux in *Torulopsis glabrata* by inhibition of oxidative phosphorylation. FEMS Yeast Res. 6: 1117–1129.

Liu, L.M., Y. Li, G.C. Du and J. Chen. 2006b. Increasing glycolytic flux in *Torulopsis glabrata* by redirecting ATP production from oxidative phosphorylation to substrate–level phosphorylation. J Appl Microbiol. 100: 1043–1053.

Loesche, W.J. 1986. Role of *Streptococcus mutans* in human dental decay. Microbiol Rev. 50: 353–380.

Lovell, S.C., A.H. Mullick and H. Muirhead. 1998. Cooperativity in *Bacillus stearothermophilus* pyruvate kinase. J Mol Biol. 276: 839–851.

Luesink, E.J., R.E.M.A. van Herpen, B.P. Grossiord, O.P. Kuipers and W.M. de Vos. 1998. Transcriptional activation of the glycolytic *las* operon and catabolite repression of the *gal* operon in *Lactococcus lactis* are mediated by the catabolite control protein CcpA. Mol Microbiol. 30: 789–798.

Magasanik, B. 1961. Catabolite repression. Cold Spring Harb Symp Quant Biol. 26: 249–256.

Mailloux, R.J., R. Beriault, J. Lemire, R. Simgh, D.R. Chenier, R.D. Hamel and V.D. Appanna. 2007. The tricarboxylic acid cycle, an ancient metabolic network with a novel twist. PLoS One. 8: e690.

Majewski, R.A. and M.M. Domach. 1990. Simple constrained-optimization view of acetate overflow in *Escherichia coli*. Biotech Bioeng. 35: 732–738.

Malcovati, M. and G. Valentini. 1982. AMP- and fructose 1, 6-biophosphate-activated pyruvate kinase from *Escherichia coli*. Methods Enzymol. 90: 170–179.

Mandelstam, J. 1962. The repression of constitutive beta-galactosidase in *Escherichia coli* by glucose and other carbon sources. Biochem J. 82: 489–493.

Marie, J., A. Kahn and P. Boivin. 1976. Pyruvate kinase isozymes in man. I. M type isozymes in adult and foetal tissues, electrofocusing and immunological studies. Hum Genet. 31: 35–45.

Martin, M.C., D. Schneider, C.J. Bruton, K.F. Chater and C. Hardisson. 1997. A *glgC* gene essential only for the first of two spacially distinct phases of glycogen synthesis in *Streptomyces coelicolor* A3(2). J Bacteriol. 179: 7784–7789.

Martinez, K., R. de Anda, G. Hernandez, A. Eacalante, G. Gosset, O.T. Ramirez and F.G. Bolivar. 2008. Coutilization of glucose and glycerol enhances the production of aromatic compounds in an *Escherichia coli* strain lacking the phosphoenolpyruvate: carbohydrate phosphotransferase system. Microb Cell Fact. 7: 1.

Mattevi, A., M. Bolognesi and G. Valentini. 1996. The allosteric regulation of pyruvate kinase. FEBS Lett. 389: 15–19.

Max-Audit, I. et al. 1984. Pattern of pyruvate kinase isozymes in erythroleukemia cell lines and in normal human erythroblasts. Blood. 64: 930–936.

Mazurek, S. 2011. Pyruvate kinase type M2: A key regulator for the metabolic budget system in tumor cells. Int J Biochem Cell Biol. 43: 969–980.

McFall, E. and B. Magasanik. 1962. Effects of thymine and of phosphate deprivation on enzyme synthesis in *Escherichia coli*. Biochim Biophys Acta. 55: 900–908.

McInerney, M.J., J.R. Sieber and R.P. Gunsalus. 2009. Syntrophy in anaerobic global carbon cycles. Curr Opin Biotechnol. 20: 623–632.

McKee, A.E., B.J. Rutherford, D.C. Chivian, E.K. Baidoo, D. Juminaga, D. Kuo, P.I. Benke, J.A. Dietrich, S.M. Ma, A.P. Arkin, C.J. Petzold, P.D. Adams, J.D. Keasling and S.R. Chhabra. 2012. Manipulation of the carbon storage regulator system for metabolite remodeling and biofuel production in *Escherichia coli*. Microb Cell Fact. 11: 79.

Meyer, D.J. and C.W. Jones. 1973. Oxidative phosphorylation in bacteria which contain different cytochrome oxidases. Eur J Biochem. 36: 144–151.

Molenaar, D., R. van Berlo, D. de Ridder and B. Teusink. 2009. Shifts in growth strategies reflect tradeoffs in cellular economics. Mol Syst Biol. 5: 323.

Monod, J. 1942. Recherches sur la croissance des cultures bacteriennes. Hermann et Cie, Paris, France.

Moon, M.W., H.L. Kim, T.K. Oh, C.S. Shin, J.S. Lee, S.J. Kim and J.K. Lee. 2005. Analyses of enzyme II gene mutants for sugar transport and heterologous expression of fructokinase gene in *Corynebacterium glutamicum* ATCC13032. FEMS Microbiol Lett. 244: 259–266.

Morán-Zorzano, M.T., A.M. Viale, F.J. Muñoz, N. Alonso-Casajús, G. Eydallin, B. Zugasti, E. Baroja-Fernández and J. Pozueta-Romero. 2007. *Escherichia coli* AspP activity is enhanced by molecular crowding and by both glucose-1, 6-bisphosphate and nucleotide-sugars. FEBS Lett. 581: 1035–1040.

Morán-Zorzano, M.T., M. Montero, F.J. Muñoz, N. Alonso-Casajús, A.M. Viale, G. Eydallin, M.T. Sesma, E. Baroja-Fernández and J. Pozueta-Romero. 2008. Cytoplasmic *Escherichia coli* ADP sugar pyrophosphatase binds to cell membranes in response to extracellular signals as the cell population density increases. FEMS Microbiol Lett. 288: 25–32.

Morelli, M.J., R.J. Allen and P.R. Wolde. 2011. Effects of macromolecular crowding on genetic networks. Biophys J. 101(12): 2882–2891.

Muirhead, H., D.A. Clayden, D. Barford, C.G. Lorimer, L.A. Fothergill-Gilmore, E. Schiltz et al. 1986. The structure of cat muscle pyruvate kinase. EMBO J. 5: 475–481.

Muller, S., F.K. Zimmermann and E. Boles. 1997. Mutant studies of phosphofruct-2-kinases do not reveal an essential role of fructose-2, 6-bisphosphate in the regulation of carbon fluxes in yeast cells. Microbiol. 143: 3055–3061.

Munoz, M.E. and E. Ponce. 2003. Pyruvate kinase: current status of regulatory and functional properties. Compara Biochem Physiol B. 135: 197–218.

Murcott, T.H.L., T. McNally, S.C. Allen, L.A. Fothergill-Gilmore and H. Muirhead. 1991. Purification, characterisation and mutagenesis of highly expressed recombinant yeast pyruvate kinase. Eur J Biochem. 198: 513–519.

Muroya, N., Y. Nagao, K. Miyazaki, K. Nishikawa and T. Horio. 1976. Pyruvate kinase isozymes in various tissues of rat, and increase of spleen-type pyruvate kinase in liver by injecting chromatins from spleen and tumor. J Biochem. 79: 203–215.

Murphy, M.P. 2009. How mitochondria produce reactive oxygen species. Biochem J. 417(1): 1–13.

Nakamura, Y., J. Takahashi, A. Sakurai, Y. Inaba, E. Suzuki, S. Nihei, S. Fujiwara, M. Tsuduki, H. Miyashita, H. Ikemoto, M. Kawachi, H. Sekiguchi and N. Kurano. 2005. Some cyanobacteria synthesize semi-amylopectin type α-polyglucans instead of glycogen. Plant Cell Physiol. 46: 539–545.

Nakashima, N. and T. Tamura. 2012. A new carbon catabolite repression mutation of *Escherichia coli*, *mlc*\*, and its use for producing isobutanol. J Biosci Bioeng. 114(1): 38–44.

Nessler, S. et al. 2003. HPr kinase/phosphorylase, the sensor enzyme of catabolite repression in Gram-positive bacteria: structural aspect of the enzyme and the complex with its protein substrate. J Bacteriol. 185: 4003–4010.

Neves, A.R., A. Ramos, H. Costa, I.I. van Swam, J. Hugenholtz, M. Kleerebezem, W. de Vos and H. Santos. 2002. Effect of different NADH oxidase levels on glucose metabolism by *Lactococcus lactis*: Kinetics of intracellular metabolite pools determined by *in vivo* nuclear magnetic resonance. Appl Environ Microbiol. 68(12): 6332–6342.

New, A.M., B. Cerulus, S.K. Govers, G. Perez-Samper, B. Zhu, S. Boogmans, J.B. Xavier and K.J. Verstrepen. 2014. Different levels of catabolite repression optimize growth in stable and variable environments. PLoS Biol. 12(1): e1001764.

Nichols, N.N., B.S. Dien and R.J. Bothast. 2001. Use of catabolite repression mutants for fermentation of sugar mixtures to ethanol. Appl Microbiol Biotechnol. 56(1-2): 120–125.

Nikel, P., J. Zhu, K. San, B.S. Mendez and G.N. Bennet. 2009. Metabolic flux analysis of *Escherichia coli creB* and *arcA* mutants reveals shared control of carbon catabolism under microaerobic growth conditions. J Bacteriol. 191: 5538–5548.

Noguchi, T., H. Inoue and T. Tanaka. 1986. The M1- and M2-type isozymes of pyruvate kinase are produced from the same gene by alterative RNA splicing. J Biol Chem. 261: 13807–13812.

Nordberg, J. and E.S. Arner. 2001. Reactive oxygen species, antioxidants, and the mammarian thioredoxin system. Free Radical Biol Med. 31(11): 1287–1312.

Ochocki, J.D. and M.C. Simon. 2013. Nutrient-sensing pathways and metabolic regulation in stem cells. J Cell Biol. 203(1): 23–33.

Ogawa, N., C.M. Tzeng, C.D. Fraley and A. Kornberg. 2000. Inorganic polyphosphate in *Viblio cholerae*: genetic, biochemical, and physiologic features. J Bacteriol. 182: 6687–6693.

Owittrim, G.W. and B. Colman. 1988. Phosphoenolpyruvate carboxylase mediated carbon flow in a cyanobacterium. Biochem Cell Biol. 66: 93–99.

Ozbudak, E.M., M. Thattai, H.N. Lim, B.I. Shiraiman and A. van Oudenaarden. 2004. Multiplicity in the lactose utilization network of *Escherichia coli*. Nature. 427: 737–740.

Padilla, L., S. Morbach, R. Kramer and E. Agosin. 2004. Impact of heterologous expression of *Escherichia coli* UDP-glucose pyrophosphorylase on trehalose and glycogen synthesis in *Corynebacterium glutamicum*. Appl Environ Microbiol. 70: 3845–3854.

Park, J.M., P. Vinuselvi and S.K. Lee. 2012. The mechanism of sugar-mediated catabolite repression of the propionate catabolic genes in *Escherichia coli*. Gene. 504: 116–121.

Park, S.Y., M.W. Moon, B. Subhadra and J.K. Lee. 2010. Functional characterization of the *glxR* deletion mutant of *Corynebacterium glutamicum* ATCC 13032: involvement of GlxR in acetate metabolism and carbon catabolite repression. FEMS Microbiol Lett. 304: 107–115.

Pasteur, L. 1857. Momoir sur la fermentation alcoolique. CR Seances Acad Aci. 45: 1032–1036.

Pawluk, A., R.K. Scopes and K. Griffiths-Smith. 1986. Isolation and properties of the glycolytic enzymes from *Zymomonas mobilis*. Biochem J. 238: 275–281.

Pawluk, A., R.K. Scopes and K. Griffiths-Smith. 1986. Isolation and properties of the glycolytic enzymes from Zymomonas mobilis. The five enzymes from glyceraldehyde-3-phosphate dehydrogenase through to pyruvate kinase. Biochem J. 238(1): 275–81.

Peng, L. and K. Shimizu. 2003. Global metabolic regulation analysis for *E. coli* K12 based on protein expression by 2DE and enzyme activity measurement. Appl Microbiol Biotech. 61: 163–178.

Peralta-Yahya, P.P., F. Zhang, S.B. del Cardayre and J.D. Keasling. 2012. Microbial engineering for the production of advanced biofuels. Nature. 488: 320–328.

Pfeiffer, T. and A. Morley. 2014. An evolutionary perspective on the Crabtree effect. Frontiers in Mol Biosci. 1: 1–6.

Pfeiffer, T. and S. Bonhoeffer. 2002. Evolutionary consequences of tradeoffs between yield and rate of ATP production. Z Phys Chem. 31: 51–63.

Pfeiffer, T. and S. Bonhoeffer. 2004. Evolution of cross-feeding in microbial populations. Am Nat. 163: E126–E135.

Pfeiffer, T., S. Schuster and S. Bonhoeffer. 2001. Cooperation and competition in the evolution of ATP-producing pathways. Science. 292: 504–507.

Podesta, F.E. and W.C. Plaxton. 1992. Plant cytosolic pyruvate kinase: a kinetic study. Biochim Biophys Acta. 1160: 213–220.

Ponce, E. 1999. Effect of growth rate reduction and genetic modifications on acetate accumulation and biomass yields in *Escherichia coli*. J Biosci Bioeng. 87: 775–780.

Ponce, E., N. Flores, A. Martinez, F. Valle and F. Bolivar. 1995. Cloning of the two pyruvate kinase isozymes structural genes from *Escherichia coli*: the relative roles of these genes in pyruvate biosynthesis. J Bacteriol. 117: 5719–5722.

Postmus, J., R. Aardema, C.G. de Koster, S. Brul and G.J. Smits. 2012. Isoenzyme expression changes in response to high temperature determine the metabolic regulation of increased glycolytic flux in yeast. FEMS Yeast Res. 12(5): 571–581.

Postmus, J., A.B. Canelas, J. Bouwman, B.M. Bakker, W. van Gulik, M.J. de Mattos, S. Brul and G.J. Smits. 2008. Quantitative analysis of the high temperature-induced glycolytic flux increase in *Saccharomyces cerevisiae* reveals dominant metabolic regulation. J Biol Chem. 283(35): 23524–23532.

Preiss, J. 1984. Bacterial glycogen synthesis and its regulation. Annu Rev Microbiol. 38: 419–458.

Preiss, J. 2009. Glycogen: biosynthesis and regulation. *In*: A. Böck, R. Curtiss III, J.B. Kaper, P.D. Karp, F.C. Neidhardt, T. Nyström, J.M. Slauch, C.L. Squires and D. Ussery (eds.). EcoSal—*Escherichia coli* and Salmonella: Cellular and Molecular Biology. ASM Press; Washington, DC.

Puech, V., M. Chami, A. Lemassu, M.A. Laneelle, B. Schiffler, P. Gounon, N. Bayan, R. Benz and M. Daffe. 2001. Structure of the cell envelope of corynebacteria: importance of the non-covalently bound lipids in the formation of the cell wall permeability barrier and fracture plane. Microbiol. 147: 1365–1382.

Ralser, M., M.M. Wamelink, S. Latkolik, E.E. Jansen, H. Lehrach and C. Jakobs. 2009. Metabolic reconfiguration precedes transcriptional regulation in the antioxidant response. Nat Biotechnol. 27(7): 604–605.

Ramos, A. et al. 2004. Effect of pyruvate kinase overproduction on glucose metabolism of *Lactococcus lactis*. Microbiol. 150: 1103–1111.

Reidl, J. and K.E. Klose. 2002. *Vibrio cholerae* and cholera: out of the water and into the host. FEMS Microbiol Rev. 26: 125–139.

Rolfe, M.D., A. Ocone, M.R. Stapleton, S. Hall, E.W. Trotter, R.K. Poole, G. Sanguinetti and J. Green. 2012. Systems analysis of transcription factor activities in environments with stable and dynamic oxygen concentrations. Open Biol. 2(7): 120091.

Romeo, T. 1996. Post-transcriptional regulation of bacterial carbohydrate metabolism: evidence that the gene product CsrA is global mRNA decay factor. Res Microbiol. 147(6-7): 505–512.

Romeo, T. 1998. Global regulation by the small RNA-binding protein CsrA and the non-coding RNA molecule CsrB. Mol Microbiol. 29(6): 1321–1330.

Romeo, T. and J. Preiss. 1989. Genetic regulation of glycogen biosynthesis in *Escherichia coli*: *In vitro* effects of cyclic AMP and guanosine 5'-diphosphate 3'-diphosphate and analysis of *in vivo* transcripts. J Bacteriol. 171: 2773–2782.

Romeo, T., M. Gong, M.Y. Liu and A.M. Brun-Zinkernagel. 1993. Identification and molecular characterization of *csrA*, a pleiotropic gene from *Escherichia coli* that affects glycogen biosynthesis, gluconeogenesis, cell size, and surface properties. J Bacteriol. 175(15): 4744–4755.

Romeo, T. and M. Gong. 1993. Genetic and physical mapping of the regulatory gene *csrA* on the *Escherichia coli* K-12 chromosome. J Bacteriol. 175(17): 5740–5741.

Roseman, S. and N.D. Meadow. 1990. Signal transduction by the bacterial phosphotransferase system. Diauxie and *crr* gene (J. Monod revisited) J Biol Chem. 265: 2993–2996.

Rowen, D.W., M. Meinke and D.C. LaPorte. 1992. *GLC3* and *GHA1* of *Saccharomyces cerevisiae* are allelic and encode the glycogen branching enzyme. Mol Cell Biol. 12: 22–29.

Russell, J.B. 2007. The energy spilling reactions of bacteria and other organisms. J Mol Microbiol Biotechnol. 13: 1–11.

Ruyter, G.J.G., P.W. Postma and K. van Dam. 1991. Control of glucose metabolism by enzyme II Glc of the phosphoenolpyruvate-dependent phosphotransferase system in *Escherichia coli*. J Bacteriol. 173: 6184–6191.

Sack, D.A., R.B. Sack and C.L. Chaignal. 2006. Getting serious about cholera. N Engl J Med. 355: 649–651.

Sakai, H., K. Suzuki and K. Imahori. 1986. Purification and properties of pyruvate kinase from *Bacillus stearothermophilus*. J Biochem. 99: 1157–1167.

Sambou, T., P. Dinadayala, G. Stadthagen, N. Barilone, Y. Bordat, P. Constant, F. Levillain, O. Neyrolles, B. Gicquel, A. Lemassu, M. Daffé and M. Jackson. 2008. Capsular glucan and intracellular glycogen of *Mycobacterium tuberculosis*: biosynthesis and impact on the persistence in mice. Mol Microbiol. 70(3): 762–74.

Sawada, K., S. Zen-in, M. Wada and A. Yokota. 2010. Metabolic changes in a pyruvate kinase gene deletion mutant of *Corynebacterium glutamicum* ATCC13032. Metab Eng.

Schaaff, I., J. Heinisch and F.K. Zimmermann. 1989. Overproduction of glycolytic enzymes in yeast. Yeast. 5: 285–290.

Schild, S., R. Tamayo, E.J. Nelson, F. Quadri, S.B. Calderwood and A. Camilli. 2007. Genes induced late in infection increase fitness of *Viblio cholera* after release into the environment. Cell Host Microbiol. 2: 264–277.

Schumacher, M.A., G.S. Allen, M. Diel, G. Seidel, W. Hillen and R.G. Brennan. 2004. Structural basis for allosteric control of the transcription regulator CcpA by the phosphoprotein HPr-Ser46-P. Cell. 118(6): 731–41.

Schumacher, M.A., G. Seidel, W. Hillen and R.G. Brennan. 2007. Structural mechanism for the fine-tuning of CcpA function by the small molecule effectors glucose 6-phosphate and fructose 1, 6-bisphosphate. J Mol Biol. 368: 1042–1050.

Scott, M., C.W. Gunderson, E.M. Mateescee, Z. Zhang and T. Hwa. 2010. Interdependence of cell growth and gene expression: origins and consequences. Science. 330: 1099–1102.

Seidel, G., M. Diel, N. Fuchsbauer and W. Hillen. 2005. Quantitative interdependence of effectors, CcpA and cre in carbon catabolite regulation of *Bacillus subtilis*. FEMS J. 272: 2566–2577.

Sekine, H., T. Shimada, C. Hayashi, A. Ishiguro, F. Tomita and A. Yokota. 2001. H+-ATPase defect in *Corynebacterium glutamicum* abolishes glutamic acid production with enhancement of glucose consumption rate. Appl Microbiol Biotechnol. 57: 534–540.

Sengupta, S., T.R. Peterson and D.M. Sabatini. 2010. Regulation of the mTOR complex 1 pathway by nutrients, growth factors, and stress. Mol Cell. 40: 310–322.

Servinsky, M.D., J.T. Kiel, N.F. Dupuy and C.J. Sund. 2010. Transcriptional analysis of differential carbohydrate utilization by *Clostridium acetobutylicum*. Microbiol. 156: 3478–3491.

Shen, C.R. et al. 2011. Driving forces enable high-titer anaerobic 1-butanol synthesis in *Escherichia coli*. Appl Environ Microbiol. 77: 2905–2915.

Shimada, T., K. Yamamoto and A. Ishihama. 2011a. Novel members of the Cra regulon involved in carbon metabolism in *Escherichia coli*. J Bacteriol. 193(3): 649–659.

Shimada, T., N. Fujita, K. Yamamoto and A. Ishihama. 2011b. Novel roles of cAMP receptor protein (CRP) in regulation of transport and metabolism of carbon . PLoS One. 6(6): e20081.

Shimizu, K. 2013. Metabolic Regulation of a Bacterial Cell System with Emphasis on *Escherichia coli* Metabolism, ISRN Biochemistry, Article ID 645983: doi:10.1155/2013/645983.

Shimizu, K. 2014. Biofuels and biochemicals production by microbes, NOVA Publ. Co., NY, USA.

Shimizu, K. 2014. Metabolic regulation of *Escherichia coli* in response to nutrient limitation and environmental stress conditions. Metabolites. 4: 1–35.

Shimizu, K. 2016. Metabolic regulation and coordination of the metabolism in bacteria in response to a variety of growth conditions. Adv Biochem Eng Biotechnol. 155: 1–54.

Siddiquee, K.A.Z., M. Arauzo-Bravo and K. Shimizu. 2004a. Metabolic flux analysis of *pykF* gene knockout *Escherichia coli* based on $^{13}$C labeled experiment together with measurements of enzyme activities and intracellular metabolite concentrations. Appl Micobiol Biotechnol. 63: 407–417.

Siddiquee, K.A.Z., M. Arauzo-Bravo and K. Shimizu. 2004b Effect of pyruvate kinase (*pykF* gene) knockout mutation on the control of gene expression and metabolic fluxes in *Escherichia coli*. FEMS Microbiol Lett. 235: 25–33.

Siebold, G. and B.J. Eikmanns. 2007. The *glgX* gene product of *Corynebacterium glutamicum* is required for glycogen degradation and for fast adaptation to hyperosmotic stress. Microbiol. 153: 2212–2220.

Siebold, G., S. Dempf, J. Schreiner and B.J. Eikmanns. 2007. Glycogen formation in *Corynebacterium glutamicum* and role of ADP-glucose pyrophosphorylase. Microbiol. 153: 1275–1285.

Siedler, S., S. Bringer, L.M. Blank and M. Bott. 2012. Engineering yield and rate of reductive biotransformation in *Escherichia coli* by partial cyclization of the pentose phosphate pathway and PTS independent glucose transport. Appl Microbiol Biotechnol. 93(4): 1459–1467.

Smith, T.L. and J. Rutter. 2007. Regulation of glucose partitioning by PAS kinase and Ugp1 phosphorylation. Mol Cell. 26: 491–499.

Smits, H.P., J. Hauf, S. Muller, T.J. Hobley, F.K. Zimmermann, B. Hahn-Hagerdal, J. Nielsen and L. Olsson. 2000. Simultaneous overexpression of enzymes of the lower part of glycolysis can enhance the fermentative capacity of *Saccharomyces cerevisiae*. Yeast. 16: 1325–1334.

Snoep, J.L., L.P. Yomano, H.V. Westerhoff and L.O. Ingram. 1995. Protein burden in Zymomonas mobilis: negative flux and growth control due to overproduction of glycolytic enzymes. Microbiol. 141: 2329–2337.

Solem, C., B.J. Koebmann and P.R. Jensen. 2003. Glyceraldehyde-3-phosphate dehydrogenase has no control over glycolytic flux in *Lactococcus lactis* MG1363. J Bicteriol. 185: 1564–1571.

Solem, C., D. Petranovic, B.J. Koebmann, I. Mijakovic and P.R. Jensen. 2010. Phosphoglycerate mutase is a highly efficient enzyme without flux control in *Lactococcus lactis*. J Mol microbial Biotechnol. 18: 174–180.

Solopova, A., J. van Gestel, F.J. Weissing, H. Bachmann, B. Teusink, J. Kok and O.P. Kuipers. 2014. Bet-hedging during bacterial diauxic shift. PNAS USA. 111(20): 7427–7432.

Sonenshein, A.L. 2005. CodY, a global regulator of stationary phase and virulence in Gram-positive bacteria. Curr Opin Microbiol. 8: 203–207.

Sonenshein, A.L. 2007. Control of key metabolic intersections in *Bacillus subtilis*. Nature Rev Microbiol. 5(12): 917–927.

Stanier, R.Y. 1951. Enzymatic adaptation in bacteria. Annu Rev Microbiol. 5: 35–56.

Suzuki, E., H. Ohkawa, K. Moriya, T. Matsubara, Y. Nagaike, I. Iwasaki, S. Fujiwara, M. Tsuzuki and Y. Nakamura. 2010. Carbohydrate metabolism in mutants of the Cyanobacterium *Synechococcus elongatus* PCC7942 defective in glycogen synthesis. Appl Environ Microbiol. 76(10): 3153–3159.

Suzuki, K., X. Wang, T. Weilbacher et al. 2002. Regulatory circuitry of the CsrA/CsrB and BarA/UvrY systems of *Escherichia coli*. J of Bacteriol. 184(18): 5130–5140.

Takata, H., T. Takaha, S. Okada, M. Takagi and T. Imanaka. 1997. Characterization of a gene cluster for glycogen biosynthesis and a heterotetrameric ADP-glucose pyrophosphorylase from *Bacillus stearothermophilus*. J Bacteriol. 179: 4689–4698.

Tanaka, Y., N. Okai, H. Teramoto, M. Inui and H. Yukawa. 2008. Regulation of the expression of phosphoenolpyruvate: carbohydrate transferase system (PTS) genes in *Corynebacterium glutamicum* R. Microbiol. 154: 264–274.

Tchieu, J.H., V. Norris, J.S. Edwards and M.H. Saier Jr. 2001. The complete phosphotransferase system in *Escherichia coli*. J Mol Microbiol. 3: 329–346.

Teramoto, H., M. Inui and H. Yukawa. 2010. Regulation of genes involved in sugar uptake, glycolysis and lactate production in *Corynebacterium glutamicum*. Future Microbiol. 5: 1475–1481.

Teramoto, H., M. Inui and H. Yukawa. 2011. Transcriptional regulators of multiple genes involved in carbon metabolism in *Corynebacterium glutamicum*. J Biotechnol. 154: 114–125.

Teusink, B. and E.J. Smid. 2006. Modelling strategies for the industrial exploitation of lactic acid bacteria. Nat Rev Microbiol. 4: 46–56.

Thattai, M. and A. van Oudenaarden. 2004. Stochastic gene expression in fluctuating environments. Genetics. 167: 523–530.

Thomas, T.D., D.C. Ellwood and V.M. Longyear. 1979. Change from homo- to heterolactic fermentation by *Streptococcus lactis* resulting from glucose limitation in anaerobic chemostat cultures. J Bacteriol. 138: 109–117.

Timmermans, J. and L. van Melderen. 2010. Post-transcriptional global regulation by CsrA in bacteria. Cell Mol Life Sci. 67(17): 2897–2908.

Toya,Y., N. Ishii, K. Nakahigashi, T. Hirasawa, T. Soga, M. Tomita and K. Shimizu. 2010. $^{13}$C-metabolic flux analysis for batch culture of *Escherichia coli* and its *pyk* and *pgi* gene knockout mutants based on mass isotopomer distribution of intracellular metabolites. Biotech. Prog. DOI 10.1002/(ISSN) 1520-6033.

Toyoda, K., H. Teramoto, M. Inui and H. Yukawa. 2011. Genome-wide identification of *in vivo* binding sites GlxR, a cyclic AMP receptor protein-type regulator in *Corynebacterium glutamicum*. J Bacteriol. 193: 4123–4133.

Traxler, M.F., D.-E. Chang and T. Conway. 2006. Guanosine 3′, 5′-bispyrophosphate coordinates global gene expression during glucose-lactose diauxie in *Escherichia coli*. PNAS USA. 103(7): 2374–2379.

Tsuge, Y., K. Uematsu, S. Yamamoto, M. Suda, H. Yukawa and M. Inui. 2015. Glucose consumption rate critically depends on redox state in *Corynebacterium glutamicum* under oxygen deprivation. Appl Microbiol Biotechnol. 99(13): 5573–5582.

Tzvetkov, M., C. Klopprogge, O. Zelder and W. Liebl. 2003. Genetic dissection of trehalose biosynthesis in *Corynebacterium glutamicum*: inactivation of trehalose production leads to impaired growth and an altered cell wall lipid composition. Microbiol. 149: 1659–1673.

Urso, M.L. and P.M. Clarkson. 2003. Oxidative stress, exercise, and antioxidant supplementation. Toxicol. 189(1-2): 41–54.

Vadlakonda, L., A. Dash, M. Pasupuleti, A.N. Mumar and P. Reddanna. 2013. Did we get Pasteur, Warburg, and Crabtree on a right note ? Frontiers in Oncology. 3: 1–4.

Valgepea, K., K. Adamberg, R. Nahku, P.J. Lahtvee, L. Arike and R. Vilu. 2010. Systems biology approach reveals that overflow metabolism of acetate in *Escherichia coli* is triggered by carbon catabolite repression of acetyl-CoA synthetase. BMC Syst. Biol. 4: 166.

Valgepea, K., K. Adamberg and R. Vilu. 2011. Decrease of energy spilling in *Escherichia coli* continuous cultures with rising specific growth rate and carbon wasting. BMC Syst Biol. 5: 106.

van Berkel, T.J., H.R. de Jonge, J.F. Koster and W.C. Hulsmann. 1974. Kinetic evidence for the presence of two forms of M2-type pyruvate kinase in rat small intestine. Biochem Biophys Res Commun. 60: 398–405.

Van den Brink, J., A.B. Canelas, W.M. van Gulik, J.T. Pronk, J.J. Heijnen, J.H. de Winde and P. Daran-Lapujade. 2008. Dynamics of glycolytic regulation during adaptation of *Saccharomyces cerevisiae* to fermentative metabolism. Appl Environ Microbiol. 74(18): 5710–5723.

Van der Heiden, M.G., L.C. Cantley and C.B. Thompton. 2009. Understanding the Warburg effect: The metabolic requirements of cell proliferation. Science. 324: 1029–1033.

van Heerden, J.H., T.W. Meike, F.J. Bruggeman, J.J. Heijnen, Y.J.M. Bollen, R. Planque, J. Hulshof, T.G. O'Toole, S.A. Wahl and B. Teusink. 2014. Lost in transition: Start-up of glycolysis yields subpopulations of nongrowing cells. Science. 343: 1245114.

Van Schaftingen, E., F.R. Opperdoes and H.G. Hers. 1985. Stimulation of *Trypanosoma brucei* pyruvate kinase by fructose 1, 6-bisphosphate. Eur J Biochem. 153: 403–406.

Vander Heiden, M.G., L.C. Cantley and C.B. Thompson. 2009. Understanding the Warburg effect: The metabolic requirements of cell proliferation. Science. 324: 1029–1033.

Varma, A. and B.O. Palsson. 1994. Stoichiometric flux balance models quantitatively predict growth and metabolic by-product secretion in wild-type *Escherichia coli* W3110. Appl Environ Microbiol. 60(10): 3724–31.

Vaulont, S., A. Munnich, J.F. Decaux and A. Kahn. 1986. Transcriptional and post-transcriptional regulation of L-type pyruvate kinase gene expression in rat liver. J Biol Chem. 261: 7621–7625.

Vazquez, A., Q.K. Beg, M.A. Demenezes, J. Ernst, Z. Bar-Joseph, A.L. Barabasi, L.G. Boros and Z.N. Oltvai. 2008. Impact of the solvent capacity constraint on *E. coli* metabolism. BMC Syst Biol. 2(7): 1752–0509.

Vemuri, G.N., E. Altman, D.P. Sangurdekar, A.B. Khodursky and M.A. Eitman. 2006. Overflow Metabolism in *Escherichia coli* during Steady-State Growth: Transcriptional regulation and effect of the redox ratio. Appl Environ Microbiol. 72: 3653–3661.

Vemuri, G.N., M.A. Eitman, J.E. McEwen, L. Olsson and J. Nielsen. 2007. Increasing NADH oxidation reduces overflow metabolism in Saccharomyces cerevisiae. PNAS 104(7): 2402–2407.

Voigt, C., H. Bahl and R.-J. Fischer. 2014. Identification of PTS$^{Fru}$ as the major fructose uptake system of *Clostridium acetobutylicum*. Appl Microbiol Biotechnol. 98: 7161–7172.

Volkmer, B. and M. Heinemann. 2011. Condition-dependent cell volume and concentration of *Escherichia coli* to facilitate data conversion for systems biology modeling. PLoS ONE. 6: e23126.

Waddell, T.G., P. Repovic, E. Mele´ndez-Hevia, R. Heinrich and F. Montero. 1997. Optimization of glycolysis: a new look at the efficiency of energy coupling. Biochem Educ. 25: 204–205.

Walker, D., W.N. Chia and H. Muirhead. 1992. Key residues in the allosteric transition of *Bacillus stearothermophilus* pyruvate kinase identified by site-directed mutagenesis. J Mol Biol. 228: 265–276.

Warner, J.B. and J.S. Lolkema. 2003. CcpA-dependent carbon catabolite repression in bacteria. Micobiol Mol Biol Rev. 67: 475–490.

Watnick, P. and R. Kolter. 2000. Biofilm, city of microbes. J Bacteriol. 182: 2675–2679.

Waygood, E.B., J.S. Mort and B.D. Sanwal. 1976. The control of pyruvate kinase of *Escherichia coli*. Binding of substrate and allosteric effectors to the enzyme activated by fructose 1, 6-bisphosphate. Biochem. 15: 277–282.

Weilbacher, T., K. Suzuki, A.K. Dubey et al. 2003. A novel sRNA component of the carbon storage regulatory system of *Escherichia coli*. Mol Microbiol. 48(3): 657–670.

Wendisch, V.H., A.A. de Graaf, H. Sahm and B.J. Eikmanns. 2000. Quantitative determination of metabolic fluxes during co-utilization of two carbon sources: comparative analyses with *Corynebacterium glutamicum* during growth on acetate and/or glucose. J Bacteriol. 182: 3088–3096.

Wilson, W.A., P.J. Roach, M. Montero, E. Baroja-Fernández, F.J. Muñoz, G. Eydallin, A.M. Viale and J. Pozueta-Romero. 2010. Regulation of glycogen metabolism in yeast and bacteria. FEMS Microbiol Rev. 34(6): 952–85.

Wolf, A., R. Kramer and S. Morbach. 2003. Three pathways for trehalose metabolism in *Corynebacterium glutamicum* ATCC 13032 and their significance in response to osmotic stress. Mol Microbiol. 49: 1119–1134.

Voit, E., A.R. Neves and H. Santos. 2006. The intricate side of systems biology. PNAS USA. 103: 9452–9457.

Warburg, O. 1956. On the origin of cancer cells. Science. 123(3191): 309–14.

Wolfe, A.J. 2005. The acetate switch. Microbiol Mol Biol Reviews. 69: 12–50.

Wu, Y., Y. Yang, C. Ren, C. Yang, S. Yang, Y. Gu and W. Jiang. 2015. Molecular modulation of pleiotropic regulator CcpA for glucose and xylose coutilization by solvent-producing *Clostridium acetobutylicum*. Metab Eng. 28: 169–179.

Xu, Y.-F. et al. 2012b. Regulation of yeast pyruvate kinase by ultrasensitive allostery independent of phosphorylation. Mol Cell. 48: 52–62.

Xu, Y.-F., D. Amador-Noguez, M.L. Reaves, X.J. Feng et al. 2012a. Ultrasentitive regulation of anapleurosis via allosteric activation of PEP carboxylase. Nature Chem Biol. 8: 562–568.

Yakandawala, N., T. Romeo, A.D. Friesen and S. Madhyastha. 2008. Metabolic engineering of *Escherichia coli* to enhance phenylalanine production. Appl Microbiol Biotechnol. 78(2): 283–291.

Yakhnin, H., P. Pandit, T.J. Petty, C.S. Baker, T. Romeo and P. Babitzke. 2007. CsrA of *Bacillus subtilis* regulates translation initiation of the gene encoding the flagellin protein (hag) by blocking ribosome binding. Mol Microbiol. 64(6): 1605–1620.

Yamamotoya, T., H. Dose, Z. Tian, A. Faure, Y. Toya, M. Honma, K. Igarashi, K. Nakahigashi, T. Soga, H. Mori and H. Matsuno. 2012. Glycogen in the primary source of glucose during the lag phase of *E. coli* proliferation. Biochim Biophys Acta. 1824: 1442–1448.

Yang, C., Q. Hua, T. Baba, H. Mori and K. Shimizu. 2003. Analysis of *E. coli* anaplerotic metabolism and its regulation mechanism from the metabolic responses to alter dilution rates and *pck* knockout. Biotechnol Bioeng. 84: 129–44.

Yao, R., H. Kurata and K. Shimizu. 2013. Effect of *cra* gene mutation on the metabolism of *Escherichia coli* for a mixture of multiple carbon sources. Adv Biosci Biotechnol. 4: 477–486.

Yao, R., Y. Hirose, D. Sarkar, K. Nakahigashi, Q. Ye and K. Shimizu. 2011. Catabolic regulation analysis of *Escherichia coli* and its *crp*, *mlc*, *mgsA*, *pgi* and *ptsG* mutants. Microb Cell Fact. 10: 67.

Ye, J., A. Mancuso, X. Tong, P.S. Ward, J. Fan, J.D. Rabinowitz and C.B. Thompson. 2012. Pyruvate kinase M2 promotes *de novo* serine synthesis to sustain mTORC1 activity and cell proliferation. PNAS USA. 109(18): 6904–6909.

Yim, H. et al. 2011. Metabolic engineering of *Escherichia coli* for direct production of 1,4-butanediol. Nat Chem Biol. 7: 445–452.

You, C., H. Okano, S. Hui, Z. Zhang, M. Kim, C.W. Gunderson, Y.P. Wang, P. Lenz, D. Yan and T. Hwa. 2013. Coordination of bacterial proteome with metabolism by cAMP signaling. Nature. 500: 301–306.

Young, J.D., A.A. Shastri, G. Stephanopoulos and J.A. Morgan. 2011. Mapping photoautotrophic metabolism with isotopically nonstationary $^{13}$C flux analysis. Metab Eng. 13: 656–665.

Yuneva, M., N. Zamboni, P. Oefner, R. Sachidanandam and Y. Lazebnik. 2007. Deficiency in glutamine but not glucose induces MYC-dependent apoptosis in human cells. J Cell Biol. 178(1): 93–105.

Zheng, X., G. Xie, A. Zhao, L. Zhao, C. Yao, N.H.L. Chiu, Z. Zhou, Y. Bao, W. Jia et al. 2011. The footprints of gut microbial–mammalian co-metabolism. J Proteome Res. 10: 5512–5522.

Zhou, Y., A. Vazquez, A. Wise, T. Warita, K. Warita, Z. Bar-Joseph and Z.N. Oltavai. 2013. Carbon catabolite repression correlates with the maintenance of near invariant molecular crowding in proliferating *E. coli* cells. BMC Stst Biol. 7: 138.

Zhou, H.X., G. Rivas and A.P. Minton. 2008. Macromolecular crowding and confinement: biochemical, biophysical, and potential physiological consequences. Ann Rev Biophys. 37: 375–397.

Zhou, J.W., L.M. Liu, Z.P. Shi, G.C. Du and J. Chen. 2009. ATP in current biotechnology: regulation, applications and perspectives. Biotechnol Adv. 27: 94–101.

Zhu, T., C. Phalakornkule, R.R. Koepsel, M.M. Domach and M.M. Ataai. 2001. Cell growth and by-product formation in a pyruvate kinase mutant of *E. coli*. Biotech Prog. 17: 624–628.

Zhuang, K., G.N. Vemuri and R. Mahadevan. 2011. Economics of membrane occupancy and respiro-fermentation. Mol Syst Biol. 7: 500.

# 5

# Metabolic Regulation in Response to Growth Environment

## ABSTRACT

Living organisms have sophisticated but well organized regulation system. It is important to understand the metabolic regulation mechanisms in relation to growth environment for the efficient design of cell factories for biofuel and biochemical production. Carbon catabolite regulation is explained in Chapters 3 and 4. Here, an overview is given for nitrogen regulation, ion, sulfur, and phosphate regulations, stringent response under nutrient starvation as well as oxidative stress regulation, redox regulation, acid shock, heat and cold shock regulations, solvent stress regulation, osmoregulation, as well as biofilm formation and quorum sensing. The coordinated regulation mechanisms are of particular interest in getting insight into the principle which governs the cell metabolism. The metabolism is controlled by both enzyme level regulation and transcriptional regulation via transcription factors such as cAMP-Crp, Cra, Csr, Fis, $P_{II}$(GlnB), NtrBC, CysB, PhoR/B, SoxR/S, Fur, MarR, ArcA/B, Fnr, NarX/L, RpoS as well as (p)ppGpp for stringent response in *Escherichia coli*, where the time scales for fast enzyme level and slow transcriptional regulations are different. Moreover, multiple regulations are coordinated by the intracellular metabolites, where fructose 1,6-bisphosphate (FBP), phosphoenol pyruvate (PEP), and acetyl CoA (AcCoA) play important roles for enzyme level regulation as well as transcriptional control, while α-ketoacids such as α-ketoglutaric acid (αKG), pyruvate (PYR), and oxaloacetate (OAA) play important roles for the coordinated regulation between carbon source uptake rate and other nutrient uptake rate such as nitrogen or sulfur uptake rate by modulation of cAMP via Cya.

**Keywords**

Catabolite regulation, nitrogen regulation, phosphate regulation, sulfur regulation, acid shock, heat shock, oxidative stress, oxygen limitation, osmoregulation, acetate overflow metabolism, stringent response, redox regulation

## 1. Introduction

The living organisms on earth survive by manipulating the cell system in response to the change in growth environment by sensing signals of both external and internal states of the cell. The complex signaling networks interconvert signals or stimuli for the cell to function properly. The transfer of information in signal transduction pathways and cascades is evolved to respond to a variety of growth environment. Metabolic network, defined as the set and topology of metabolic biochemical reactions within a cell, plays an essential role for the cell to survive, where it is under organized control. The set of enzymes changes dynamically in accordance with the change in growth environment and the cell's state. It is important to properly understand how it is managed by the cell by coordinating the metabolism in response to the change in growth condition.

Deep understanding on the metabolic regulation mechanism is essential for manipulating and redesigning the metabolism, and it is critical to understand the basic principles which govern the cell metabolism. Such principles may be in common to various organisms, or some set of organisms, while some are the specific to the organism of concern.

The metabolic regulation may be made by enzyme level and also by the transcriptional regulation together with global regulators or transcription factors. Moreover, some specific metabolites are also involved for the coordinated regulation of the metabolism. Here, brief explanation is given for the metabolic regulation of microbes in response to a variety of growth environment.

## 2. Nitrogen Regulation

Next to carbon (C) source metabolism, nitrogen (N) metabolism is important to understand the cell metabolism. The N-regulation is controlled by $\sigma^{54}$ encoded by *rpoN*. The main players in the hierarchical network for nitrogen metabolism and regulation are the ammonia transporter AmtB and a glutamine transporter GlnHPQ, metabolic pathways such as glutamate dehydrogenase (GDH) encoded by *gdhA*, glutamine synthetase (GS) encoded by *glnA*, and glutamate synthase (GOGAT) encoded by *gltBD*, the two bifunctional enzymes such as adenylyl transferase/adenylyl-removing enzyme (ATase) and uridylyl transferase/uridylyl removing enzyme (UTase), the two-component regulatory system composed of the histidine protein kinase, nitrogen regulator II ($NR_{II}$) encoded by *glnL* and the response regulator I ($NR_I$) encoded by *glnG*, three global transcriptional regulators such as nitrogen assimilation control protein (Nac), leusine-responsive regulatory protein (Lrp), and Crp, the glutaminases, and the nitrogen-phosphotransferase system in *Escherichia coli* (van Heeswijk et al. 2013).

N-source such as ammonia ($NH_3$)/ammonium ($NH_4^+$) is predominantly assimilated at glutamate dehydrogenase (GDH) reaction, where α-ketoglutarate (αKG) is converted to glutamic acid (Glu), where NADPH is required for this reaction (Fig. 1). Then glutamate is converted to glutamine (Gln) by glutamate synthetase (GS) reaction, where $NH_3$/$NH_4^+$ and ATP are required for this reaction. Thus, the flux goes from αKG via Glu to Gln, and thus Gln accumulates under excess ammonia condition. Under N-limitation, the expression of *gdhA*, which encodes GDH, is repressed by Nac, and thus Gln concentration decreases, and αKG accumulates, where Glutamate is formed from Gln by glutamate synthase (GOGAT) reaction (Fig. 1). Namely, under N-limitation, GS/GOGAT cycle plays an important role.

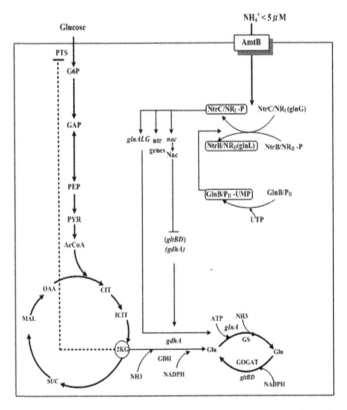

**Figure 1.** Ammonia transport via AmtB, ammonia accimilation, and the effect of αKG on the glucose PTS under N-limitation.

Intracellular ammonium is assimilated into biomass in two steps: Namely, it is first captured in the form of glutamic acid using carbon skeleton of αKG via GS/GOGAT cycle. Then N-group in glutamate is transferred by amino-transferase reactions to synthesize other amino acids thus incorporating into biomass, while recycling the carbon skeleton back to αKG (Kim et al. 2012). The αKG pool, which integrates imbalance between the ammonium assimilation flux and incorporation flux to biomass activates AmtB (Gruswitz et al. 2007, Radchenko et al. 2010, Truan et al. 2010) via GlnK. If ammonia level drops, then the rate of ammonia assimilation will drop immediately which results in αKG accumulation (Yuan et al. 2009). When extracellular ammonia concentration is low around 5 μM or less, ammonia is captured and transported into the cell via AmtB, and is converted to glutamine by GS, and UTase uridylylates both GlnK and GlnB (Ninfa et al. 2000) (Fig. 2). When extracellular $NH_4^+$ concentration is more than 50 μM, glutamine pool rises, and UTase deuridylylates GlnK and GlnB. Then GlnK complexes with AmtB, thereby inhibiting the transport via AmtB, and ammonia may enter by diffusion. $P_{II}$ (GlnB) interacts with NtrB ($NR_{II}$) and activates its phosphatase activity leading to dephosphorylation of NtrC ($NR_I$), and NtrC-dependent gene expression ceases (Fig. 2) (Ninfa et al. 2000).

Lrp regulates the expression of the genes involved in catabolism and anabolism of amino acids (AAs). In particular, leusine indicates AA sufficiency, and it is affected by Lrp, where Lrp does not restrict to leucine but the other AAs such as isoleucine, histidine, and threonine. Lrp may activate *gltBD* and pyridine nucleotide transhydrogenase (van Heeswijk et al. 2013).

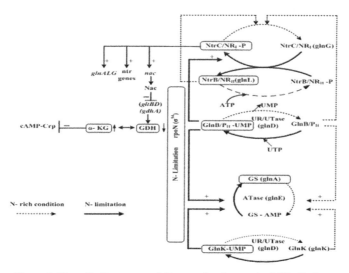

**Figure 2.** Overall nitrogen regulation mechanism under N-limitation.

When arginine is abundant, the transcription factor ArgR binds to arginine to repress arginine biosynthesis enzymes (Gerosa et al. 2013), and activates arginine degradation enzymes (Kiupakis et al. 2002). This regulation is also subject to the NtrC regulation.

The GlnHPQ enables active transport of glutamine into the cell with higher specificity, where *glnH* is the structure gene for the periplasmic binding protein, *glnP* gene codes for the membrane bound glutamine permease, and *glnQ* codes for the ATP hydrolyzing component of ABC transporter system (van Heeswijk et al. 2013).

Two of the major signal transduction systems of N and C metabolisms are identified as $P_{II}$ (GlnB) and PTS. Because of the important roles in the regulatory functions, $P_{II}$ and PTS can be regarded as the central processing units of N and C metabolisms, respectively. The $P_{II}$ protein senses $\alpha$KG and ATP, thus link the state of the central carbon and energy metabolism for the control of N assimilation (Commichau et al. 2006). Nitrogen (N) assimilation is regulated by $P_{II}$-Ntr system together with Crp for the coordinative modulation between C and N assimilation in *E. coli* (Mao et al. 2007, Kumar and Shimizu 2010). The C and N metabolisms may be linked by energy metabolism, where $P_{II}$ controls N assimilation by acting as a sensor of adenylate energy charge. Moreover, $\alpha$KG serves as a cellular signal of C and N status, and strongly regulates $P_{II}$ functions (Jiang and Ninfa 2007). Gln and $\alpha$KG are the signal metabolites for nitrogen and carbon status, respectively, and these signals regulate GS adenylylation state and nitrogen regulator I ($NR_I$ or NtrC) phosphorylation state (Ninfa and Jiang 2005). Nitrogen shortage is reflected by the reduced Gln levels and increased $\alpha$KG level (Yuan et al. 2009, Brauer et al. 2006). This ratio is nearly constant under C-limitation, where this constant ratio is the result of tight regulation of ammonia assimilation to match exactly the carbon uptake rate. This ratio is insensitive to variations in protein levels of the core circuit and to the N-utilization rate, and this robustness depends on bifunctional enzyme adenylyl transferase (Hart et al. 2011).

During N-limitation, a sudden increase in nitrogen availability results in immediate increase in glucose uptake, and $\alpha$KG plays an important role for this, where $\alpha$KG directly reduces the glucose uptake under N-limitation by inhibiting EI of PTS (Fig. 3) (Doucette et al. 2011). This implies the followings:

(1) $\alpha$KG inhibition of sugar uptake is for all PTS sugars by inhibiting EI but not carbohydrate specific E II,

(2) this is performed without perturbing the concentrations of the glycolytic intermediates such as G6P, PEP, and PYR,

(3) inhibition of EI by $\alpha$KG leads to reduced amount of phosphorylated E II A$^{Glc}$ and decreases cAMP level, where the effect of $\alpha$KG on cAMP production is caused by the difference in E II A$^{Glc}$ phosphorylation rather than a difference in substrate availability (Doucette et al. 2011).

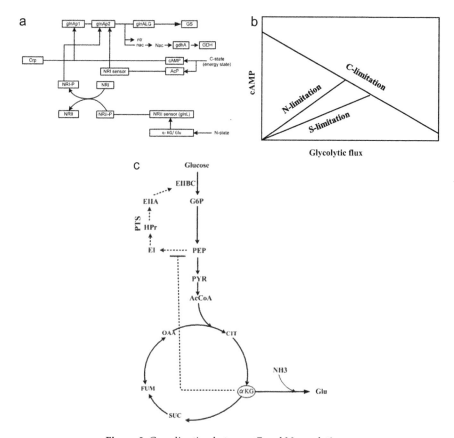

**Figure 3.** Coordination between C and N-regulations.

Not only αKG but also other α-ketoacids such as OAA and PYR play also the similar roles, and affect not only PTS but also the cAMP level by Cya (You et al. 2013). Moreover, αKG is a promiscuous enzymatic regulator that competitively inhibits citrate synthase (CS) of the TCA cycle and 3PG dehydrogenase for serine biosynthesis, and further controls aspartate production by product inhibition of transaminase under N-limitation (Doucette et al. 2011). αKG noncompetitively inhibits EI and Pps, while PtsP (EI homolog in the nitrogen PTS) is insensitive to αKG.

## 3. Sulfur Regulation

Under sulfur (S)-limitation, at least three metabolites such as sulfide, the reduction product of sulfate used for cystein biosynthesis, N-acetylserine, the only precursor of cysteine, and adenosine 5′-phosphosulphate (APS), the first intermediate in sulfate assimilation are involved for the metabolic regulation (Bykowski et al. 2002, Kredich 1996). Under S-limitation, the concentrations of sulfide and APS decrease, while N-acetylserine pool increases. The two regulators CysB and Cbl mediate homeostatic responses to S-limitation, where these responses help *E. coli* to scavenge trace amounts of cysteine and sulfate, preferred S sources, or the alternative S sources such as glutathione and various alkaline sulfonate including taurin. S-limitation affects methionine metabolism, synthesis of FeS clusters, and oxidative stress.

Like NtrC for N-regulation, CysB is the primary regulator for homeostatic responses to S, and it is required for the synthesis of Cbl (Iwanicka-Nowicka and Hryniewicz 1995). CysB is positively controlled by N-acetylserine and negatively controlled by sulfide or thiosulfate (Kredich 1996), and Cbl is negatively controlled by APS (Bykowski et al. 2002). It is of interest that *cbl* gene is transcribed from *nac* promoter under N-limitation (Zimmer et al. 2000). The *ddp* operon is activated by NtrC, and there might be a cross regulation between S-limitation and N-limitation (Gyaneshwar et al. 2005).

## 4. Phosphate Regulation

The phosphate (P) metabolism is also quite important from the energy generation and phosphorelay regulation points of view. The phosphorous compounds serve as major building blocks of many biomolecules and have important roles in signal transduction (Wanner 1996). Depending on the concentration of environmental phosphate, *E. coli* controls phosphate metabolism through Pho regulon, which forms a global regulatory circuit involved in a bacterial phosphate management (Wanner 1996, 1993). The PhoR/PhoB two-component system plays important roles in detecting and responding to the changes of the environmental phosphate concentration (Baek and Lee 2007). Namely, under phosphate limitation, the phosphate is transferred by an ABC transporter composed of PstSCAB for the high-affinity capture of $P_i$, and the phosphate is then transferred to PhoR (PhoR-P), and in turn PhoB is phosphorylated by PhoR. The phosphorylated PhoB acts as the response regulator, and regulates Pho-Box genes such as *eda, phnCDEFGHIJKLMNOP, phoA, phoR/B, phoE, phoH, psiE, pstSCAB, phoU,* and *ugpBAECQ* (van Dien and Keasling 1998). When $P_i$ is rich or in excess, $P_i$ is taken up by the low affinity transporter Pit, and PhoR, Pst, and PhoU together turn off the Pho regulon by dephosphorylating PhoB. The sensor protein CreC (PhoM) can phosphorylate PhoB, while acetyl phosphate can also directly phosphorylate PhoB (Wanner 1993). The overall regulation mechanism is complex, and it is not so clear how the phosphate limitation affects the metabolism (Marzan and Shimizu 2011).

The promoters of the Pho genes are recognized by $\sigma^D$-associated RNA polymerase. A mutation in *rpoS* significantly increases the level of AP (alkaline phosphatase) activity, and the overexpression of $\sigma^S$ inhibits it. The Pho regulon is thus evolved to maintain a tradeoff between cell nutrition and cell survival during $P_i$-starvation (Spira et al. 1995).

## 5. Metal Ion Regulation and Oxidative Stress Regulation

Iron is ubiquitous and the fourth most abundant element on earth, and supports the metabolism of living organisms on the planet (Frey and Reed 2012). Iron is involved in the formation and destruction of ROSs such as superoxide ($O_2^-$), peroxidase ($H_2O_2$ and ROOH) and free radicals ($\bullet$ OH and $\bullet$ OR) usually generated as toxic by-products of aerobic metabolism in a cascade of monovalent reductions from molecular oxygen. Although certain amounts of iron and ROSs are required for the cell to survive, the excess amounts cause stress to the cell leading to the cell death (Dixon and Stockwell 2014).

The metal ion levels are often sensed by metal-sensing regulatory RNA, which encodes metal-sensing proteins involved in the transport and storage of intra-cellular metals (Dann et al. 2007, Helmann 2007). In the native environment, the cell continuously faces iron deficiency, where metal ion functions as cofactor in many of the cellular constituents such as flavoproteins, and therefore, the cell furnishes the mechanism for iron uptake and storage system (Zheng et al. 1999, McHugh et al. 2003). However, excess iron causes toxicity by catalyzing the formation of reactive free radicals through Fenton/Haber-Weiss reaction (Storz and Imlay 1999). In combination with inability to convert NADH to $NAD^+$ in the respiration, a decrease in endogenous $O_2^-$ causes reductive stress, and in turn activates **Fur (ferric uptake regulator)** (Jovanovic et al. 2006). Fur generally represses ion transport and ion siderophore biosynthetic genes when complexed with ferrous ion. Under ion limitation, ion dissociates from Fur, where Fur requires binding to $Fe^{2+}$ to become active. $O_2^-$ deactivates Fur after its conversion to $H_2O_2$ by superoxide dismutase (SOD) through **Fenton reaction** ($H_2O_2 + Fe^{2+} \rightarrow HO\bullet + OH^- + Fe^{3+}$) (Blanchard et al. 2007). Therefore, a decrease in endogenous $O_2^-$ increases the availability of $Fe^{2+}$, through a decrease in $H_2O_2$ level, and in effect activates Fur (Brynildsen and Liao 2009). Namely, Fur senses the reductive stress and protects Fe-S clusters to be safe from damage by **reactive oxygen species (ROSs)**. It is essential for the cell to use iron economically, and this is attained by siderophore synthesis and iron transport regulation (Braun et al. 1998). Iron transport and siderophores (e.g., enterobactin) pathway genes such as *talB* and *entF* are repressed by Fur (Azpiroz and Lavina 2004, Semsey et al. 2006, Hantash et al. 1997), and enterobactin may be produced in *fur* mutant *E. coli* (Kumar and Shimizu 2011). There are functional interactions between carbon and ion utilization via Crp and Fur, where many ion transport genes and several catabolic genes are subject to dual control (Zhang et al. 2005).

The widely conserved **multiple antibiotic resistant regulator (MarR)** family of transcription factors modulate bacterial detoxification in response to antibiotics such as fluoroquinolones, $\beta$-lactams such as ampicilin, tetracycline, and chloramphenicol, as well as toxic chemicals, and synthesis of virulence determinants, etc. (Perera and Grove 2010). MarR senses copper (II) as a signal to cope with stress caused by antibiotics, etc., where copper (II) oxidizes a cystein residue on MarR to generate disulfide bonds between two MarR dimers, thereby inducing tetramer formation and dissociation of MarR from its cognate promoter DNA (Hao et al. 2014).

The microbial cell responds to oxidative stress by inducing antioxidant proteins such as SOD and catalase, where those are regulated by OxyR and SoxR/S (Pomposiello and Demple 2001). SoxR is a member of the MarR family of metal-binding transcription factors, and it exists in solution as a homodimer with each subunit containing a [2Fe-2S] cluster. These clusters are in the reduced state in inactivated SoxR, and their oxidation activates SoxR as a powerful transcription factor (Gaudu and Weiss 1996). The active form of SoxR activates transcription of *soxS* gene, where SoxS belongs to the AraC/XylS family.

Although the respiration is universal among all aerobic organisms, inefficient electron transfer via the respiratory complexes results in one electron reduction of diatomic oxygen, a phenomenon known to generate toxic **reactive oxygen species (ROSs)** (Mailloux et al. 2007). Since NADPH plays an important role for detoxification of ROSs, some prokaryotic microorganisms such as *E. coli* produce NADPH at ICDH in the TCA cycle together with the modulation at G6PDH, 6PGDH, and possibly at Mez.

The $\alpha$KG is a key participant in the detoxification of ROSs with concomitant formation of succinate, where succinate is a biomarker for oxidative stress. Moreover, NADPH producing ICDH is activated, while NADH producing KGDH is deactivated in *Pseudomonas fluorescens*. This indicates that for both prokaryotic and eukaryotic cells, the TCA cycle acts both as a scavenger and generator of ROS, and its modulation is important for regulating ROSs. The TCA cycle can both regulate their formation and decomposition, where the concomitant accumulation of succinate may act as a potent signal for this (Mailloux et al. 2007).

The proper understanding on the regulation of ROS homeostasis gives a way for medical applications. Namely, iron- and ROS-dependent cell death may be considered for cancer treatment. As mentioned above for bacteria, high NADPH production with low ROS levels is essential for tumor cell proliferation and survival (Denicola et al. 2011, Son et al. 2013, Maddocks et al. 2013). NADPH is required for glutathione homeostasis, which indicates that tumor cells require a highly reduced environment for survival. Therefore, one idea for pushing cancer cells to sentence or death is the decrease of the glutathione levels and/or the increase of the oxidative stress levels (Trachootham et al. 2009).

## 6. Redox State Regulation

Global regulators such as **Fnr (fumarate nitrate reduction)**, **Arc (anoxic respiration control)** system, and **Nar (nitric acid reduction)** are mainly responsible for the regulation under oxygen limitation and other electron acceptors in the culture environment, where Fnr directly senses molecular oxygen, and plays a role under anaerobic condition (Gunsalus 1992), in coordination with ArcA/B system, where Fnr activates *arcA* gene expression. Under oxygen limitation, Fnr binds a $[4Fe-4S]^{2+}$ cluster and becomes a transcriptionally active dimeric form. Molecular oxygen can oxidize the ion-sulfer cluster of the corresponding region, resulting in monomerization of the protein and subsequent loss of its ability to bind DNA (Salmon et al. 2003). The ArcA/B system plays a role under both anaerobic and micro-aerobic conditions (Alexeeva et al. 2003, Zhu et al. 2006), where it is composed of ArcA, the cytosolic response regulator, and ArcB, the membrane bound sensor kinase. The ArcA/B two-component system responds to the redox state of the membrane-associated redox carriers such as quinones in the respiratory chain (Georgellis et al. 2001, Malpica et al. 2004). The quinones pool decreases under oxygen limitation, and causes ArcB to be self-phosphorylated (ArcB-P), and then ArcB-P transphosphorylates ArcA (Fig. 4) (Constantinidou et al. 2006). The ArcA-P then represses the expression of the TCA cycle and the glyoxylate shunt genes (Appendix A). Moreover, the genes which encode the primary dehydrogenases such as *glpD*, *lctPRD*, *aceE,F* and *lpdA* are also repressed by ArcA (Appendix A). The *cyoABCDE* operon is represses by both ArcA and Fnr, while *cydAB* operon is activated by ArcA and repressed by Fnr (Fig. 4) (Kessler and Knappe 1996).

The expression of *pfl* genes which encode pyruvate formate lyase, Pfl is activated by ArcA and Fnr, whereas the expression of *aceE,F*, and *lpdA* which encode PDHc is repressed by ArcA under oxygen limitation (Fig. 4). The formate can be excreted via Foc, or converted to hydrogen via formate dehydrogenase, $FDH_H$ and formate hydrogen lyase, Fhl and deletion of FocAB, $FDH_N$ and hydrogenase Hyd (Fig. 5) (Toya et al. 2012, Maeda et al. 2007). Moreover, the flux from PYR to AcCoA is blocked in *pfl* mutants (Δ*pflA* or Δ*pflB*), and pyruvate exclusively goes to lactate formation via LDH reaction (Zhu and Shimizu 2004, 2005). Moreover, Fnr activates *frd* gene expression, while repressing *sdh* gene expression, resulting in branched pathways for TCA cycle under anaerobic condition.

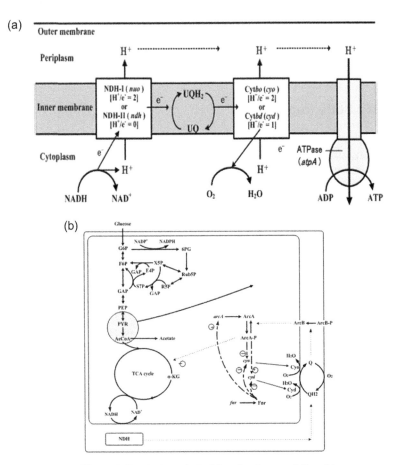

**Figure 4.** Respiratory chain (a) and redox regulation (b).

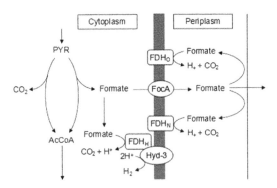

**Figure 5.** The formate pathway of *E. coli* (Toya et al. 2012).

As mentioned before, the TCA cycle activity is repressed as the glucose consumption rate was increased due to lower level of cAMP-Crp, which in turn causes acetate overflow metabolism. This also occurs by the higher redox ratio (Vemuri et al. 2006). This phenomenon can be relaxed by activating TCA cycle by *arcA/B* genes knockout (Alexeeva et al. 2003, Toya et al. 2012, Valgepea et al. 2014). The activated TCA cycle produces more NADH, and allosterically inhibits CS and ICDH activities (Nizam et al. 2009). Thus, the NADH oxidation by the expression of *nox* gene coding for NADH oxidase, NOX in the *arc*A mutant further reduces the acetate formation, resulting in the increased recombinant protein production (Vemuri et al. 2006), while nicotinic acid and Na nitrate may also activate TCA cycle (Nizam and Shimizu 2008). The activation of the TCA cycle causes the decrease in the cell yield due to higher production of $CO_2$ in the TCA cycle.

Many bacteria utilize oxygen as the terminal electron acceptor, but they can switch to other acceptors such as nitrate under oxygen limitation. The reducing equivalents such as NADH are reoxidized in the respiratory chain, where oxygen, nitrate, fumarate, and dimethyl sulfoxide can be the electron acceptors. Nar plays a role when nitrate is present under oxygen limitation. Nar belongs to the two-component redox regulation systems, where it comprises a membrane sensor (NarX) that acts as a kinase causing phosphorylation of the regulator (NarL) under certain conditions (Constantinidou et al. 2006). The Nar system activates such genes as nitrate reduction encoding nitrate and nitrite reductases, and represses such genes as *frd* genes for fumalate reductase.

## 7. Acid Shock Response

The cell such as *E. coli* has the regulation systems in response to acidic condition (Foster 2004, Stincone et al. 2011, Richard and Foster 2004). Some of these depend on the available extracellular amino acids such as glutamate, arginine, and lysine, where the intracellular proton is consumed by the reductive decarboxylation of the amino acid followed by the excretion of γ-amino butyric acid (GABA) from cytoplasm to the periplasm by the anti-porter that also imports the original amino acid (Foster 2004). *E. coli* is acid resistant by glutamate decarboxylase system, where *gadA* and *gadB* which encode glutamate decarboxylase isozymes and *gadC* which encodes glutamate/GABA anti-porter (Fig. 6). Glutamate decarboxylase is activated in response to acid, osmotic, and stationary phase signals. The GADAB form a glutamate-dependent acid response system, where the process of decarboxylation consumes an intracellular proton and helps maintain pH homeostasis. There are other similar acid resistant systems in case arginine instead of glutamate is used, like arginine decarboxylase, where

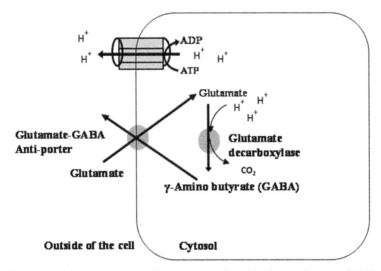

**Figure 6.** Acid shock regulation by amino acid decarboxylase and reversed PMF.

the antiporter is AdiC in this case (Gong et al. 2003, Iyer et al. 2003), and in the case of using lysine by lysine decarboxylase (Iyer et al. 2003). The cells grown in the media rich in amino acids such as LB are acid resistant because of the above mechanism (Foster 2004).

Moreover, ATPase can be involved in acid regulation system (Richard and Foster 2004), where ATPase is usually utilized for the protons in the periplasm move into the cytosol across the cell membrane producing ATP from ADP and $P_i$ by the negative **proton motive force (PMF)**. Since the basic problem of acid stress is the accumulated proton in the cytosol, this proton may be pumped out through ATPase by hydrolyzing ATP with reversed proton move due to positive PMF at low pH such as pH 2 or 3 (Richard and Foster 2004). Without amino acid in the media, this acid response system is activated by utilizing ATPase (Richard and Foster 2003, Martin-Galiano et al. 2001), where the positive PMF pumps out extra protons ($H^+$) out of the cytoplasm using ATP (Fig. 6) (Richard and Foster 2003). This proton homeostasis by PMF is conserved in large class of organisms.

RpoS which increases at the late growth phase and the stationary phase as well as Crp are involved in acid resistance (Foster 2004, Castanie-Cornet and Foster 2001). As implied by the involvement of Crp, the acid resistant system is repressed when glucose is present. The acidic pH lowers cAMP levels in exponentially growing cells in the minimal glucose medium. Since cAMP-Crp represses RpoS, this may elevate RpoS, and increases the expression of *gadX*. The overall regulation system seems to be quite complex involving EvgS/A, B1500, PhoQ/PhoP, GadX, GadW, etc. (Marzan and Shimizu 2011).

OmpR may be a key regulator for acid adaptation, and thus *ompR* mutant is sensitive to acid exposure (Greenberg et al. 1990). The acid-inducible *asr* gene is regulated by PhoR/B, and thus *phoR/phoB* deletion mutant fails to induce *asr* gene expression (Suziedeliene et al. 1999).

In order to keep pH constant, alkali such as NaOH is supplied during the cell growth in practice, which results in the increase in sodium ion ($Na^+$), where *nhaA* gene encoding $Na^+/H^+$ anti-porter membrane protein and *nhaR* gene encoding the NhaA regulatory protein will be over-expressed in *pflB* mutant, showing performance improvement for lactate fermentation (Wu et al. 2013).

## 8. Heat Shock Stress Response

The organisms respond to a sudden temperature up-shift by increasing the synthesis of a set of proteins. This phenomenon is called as **heat shock response**, where this does not restrict to the temperature up-shift, but also other stresses such as solvent stress. The heat shock proteins play important roles in the assembly and disassembly of macromolecular complex such as GroE, for the intracellular transport such as Hsp70, for the transcription such as $\sigma^{70}$, for the proteolysis such as Lon, and for the translation such as lysyl tRNA synthetase. The heat shock response in *E. coli* is mediated by $\sigma^{32}$ encoded by *rpoH*. Among them, *groEL, dnaK*, and *htpG* encode major chaperones such as Hsp60, Hsp70, and Hsp90. ClpP, Lon, and HtrC are involved in the proteolysis. DnaK, DnaJ, GrpE, and RpoH are involved in

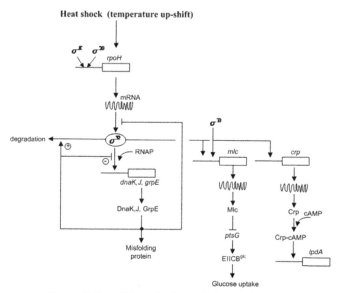

**Figure 7.** Overall heat shock regulation mechanism.

the autoregulation of heat shock response. DnaK prevents the formation of inclusion bodies by reducing aggregation and promotion of proteolysis of misfolded proteins. A bichaperone system involving DnaK and ClpB mediates the solubilization or disaggregation of proteins. GroEL plays an important role for the protein transit between soluble and insoluble protein fractions, and participates positively in disaggregation and inclusion body formation. Small heat shock proteins such as IbpA and IbpB protect heat-denatured proteins from irreversible changes in association with inclusion bodies (Kitagawa et al. 2002, Sorensen and Mortenses 2005).

Hoffmann et al. (2002) investigated the metabolic adaptation of *E. coli* during temperature induced recombinant protein production, and showed that cAMP-Crp-controlled LpdA of pyruvate dehydrogenase complex (PDHc) and SdhA in the TCA cycle are highly induced. Namely, the TCA cycle is activated due to increased level of cAMP-Crp at higher temperature. In *E. coli*, heat shock protein synthesis rates peak at about 5–10 min after the temperature up-shift and then decline to a new steady-state level (Tilly et al. 1986). The $\sigma^{70}$ is itself a heat shock protein and the increase in its concentration after heat shock may contribute to its decline in heat shock protein synthesis. DnaK contributes to the shutoff of the high level synthesis of heat shock proteins (Tilly et al. 1989). The heat shock activates *crp* gene expression, and in turn Crp activates *mlc* gene which codes for Mlc (Shin et al. 2001), and thus the glucose consumption rate decreases (Hasan and Shimizu 2008), which also causes cAMP level to be increased (Fig. 7).

Acetate production is affected by higher temperature. Transcription of *acs* gene occurs from two $\sigma^{70}$-dependent promoters such as distal promoter *acs* $P_1$ and proximal promoter *acs* $P_2$ (Kumari et al. 2000, Browning et al. 2002). The cAMP-Crp binds two sites within the *acs* regulatory region. However, Fis and Ihf independently modulate Crp-dependent activation of *acs* $P_2$ transcription (Beatty et al. 2003, Browning et al. 2004).

The respiration is activated during the temperature up-shift (Hoffmann et al. 2002), and *sod* is induced in response to the oxidative stress imposed by dioxygen or by the redox active compounds such as viologens or quinones caused by the temperature up-shift (Fridovich 1987). This phenomenon may be also caused by the activated TCA cycle.

## 9. Cold Shock Response

Upon temperature down-shift from 37°C to 15°C, the major cold shock proteins such as CspA, CspB and CspC are induced, where cold shock proteins are able to bypass the inhibitory effect of the antibiotics such as kanamycin and chloramphenicol (Etchegaray and Inoue 1999). Although thermoregulatory mechanism is not well understood, the adaptation of the cell to low temperature such as 20°C–23°C requires coordinated and multifunctional response, where RpoS and the small regulatory RNA DsrA are involved in both cold shock and biofilm formation genes (White-Ziegler et al. 2008) as well as flagella biosynthesis and motility genes (Kim et al. 2005).

## 10. Solvent Stress Regulation

The biofuels production by microorganisms has been paid recent attention. However, many biofuels are toxic to microorganisms, and reduce the cell viability through damage to the cell membrane and interference with essential physiological processes. Several attempts have been made to improve the tolerance to biofuels, where biofuel export systems, heat shock proteins, and membrane modifications have been considered (Dunlop 2011). The effect of biofuels on the cell is through hydrophobicity of the cytoplasmic membrane, where the accumulation of solvent in the cytoplamic membrane increases permeability of membrane, diminishes energy transduction, interferes with membrane protein function, and increases fluidity (Dunlop 2011, Isken and Bont 1998, Ramos et al. 2002, Nicolaou et al. 2010). This may cause the release of ATP, ions, and phospholipids, RNA, and proteins, and thus the cell growth is depressed due to disturbances on ATP production by diminished PMF. Moreover, the increase in fluidity affects the nutrient transport as well as energy transduction.

Toxicity levels vary depending on the microbes and the types of biofuels and biochemicals. In general, longer chain alcohols are more toxic than short chain alcohols. Efflux pumps are membrane transporters that recognize and export toxic compounds from the cell by PMF, where this is important for the cell to survive by exporting bile salts, antimicrobial drugs, and solvents. The *acrAB-tolC* pump in *E. coli* provides tolerance to hexane, heptanes, octane, and nonane (Takatsuka et al. 2010). Efflux pumps are effective for increasing tolerance and production of biofuels, in particular, for long chain alcohols, but those may not be effective for exporting short chain alcohols such as 1-propanol and isobutanol (Ankarloo et al. 2010).

The heat shock proteins are up-regulated in response to short chain alcohols, and heat shock and protein refolding genes such as *rpoH, dnaJ, htpG,* and *ibpAB* are up-regulated (Rutherford et al. 2010), while *groESL, dnaKJ, hsp18, hsp90* are up-regulated in *Chrostridium acetobutylicum* (Tomas et al. 2004). Over-expression of heat shock proteins may increase tolerance against biofuels (Ficco et al. 2007, Reyes et al. 2011).

In general, solvents disrupt the cell membrane structure and have a strong impact on physiological function, and eventually leading to the cell death (Sikkema et al. 1995). To overcome this problem, solvent tolerant microbes change the composition of the fatty acids from *cis* to *trans*-unsaturated fatty acids catalyzed by *cis-trans* isomerase (*cti*), thus decreasing membrane fluidity, preventing the entry of solvents into the cell (Holtwick et al. 1997, Kiran et al. 2004). In addition, modifications to phospholipid headgroups or phospholipid chain length increase solvent tolerance (Piper 1995).

In relation to solvent stresses caused by the accumulation of biofuels in the culture broth, the primary role to protect the cell from such stress is made by outer membrane porin proteins. Since cytosolic membrane is also under stress condition, respiration and membrane proteins as well as general stress response mechanism are affected (Nicolaou et al. 2010). ROS highly increase in response to the stress caused by *n*-butanol in *E. coli* (Rutherford et al. 2010).

## 11. Osmoregulation

The bacterial cell is surrounded by the cell envelope, where the plasma membrane is responsible for the transport of ions such as $H^+$, $Na^+$, $K^+$, and various substrates or nutrients, and metabolites to maintain homeostasis. The bacterial cells exchange such components together with energy and information with their surroundings by the appropriate sensing and responding mechanisms (Kramer 2010). Under osmotic stress condition, a number of transport systems for ions such as $K^+$, and compatible solutes such as proline betain and the precursor choline are activated (Wood 2011). The typical two-component histidine kinase/response regulator system such as KdpD/KdpE is ubiquitous in various bacteria (Walderhaug et al. 1992), where it regulates *kdpFABCDE* operon including the Kdp ATPase, and active $K^+$ uptake system. Namely, KdpD/KdpE system responds to $K^+$ limitation and salt stress (Sugiura et al. 1994, Hamann et al. 2008, Heermann and Jung 2010). As also mentioned before, EnvZ/OmpR two-component system regulates the expression of the porin genes such as *ompC* and *ompF* encoding outer membrane porins in relation to osmolarity.

The cytoplasmic or inner membrane is impermeable to most large and polar solutes, while these are compensated for by freely diffusing water molecules, and thus the transmembrane concentration gradients are developed for such compounds. The resulting changes in cellular volume and turgor pressure exert strong mechanical force on the cytoplasmic membrane and associated proteins, and preclude the cell growth (Sevin and Sauer 2014). To cope with osmotic stress, bacteria adapt their intracellular osmolarity (Kramer 2010), or increase the cell wall stability (Piuri et al. 2005). The salt stress tolerance is mediated by flux control of water across the cell membrane, adjustments of intracellular potassium levels, synthesis of disaccharide traharose and/or transport of small molecule osmoprotectants (Clarke et al. 2014).

In principle, bacterial cells respond to environmental or growth conditions by immediate protein or enzyme level regulation, and by slow gene transcriptional regulation via transcription factors. In summary, in response to sudden changes in osmotic pressure, *E. coli* controls in and outflux of water and other small molecules by activating aquaporins as an immediate response (Calamita et al. 1995). It regulates intracellular potassium concentrations by adjusting the potassium transporters such as Kup, KdpFABC or TrkA for transient adaptation to short term osmotic stress (Lamins et al. 1981). In the case of prolonged osmotic stress, *E. coli* takes up the osmotolerants such as glycine betain and proline from the environment via ABC transporter encoded by *proVWX*, or synthesizes glycine betain from the extracellular precursor choline (Record et al. 1998, Grothe et al. 1986, von Weyman et al. 2001). If no extracellular compatible solutes are available, *E. coli* induces expression of trehalose 6-phosphate synthase (OtsA) and phosphatase (OtsB) to produce high intracellular concentrations of the nonreducing disaccharide trehalose from the precursors such as UDP-glucose and G6P in response to long-term resistance to sustained osmotic stress (Giaever et al. 1998, Elbein et al. 2003, Klein et al. 1991).

There is indeed an interaction between trehalose and membrane lipid head groups, but this effect is insufficient to fully account for the resistance of membrane against strong osmotic stress. Upon osmotic stress, bacteria adjust their intracellular osmolarity and modify their cell-wall structure (Sevin and Sauer 2014). For this, polyisoprene lipids may also contribute to osmoprotection by increasing resistance to high-salt conditions in the cytoplasmic membrane and in the membrane bilayers of liposomes in *E. coli* (Sevin and Sauer 2014). Coenzyme Q functions as an electron and proton carrier in aerobic respiration, and has an additional crutial role as a chain breaking antioxidant (Turunen et al. 2004). The long polyisoprenyl tail of $CoQ_n$ functions to anchor this lipid in the membranes of cells (Fig. 8), where n designates the number of five carbon isoprene units such as $CoQ_6$ in *S. cerevisiae*, $CoQ_8$ in *E. coli*, and $CoQ_{10}$ in human (Clarke et al. 2014). In *E. coli*, $CoQ_8$ level becomes significantly high in response to high-salt condition (Fig. 8) (Sevin and Sauer 2014).

## 12. Biofilm, Motility by Flagella, and Quorum Sensing

Biofilm formation is one of the important microbial survival strategies, where biofilm development involves attachment of bacteria to surfaces and cell-cell adhesion to form microcolonies. This is useful for the cell to protect against predetors and antibiotics (Wang et al. 2005). The attachment of bacteria to abiotic and biotic surfaces is made by motility, proteinaceous adhesion, and a cell-bound polysaccharide such as PGA (poly-$\beta$-1,6-N-acetyl-D-glucosamine), where PGA is a cell-bound exopolysaccharide

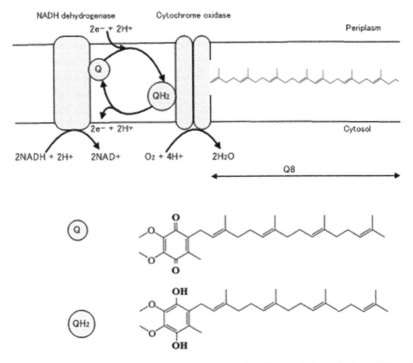

**Figure 8.** Electron transport carrier quinon and quinol in the respiratory chain, and the role of osmoprotection.

adhesion (Wang et al. 2005). As mentioned before, Csr plays important roles for biofilm formation, where *pga* operon involved in PGA formation and excretion is negatively regulated by CsrA. CsrA also negatively regulates c-di-GMP, a second messenger involved in biofilm formation and motility (Hengge 2009). Curli are extracellular proteinaceous structures extending from the cell surface for attachment during biofilm development (Barnhart and Chapman 2006). Curli filaments are activated by CsgD, where it is inversely correlated with flagella synthesis. The master regulator of flagella synthesis is $FlhD_2C_2$, which activates the genes involved in motility and chemotaxis (Thomason et al. 2012). McaS (multi-cellular adhesion sRNA) represses CsgD expression, while activates FlhD and PgaA (Thomason et al. 2012), and thus regulates the synthesis of curli flagella and polysaccharide. Moreover, biofilm formation is under catabolite repression by cAMP and Crp (Jackson et al. 2002).

Quorum sensing is a cell-to-cell communication (Waters and Bassler 2005), where the signal molecules are **acyl-homoserine lactones (AHL)** synthesized by LuxI-type enzyme. At high cell density cultivations, LuxR-type regulator plays a role for the positive feedback in association with AHL when its concentration exceeds a threshold level (Timmermans and van

Melderen 2010). The quorum sensing is the sensing of cell density, where in *E. coli*, CyaR represses *luxS* gene which encodes autoinducer-2 synthase (Delay and Gottesman 2009).

## 13. Concluding Remarks

As seen above, the global regulators are responsive to the specific stimuli. Examples of such pleiotropic TFs in *E. coli* are Crp, a primary sensor for C-availability, NtrBC, a sensor for N-availability, PstSCAB and PhoR, the sensor for P-availability, CysB, the sensor for S-availability, and Fur, the sensor for ion availability. Functional interactions among such regulators must coordinate the activities of the metabolon so that the supply of one type of nutrient matches the supply of other nutrients (Gutierrez-Rios et al. 2003). Thus, multiple links between C and N metabolism has been identified (Maheswaran and Forchhammer 2003). Other functional links between C and S metabolism (Quan et al. 2002), and between C and ion metabolism (De Lorenzo et al. 1998, Zhang et al. 2005) have been identified. Moreover, the links between S and N limitations have been also identified (Gyaneshwar et al. 2005).

In general, bacteria in nature live far away from the optimal growth condition, where multiple stresses are imposed on the cell. Therefore, the cell must have the ability to sense, integrate, and respond to the variety of stresses for survival. Although little is known about "cross-stress" protection, cross stress dependencies are ubiquitous, highly interconnected, and may emerge within short time frames (Dragosits et al. 2013). In fact, high degree of overlap was observed in the transcriptional profiling for different stresses such as starvation, osmotic and acidic stresses (Weber et al. 2005), as well as starvation and heat shock or oxidative stress (Jenkins et al. 1988, Jenkins et al. 1991), where high osmolarity and high temperature induces the oxidative stress regulons such as SoxRS and OxyR (Gunasekera et al. 2008, White-Ziegler et al. 2008). The responses to n-butanol share the same high overlap with those in heat shock, oxidative, and acidic stresses (Rutherford et al. 2012).

As mentioned in this article, the specific metabolites such as FBP, PEP, PYR, OAA, AcCoA, and αKG in the main metabolic pathways play important roles for metabolic regulation. This implies that these metabolites play roles for the coordinated and integrated metabolic regulation. The regulation system ranges from relatively rapid interactions such as enzyme level regulation by allosteric binding of the specific molecules or post-translational modification to slow interactions such as transcriptional regulation via transcription factors. It is important to get deep insight into the whole cellular metabolic systems not only by molecular biology, biochemistry, but also by systems biology approach, and apply this for the efficient metabolic engineering.

Among intracellular metabolites, $\alpha$-ketoacids such as $\alpha$KG, OAA, and PYR turn to be mater regulators for catabolite regulation and co-ordination of different regulations (Rabinowitz and Silhavy 2013). Namely, when favoured carbon sources are depleted, $\alpha$-ketoacid levels fall, and cAMP increases to stimulate other carbon catabolite machinery. When preferred nutrients are abundant, the cell growth rate becomes higher with lower cAMP level, while if they are scarce, the cell growth rate declines with higher cAMP level. This change in growth rate is accompanied by a change in cellular composition, where ribosomes are needed for rapid protein production at higher growth rate, while more metabolic enzymes for nutrient assimilation (catabolism) is needed at lower cell growth rate (Klumpp et al. 2009, Scott et al. 2010). There is a linear relationship between the total protein composition of a cell and its growth rate, where this can be extended beyond ribosomes to metabolic enzymes (Hui et al. 2015). Under N- or S-limitation or other nutrient limitation, $\alpha$keto acids such as $\alpha$KG accumulates, and inhibits carbon assimilation, where there is less need for carbon-catabolic enzymes, and more demand for those involved in such nutrient assimilation. When anabolic nutrients are in excess, the $\alpha$KG concentration decreases, cAMP level increases and the carbon catabolic enzymes increases to accelerate carbon assimilation. In the end, the physiological function of cAMP signaling goes beyond simply enabling hierarchical utilization of carbon sources, but also controls the function of the proteome (You et al. 2013, Rabinowitz and Silhavy 2013). The energy level also affects carbon uptake rate (Chubukov et al. 2014, Koebman et al. 2002).

As mentioned before, cAMP-Crp and Cra play important roles for carbon catabolite regulation in *E. coli*, where either PTS or non-PTS sugars are ranked for assimilation in a hierarchy (Luo et al. 2014, Yao et al. 2012, Aidelberg et al. 2014). This may be caused by the cAMP-Crp level and the promoter activities of the corresponding promoters of the transporters (Luo et al. 2014, Yao et al. 2012, Aidelberg et al. 2014).

The roles of cAMP-Crp are not only limited to carbon catabolite regulation. Among many transcription factors, Crp plays significantly important roles in the wide range of regulations such as osmoregulation (Balsalobre et al. 2006), osmotolerance (Zhang et al. 2012), oxidative stress (Basak and Jiang 2012), acid tolerance (Basak et al. 2013), acetate tolerance (Chong et al. 2013a), ethanol tolerance (Chong et al. 2013b), and butanol tolerance (Zhang et al. 2012) as well as catabolite regulation. Therefore, it may be of interest to modulate such transcription factors (Khankal et al. 2009, Gosset et al. 2004) for the development of next generation cell factories.

It is of surmount interest to understand how the cell growth rate is regulated, since such information gives us a hint for improving the cell growth rate, and thus increasing the protein or metabolite production. For this, it is important to recognize at which regulation levels affect the cell growth rate, where post-transcriptional control of protein abundances and post-translational control of flux rates are dominated (Valgepea et al. 2014).

Moreover, it is also important to understand the effect of the specific pathway gene mutation on the metabolic regulation in addition to the effect of growth condition. In fact, [13]C-metabolic flux analysis has been extensively employed for the metabolic flux distributions of pathogens such as *Mycobacterium tuberculosis* (Baste et al. 2013) and the specific gene knockout mutant *E. coli* (Ishii et al. 2007, Shimizu 2004, Shimizu 2013), where the flux information is located on top of different levels of information, manifested as the result of metabolic regulation, and central to understanding the metabolism. It is, therefore, of interest to investigate the effect of the specific gene knockout on the metabolism as well in view of metabolic regulation for the design of next generation cell factories.

## References

Aidelberg, G., B.D. Towbin, D. Rothschild, E. Dekel, A. Bren and U. Alon. 2014. Hierarchy of non-glucose sugars in *Escherichia coli*. BMC Syst Biol. 8: 133.

Alexeeva, S., K.J. Hellingwerf and M.J.T. de Mattos. 2003. Requirement of ArcA for redox regulation in *Escherichia coli* under microaerobic but not anaerobic or aerobic conditions. J Bacteriol. 185(1): 204–209.

Ankarloo, J., S. Wikman and I.A. Nicholls. 2010. *Escherichia coli mar* and *acrAB* mutants display no tolerance to simple alcohols. Int J Mol Sci. 11: 1403–1412.

Azpiroz, M.F. and M. Lavińa. 2004. Involvement of enterobactin synthesis pathway in production of Microcin H47. Antimicrob Agents Chemother. 48: 1235–1241.

Baek, J.H. and S.Y. Lee. 2007. Transcriptome analysis of phosphate starvation response in *Escherichia coli*. J Microbiol Biotechnol. 17(2): 244–252.

Balsalobre, C., J. Johansson and B.E. Uhlin. 2006. Cyclic AMP-dependent osmoregulation of crp gene expression in *Escherichia coli*. J Bacteriol. 188(16): 5935–5944.

Barnhart, M.M. and M.R. Chapman. 2006. Curli biogenesis and function. Annu Rev Microbiol. 60: 131–147.

Basak, S., H. Geng and R. Jiang. 2013. Rewiring global regulator cAMP receptor protein (CRP) to improve *E. coli* tolerance towards low pH. J Biotechnol, in press.

Basak, S. and R. Jiang. 2012. Enhancing *E. coli* tolerance towards oxidative stress via engineering its global regulator cAMP receptor protein (CRP). PLOS One. 7(12): e51179.

Baste, D.J.V., K. Nol, S. Niedenfuhr, T.A. Mendum, N.D. Hawkins, J.L. Ward, M.H. Beale, W. Wiechert and J. McFadden. 2013. [13]C-Flux spectral analysis of host-pathogen metabolism reveals a mixed diet for intracellular mycobacterium tuberculosis. Chem and Biol. 20: 1–10.

Beatty, C.M., D.F. Browning, S.J.W. Busby and A.J. Wolfe. 2003. Cyclic AMP receptor protein-dependent activation of the *Escherichia coli acs* P2 promoter by a synergistic class III mechanism. J Bacteriol. 185(17): 5148–5157.

Blanchard, J.R., W.Y. Wholey, E.M. Conlon and P.J. Pomposiello. 2007. Rapid changes in gene expression dynamics in response to superoxide reveal SoxRS dependent and independent transcriptional network. PLoS One. 2: e1186.

Brauer, M.J., J. Yuan, B.D. Bennet, W. Lu, E. Kimball, D. Botstein and J.D. Rabinowitz. 2006. Conservation of the metabolomic response to starvation across two divergent microbes. PNAS USA. 103: 19302–19307.

Braun, V., K. Hantke and W. Koster. 1998. Bacterial ion transport: mechanisms, genetics, and regulation. pp. 67–145. *In*: A. Sigel and H. Sigel (eds.). Metal Ions in Biological Systems, Marcel Dekker, New York, NY.

Browning, D.F., C.M. Beatty, A.J. Wolfe, J.A. Cole and S.J.W. Busby. 2002. Independent regulation of the divergent *Escherichia coli nrfA* and *acs* P1 promoters by a nucleoprotein assembly at a shared regulatory region. Mol Microbiol. 43(3): 687–701.

Browning, D.F., C.M. Beatty, E.A. Sanstad, K.E. Gunn, S.J.W. Busby and A.J. Wolfe. 2004. Modulation of CRP-dependent transcription at the *Escherichia coli acs* P2 promoter by nucleoprotein complexes: anti-activation by the nucleoid proteins FIS and IHF. Mol Microbiol. 51(1): 241–254.

Brynildsen, M.P. and J.C. Liao. 2009. An integrated network approach indentifies the isobutanol response network of *Escherichia coli*. Mol Syst Biol. 5: 34.

Bykowski, T., J.R. van der Ploeg, R. Iwanicka-Nowicka and M.M. Hryniewicz. 2002. The switch from inorganic to organic sulphur assimilation in *Escherichia coli*: adenosine 5′-phosphosulphate (APS) as a signalling molecule for sulphate excess. Mol Microbiol. 43: 1347–1358.

Calamita, G., W.R. Bishai, G.M. Preston, W.B. Guggino and P. Agre. 1995. Molecular cloning and characterization of AqpZ, a water channel from *Escherichia coli*. J Biol Chem. 270: 29063–29066.

Castanie-Cornet, M.P. and J.W. Foster. 2001. *Escherichia coli* acid resistance: cAMP receptor protein and a 20 bp *cis*-acting sequence control pH and stationary phase expression of the *gadA* and *gadBC* glutamate decarboxylase genes. Microbiol. 147: 709–715.

Chong, H., J. Yeow, I. Wang, H. Song and R. Jiang. 2013a. Improving acetate tolerance of *Escherichia coli* by rewiring its global regulator cAMP receptor protein (CRP). PLOS One. 8(10): e77422.

Chong, H., L. Huang, J. Yeow, I. Wang, H. Zhang, H. Song and R. Jiang. 2013b. Improving ethanol tolerance of *Escherichia coli* by rewiring its global regulator cAMP receptor protein (CRP). PLOS One. 8(2): e57628.

Clarke, C.F., A.C. Rowat and J.W. Gober. 2014. Is CoQ a membrane stabilizer ? Nature Chem Biol. 10: 242–243.

Commichau, F.M., K. Forchhammer and J. Stulke. 2006. Regulatory links between carbon and nitrogen metabolism. Curr Opin Microbiol. 9(2): 167–172.

Constantinidou, C., J.L. Hobman, L. Grifiths, M.D. Patel, C.W. Penn, J.A. Cole and T.W. Overton. 2006. A reassessment of the FNR regulon and transcriptomic analysis of the effects of nitrate, nitritre, NarXL, and NarQP as *Escherichia coli* K12 adapts from aerobic to anaerobic growth. J Biol Chem. 281(8): 4802–4815.

Dann, C.E. III, C.A. Wakeman, C.L. Sieling, S.C. Baker, I. Irnov and W.C. Winker. 2007. Structure and mechanism of a metal-sensing regulatory RNA. Cell. 130: 878–892.

De Lay, N. and S. Gottesman. 2009. The Crp-activated small noncoding regulatory RNA CyaR (RyeE) links nutritional status to group behavior. J Bacteriol. 191: 461–476.

De Lorenzo, V., M. Herrero, F. Giovannini and J.B. Neilands. 1998. Fur (ferric uptake regulation) protein and CAP (catabolite-activator protein) modulate transcription of *fur* gene in *Escherichia coli*. Eur J Biochem. 173: 537–546.

DeNicola, G.M. et al. 2011. Oncogene-induced Nrf2 transcription promotes ROS detoxification and tumorigenesis. Nature. 475: 106–109.

Dixon, S.J. and B.R. Stockwell. 2014. The role of iron and reactive oxygen species in cell death. Nature Chem Biol. 10: 9–17.

Doucette, C.D., D.J. Schwab, N.S. Wingreen and J.D. Rabinowitz. 2011. Alpha-ketoglutarate coordinates carbon and nitrogen utilization via enzyme I inhibition. Nat Chem Biol. 7: 894–901.

Dragosits, M., V. Mozhayskiy, S. Quinones-Soto, J. Park and I. Tagkopoulos. 2013. Evolutionary potential, cross-stress behavior and the genetic basis of acquired stress resistance in *Escherichia coli*. Mol Syst Biol. 9: 643.

Dunlop, M.J. 2011. Engineering microbes for tolerance to next-generation biofuels. Biotechnol Biofuels. 4: 32.

Elbein, A.D., Y.T. Pan, I. Pastuszak and D. Carroll. 2003. New insights on trehalose: a multifunctional molecule. Glycobiology. 13: 17R–27R.

Etchegaray, J.-P. and M. Inoue. 1999. CspA, CspB, and CspG, major cold shock proteins of *Escherichia coli*, are induced at low temperature under conditions that completely block protein synthesis. J Bacteriol. 181(6): 1827–1830.

Fiocco, D., V. Capozzi, P. Goffin, P. Hols and G. Spano. 2007. Improved adaptation to heat, cold, and solvent tolerance in Lactobacillus plantarum. Appl Microbiol Biotechnol. 77: 909–915.

Foster, J.W. 2004. *Escherichia coli* acid resistance: tales of an amateur acidophile. Nature Rev Microbiol. 2(11): 898–907.

Frey, P.A. and G.H. Reed. 2012. The ubiquity of iron. ACS Chem Biol. 7: 1477–1481.

Gaudu, P. and B. Weiss. 1996. SoxR, a [2Fe–2S] transcription factor, is active only in its oxidized form. PNAS USA. 93(19): 10094–10098.

Georgellis, D., O. Kwon and E.C.C. Lin. 2001. Quinones as the redox signal for the arc two-component system of bacteria. Science. 292(5525): 2314–2316.

Gerosa, L., K. Kochanowski, M. Heinemann and U. Sauer. 2013. Dissecting specific and global transcriptional regulation of bacterial gene expression. Mol Syst Biol. 9: 658.

Giaever, H.M., O.B. Styrvoid, I. Kaasen and A.R. Strom. 1998. Biochemical and genetic characterization of osmoregulatory trehalose sysnthesis in *Escherichia coli*. J Bacteriol. 170: 2841–2849.

Gong, S., H. Richard and J.W. Foster. 2003. YjdE (AdiC) is the arginine: agmatine antiporter essential for arginine-dependent acid resistance in *Escherichia coli*. J Bacteriol. 185(15): 4402–4409.

Gosset, G., Z. Zhang, S. Nayyar, W.A. Cuevas and M.H. Saier Jr. 2004. Transcriptome analysis of Crp-dependent catabolite control of gene expression in *Escherichia coli*. J Bacteriol. 186(11): 3516–3524.

Greenberg, J.T., P. Monach, J.H. Chou, P.D. Josephy and B. Demple. 1990. Positive control of a global antioxidant defense regulon activated by superoxide-generating agents in *Escherichia coli*. PNAS USA. 87(16): 6181–6185.

Grothe, S., R.L. Krogsrud, D.J. McClellan, J.L. Milner and J.M. Wood. 1986. Proline transport and osmotic response in *Escherichia coli* K-12. J Bacteriol. 166: 253–259.

Gruswitz, F., J. O'Connell and R.M. Stroud. 2007. Inhibitory complex of the transmembrane ammonia channel, AmtB, and the cytosolic regulatory protein, GlnK, at 1.96 A. PNAS USA. 104: 42–47.

Gunasekera, T., L. Csonka and O. Paliy. 2008. Genome-wide transcriptional responses of *Escherichia coli* K-12 to continuous osmotic and heat stresses. J Bacteriol. 190: 3712–3720.

Gunsalus, R.P. 1992. Control of electron flow in *Escherichia coli*: coordinated transcription of respiratory pathway genes. J Bacteriol. 174(22): 7069–7074.

Gutierrez-Rios, R.M., D.A. Rosenblueth, J.A. Loza, A.M. Huerta, J.D. Glasner, F.R. Blattner and J. Collado-Vides. 2003. Regulatory network of *Escherichia coli*: consistency between literature knowledge and microarray profiles. Genome Res. 13: 2435–2443.

Gyaneshwar, P., O. Paliy, J. McAuliffe, D.L. Popham, M.I. Jordan and S. Kustu. 2005. Sulfur and nitrogen limitation in *Escherichia coli* K-12: specific homeostatic responses. J Bacteriol. 187(3): 1074–1090.

Hamann, K., P. Zimmann and K. Altendorf. 2008. Reduction of turgor is not the stimulus for the sensor kinase KdpD of *Escherichia coli*. J Bacteriol. 190: 2360–2367.

Hantash, F.M., M. Ammerlaan and C.F. Earhart. 1997. Enterobactin synthase polypeptides of *Escherichia coli* are present in an osmotic-shocksensitive cytoplasmic locality. Microbiol. 143: 147–156.

Hao, Z., H. Lou, R. Zhu, J. Zhu, D. Zhang, B.S. Zhao, S. Zeng, X. Chen, J. Chan, C. He and P.R. Chen. 2014. The multiple antibiotic resistance regulator MarR is a copper sensor in *Escherichia coli*. Nature Chem Biol. 10: 21–28.

Hart, Y., D. Madar, J. Yuan, A. Bren, A.E. Mayo, J.D. Rabinowitz and U. Alon. 2011. Robust control of nitrogen assimilation by a bifunctional enzyme in *E. coli*. Moleculae Cell. 41: 117–127.

Hasan, C.M. and K. Shimizu. 2008. Effect of temperature up-shift on fermentation and metabolic characteristics in view of gene expressions in *Escherichia coli*. Microb Cell Fact. 7: 35.

Heermann, R. and K. Jung. 2010. The complexity of the 'simple' two-component system KdpD/ KdpE in *Escherichia coli*. Microbiol Lett. 304: 97–106.

Helmann, J.D. 2007. Measuring metals with RNA. Mol Cell. 27: 859–860.

Hengge, R. 2009. Principles of c-di-GMP signalling in bacteria. Nat Rev Microbiol. 7: 263–273.

Hoffmann, F., J. Weber and U. Rinas. 2002. Metabolic adaptation of *Escherichia coli* during temperature-induced recombinant protein production: 1. Readjustment of metabolic enzyme synthesis. Biotechnol Bioeng. 80(3): 313–319.

Holtwick, R., F. Meinhardt and H. Keweloh. 1997. Cis-trans isomerization of unsaturated fatty acids: cloning and sequencing of the cti gene from Pseudomonas putida P8. Appl Environ Microb. 63: 4292–4297.

Hui, S., J.M. Silverman, S.S. Chen, D.W. Erickson, M. Basan, J. Wang, T. Hwa and J.R. Williamson. 2015. Quantitative proteomic analysis reveals a simple strategy of global resource allocation in bacteria. Mol Syst Biol. 11: 784.

Isken, S. and J.A.M. de Bont. 1998. Bacteria tolerant to organic solvents. Extremophiles. 2: 229–238.

Iwanicka-Nowicka, R. and M.M. Hryniewicz. 1995. A new gene, *cbl*, encoding a member of the LysR family of transcriptional regulators belongs to *Escherichia coli cys* regulon. Gene. 166: 11–17.

Iyer, R., C. Williams and C. Miller. 2003. Arginin-agmatine antiporter in extreme acid resistance in *Escherichia coli*. J Bacteriol. 185: 6556–6561.

Jackson, D.W., J.W. Simecka and T. Romeo. 2002. Catabolite repression of *Escherichia coli* biofilm formation. J Bacteriol. 184: 3406–3410.

Jenkins, D., E. Auger and A. Matin. 1991. Role of RpoH, a heat shock regulator protein, in *Escherichia coli* carbon starvation protein synthesis and survival. J Bacteriol. 173: 1992–1996.

Jenkins, D., J. Schultz and A. Matin. 1988. Starvation-induced cross protection against heat or $H_2O_2$ challenge in *Escherichia coli*. J Bacteriol. 170: 3910–3914.

Jiang, P. and A.J. Ninfa. 2007. *Escherichia coli* PII signal transduction protein controlling nitrogen assimilation acts as a sensor of adenylate energy charge *in vitro*. Biochem. 46(45): 12979–12996.

Jovanovic, G., L.J. Lloyde, M.P. Stumpf, A.J. Mayhew and M. Buck. 2006. Induction and function of the phase shock protein extracytoplasmic stress response in *Escherichia coli*. J Biol Chem. 281: 21147–21161.

Kessler, D. and J. Knappe. 1996. Anaerobic dissimilation of pyruvate. pp. 199–205. *In*: F.C. Neidhardt, R. Curtiss, J.I. Ingraham et al. (eds.). *E. coli* and Salmonella: Cellular and Molecular Biology, Vol. 1. ASM Press, Washington, DC, USA, 2nd Edition.

Khankal, R., J.W. Chin, D. Ghosh and P.C. Cirino. 2009. Transcriptional effects of CRP* expression in *Escherichia coli*. J Biol Eng. 3: 13.

Kim, M., Z. Zhang, H. Okano, D. Yan, A. Groisman and T. Hwa. 2012. Need-based activation of ammonium uptake in *Escherichia coli*. Mol Syst Biol. 8: 616.

Kim, Y.-H., K.Y. Han, K. Lee and J. Lee. 2005. Proteome response of *Escherichia coli* fed-batch culture to temperature downshift. Appl Microbiol Biotechnol. 68: 786–793.

Kiran, M., J. Prakash, S. Annapoorni, S. Dube, T. Kusano, H. Okuyama, N. Murata and S. Shivaji. 2004. Psychrophilic *Pseudomonas syringae* requires transmonounsaturated fatty acid for growth at higher temperature. Extremophiles. 8: 401–410.

Kitagawa, M., M. Miyakawa, Y. Matsumura and T. Tsuchido. 2002. *Escherichia coli* small heat shock proteins, IbpA and IbpB, protect enzymes from inactivation by heat and oxidants. Eur J Biochem. 269(12): 2907–2917.

Kiupakis, A.K. and L. Reitzer. 2002. ArgR-independent induction and ArgR-dependent superinduction of the *astCADBE* operon in *Escherichia coli*. J Bacteriol. 184(11): 2940–2950.

Klein, W., U. Ehmann and W. Boos. 1991. The repression of trehalose transport and metabolism in *Escherichia coli* by high osmolarity is mediated by trehalose-6-phosphate phosphatase. Res Microbiol. 142: 359–371.

Klumpp, S., Z. Zhang and T. Hwa. 2009. Growth rate-dependent global effects on gene expression in bacteria. Cell. 139: 1366–1375.

Koebman, B.J., H.V. Westerhoff, J.L. Snoep, D. Nilsson and P.R. Jensen. 2002. The glycolytic flux in *Escherichia coli* is controlled by the demand for ATP. J Bacteriol. 184: 3909.

Kramer, R. 2010. Bacterial stimulus perception and signal transduction: response to osmotic stress. Chem Res. 10: 217–229.

Kredich, N.M. 1996. Biosynthesis of cysteine. pp. 514–527. *In*: F.C. Neidhardt, R. Curtiss III, J.L. Ingraham, E.C.C. Lin, K.B. Low, B. Magasanik, W.S. Reznikoff, M. Riley, M. Schaechter and H.H. Umbarger (eds.). *Escherichia coli* and *Salmonella*: Cellular and Molecular Biology, 2nd Ed., Vol. 1. ASM Press, Washington, D.C.

Kumar, R. and K. Shimizu. 2010. Metabolic regulation of *Escherichia coli* and its *gdhA*, *glnL*, *gltB*, *D* mutants under different carbon and nitrogen limitations in the continuous culture. Microb Cell Fact. 9: 8.

Kumar, R. and K. Shimizu. 2011. Transcriptional regulation of main metabolic pathways of cyoA, cydB, fnr, and fur gene knockout *Escherichia coli* in C-limited and N-limited aerobic continuous cultures. Microb Cell Fact. 10: 3.

Kumari, S., C.M. Beatty, D.F. Browning et al. 2000. Regulation of acetyl coenzyme A synthetase in *Escherichia coli*. J Bacteriol. 182(15): 4173–4179.

Lamins, L.A., D.B. Rhoads and W. Epstein. 1981. Osmotic control of *kdp* operon expression in *Escherichia coli*. PNAS USA. 78: 464–468.

Luo, Y., T. Zhang and H. Wu. 2014. The transport and mediation mechanisms of the common sugars in *Escherichia coli*. Biotechnol Adv. 32: 905–919.

Maddocks, O.D. et al. 2013. Serine starvation induces stress and p53-dependent metabolic remodeling in cancer cells. Nature. 493: 542–546.

Maeda, T., V. Sanchez-Torres and T.K. Wood. 2007. Enhanced hydrogen production from glucose by metabolically engineered *Escherichia coli*. Appl Microbiol Biotechnol. 77: 879–890.

Maheswaran, M. and K. Forchhammer. 2003. Carbon-source-dependent nitrogen regulation in *Escherichia coli* is mediated through glutamine-dependent GlnB signaling. Microbiol. 149: 2163–2172.

Mailloux, R.J., R. Bériault, J. Lemire, R. Singh, R.R. Chénier, R.D. Hamel and V.D. Appanna. 2007. The tricarboxylic acid cycle, an ancient metabolic network with a novel twist. Plos One. 2(8): e690.

Malpica, R., B. Franco, C. Rodriguez, O. Kwon and D. Georgellis. 2004. Identification of a quinone-sensitive redox switch in the ArcB sensor kinase. PNAS USA. 101(36): 13318–13323.

Mao, X.J., Y.X. Huo, M. Buck, A. Kolb and Y.P. Wang. 2007. Interplay between CRP-cAMP and PII-Ntr systems forms novel regulatory network between carbon metabolism and nitrogen assimilation in *Escherichia coli*. Nucleic Acids Res. 35(5): 1432–1440.

Martin-Galiano, A.J., M.J. Ferrandiz and A.G. de La Campa. 2001. The promoter of the operon encoding the F0F1 ATPase of *Streptococcus pneumonia* is inducible by pH. Mol Microbiol. 41: 327–338.

Marzan, L.W. and K. Shimizu. 2011. Metabolic regulation of *Escherichia coli* and its *phoB* and *phoR* gene knockout mutants under phosphate and nitrogen limitations as well as acidic condition. Microb Cell Fact. 10: 39.

McHugh, J.P., F. Rodríguez-Quiñones, H. Abdul-Tehrani, D.A. Svistunenko, R.K. Poole, C.E. Cooper and S.C. Andrews. 2003. Global iron-dependent gene regulation in *Escherichia coli*. J Biol Chem. 278: 29478–29486.

Nicolaou, S.A., S.M. Gaida and E.T. Papoutsakis. 2010. A comparative view of metabolite and substrate stress and tolerance in microbial bioprocessing: from biofuels and chemicals, to biocatalysis and bioremediation. Metab Eng. 12: 307–331.

Ninfa, A.J., P. Jiang, M.R. Atkinson and J.A. Peliska. 2000. Integration of antagonistic signals in the regulation of nitrogen assimilation in *Escherichia coli*. Curr Topics Cell Regul. 36: 31–75.

Ninfa, J. and P. Jiang. 2005. PII signal transduction proteins: sensors of $\alpha$-ketoglutarate that regulate nitrogen metabolism. Curr Opin Microbiol. 8(2): 168–173.

Nizam, S.A., J. Zhu, P.Y. Ho and K. Shimizu. 2009. Effects of *arcA* and *arcB* genes knockout on the metabolism in *Escherichia coli* under aerobic condition. Biochem Eng J. 44(2-3): 240–250.

Nizam, S.A. and K. Shimizu. 2008. Effects of *arc* A and *arc* B genes knockout on the metabolism in *Escherichia coli* under anaerobic and microaerobic conditions. Biocheml Eng J. 42: 229–236.

Perera, I.C. and A. Grove. 2010. Molecular mechanisms of ligand-mediated attenuation of DNA binding by MarR family transcriptional regulators. J Mol Cell Biol. 2: 243–254.

Piper, P. 1995. The heat-shock and ethanol stress responses of yeast exhibit extensive similarity and functional overlap. FEMS Microbiol Lett. 134: 121–127.

Piuri, M., C. Sanchez-Rivas and S.M. Ruzal. 2005. Cell wall modifications during osmotic stress in *Lactobacullus casei*. J Appl Microbiol. 98: 84–95.

Pomposiello, P.J. and B. Demple. 2001. Redox-operated genetic switches: the SoxR and OxyR transcription factors. Trends in Biotechnol. 19(3): 109–114.

Privalle, C.T. and I. Fridovich. 1987. Induction of superoxide dismutase in *Escherichia coli* by heat shock. PNAS USA. 84(9): 2723–2726.

Quan, J.A., B.L. Schneider, I.T. Paulsen, M. Yamada, N.M. Kredich and M.H. Saier, Jr. 2002. Regulation of carbon utilization by sulfur availability in *Escherichia coli* and *Salmonella typhimurium*. Microbiol. 148: 123–131.

Rabinowitz, J. and T.J. Silhavy. 2013. Metabolite turns master regulator. Nature. 500: 283–284.

Radchenko, M.V., J. Thornton and M. Merrick. 2010. Control of AmtB-GlnK complex formation by intracellular levels of ATP, ADP, and 2-oxoglutarate. J Biol Chem. 285: 31037–31045.

Ramos, J.L., E. Duque, M.-T. Gallegos, P. Godoy, M.I. Ramos-Gonzalez, A. Rojas, W. Teran and A. Segura. 2002. Mechanisms of solvent tolerance in Gram-negative bacteria. Annu Rev Microbiol. 56: 743–768.

Record, Jr M.T., E.S. Courtenary, D.S. Cayley and H.J. Guttman. 1998. Responses of *E. coli* to osmotic stress: large changes in amounts of cytoplasmic solutes and water. Trends Biochem Sci. 23: 143–148.

Reyes, L.H., M.P. Almario and K.C. Kao. 2011. Genomic library screens for genes involved in n-butanol tolerance in *Escherichia coli*. PLoS One. 6: e17678.

Richard, H.T. and J.W. Foster. 2003. Acid resistance in *Escherichia coli*. Adv Appl Microbiol. 52: 167–186.

Richard, H.T. and J.W. Foster. 2004. *Escherichia coli* glutamate- and arginine-dependent acid resistance systems increase internal pH and reverse transmembrane potential. J Bacteriol. 86(18): 6032–6041.

Rutherford, B., R. Dahl, R. Price, H. Szmidt, P. Benke, A. Mukhopadhyay and J. Keasling. 2012. Functional genomic study of exogenous n-butanol stress in *Escherichia coli*. Appl Environ Microbiol. 76: 1935–1945.

Rutherford, B.J., R.H. Dahl, R.E. Price, H.L. Szmidt, P.I. Benke, A. Mukhopadhyay and J.D. Keasling. 2010. Functional genomic study of exogenous n-butanol stress in *Escherichia coli*. Appl Environ Microbiol. 76: 1935–1945.

Salmon, K., S.P. Hung, K. Mekjian, P. Baldi, G.W. Hatfield and R.P. Gunsalus. 2003. Global gene expression profiling in *Escherichia coli* K12: The effects of oxygen availability and FNR. J Biol Chem. 278(32): 29837–29855.

Scott, M., C.W. Gunderson, E.M. Mateescu, Z. Zhang and T. Hwa. 2010. Interdependence of cell growth and gene expression: origins and consequences. Science. 330: 1099–1102.

Semsey, S., A.M. Andersson, S. Krishna, M.H. Jensen, E. Masse and K. Sneppen. 2006. Genetic regulation of fluxes: iron homeostasis of *Escherichia coli*. Nucleic Acids Res. 34: 4960–4967.

Sevin, D.C. and U. Sauer. 2014. Ubiquinone accumulation improves osmotic-stress tolerance in *Escherichia coli*. Nature Chem Biol. 10: 266–272.

Shimizu, K. 2004. Metabolic flux analysis based on $^{13}C$-labeling experiments and integration of the information with gene and protein expression patterns. Adv Biochem Eng Biotechnol. 91: 1–49.

Shimizu, K. 2013. Bacterial Cellular Metabolic Systems, Woodhead Publ. Co., Oxford, UK.

Shin, D., S. Lim, Y.J. Seok and S. Ryu. 2001. Heat shock RNA polymerase (Eσ 32) is involved in the transcription of mlc and crucial for induction of the Mlc regulon by glucose in *Escherichia coli*. J Biol Chem. 276(28): 25871–25875.

Sikkema, J., J.A.M. de Bont and B. Poolman. 1995. Mechanisms of membrane toxicity of hydrocarbons. Microbiol Rev. 59: 201–222.

Son, J. et al. 2013. Glutamine supports pancreatic cancer growth through a KRAS-regulated metabolic pathway. Nature. 496: 101–105.

Sørensen, H.P. and K.K. Mortensen. 2005. Soluble expression of recombinant proteins in the cytoplasm of *Escherichia coli*. Microb Cell Fact. 4: 1.

Spira, B., N. Silberstein and E. Yagil. 1995. Guanosine 3′, 5′-bispyrophosphate (ppGpp) synthesis in cells of *Escherichia coli* starved for $P_i$. J Bacteriol. 177(14): 4053–4058.

Stincone, A., N. Daudi, A.S. Rahman et al. 2011. A systems biology approach sheds new light on *Escherichia coli* acid resistance. Nucleic Acids Research. 39(17): 7512–7528.

Storz, G. and J.A. Imlay. 1999. Oxidative stress. Curr Opin Microbiol. 2: 188–194.

Sugiura, A., K. Hirokawa, K. Nakashima and T. Mizuno. 1994. Signal-sensing mechanisms of the putative osmosensor KdpD in *Escherichia coli*. Mol Microbiol. 14: 929–938.

Suziedeliene, E., K. Suziedelis, V. Garbenciute and S. Normark. 1999. The acid-inducible asr gene in *Escherichia coli*: transcriptional control by the phoBR operon. J Bacteriol. 181(7): 2084–2093.

Takatsuka, Y., C. Chen and H. Nikaido. 2010. Mechanism of recognition of compounds of diverse structures by the multidrug efflux pump AcrB of *Escherichia coli*. PNAS USA. 107: 6559–656.

Thomason, M.K., F. Fontaine, N. De Lay and G. Storz1. 2012. A small RNA that regulates motility and biofilm formation in response to changes in nutrient availability in *Escherichia coli*. Mol Microbiol. 84(1): 17–35.

Tilly, K., J. Erickson, S. Sharma and C. Georgopoulos. 1986. Heat shock regulatory gene *rpoH* mRNA level increases after heat shock in *Escherichia coli*. J Bacteriol. 168(3): 1155–1158.

Tilly, K., J. Spence and C. Georgopoulos. 1989. Modulation of stability of the *Escherichia coli* heat shock regulatory factor σ32. J Bacteriol. 171(3): 1585–1589.

Tomas, C., J. Beamish and E. Papoutsakis. 2004. Transcriptional analysis of butanol stress and tolerance in Clostridium acetobutylicum. J Bacteriol. 186: 2006–2018.

Toya, Y., K. Nakahigashi, M. Tomita and K. Shimizu. 2012. Metabolic regulation analysis of wild-type and *arcA* mutant *Escherichia coli* under nitrate conditions using different levels of omics data. Mol Biosyst. 8: 2593–2604.

Trachootham, D., J. Alexandre and P. Huang. 2009. Targeting cancer cells by ROS-mediated mechanisms: a radical therapeutic approach ? Nature Rev Drug Discov. 8: 579–591.

Truan, D., L.F. Huergo, L.S. Chubatsu, M. Merrick, X.D. Li and F.K. Winkler. 2010. A new P(II) protein structure identifies the 2-oxoglutarate binding site. J Mol Biol. 400: 531–539.

Turunen, M., J. Olsson and G. Dallner. 2004. Metabolism and function of coenzyme Q. Biochim Biophys ACTA Biomembranes. 1660: 171–199.

Valgepea, K., K. Adamberg, A. Seiman and R. Vilu. 2014. *Escherichia coli* achieves faster growth by increasing catalytic and translation rates of proteins. Mol BioSys. 9: 2344–2358.

Van Dien, S.J. and J.D. Keasling. 1998. A dynamic model of the *Escherichia coli* phosphate-starvation response. J Theoret Biol. 190(1): 37–49.

Van Heeswijk, W.C., H.V. Westerhoff and F.C. Boogerd. 2013. Nitrogen assimilation in *Escherichia coli*: putting molecular data into a systems perspective. Microbiol Mol Biol Rev. 77(4): 628–695.

Vemuri, G.N., E. Altman, D.P. Sangurdekar, A.B. Khodursky and M.A. Eiteman. 2006. Overflow metabolism in *Escherichia coli* during steady-state growth: transcriptional regulation and effect of the redox ratio. Appl Environ Microbiol. 72(5): 3653–3661.

Vemuri, G.N., M.A. Eiteman and E. Altman. 2006. Increased recombinant protein production in *Escherichia coli* strains with overexpressed water-forming NADH oxidase and a deleted ArcA regulatory protein. Biotechnol Bioeng. 94(3): 538–542.

Von Weyman, N., A. Nyyssola, T. Reinikainen, M. Leisola and H. Ojamo. 2001. Improved osmotolerance of recombinant *Escherichia coli* by *de novo* glycine betain biosynthesis. Appl Microbiol Biotechnol. 55: 214–218.

Walderhaug, M.O., J.W. Polarek, P. Voelkner, J.M. Daniel, J.E. Hesse, K. Altendorf and W. Epstein. 1992. KdpD and KdpE, proteins that control expression of the *kdpABC* operon, are members of the two-component sensor-effector class of regulators. J Bacteriol. 174: 2152–2159.

Wang, X., A.K. Dubey, K. Suzuki, C.S. Baker, P. Babitzke and T. Romeo. 2005. CsrA post-transcriptionally represses *pgaABCD*, responsible for synthesis of a biofilm polysaccharide adhesin of *Escherichia coli*. Mol Microbiol. 56: 1648–1663.

Wanner, B.L. 1993. Gene regulation by phosphate in enteric bacteria. J Cellr Biochem. 51(1): 47–54.

Wanner, B.L. 1996. Phosphorus assimilation and control of the phosphate regulon. pp. 1357–1381. *In*: F.C. Neidhardt, I.I.I. Curtiss, R.J.L. Ingraham et al. (eds.). *Escherichia coli and Salmonella*: Cellular and Molecular Biology. ASM Press, Washington, DC, USA.

Waters, C.M. and B.L. Bassler. 2005. Quorum sensing: cell-to-cell communication in bacteria. Annu Rev Cell Dev Biol. 21: 319–346.

Weber, H., T. Polen, J. Heuveling, V. Wendisch and R. Hengge. 2005. Genome-wide analysis of the general stress response network in *Escherichia coli*: sigmaS-dependent genes, promoters, and sigma factor selectivity. J Bacteriol. 187: 1591–1603.

White-Ziegler, C.A., S. Um, N.M. Perez, A.L. Berns, A.J. Malhowski and S. Young. 2008. Low temperature (23°C) increases expression of biofilm-, cold shock-, and RpoS-dependent genes in *Escherichia coli* K-12. Microbiol. 154: 148–166.

Wood, J.M. 2011. Bacterial osmoregulation: a paradigm for the study of cellular homeostasis. Ann Rev Microbiol. 65: 215–238.

Wu, X., R. Altman, M.A. Eiteman and E. Altman. 2013. Effect of overexpressing nhaA and nhaR on sodium tolerance and lactate production in *Escherichia coli*. J Biol Eng. 7: 3.

Yao, R. and K. Shimizu. 2012. Recent progress in metabolic engineering for the production of biofuels and biochemical from renewable sources with particular emphasis on catabolite regulation and its modulation. Process Biochem. 48(9): 1409–1417.

You, C., H. Okano, S. Hui, Z. Zhang, M. Kim, C.W. Gunderson, Y.-P. Wang, P. Lenz, D. Yan and T. Hwa. 2013. Coordination of bacterial proteome with metabolism by cyclic AMP signaling. Nature. 500: 301–306.

Yuan, J., C.D. Doucette, W.U. Fowler, X.-J. Feng, M. Piazza, H.A. Rabitz, N.S. Wingreen and J.D. Rabinowitz. 2009. Metabolomics-driven quantitative analysis of ammonia assimilation in *E. coli*. Mol Syst Biol. 5: 302.

Zhang, H., H. Chong, C.B. Ching, H. Song and R. Jiang. 2012. Engineering global transcription factor cyclic AMP receptor protein of *Escherichia coli* for improved 10 butanol tolerance. Appl Microbiol Biotechnol. 94: 1107–1117.

Zhang, H., H. Chong, C.B. Ching and R. Jiang. 2012. Random mutagenesis of global transcription factor cAMP receptor protein for improved osmotolerance. Biotechnol Bioeng. 109(5): 1165–1172.

Zhang, Z., G. Gosset, R. Barabote, C.S. Gonzalez, W.A. Cuevas and M.H. Saier Jr. 2005. Functional interactions between the carbon and iron utilization regulators, Crp and Fur, in *Escherichia coli*. J Bacteriol. 187(3): 980–990.

Zheng, M., B. Doan, T.D. Schneider and G. Storz. 1999. OxyR and SoxRS regulation of fur. J Biotechnol. 181: 4639–4643.

Zhu, J. and K. Shimizu. 2004. The effect of *pfl* genes knockout on the metabolism for optically pure d-lactate production by *Escherichia coli*. Applied Microbiol Biotechnol. 64: 367–75.

Zhu, J. and K. Shimizu. 2005. Effect of a single-gene knockout on the metabolic regulation in *E. coli* for d -lactate production under microaerobic conditions. Metab Eng. 7: 104–15.

Zhu, J., S. Shalel-Levanon, G. Bennett and K.Y. San. 2006. Effect of the global redox sensing/regulation networks on *Escherichia coli* and metabolic flux distribution based on C-13 labeling experiments. Metab Eng. 8(6): 619–627.

Zimmer, D.P., E. Soupene, H.L. Lee, V.F. Wendisch, B.J. Khodursky, R. Peter, A. Bender and S. Kustu. 2000. Nitrogen regulatory protein controlled genes of *Escherichia coli*: scavenging as a defense against nitrogen limitation. PNAS USA. 97: 14674–14679.

# 6

# Metabolic Engineering for the Production of a Variety of Biofuels and Biochemicals

## ABSTRACT

Metabolic engineering strategies for the production of a variety of biofuels and biochemicals by microorganisms are explained. The organic acid fermentations such as pyruvate, lactate, citric acid, malic acid, fumaric acid, and succinate fermentations are explained with pathway modification of *Escherichia coli*, *Corynebacterium glutamicum*, and others. As for biofuels production, hydrogen, ethanol, 1-butanol, 1-propanol, isobutanol, and isopropanol fermentations are briefly explained for engineered *Saccharomyces cerevisiae*, *Zymomonas* sp., *E. coli*, and *Corynebacterium* sp., etc. Metabolic engineering of *E. coli* and *Corynebacterium* sp. for shikimic acid and amino acids fermentations is also explained. The production of isoprenoids by microorganisms is also explained. The production of diols such as butandiol and propandiol from glycerol is also explained focusing on the use of engineered *E. coli*.

### Keywords

Metabolic engineering, cell factory, biofuels, biochemicals, ethanol fermentation, succinate fermentation, lactate fermentation, butanol, propanol, fatty acids, higher alcohols, shikimic acid, amino acid, propandiol, butandiol, hydrogen fermentation, consolidated bioprocess (CBP)

## 1. Introduction

One of the successful examples for metabolic engineering practice in industry may be the production of natural products, particularly active pharmaceutical ingredients, some of which are too complex to be chemically synthesized and yet a value that justifies the cost for developing a genetically engineered microorganisms (Keasling 2010). Such examples are alkaloids and flavonoids found primarily in and derived from plants and widely used as drugs (Hawkins and Smalke 2008, Pandey et al. 2016). Other examples are polyketides and nonribosomal peptides, which are used as veterinary agents and agrochemicals. Some of the most valuable molecules have been produced with engineered industrial host organisms (Pfeifer et al. 2001). Isoprenoids can be used as fragrances and essential oils, nutraceuticals, and pharmaceuticals, and many isoprenoids such as isoprene, carotenoids and terpenes have been produced by microorganisms. Other metabolites have also been produced by metabolic engineering.

In contrast to the above chemicals, solvents and polymer precursors are rarely produced by microorganisms in industry due to higher cost as compared to their production from inexpensive petroleum by chemical catalysis. Most popular examples are the production of transportation fuels such as ethanol, butanol, isobutanol, propanol, isopropanol, and 1,3-propanediol by recombinant microorganisms. Large branched-chain alcohols can be also produced by engineered microorganisms, where those may be considered to be better fuels than ethanol and butanol, and can be also used to produce commodity chemicals. As the next generation biofuels, the isopentenols such as isoprenol and prenol can be produced by engineered *E. coli*. Metabolic engineering can be also applied to produce hydrocarbons with properties similar to those petroleum-derived fuels, where linear hydrocarbons such as alkanes, alkenes and esters, typical of diesel and jet fuel have been produced by way of fatty acid synthetic pathway.

In industry, a limited number of platform cell factories have been employed for the production of a wide range of fuels and chemicals (Papagianni 2012, Peralta-Yohya et al. 2012, Keasling 2010). The most popular platform cell factories may be *Escherichia coli* due to the advantages of high cell growth rate and well-known physiolosical characteristics (Huffer et al. 2012, Chen et al. 2013, Vickers et al. 2012, Clomburg and Gonzalez 2010). In the case of bioethanol production, much interest has centered on *S. cerevisiae* in view of tolerance against ethanol and various other stresses (Kondo et al. 2012, Hasunuma et al. 2011). *Zymomonas mobilis* and its related microorganisms have also been used for ethanol fermentation and others. *Chrostridium acetobutyricum* has been considered for aceton-butanol-ethanol fermentation and others. The gram-positive soil bacteria such as *Corynebacterium glutamicum* and the related species have also been used in industrial production of L-amino acids, while the potentials of biofuels and biochemicals production by *Corynebacterium* sp. have also been shown. *Pseudomonas putida* can overcome the toxicity using efflux pumps, and *Bacillus subtilis* can change its cell-wall composition in response to solvent toxicity. *Lactobacillus brevis* digests hexose and pentose sugars with high tolerance to high concentration of solvents such as butanol.

Recent progress in metabolic engineering, synthetic biology, together with systems biology allows the design and construction of efficient microbial cell factories for producing biochemicals with higher performances in terms of yield, productivity, and titer, thus enabling cost reduction and make microbial production of biochemicals to be feasible in industrial scale production.

Figure 1 shows the fermentative pathways for the production of biofuels and biochemicals from various carbon sources, and non-fermentative pathways for the production of fatty acids and isoprenoid, etc. (Yao and Shimizu 2013).

For metabolic engineering of a microbial cell, the following factors must be taken into account (Keasling 2010, Yadav et al. 2012):

(1) available raw materials such as cellulosic biomass, and the available carbon sources such as glucose, xylose, fructose, glycerol, arabinose, galactose, mannose, sorbitol, mannitol, sucrose, maltose, lactose, etc. obtained after pretreatment of cellulosic biomass,

(2) appropriate host microbes such as *Escherichia coli, Saccharomyces cerevisiae, Corynebacterium glutamicum, Bacillus subtilis, Pichia pastoris, Zymomonas mobilis, Lactococcus lactis, Lactobacillus* sp., *Pseudomonas putida, Clostridium acetobutyricum, Streptomyces* sp., photosynthetic bacteria such as Cyanobacteria and algae, etc.,

(3) metabolic pathways available in the microbes chosen, or introduction of heterologous pathways available from other organisms for the conversion of the available carbon source(s) to the target metabolite(s) with the corresponding genes which encode the enzymes in the pathways as well as its regulators or transcription factors,

(4) metabolic engineering strategy for debottlenecking the limiting pathways, and ways to maximize the yields, titers, and productivities together with robustness against stresses,

(5) modulation of enzymatic (allosteric) regulation and transcriptional regulation by the transcription factors,

(6) enhancement of the substrate uptake rate (for the improvement of the production rate or the productivity), and co-consumption of multiple sugars, as well as reduction or deletion of the undesirable by-product pathways (for the improvement of yield),

(7) enhancement of the precursor metabolites production together with the production of necessary cofactors such as NAD(P)H, efficient induction strategy for the heterologous protein production by recombinant DNA technology, and

(8) modulation of transporters for excreting the target metabolite(s) to the culture broth.

In the present chapter, metabolic engineering practice is explained for the production of a variety of biofuels and biochemicals by microbes.

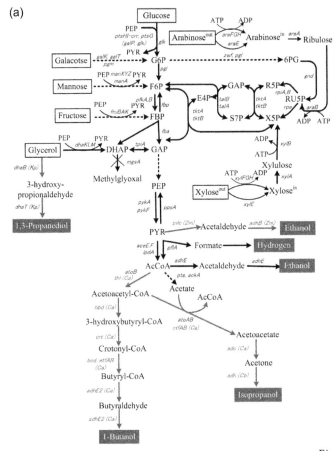

*Fig. 1 contd. ...*

*... Fig. 1 contd.*

(b)

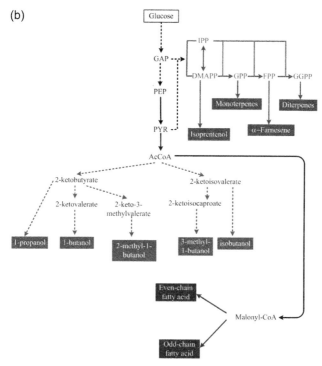

**Figure 1.** An overview on the metabolic pathways for a variety of useful metabolite production. (a) Fermentative pathways for the production of biofuels in *E. coli*. *E. coli* genes unless otherwise noted in the parenthesis as follows: *Clostridium acetobutylicum*, *Ca*; *Clostridium Beijerinckii*, *Cb*; *Zymomonas mobilis*, *Zm*; *Klebsiella pneumoniae*, *Kp*; *adc*, acetoacetate dehydrogenase (*Ca*); *adhE2*, secondary alcohol dehydrogenase (*Ca*); *bcd*, butyryl-CoA dehydrogenase (*Ca*); *crt*, crotonase (*Ca*); *ctfAB*, acetoacetyl-CoA transferase (*Ca*); *etfAB*, electron transfer flavoprotein (*Ca*); *hbd*, β-hydroxybutyryl-CoA dehydrogenase (*Ca*); *thl*, acetyl-CoA acyltransferase (*Ca*); *adh*, secondary alcohol dehydrogenase (*Cb*); *adhB*, alcohol dehydrogense (*Zm*); *pdc*, pyruvate decarboxylase (*Zm*); *dhaB*, glycerol dehydratase (*Kp*); *dhaT*, 1, 3-propanediol oxidoreductase (*Kp*). (b) Nonfermentative pathways for the production of alcohols, fatty acid pathways, fatty-acid derived molecules, and isoprenoid pathways in *E. coli*. Broken lines represent multiple reaction steps (Yao and Shimizu 2013).

## 2. Typical Fermentation

Microbial production of organic acids is a promising approach for building block (bulk) chemicals from renewable sources (Goldberg et al. 2006, Sauer et al. 2008, Liu et al. 2016). Bulk chemicals are produced in general at high volume with low cost, and are used directly or as platform for the production of a variety of chemicals. Although most of the bulk chemicals have been produced via the petrochemical processes as shown in Table 1 (Sauer et al.

Table 1. Potential of microbial organic acid production*.

| Compounds | Organic acids | Annual production (A.P.) (t) | A.P. by microbes (t) | References |
|---|---|---|---|---|
| C2 | Acetic acid | 7,000,000 | 190,000 | |
| | Oxalic acid | 124,000 | | |
| | | | | |
| C3 | Acrylic acid | 4,200,000 | | |
| | Lactic acid | 150,000 | 150,000 | |
| | Propionic acid | 130,000 | | |
| | | | | |
| C4 | Butyric acid | 50,000 | | |
| | Fumaric acid | 180,000 | | |
| | Malic acid | 10,000 | | |
| | Succinic acid | 16,000 | | |
| | | | | |
| C5 | Itaconic acid | 80,000 | 80,000 | Okabe et al. 2009 |
| | Levulic acid | 450 | | |
| | | | | |
| C6 | Adipic acid | 2,800,000 | | Burridge et al. 2011 |
| | Ascorbic acid | 80,000 | | |
| | Citric acid | 1,600,000 | 1,600,000 | |
| | Gluconic acid | 87,000 | 87,000 | |
| More than C6 | | | | |
| | Terephthalic acid | 71,000,000 | | Mirasol 2011 |

* Updated from Table 1 of Sauer et al. 2008.

2008), microbial production is believed to be attractive and promising from the long range future perspectives, due to energy shortage, environmental protection, etc. for the realization of the sustainable low carbon society as mentioned in Chapter 1.

Figure 2 shows the central metabolic pathways of *E. coli* cultivated under anaerobic condition. The organic acids such as pyruvic acid, lactic acid, succinic acid, fumaric acid, maleic acid, and citric acid produced as the intermediates in the central metabolism can be used for the industrial applications such as food, beverages, cosmetics, pharmaceuticals, chemicals, and biopolymers. Here, the typical metabolic engineering strategies for the production of some of the above metabolites are explained, where Fig. 3 shows the basic idea of metabolic engineering for the production of lactate, succinate, and ethanol.

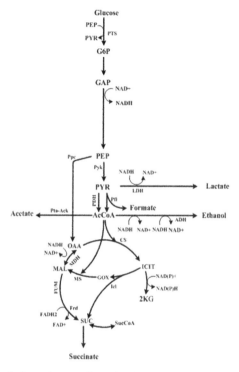

**Figure 2.** Metabolic pathways of *E. coli* cultivated under anaerobic condition.

## 2.1 Pyruvic acid fermentation

Pyruvic acid is currently manufactured for use as a food additive or a nutriceutical supplement (Li et al. 2001). Pyruvic acid can be also used as a starting material for the synthesis of amino acids (Ingram et al. 1987). Pyruvate is currently produced commercially by both chemical and microbial processes. Microbial pyruvate production is based primarily on two microorganisms, a multi-vitamin auxotroph of the yeast *Torulopsis glabrata* (Li et al. 2001, Yonehara and Miyata 1995, Miyata and Yonehara 1996) and a lipoic acid auxotroph of *E. coli* containing a mutation in the $F_1$ATPase component of $(F_0F_1)$ H⁺-ATP synthase (Tomar et al. 2003, Yokota et al. 1994). Both strains require precise regulation of the media composition during fermentation and the use of complex supplements.

The basic idea for pyruvate fermentation is to block the pyruvate consuming pathways such as pyruvate dehydrogenase (PDH) under aerobic condition. Tomar et al. (2003) considered the mutations in *ppc, aceF* and *adhE*, while Causey et al. (2004) engineered *E. coli* TC44, a derivative of W3110 for such mutation as $\Delta atpFH$, $\Delta adhE$, $\Delta sucA$ to minimize the ATP yield, the cell growth rate, and $CO_2$ production, where pyruvate was produced under aerobic condition. The reason why the respiratory chain and ATP synthesis pathways were disrupted may be to increase the glycolytic flux as mentioned in Chapter 4. In addition, the production of acetate can be reduced by the mutation of acetate producing pathway genes such as *poxB*::FRT (FLP recognition target) and Δ*ackA*. The production of other metabolites may be reduced by blocking the corresponding fermentation pathways such as Δ*focA-pflB* for formate production, Δ*frdBC* for succinate production, Δ*ldhA* for lactate production, and Δ*adhE* for ethanol production. The resulting strain converted glucose to pyruvate with the yield of 75% (g-pyruvate/g-glucose), the productivity of 1.2 g/l.h, and the maximum titer of 749 mM, where the glycolytic flux was more than 50% faster than that of the wild type W3110 under fully aerobic condition (Causey et al. 2004). The performance improvement may be made by the repetitive fed-batch operation with *in situ* product recovery by electrodyalysis (Zelic et al. 2004). However, the scale-up may be the problem for its application in practice.

The basic idea of pyruvate production is to block all the consuming pathways such as pyruvate dehydrogenase (PDH), pyruvate formate lyase (Pfl), pyruvate oxidase (Pox), PEP synthase (Pps), and lactate dehydrogenase (LDH), where these can be attained by the mutations in *aceEF*, *pfl*, *poxB*, *pps*, and *ldhA* genes. In this mutant, however, AcCoA is not formed from pyruvate, and thus the TCA cycle intermediates are not produced, resulting in the depletion of some of the precursors for the cell synthesis. This may be overcome by adding acetate, and the acetate limiting cultivation may be considered, giving the glycolytic flux of 1.60 h$^{-1}$, pyruvate productivity of 1.11 g/l.h, the yield of 0.70 g/g-glucose in the continuous culture at the dilution rate of 0.15 h$^{-1}$ (Zhu et al. 2008). The glycolytic flux can be further increased by modulating the respiratory chain as Δ*atpFH* and the transcription factor as Δ*arcA* to 2.38 h$^{-1}$. A fed-batch operation by the exponential feeding rate at 0.15 h$^{-1}$ achieved the titer of 90 g/l, the productivity of 2.1 g/l.h, and the yield of 0.68 g/g-glucose (Zhu et al. 2008) (see also Table 2).

### *2.2 Lactic acid fermentation*

Lactic acid is widely used in food industry and has attracted interest for the production of biodegradable plastics such as poly-lactic acid (PLA) (Jung et al. 2010, Sauer 2008, Pang et al. 2010, Datta et al. 1995, Martinez et al. 2013, Upadhyaya et al. 2014), where much attention has been focused on utilizing the lignocellulosic biomass or waste to reduce the cost for industrialization (Hofvendahl and Hahn-Hagerdal 2000, John et al. 2007, Okano et al. 2010, Abdel-Rahman et al. 2013, Wang et al. 2015, Zhang et al. 2016). There are two types of isomers such as L- and D-lactic acids depending on the chiral-specific L or D-lactate dehydrogenase (LDH), where the optical purity is

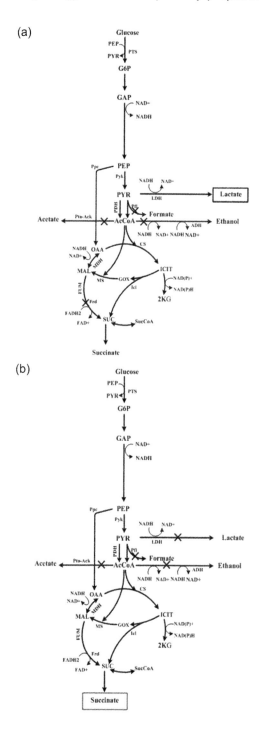

*Fig. 3 contd. ...*

*...Fig. 3 contd.*

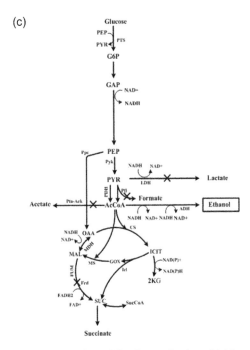

**Figure 3.** Metabolic engineering of *E. coli* for the production of (a) lactate, (b) succinate, and (c) ethanol.

critical, since the purification is difficult in practice, and the impurities cause the change in crystallinity and thus affect the thermal properties (Narayanan et al. 2004, Wee et al. 2006, Matsumoto and Taguchi 2013). In the past, lactic acid bacteria such as *Lactobacillus* sp. and *Lactococcus* sp. have been employed for lactic acid fermentation (Benthin and Villadsen 1995, Kyla-Nikkila et al. 2000). However, lactic acid bacteria suffer from several drawbacks such as requiring complex nutrients, an optically impure product, a mixed acid fermentation, and poor ability to utilize pentoses (Wendisch et al. 2006).

Metabolically engineered *E. coli* has therefore been attracted much attention. In *E. coli*, D-lactate is produced from pyruvate together with NADH by LDH encoded by *ldhA*, where the redox balance is important to drive this pathway for lactate production under anaerobic conditions. *E. coli* has been successfully manipulated to produce optically pure D- or L-lactate, indicating the potential for large scale industrial production (Okano et al. 2010, Sauer et al. 2008, Wendisch et al. 2006, Yu et al. 2011, Chen et al. 2013, Upadhyaya et al. 2014, Tsuge et al. 2016).

Under micro-aerobic or anaerobic conditions, the expression of *aceE*, *F* (and *lpdA*) genes which code for pyruvate dehydrogenase complex (PDHc) is repressed, while the expression of *pflAB-foc-CD* and *frdABCD* genes is activated by the transcription factors such as ArcA and Fnr in *E. coli* (Appendix A and Chapter 5), and the metabolism shows the mixed acid fermentation. Table 3 shows the fermentation characteristics of *E. coli* wild type (BW25113) and its single-gene knockout mutants such as Δ*pflA*, Δ*pta*, Δ*ppc*, Δ*adhE*, and Δ*pykF* mutants cultivated under micro-aerobic conditions, which indicates that D-lactate is overproduced in Δ*pflA*, Δ*pta*, and Δ*ppc* mutants, while succinate is overproduced in *pykF* mutant (Zhu and Shimizu 2005). Among the possible gene knockout, *pflA* gene knockout shows exclusive D-lactate formation, where pyruvate is accumulated (Table 4), and the enzyme activities of GAPDH and LDH are higher (Table 5), which together with higher NADH/NAD$^+$ ratio (Table 4) gave higher D-lactate formation flux (Table 6) (Zhu and Shimizu 2004, 2005). The similar phenomenon is also seen for d*pflB* mutant, whereas the effect of *pflC* or *pflD* gene knockout on Pfl activity as well as D-lactate production is small (Zhu and Shimizu 2004). The Foc encoded by *foc* gene is the membrane embedded protein for the transport of formate.

Moreover, Δ*pta* and Δ*ppc* mutants also overproduced D-lactate under micro-aerobic conditions. In the case of r*pta* mutant, pyruvate as well as PEP accumulates (Table 4), and GAPDH and LDH activities are also higher (Table 5), which together with relatively higher NADH/NAD$^+$ ratio gave higher D-lactate production (Table 6). Although Δ*ppc* mutant may also contribute for D-lactate production by reducing the succinate production, the cell growth rate (Table 3) and the specific glucose consumption rate (Table 6) are low. The brief metabolic characteristics are illustrated in Fig. 4 (Matsuoka and Shimizu 2015), where *ldhA* mutant enhances ethanol and acetate formation (Kabir and Shimizu 2006). The increase in the glucose uptake rate causes the activation of PTS, and then causes the decrease in PEP concentration, and increase in pyruvate concentration, and thus the activation of PTS is preferred for lactate fermentation, while less preferred for succinate fermentation.

The fed-batch cultivation of Δ*pta* Δ*ppc* mutant produced more than 62.2 g/*l* of D-lactate with the productivity of 1.04 g/*l*.h, where the cultivation was started with aerobic condition with initial glucose concentration of 50 g/*l* until 10 h to enhance the cell growth, followed by the anaerobic conditions to produce lactate (Chang et al. 1999). Since succinate, acetate, and ethanol are also co-produced, further gene knockout mutation such as Δ*frdABCD* Δ*ackA* Δ*adhE* together with Δ*pflB* gave D-lactate production with the yield of 96% of the theoretical yield, titer of 539 mM, and the productivity of 5.98 mmol/*l*.h with the optical purity of 99% (Zhou et al. 2003). However, this strain showed some inefficient metabolism at higher glucose concentration (more than about 50 g/*l*), and therefore, another strain SZ132 was constructed from an ethanologenic KO11, where 100 g/*l* of glucose could be rapidly assimilated to produce D-lactate of over 90 g/*l* in

**Table 2.** Biofuel and biochemical production by engineered microorganisms at different culture conditions.

| Product | Strains | Genotype (mutation) | Carbon source | Cultivation method | Titer [g/l] | Yield [g/g] | Productivity [g/l.h] | Reference |
|---|---|---|---|---|---|---|---|---|
| **C0** | | | | | | | | |
| Hydrogen | *E. coli* | | Formate | Immobilized | | 1.0*a) | 2.4*b) | Seol et al. 2011 |
| | *E. coli* | *ΔhycA fhlA⁺* | Formate | | | | 23.6*c) | Yoshida et al. 2005 |
| | *E. coli* | *ΔhycA.ΔhyaAB.ΔhybBC.ΔldhA.ΔfrdAB* | Glucose | | | | 31.3*d) | Kim et al. 2009 |
| **C2** | | | | | | | | |
| Ethanol | *S. cerevisiae* | | Corn stover hydrolysate | Batch | 40 | 0.46 | 0.8 | Lau and Dale 2009 |
| | *S. cerevisiae* | | Cellobiose, xylose, glucose | Batch | 48 | 0.37 | 0.8 | Ha et al. 2011 |
| | *Z. mobilis* | | Glucose and xylose | Batch | 62 | 0.46 | 1.29 | Johachimsthat et al. 1999 |
| | *E. coli* | *fucO, ΔyqhD* | Xylose | Batch | ~45 | 0.48 | 0.62 | Wang et al. 2011 |
| Acetic acid | *A. aceti* | | Ethanol | Aerobic batch | 111.7 | | 0.6 | Nakano et al. 2006 |
| **C3** | | | | | | | | |
| Pyruvic acid | *E. coli* | *atpFH, adhE, sucA, ackA, poxB, focA-pflB, frd, ldhA* | Glucose | Aerobic batch | 0.749 | 0.75 | 1.2 | Causey et al. 2004 |
| | *E. coli* | *ΔaceEF, Δpfl, ΔpoxB, Δpps, ΔldhA, ΔatpFH, ΔarcA* | Glucose (+acetate) | Microaerobic chemostat (2.5 l) | 90 | 0.68 | 2.1 | Zhu et al. 2008 |
| Propionic acid | *P. acidipropionici* | | Glycerol | Fed-batch with fibrous bed bioreactor | 106 | 0.56 | 0.035 | Zhang and Yang 2009 |

| Product | Organism | Genetic modification | Substrate | Condition | Titer | Yield | Productivity | Reference |
|---|---|---|---|---|---|---|---|---|
| Lactic acid | E. coli B0013 | ack-pta, pps, pflB, dld, poxB, adhE, frdA | Glucose | Aerobic +O2 limit Fed-batch (71) | 122.8 | 0.866 | 5.5 | Zhou et al. 2012a |
| | Sporolactobacillus | | Glucose | Fed-batch with 40 g/l of peanat meal | 207 | 0.93 | 3.8 | Wang et al. 2011 |
| | E. coli | | Glucose | Fed-batch | 138 | 0.86 | 3.5 | Zhu et al. 2007 |
| | C. glutamicum | ΔldhA/pCRB204 (L. delbluckii ldhA) | | Fed-batch | 120 | 1.73 *f) | 4.0 | Okino et al. 2008 |
| 3-HP | K. pneumonia | | Glycerol | Fed-batch | 16 | | 0.01 | Ashok et al. 2011 |
| | E. coli | | Glycerol | Fed-batch | 38.7 | 0.34 | 0.53 | Rathnasingh et al. 2009 |
| | E. coli BL21 | PQE-80L-mcr pACYCDuet-1r-accADBC-birA-pmtAB | Glucose | Aerobic batch | 1.19 | 0.016 | | Rathnasingh et al. 2012 |
| | E. coli BL21 | ΔglpK,ΔyqhD, PELDRR L. brevis DhaB-DhaE pCPa72, P. aeruginosa PSALDH | Glycerol | Aerobic fed-batch | 57.3 | 0.88 | | Kim et al. 2014 |
| Propanol | E. coli | | Glucose | Flask | 3.9 | | 0.04 | Atsumi and Liao 2008 |
| Iso-propanol | C. acetobutyricum | | Glucose | Anaerobic flask | 5.1 | | | Lee et al. 2012 |
| | E. coli | | Glucose | Fed-batch | 13.6 | 0.15 | 0.28 | Jojima et al. 2008 |
| 1,2-PDO | C. thermosaccharolyticum | | Glucose | Anaerobic batch | 9.1 | 0.2 | 0.35 | Sanchez-Riera et al. 1987 |

*Table 2 contd. ...*

*...Table 2 contd.*

| Product | Strains | Genotype (mutation) | Carbon source | Cultivation method | Titer [g/l] | Yield [g/g] | Productivity [g/l.h] | Reference |
|---|---|---|---|---|---|---|---|---|
| 1,3-PDO | E. coli | ΔackΔpta ΔldhA ΔdhaK. pTHKLcfgldA.mgsA.yqhD | Glycerol | Batch | 5.6 | 0.21 | 0.077 | Clomburg and Gonzalez 2011 |
| | C. acetobutyricum | | Glycerol | Anaerobic fed-batch | 83.6 | 0.54 | 1.7 | Gonzalez-Pajuelo et al. 2005 |
| | E. coli | | Glucose | Fed-batch | 135 | 0.51 | 3.5 | Nakamura and Whited 2003 |
| **C4** | | | | | | | | |
| Butyric acid | C. tyrobutyricum | Δack/Δpta | Glucose | Fed-batch | 32.5–41.7 | 0.38–0.42 | 0.24–0.68 | Zhu et al. 2003, Liu et al. 2006 |
| | C. tyrobutyricum | | Glucose | Repeated fed-batch immobilized in a fibrous bed bioreactor | 86.9 | 0.46 | 1.1 | Jiang et al. 2011 |
| Succinic acid | Engineered rumen bacteria | | Glucose | Anaerobic fed-batch | 52–106 | 0.76–0.88 | 1.8–2.8 | Guettler et al. 1996a,b, Lee et al. 2006 |
| | E. coli | | Glucose | Fed-batch | 73–87 | 0.8–1.0 | 0.7–0.9 | Jantaa et al. 2008a,b, Thakker et al. 2012 |
| | E. coli K12 | ΔpflΔldhΔptsGpTrc99A-R.etlipyc | Glucose+Na2CO3 | Two-stage fermentation | 99.2 | | | Chatterjee et al. 2001, Vemuri et al. 2002 |
| | C. glutamicum | Δcat.Δpqo.Δpta-ack. ΔldhA.ppc.aceAB. gltA. sucE | | Anaerobic fed-batch | 109 | | 1.11 | Zhu et al. 2014 |

| | | | | | | | | |
|---|---|---|---|---|---|---|---|---|
| Malic acid | C. glutamicum | Δcat, Δpqo, Δpta-ack:: Ppyc ΔldhA | Glucose | Fed-batch | 140–146 | 0.92–1.1 | 1.9–2.5 | Litsanov et al. 2012, Okino et al. 2008 |
| | A. flavus | | Glucose | Fed-batch | 113 | 0.95 | 0.59 | Battat 1991 |
| | A. oryzae NRR3488 | pyc+,mdh+,C4T318+ | Glucose | Batch | 154 | 1.38*f) | 0.94 | Brown et al. 2013 |
| | S. cerevisiae | | Glucose | Fed-batch | 59 | 0.31 | 0.19 | Zelle et al. 2008 |
| | E. coli | ΔldhAΔackAΔadhEΔpflB ΔmgsAΔpoxBΔfrdBCΔsfc AΔmaeBΔfumABC | Glucose | Two-step fermentation | 33.9 | 0.47 | 1.06 | Zhang et al. 2011 |
| | E. coli | Δpta, pck (M. Succiniciproducens) | Glucose | Batch | 9.25 | 0.42 | 0.74 | Moon et al. 2008 |
| Fumaric acid | R. arrhizus NRRL 2582 | | Glucose | Batch | 97.7 | 0.81 | 1.02 | Kenealy et al. 1986 |
| | R. arrhizus NRRL 1526 | | Glucose | Two-stage DO | 56.2 | 0.54 | 1.3 | Fu et al. 2010 |
| | E. coli W3110 | ΔiclRΔfumCABΔiackAΔpt sGΔaspAΔiacI, galP::Prrc pTac15kppc | Glucose+Na2CO3 +L-asprtic acid | Fed-batch | 28.2 | 0.389 | | Song et al. 2013 |
| GABA | L. brevis NCL912 | | Glucose and glutamate | Fed-batch | 103.7 | | | Li et al. 2010 |
| | C. glutamicum | | Glucose | Batch | 2.2 | | 0.01 | Shi and Li 2011 |
| 1-Butanol | C. acetobutyricum | Δbuk | Glucose | Anaerobic batch | 16.7 | | 0.31 | Harris et al. 2000 |
| | E. coli | | Glucose | Batch | 14–15 | 0.33–0.36 | 0.20–0.29 | Dellomonaco et al. 2011, Shen et al. 2011 |
| Iso-butanol | E. coli | | Glucose | Batch | 20 | | | Atsumi et al. 2008b |

*Table 2 contd. ...*

...*Table 2 contd.*

| Product | Strains | Genotype (mutation) | Carbon source | Cultivation method | Titer [g/l] | Yield [g/g] | Productivity [g/l.h] | Reference |
|---|---|---|---|---|---|---|---|---|
| | *C. glutamicum* | | Glucose | Fed-batch | 13 | 0.2 | 0.33 | Blombach et al. 2011 |
| 1,4-BDO | *E. coli* | | Glucose | Microaerobic fed-batch | 18 | | 0.15 | Yim et al. 2011 |
| 2,3-BDO | *K. pneumonia* SDM | | Glucose | Fed-batch | 150 | 0.48 | 4.21 | Ma et al. 2009 |
| | *S. macescens* | | Glucose | Fed-batch | 152 | 0.46 | 2.67 | Zhang et al. 2010 |
| Putresine | *E. coli* | | Glucose | Fed-batch | 24.2 | | 0.75 | Qian et al. 2009 |
| **C5** | | | | | | | | |
| Itaconic acid | *A. terreus* IFO 6365 | | Glucose and Corn steep | Flask and 100L batch | 82–85 | 0.54 | 0.57 | Yahiro et al. 1995 |
| | *A. terreus* | | Glucose | Batch | 146 | 0.81*f) | 0.48 | Havekerl et al. 2014 |
| | *A. terreus* | | Glucose | Batch with pH control | 129 | | 1.15 | Havekerl et al. 2014 |
| | *A. niger* | *cadA,mttA,mfsA,citB* | Glucose | Batch | 26.2 | | 0.35 | Hossain et al. 2016 |
| | *E. coli* | | Glucose | Flask, batch | 6 | 0.61 | 0.06 | Liao et al. 2010 |
| | *E. coli* MG1655 | *ΔaceAΔsucCDΔpykAΔpy kFΔptaΔPicd::cam BBa_ J23115,* pCadCS | Glucose | Fed-batch | 32 | 0.68*f) | 0.45 | Harder et al. 2016 |
| 3-HV | *P. putida* | | Glucose and Levulinic acid | Flask, batch | 5.3 | | | |
| | *E. coli* | | Glucose and threonine | Flask, batch | 1.3 | | | Tseng et al. 2010 |
| | *E. coli* | | Glucose | Flask, batch | 0.81 | | | Tseng et al. 2010 |

| Product | Organism | Genes | Substrate | Mode | | | | Reference |
|---|---|---|---|---|---|---|---|---|
| 1-Pentanol | *E. coli* | | Glucose | | 0.5 | | | Zhang et al. 2008 |
| 2-M1B | *E. coli* | | Glucose | | 1.25 | | 0.17 | Cann and Liao 2008 |
| 3-M1B | *E. coli* | | Glucose | | 1.28 | | 0.11 | Connor and Liao 2008 |
| Xylitol | *C. tropicalis* | | Xylose | O2-limited with cell recycle | 1.82 | 0.85 | 12 | Granstrom et al. 2007a,b |
| | *E. coli* | | Glucose and xylose | Fed-batch | 38 | | | Cirino et al. 2006 |
| Cadaverine | *E. coli* | | Glucose | Fed-batch | 9.61 | | 0.12 | Qian et al. 2011 |
| **C6** | | | | | | | | |
| Citric acid | *A. niger* | | Glucose | Batch | 46 | | 0.27 | Ruijter et al 2000 |
| | *A. niger* | | Beet, Cane molasse | Shake flask | 113.5 | 0.71 | | Lotfy et al. 2007 |
| | *A. niger* | | Cane molasse | | 114 | 0.76 | 0.61 | Ikram-ul et al. 2004 |
| | *A. niger* | | Glucose | | | 0.9 | 0.025 | de Jongh et al. 2008 |
| | *Y. lipolytica* | ΔACL1. ICL1. INO1 | Glucose | Batch | 84 | | 0.39 | Liu et al. 2013 |
| | *Y. lipolytica* | SUC2.ICL1 | Glucose | Batch | 140 | | 0.73 | Forster et al. 2007 |
| | *Y. lipolytica* | | n-Paraffine | | 40 | 0.99 | 0.1 | Crolla and Kennedy 2004 |
| | *Y. lipolytica* | | Fatty acid, glucose | | 42.9 | 0.56 | | Papanikolaou et al. 2006 |
| | *Y. lipolytica* | | Sucrose | | 140 | 0.82 | 0.73 | Wang et al. 2015 |

*Table 2 contd. ....*

...*Table 2 contd.*

| Product | Strains | Genotype (mutation) | Carbon source | Cultivation method | Titer [g/l] | Yield [g/g] | Productivity [g/l.h] | Reference |
|---|---|---|---|---|---|---|---|---|
| Glucaric acid | *E. coli* | | Glucose | Flask | 2.5 | | | Moon et al. 2010 |
| | *E. coli* | Direct evolution for MIOX activity increase | *myo*-inositol (10.8 g/l) | | 4.85 | | | Shiue and Prather 2014 |
| Muconic acid | *Pseudomonas* sp. | | Benzoate | Flask | 7.2 | | (11 h) | Xie et al. 2014b |
| | *P. putida* | Random mutagenesis | Toluene | | 45 | | (4 d) | Chua and Hsieh 1990 |
| | *P. putida* | | Benzoate | Fed-batch | 18.5 | 1 | | van Duuren et al. 2012 |
| | *E. coli* | *aroZ, aroY, catA, aroE* | Glucose | Fed-batch | 36.8 | 22*d) | | Niu et al. 2002 |
| | *E. coli* | *aroZ, aroY, catA, aroE* | Glucose | Fed-batch | 59.2 | | (48 h) | Bui et al. 2011, 2013 |
| Anthranilic acid | *E. coli* | | Glucose | Fed-batch | 14 | 0.2 | 0.41 | Balderas-Hernandez et al. 2009 |
| Phenol | *P. putida S12* | | Glucose | Flask | 0.14 | | 0.006 | Wierckx et al. 2005 |
| Catechol | *P. putida ML2* | | 3-Dehydroshikimate (DHS) | | 4.2 | | 0.12 | Wang et al. 2001 |
| | *E. coliW3110* | *trp9923.PTS⁻, tktA, aroG^{fbr}, ant3(P.aeruginosa)* | Glucose | fed-batch | 4.47 | 0.16 | | Balderas-Hernandez et al. 2014 |

*a) mol H2/mol formate, *b) H2 l/l.h, *c) H2 g/l.h, *d) mmol H2/gDCW.h, *e) Optical purity, *f) mol/mol glucose
3HP: 3-Hydroxy propionate, 3HV: 3-Hydroxyvalerate, PDO: Propanediol, BDO: Butanediol, M1B: Methyl-1-butanol.

rich medium containing betain as a protective osmolyte (Zhou et al. 2005). In general, the defined synthetic medium is preferred from the point of view of product purification. Because this strain showed poor performance in minimal salts medium, several strategies have been attempted by evolution in conjunction with genetic manipulations to overcome such a problem, where the resulting strain could produce D-lactate with the titer of 110 g/l, the yield of 0.95 g/g-glucose in mineral salts medium supplemented with 1 mM of betain (Zhou et al. 2006). Unfortunately, the chiral impurity was 5%, which requires additional cost for purification. A higher titer of D-lactate (138 g/l) with the yield of 0.99 g/g and the productivity of 6.3 g/l.h could be attained by such gene mutation as $\Delta aceEF$, $\Delta pfl$, $\Delta poxB$, $\Delta pps$, and $\Delta frd\ ABCD$ (*E. coli* ALS974), and by the two step process operation of aerobic and anaerobic conditions to enhance the cell growth rate during aerobic period (14 h), and thus to enhance the productivity of lactate using the defined medium (Zhu et al. 2007). The overall productivity was 3.5 g/l.h and the overall yield of 0.86 g/g-glucose could be attained by the fed-batch fermentation, where acetate has to be supplemented, and controlled at certain level during aerobic condition. Further gene mutation such as $\Delta ackA$-*pta*, $\Delta pps$, $\Delta pflB$, $\Delta dld$, $\Delta poxB$, $\Delta adhE$, and $\Delta frdA$ (*E. coli* B0013) yielded an overall productivity of 5.5 g/l.h, the yield of 0.86 g/g-glucose, and the titer of 122.8 g/l, where the additional knockout of the gene *dld* encoding the D-lactate dehydrogenase which converts D-lactate to pyruvate was made (Pratt et al. 1979), and the lactate fermentation was coupled with the cell growth (Zhou et al. 2012a).

As for the two-step fermentation for the cell growth and lactate production phases, the chromosomal native promoter of D-lactate dehydrogenase gene may be replaced by the temperature sensitive $\lambda p_R$ and $\lambda p_L$ promoters, where the two-step fermentation can be made by changing

**Table 3.** Comparison of the fermentation results of different mutants cultivated under anaerobic condition.

| Strain | Specific growth rate μ (h⁻¹) | Yield on glucose (g/g) [%] | | | | | | |
|---|---|---|---|---|---|---|---|---|
| | | **Biomass** | **Lactate** | **Acetate** | **Formate** | **Succinate** | **Ethanol** | **Pyruvate** |
| Wild | 0.85 ± 0.08 | 30.1 ± 0. | 1.2 ± 0.0 | 24.3 ± 0.1 | 30.0 ± 0.2 | 5.0 ± 0.2 | 14.7 ± 0.3 | 0.2 ± 0.0 |
| $\Delta pfl$ A | 0.18 ± 0.04 | 8.1 ± 0.1 | 72.5 ± 0.0 | 1.1 ± 0.1 | 0.0 ± 0.2 | 0.0 ± 0.0 | 1.0 ± 0.0 | 1.1 ± 0.0 |
| $\Delta pta$ | 0.24 ± 0.04 | 6.9 ± 0.1 | 51.3 ± 0.0 | 1.6 ± 0.1 | 2.1 ± 0.2 | 4.5 ± 0.1 | 0.0 ± 0.0 | 0.0 ± 0.0 |
| $\Delta ppc$ | 0.01 ± 0.02 | 4.6 ± 0.1 | 32.2 ± 0.0 | 13.5 ± 0.1 | 9.4 ± 0.2 | 0.3 ± 0.0 | 6.2 ± 0.1 | 4.6 ± 0.0 |
| $\Delta adh$ E | 0.03 ± 0.02 | 11.4 ± 0.1 | 9.7 ± 0.0 | 42.4 ± 0.1 | 12.1 ± 0.2 | 3.2 ± 0.1 | 0.2 ± 0.0 | 0.0 ± 0.0 |
| $\Delta pyk$ F | 0.48 ± 0.05 | 13.0 ± 0.1 | 2.1 ± 0.0 | 34.0 ± 0.1 | 22.9 ± 0.2 | 6.9 ± 0.2 | 12.0 ± 0.2 | 0.0 ± 0.0 |

**Table 4.** Comparison of enzyme activities for different mutants cultivated under anaerobic condition.

| | Enzyme activities (mmol/min. mg protein) | | | | | | | | |
|---|---|---|---|---|---|---|---|---|---|
| | G6PDH | 6PGDH | GAPDH | Pyk | LDH | Ppc | Pfl | Ack | ADH |
| wt | 0.152 ± 0.023 | 0.18 ± 0.002 | 0.008 ± 0.001 | 0.657 ± 0.102 | 0.503 ± 0.120 | 0.008 ± 0.003 | 0.071 ± 0.012 | 0.038 ± 0.006 | 0.006 ± 0.000 |
| Δpfl A | 0.252 ± 0.061 | 0.079 ± 0.010 | 0.076 ± 0.011 | 0.726 ± 0.170 | 2.093 ± 0.512 | 0.343 ± 0.122 | 0.004 ± 0.004 | 4.088 ± 0.200 | 0.011 ± 0.002 |
| Δpta | 0.187 ± 0.027 | 0.001 ± 0.001 | 0.060 ± 0.008 | 0.512 ± 0.006 | 1.621 ± 0.306 | 0.005 ± 0.001 | 0.013 ± 0.001 | 3.465 ± 0.195 | 0.005 ± 0.001 |
| Δppc | 0.017 ± 0.002 | 0.000 ± 0.000 | 0.014 ± 0.001 | 0.183 ± 0.060 | 0.164 ± 0.020 | 0.003 ± 0.002 | 0.025 ± 0.006 | 0.629 ± 0.035 | 0.040 ± 0.009 |
| Δadh E | 0.013 ± 0.002 | 0.004 ± 0.001 | 0.008 ± 0.003 | 0.108 ± 0.021 | 0.000 ± 0.000 | 0.005 ± 0.001 | 0.043 ± 0.001 | 0.536 ± 0.035 | 0.000 ± 0.000 |
| Δpyk F | 0.052 ± 0.006 | 0.017 ± 0.003 | 0.020 ± 0.005 | 0.009 ± 0.003 | 0.069 ± 0.030 | 0.013 ± 0.001 | 0.050 ± 0.004 | 0.293 ± 0.016 | 0.001 ± 0.001 |

**Table 5.** Comparison of the intracellular metabolite concentrations in different mutants cultivated under anaerobic condition.

| | Metabolite concentration (mM/gDCW) | | | | | | | | | | | |
|---|---|---|---|---|---|---|---|---|---|---|---|---|
| | G6P | FDP | PEP | PYR | AcCoA | ATP | ADP | AMP | ATP/ | NADH | NAD | NADH/ |
| BW25113 | 0.05 ± 0.01 | 4.59 ± 0.02 | 0.32 ± 0.12 | 8.21 ± 0.01 | 0.07 ± 0.00 | 3.06 ± 0.17 | 0.38 ± 0.01 | 0.21 ± 0.01 | 0.84 | 0.018 ± 0.002 | 0.143 ± 0.001 | 0.126 |
| Δpfl A | 1.09 ± 0.01 | 16.41 ± 0.11 | 0.12 ± 0.04 | 25.49 ± 0.01 | 0.05 ± 0.00 | 1.69 ± 0.11 | 0.20 ± 0.01 | 0.20 ± 0.01 | 0.81 | 0.056 ± 0.001 | 0.060 ± 0.002 | 0.933 |
| Δpta | 0.54 ± 0.03 | 15.08 ± 0.03 | 1.27 ± 0.20 | 21.83 ± 0.55 | 0.01 ± 0.00 | 1.82 ± 0.16 | 1.17 ± 0.08 | 1.12 ± 0.17 | 0.44 | 0.047 ± 0.001 | 0.129 ± 0.007 | 0.68 |
| Δppc | 0.84 ± 0.03 | 24.46 ± 0.04 | 0.96 ± 0.07 | 38.72 ± 2.26 | 0.05 ± 0.02 | 4.19 ± 0.07 | 3.71 ± 1.14 | 1.16 ± 0.43 | 0.53 | 0.018 ± 0.004 | 0.170 ± 0.010 | 0.106 |
| Δadh E | 0.32 ± 0.01 | 11.41 ± 0.03 | 0.46 ± 0.08 | 18.40 ± 0.25 | 0.01 ± 0.00 | 1.38 ± 0.04 | 0.69 ± 0.02 | 0.46 ± 0.04 | 0.55 | 0.010 ± 0.006 | 0.058 ± 0.009 | 0.167 |
| Δpyk F | 1.15 ± 0.02 | 30.90 ± 0.90 | 3.67 ± 0.06 | 46.96 ± 1.98 | 0.08 ± 0.02 | 2.65 ± 0.10 | 2.27 ± 0.05 | 1.01 ± 0.08 | 0.47 | 0.002 ± 0.005 | 0.177 ± 0.008 | 0.01 |

**Table 6.** Effect of a single-gene knockout on the flux distribution of a cell cultivated under anaerobic condition.

| | Fluxes (mmol/gDCW/h) | | | | | | | | | | | |
|---|---|---|---|---|---|---|---|---|---|---|---|---|
| | $v_1$ | $v_2$ | $v_3$ | $v_4$ | $v_5$ | $v_6$ | $v_7$ | $v_8$ | $v_9$ | $v_{10}$ | $v_{11}$ | $v_{12}$ |
| Wild | 6.30 ± 0.06 | 5.90 ± 0.07 | 11.73 ± 0.10 | 3.85 ± 0.09 | 1.39 ± 0.09 | 0.42 ± 0.06 | 0.09 ± 0.06 | 7.28 ± 0.35 | 1.97 ± 0.23 | 4.47 ± 0.12 | 3.52 ± 0.11 | 3.94 ± 0.10 |
| Δ*pfl* A | 8.15 ± 0.06 | 7.58 ± 0.07 | 14.93 ± 0.10 | 5.71 ± 0.10 | 0.78 ± 0.10 | 0.18 ± 0.11 | 11.69 ± 0.7 | 0.00 ± 0.00 | 1.64 ± 0.11 | 0.46 ± 0.2 | 0.45 ± 0.10 | 1.62 ± 0.11 |
| Δ*pta* | 6.51 ± 0.06 | 5.79 ± 0.06 | 11.27 ± 0.10 | 3.03 ± 0.10 | 1.37 ± 0.10 | 0.77 ± 0.10 | 7.36 ± 0.83 | 0.70 ± 0.06 | 0.92 ± 0.10 | 0.61 ± 0.3 | 0.31 ± 0.12 | 1.26 ± 0.12 |
| Δ*ppc* | 1.59 ± 0.02 | 1.48 ± 0.02 | 2.90 ± 0.03 | 1.12 ± 0.02 | 0.14 ± 0.02 | 0.00 ± 0.03 | 1.13 ± 0.1 | 0.62 ± 0.07 | 0.66 ± 0.03 | 0.72 ± 0.08 | 0.45 ± 0.06 | 0.20 ± 0.02 |
| Δ*adh* E | 1.10 ± 0.02 | 1.04 ± 0.02 | 2.05 ± 0.03 | 0.76 ± 0.04 | 0.15 ± 0.03 | 0.07 ± 0.02 | 0.23 ± 0.03 | 0.53 ± 0.06 | 1.03 ± 0.05 | 1.45 ± 0.05 | 0.00 ± 0.02 | 0.27 ± 0.04 |
| Δ*pyk* F | 9.02 ± 0.07 | 7.05 ± 0.07 | 13.21 ± 0.07 | 0.00 ± 0.00 | 3.21 ± 0.33 | 1.54 ± 0.23 | 0.13 ± 0.2 | 7.82 ± 0.29 | 2.56 ± 0.43 | 7.75 ± 0.4 | 2.93 ± 0.31 | 2.48 ± 0.23 |

the culture temperature from low (30°C) for the cell growth to high (42°C) for the induction of the *ldh* gene. This temperature shifting fed-batch fermentation yielded the titer of D-lactate of 122.8 g/*l* with oxygen limited productivity of 6.73 g/*l*.h (Zhou et al. 2012b).

Tables 7 and 8 show the fermentation characteristics of *pflA* mutant cultivated under micro-aerobic and anaerobic (AN) conditions using different carbon sources such as glucose, gluconate, pyruvate, fructose, and glycerol (Zhu and Shimizu 2005), which indicate that glucose is the best for the substrate consumption rate and the lactate production rate. Although pyruvate accumulates for the cases of using gluconate, pyruvate, and fructose, NADH/NAD$^+$ ratio is lower, and GAPDH and LDH activities are lower (Table 9) as compared to the case of using glucose (Table 10), resulting in less production of lactate (Table 7). Acetate production is higher for the case of using gluconate and pyruvate (Table 7), where ATP is produced at Ack. The reason why ATP production is lower when either gluconate or pyruvate was used as a carbon source (Table 7) may be due to low glycolytic flux (Table 7). Although NADH/NAD$^+$ ratio is significantly higher when glycerol was used as a carbon source (Table 9), the substrate uptake rate is significantly low (Table 7).

Since glycerol is the byproduct of biodiesel and bioethanol production, and is abundant and inexpensive carbon source, it is preferred to utilize

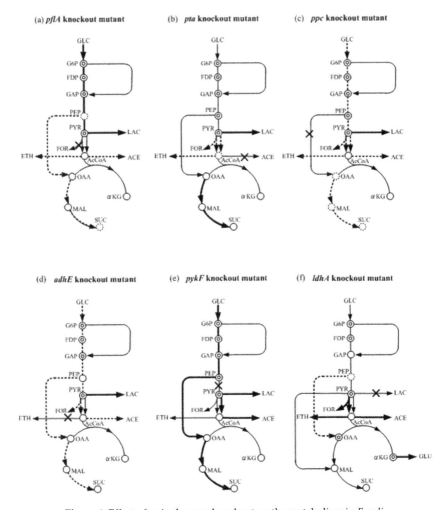

**Figure 4.** Effect of a single-gene knockout on the metabolism in *E. coli*.

such substrate for the production of biochemicals and biofuels (Khanal et al. 2008, Rausch and Belyea 2006, Yazdani and Gonzalez 2008, da Silva et al. 2009). Unlike sugars, however, when glycerol was used as a carbon source, its uptake rate is quite low. In particular, only very small amount is consumed by *E. coli* under micro-aerobic or anaerobic conditions (Zhu and Shimizu 2005, Durnin et al. 2009, Murarka et al. 2008).

Glycerol may be converted to dihydroxyacetone phosphate (DHAP) in the glycolysis by two routes, where one fermentative route converts glycerol to dihydroxyacetone (DHA) by glycerol dehydrogenase (GLDH) encoded by *gldA* and then DHA to DHAP by DHA kinase (DHAK) encoded

**Table 7.** The specific carbon source uptake rates and the specific product formation rates for the *pfl*A mutant using different carbon sources under anaerobic condition.

| Carbon source | q *[mmol gDCW$^{-1}$ h$^{-1}$ | | | | | |
|---|---|---|---|---|---|---|
| | Substrate | Lactate | Succinate | Formate | Acetate | Ethanol |
| Glucose | 8.15 | 11.01 | 0 | 0 | 0.27 | 0.2 |
| Gluconate | 6.57 | 8.22 | 0.21 | 0 | 5.63 | 0.32 |
| Pyruvate | 6.87 | 1.78 | 0.25 | 0 | 4.58 | 0.11 |
| Fructose | 3.21 | 4.42 | 0 | 0 | 0.23 | 0 |
| Glycerol | 1.33 | 0.83 | 0 | 0 | 0.05 | 0 |

**Table 8.** The yields (*Y*) of cell mass and metabolites for different carbon sources for *pfl* A mutant grown under anaerobic condition. AN: The *pfl* A mutant grown on glucose under anaerobic condition. The culture vessel was filled with $CO_2$.

| Carbon source | Yields on carbon source (g g$^{-1}$)% | | | | | |
|---|---|---|---|---|---|---|
| | $Y_{x/s}$ | $Y_{lactate/s}$ | $Y_{acetate/s}$ | $Y_{ethanol/s}$ | $Y_{succinate/s}$ | $Y_{formate/s}$ |
| Glucose | 8.1 | 72.5 | 1.1 | 3.8 | 0 | 0 |
| Gluconate | 8.1 | 57.4 | 26.2 | 1.1 | 1.9 | 0 |
| Pyruvate | 5.9 | 26.5 | 45.4 | 0.2 | 1.3 | 0 |
| Fructose | 9.7 | 69 | 3.7 | 0 | 0 | 0 |
| Glycerol | 0 | 61.1 | 2.6 | 0 | 0 | 0 |
| Glucose (AN) | 6.7 | 48.2 | 0.6 | – | 3.6 | 0 |

**Table 9.** Enzyme activities for *pfl* A mutant grown on different carbon sources under anaerobic condition.

| Carbon source | Enzyme activity (mmol (mg protein)$^{-1}$min$^{-1}$) | | | | | | | |
|---|---|---|---|---|---|---|---|---|
| | G6PDH | 6PGDH | GAPDH | Pyk | LDH | Ppc | Ack | ADH |
| Glucose | 0.252 ± 0.019 | 0.080 ± 0.003 | 0.076 ± 0.003 | 0.73 ± 0.04 | 2.09 ± 0.12 | 0.343 ± 0.032 | 4.09 ± 0.77 | 0.010 ± 0.001 |
| Gluconate | 0.449 ± 0.02 | 0.244 ± 0.009 | 0.126 ± 0.005 | 0.90 ± 0.04 | 4.38 ± 0.42 | 0.084 ± 0.010 | 38.15 ± 2.5 | ND |
| Pyruvate | 0.264 ± 0.019 | ND | ND | ND | 0.42 ± 0.04 | ND | 39.82 ± 2.8 | ND |
| Fructose | 0.170 ± 0.019 | 0.066 ± 0.004 | 0.030 ± 0.006 | 0.88 ± 0.09 | 0.6 ± 0.07 | 0.013 ± 0.001 | 0.036 ± 0.006 | 0.015 ± 0.002 |
| Glycerol | 0.141 ± 0.013 | 0.098 ± 0.005 | 0.026 ± 0.002 | 0.44 ± 0.03 | 0.52 ± 0.10 | 0.025 ± 0.002 | 0.072 ± 0.009 | 0.006 ± 0.001 |

**Table 10.** Intracellular metabolite concentrations in the *pfl* A mutant grown on different carbon sources under anaerobic condition.

| Metabolite | Intracellular concentration [µmol gDCW$^{-1}$] | | | | |
|---|---|---|---|---|---|
| | Glucose | Gluconate | Pyruvate | Fructose | Glycerol |
| G6P | 1.093 ± 0.006 | 0.123 ± 0.001 | 0.120 ± 0.001 | 0.240 ± 0.001 | 0.045 ± 0.003 |
| FDP | 16.41 ± 0.11 | 27.08 ± 0.10 | 164.9 ± 2.7 | 28.92 ± 0.04 | 33.02 ± 0.04 |
| PEP | 0.12 ± 0.04 | 0.22 ± 0.11 | 0.50 ± 0.13 | 0.45 ± 0.13 | 0.10 ± 0.05 |
| PYR | 25.49 ± 0.01 | 41.81 ± 0.70 | 250.6 ± 0.85 | 47.22 ± 0.50 | 52.98 ± 0.30 |
| AcCoA | 0.050 ± 0.001 | 0.003 ± 0.001 | 0.039 ± 0.004 | 0.132 ± 0.035 | 0.037 ± 0.001 |
| ATP | 0.052 ± 0.012 | 0.008 ± 0.004 | 0.034 ± 0.011 | 0.027 ± 0.006 | 0.056 ± 0.026 |
| ADP | 0.195 ± 0.010 | 0.143 ± 0.003 | 0.835 ± 0.011 | 0.458 ± 0.092 | 0.531 ± 0.091 |
| AMP | 0.197 ± 0.004 | 0.226 ± 0.007 | 0.932 ± 0.049 | 0.127 ± 0.013 | 0.499 ± 0.006 |
| NADH | 0.056 ± 0.001 | 0.032 ± 0.002 | 0.016 ± 0.002 | 0.092 ± 0.012 | 0.085 ± 0.020 |
| NAD+ | 0.060 ± 0.002 | 0.046 ± 0.001 | 0.032 ± 0.001 | 0.244 ± 0.003 | 0.037 ± 0.001 |
| ATP/AMP | 0.264 | 0.035 | 0.036 | 0.213 | 0.112 |
| NADH/ NAD+ | 0.93 | 0.7 | 0.5 | 0.38 | 2.3 |

by *dhaKLM*, while another respiratory route converts glycerol to glycerol 3-phosphate (GL3P) by glycerol kinase (GlpK) encoded by *glpK*, and then GL3P is converted to DHAP by aerobic GL3P dehydrogenase (GL3PDH) encoded by *glpD* (Durnin et al. 2009) (Fig. 5).

The over-expression of *gldA-dhaKLM* (activation of GLDH-DHAK route) activates *adhE* gene expression, and enhances the generation of the reducing equivalents in the form of NADH, which activates alcohol/acetaldehyde dehydrogenase (Leonardo et al. 1993, 1996), and in turn increase the ethanol formation, while reducing the acetate formation since both ethanol and acetate are formed from AcCoA in *E. coli* (Mazumdar et al. 2010). Moreover, DHAK utilizes phosphotransferase system (PTS), where DHAK consists of three subunits such as DhaK which contains the DHA binding site, DhaL which contains ADP as cofactor for the double displacement of phosphate from DhaM to DHA, and DhaM which provides a phospho histidine relay between the PTS and DhaL::ADP, where *dhaKLM* operon is repressed by the transcription factor DhaR (Bachler et al. 2005). DHAK thus uses PEP as a phosphate donor in the phosphorylation of DHA (Bachler et al. 2005), and therefore PEP pool size decreases, while pyruvate pool size increases as this pathway is activated. This implies the decreased succinate formation from PEP, and the increased formation of lactate (Mazumdar et al. 2010).

On the other hand, the over-expression of the respiratory pathway for glycerol uptake causes the decrease in ethanol production, while

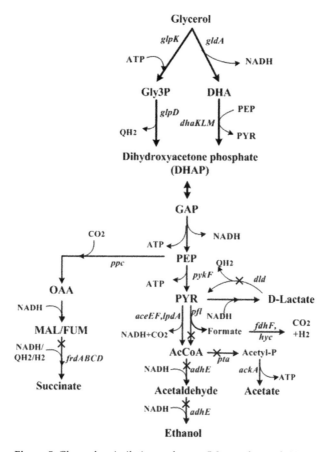

**Figure 5.** Glycerol assimilation pathways (Mazumdar et al. 2010).

significantly increases D-lactate production (Mazumdar et al. 2010). This pathway involves a respiratory GLPDH that donates electrons directly to quinol/quinone (Schweizer and Larson 1987, Walz et al. 2002), where the electrons are then transferred to oxygen via cytochrome oxidases under aerobic condition. Under micro-aerobic or anaerobic conditions, NADH produced at GAPDH is balanced with the consumption at LDH with the ATP production at Pyk by the substrate level phosphorylation.

The double mutant (Δ*pflB*Δ*frdA*) (LA01) and the triple mutant (Δ*adhE*Δ*pta*Δ*frdA*) (LA02) over-produced D-lactate, where the latter strain consumed about 40 g/l of glycerol and produced about 30 g/l of D-lactate in about 72 h with the yield of 0.80 g D-lactate/g glycerol. The chiral purity was 99.9% (Mazumdar et al. 2010). Noting that D-lactate produced may be consumed by the respiratory D-lactate dehydrogenase encoded by *dld* to yield pyruvate (Pratt et al. 1979), this *dld* gene (Santos et al. 1982) may be disrupted to improve the yield. Moreover, as mentioned before, it is preferred to activate *glpK-glpD* in LA01 and LA02 background with *dld* gene knockout. The resulting strain LA02Δ*dld* (pZglpKglpD) produced 45 g/l of D-lactate from 60 g/l of glycerol (initial concentration of 40 g/l with fed batch of 20 g/l) in 84 h with the yield of 0.83 g/g glycerol, where small amount of acetate is also produced (Mazumdar et al. 2010).

It is also of practical interest to produce lactate from inexpensive sucrose obtained from sugarcane and/or beet molasses. Sucrose utilization operon is formed as *cscBKAR*, where CscR encoded by *cscR* represses the sucrose utilization operon which encodes for sucrose permease (CscB), sucrose 6-phosphate hydrolase (CscA) and fructokinase (CscK) (Fig. 6) (Wang et al. 2012). Sucrose is first converted to glucose 1-phosphate (G1P) and fructose by sucrose 6-phosphate hydrolase, where G1P is then converted

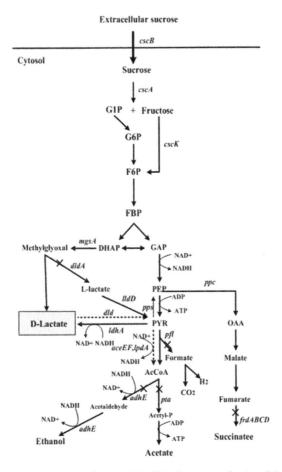

**Figure 6.** Sucrose accimilating pathway in *E. coli* for lactate production (Wang et al. 2012).

Metabolic Engineering for the Production of a Variety of Biofuels and Biochemicals

to G6P in the glycolysis by Pgm, while fructose is converted to F6P by fructokinase. The theoretical yield is 1.05 g D-lactate/g sucrose based on the stoichiometry that 4 moles of D-lactate can be produced from one mole of sucrose (Fig. 6). In *E. coli*, there are three lactate dehydrogenase genes such as *ldhA*, *dld*, and *lldD*, where *ldhA* is dominant, enabling the conversion of pyruvate to D-lactate under anaerobic conditions (Clark 1989), while some amount (about 5%) of L-lactate may be also formed (Grabar et al. 2006). This contamination may be removed by *dldA* gene knockout to some extent (Fig. 6). By enhancing the sucrose utilization by ΔcscR, and blocking the byproduct formation pathway genes such as Δ*dldA*, Δ*frdBC*, Δ*adhE*, Δ*pta*, and Δ*pflB*, 85 g/l of D-lactate was produced from 100 g/l of sucrose in 72–84 h, with the productivity of about 1 g/l.h, yield of 85%, and the optical purity of 98.3% with a minor by-prduct of acetate (4 g/l). The impurity of D-lactic acid caused by L-lactate isomer may be due to *dld-lldD* pathway, which encodes Dld which converts D-lactate to pyruvate, and LldD which converts L-lactate to pyruvate (Wang et al. 2012).

A mixture of glucose and xylose can be converted to lactate by the mixed culture of *E. coli* strains, one of which is unable to consume glucose, while the other is unable to consume xylose. The resulting mixed culture gave 37 g/l of lactate and a yield of 0.88 g/g (Eitmann et al. 2008, 2009). The problem of the mixed culture may be the difficulty in controlling each population in practice, where the optical density monitors the total cell density.

*Escherichia coli* do not produce L-lactate in a noticeable amount, but can uptake it via L-lactate dehydrogenase. L-lactate can be produced in *E. coli* by (1) removing *ldhA* gene to prevent D-lactate formation, (2) removing *lldD* gene to prevent the conversion of L-lactate to pyruvate, and (3) expressing heterologous L-lactate dehydrogenase gene to produce L-lactate from pyruvate, where this gene may be controlled by the native promoter of *ldhA* gene (Fig. 7). An attempt has been made by introducing and over-expressing L-lactate dehydrogenase gene from *Lactobacillus casei* with deletion of *pta* and *ldhA* genes. The resulting strain gave the titer of 45 g/l of L-lactate (Chang et al. 1999). The exclusive L-lactate production may be made by the above operations to such mutants as Δ*ack* Δ*pta* Δ*pps* Δ*pflB* Δ*dld* Δ*poxB* Δ*adhE* Δ*frdA* (B0013-070) (Zhou et al. 2011). L-LDH gene may be introduced from *Bacillus coagulans*, *Streptococcus bovis*, or *Lactobacillus casei* (Niu et al. 2014). In particular, L-lactate producing strain (B0013-070, Δ*ldhA::diflldD::Pldh-ldh*$_{Bcoa}$) (090B3) produced 144.2 g/l of L-lactate with no more than 1.2 g/l of by-products such as acetate, pyruvate, and succinate in 30 hours of fermentation using 25 l bioreactor, where the two-step fed-batch cultivation was made since the optimal temperatures for the cell growth and L-LDH activity are different. The cultivation was started by the aerobic batch culture with initial glucose concentration of 30 g/l for the cell growth during 10 hours at 30°C until the $OD_{600}$ becomes about 30 (about 11.4 g/l

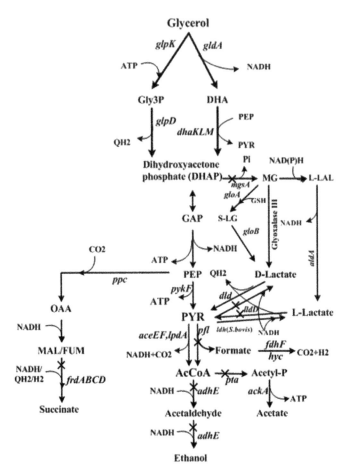

**Figure 7.** D-lactate and L-lactate forming pathways (Mazumdar et al. 2013). Pathways involved in the microaerobic utilization of glycerol and the synthesis of fermentation products in native and engineered *E. coli*. Genetic modifications supporting the metabolic engineering strategies employed in this work are illustrated by thicker lines (overexpression of *E. coli gldA-dhaKLM* and *glpK-glpD* and *S. bovis ldh*) or cross bars (disruption of *pflB, pta, adhE, frdA, ldhA, mgsA* and *lldD*). Relevant reactions are represented by the names of the gene(s) coding for the corresponding enzymes (*E. coli* genes/enzymes unless otherwise specified in parenthesis): *aceEF-lpdA*, pyruvate dehydrogenase complex; *adhE*, acetaldehyde/alcohol dehydrogenase; *ackA*, acetate kinase; *aldA*, aldehyde dehydrogenase A; *dhaKLM*, dihydroxyacetone kinase; *dld*, respiratory D-lactate dehydrogenase; *fdhF*, formate dehydrogenase, part of fomate hydrogenlyase complex; FrdABCD, fumarate reductase; *gldA*, glycerol dehydrogenase; *gloA*, glyoxalase I; *gloB*, glyoxalase II; *glpD*, aerobic glycerol-3-phosphate dehydrogenase; *glpK*, glycerol kinase; *hycB-I*, hydrogenase 3, part of formate hydrogenlyase complex; *ldh*, fermentative L-lactate dehydrogenase (*S. bovis*); *ldhA*, fermentative D-lactate dehydrogenase; *lldD*, respiratory L-lactate dehydrogenase; *mgsA*, methylglyoxal synthase; *pflB*, pyruvate formate-lyase; pta, phosphate acetyltransferase; *pykF*, pyruvate kinase. Abbreviations: DHA, dihydroxyacetone; DHAP, DHA phosphate; G-3-P, glycerol-3-phosphate; PEP, phosphoenolpyruvate; $P_i$, inorganic phosphate; PYR, pyruvate; $QH_2$, quinols; S-LG, S-lactoylglutathione.

of DCW), and then switched to anaerobic condition to enhance L-lactate production at 42°C, where the fed-batch feeding of the glucose was made during 20 hours until 30 h. The resulting L-lactate productivity and the yield were 6.77 g/*l*.h and 97% (g/g glucose), respectively (Niu et al. 2014).

As mentioned before, LA02Δ*dld* (pZ*glpKglpD*) over-produced D-lactate from glycerol. This strain was further engineered to produce L-lactate from glycerol by replacing the native D-lactate specific dehydrogenase gene *ldhA* by L-lactate dehydogenase (L-LDH) gene *ldh* obtained from *Streptococcus bovis*, where this gene was chromosomally integrated into LA02 (Δ*ptaΔadhEΔfrdA*) strain. Moreover, *mgsA* gene which encodes the methylglyoxal bypass pathways was removed to avoid the synthesis of a racemic mixture of D- and L-lactate, and also prevent the toxic metabolite such as methylglyoxal, yielding LA20 (Δ*ptaΔadhEΔfrdAΔmgsAΔldhA ldh⁺*).


The overexpression of the respiratory route of glycerol assimilation (GlpK/GlpD) and the disruption of the native L-lactate dehydrogenase in LA20 strain giving LA20Δ*dld* (pZ*glpKglpD*), which produced 50 g/*l* of L-lactate from 56 g/*l* of crude glycerol with the yield of 93% of the theoretical yield, with the optical purity of 99.9% and chemical purity of 97%, where 25-ml falsk was used with 200 RPM at 37°C. The calcium carbonate (5% wt/wt) was used to buffer the pH, where the fermentation was started with 40 g/*l* of crude glycerol (83.3% of glycerol), and another 20 g/*l* of glycerol was fed at 48 h, and the fermentation was continued until 84 h (Mazumdar et al. 2013).

A *ptsG*-negative but glucose-positive *E. coli* strain having heterologous L-lactate dehydrogenase gene with disruption of *pfl* and d-*ldh* genes was constructed, giving a yield of 0.77 g/g-sugar mixture of glucose and xylose, the titer of more than 800 mM without the production of acetate, ethanol, formate, and D-lactate (Dien et al. 2002).

Metabolically engineered *E. coli* strains that lack the genes encoding Pfl, Frd, ADH, and Ack, and possessing either D-LDH (strain SZ63) or L-LDH activity (strain SZ85) (Zhou et al. 2003) have been constructed. Introduction of the genes encoding an invertase, a fructokinase, an anion symport protein and a repressor protein to *E. coli* SZ63, for D-lactate fermentation, enabled the production of more than 500 mM D-lactate from sucrose and molasses with a yield of 0.95 g/g (Shukla et al. 2004). The inactivation of NADH dehydrogenase genes such as *ndh* and *nuo* in *adhE* and *pta-ackA* mutants led to a positive effect on the production of D-lactate (Yun et al. 2005).

As stated above, most of the lactic acid fermentation is made under micro-aerobic or anaerobic conditions, where by-product formation pathway genes must be knocked out to avoid mixed acid fermentation. The mixed acid formation can be avoided by the aerobic cultivation of the cell, where aerobic/anaerobic metabolisms are regulated by the redox control transcription factors such as ArcA/B and Fnr in relation to the electron transfer to oxygen by cytochrome oxidases. Removal of cytochrome oxidases (encoded by *cyoABCD*, *cydAB*, and *cbdAB* (*appBC*)) from *E. coli* K12 MG1655 reduces the oxygen uptake rate by about 85% (Portnoy et al. 2008). Although this knockout strain (ECOM3) was initially unable to grow on M9 minimal medium, the cell growth could be recovered up to that of the wild type strain cultivated under anoxic conditions (0.45 $h^{-1}$) after 60 days of adaptive evolution (ECOM31) (Portnoy et al. 2008). The specific glucose consumption rate of ECOM31 could be increased from 9.02 mmol/gDCW.h to 18.89 mmol/gDCW.h, while the oxygen uptake rate (OUR) decreased from 14.92 mmol/gDCW.h to 5.61 mmol/gDCW.h as compared to the wild type strain. The specific D-lactic acid production rate becomes 36.36 mmol/gDCW.h (D-lactate yield of 0.8 g/g glucose) from 0.40 mmol/gDCW.h, while byproduct (acetate) formation could be reduced from 3.4 mmol/gDCW.h to 0.50 mmol/gDCW.h (Portnoy et al. 2008). This strain shows 21-fold higher expression of *ldhA* gene, while the expression of *pflA* gene is low, and the expression of *aceF* is the similar as compared to the wild type strain (MG1655) (Portnoy et al. 2008). Significant up-regulation is seen for quinol monooxygenase gene *ygiN* as well as super oxide dismutase gene *sodAB*, indicating the remaining oxygen uptake, where the ubiquinol/ubiquinon cycle can be formed by the flux coupling between NADH dehydrogenase (*nuo/ndh* operons) and the reaction catalyzed by YgiN, which can potentially react with ubiquinol (UQOH) molecule and oxidize it to the ubiquinone (UQ) form through coupling of this oxidation reaction with reduction of the molecular oxygen (Fig. 8) (Adams and Jia 2005). Further removal of *ygiN* gene, therefore, almost completely eliminate oxygen uptake (Portnoy et al. 2008).

Removal of three cytochrome oxidases (*cyoABCD*, *cydAB*, *cbdAB* (*appBD*)) and quinol mono-oxigenase (*ygiN*) results in the reduction of oxygen uptake rate by nearly 98%, and the activation of anaerobic respiration under oxic condition, where the content of quinone pool shifts from ubiquinones (UQs) to menaquinones (MQs), which in turn activates ArcA, which causes the repression of the TCA cycle activity (Portnoy et al. 2010). In this case, the specific growth rate decreased from 0.71 $h^{-1}$ (K12 MG1655) to 0.32 $h^{-1}$, while the specific glucose uptake rate increased from 9.02 mmol/gDCW.h to 26.4 mmol/gDCW.h, and the lactate production rate was nearly 48.6 mmol/gDCW.h (0.98 g/g glucose) without acetate formation (3.37 mmol/gDCW.h in the wild type strain) (Portnoy et al. 2010). The reason why the specific glucose uptake rate was increased may be the disruption of the cytochromes in the respiratory chain as mentioned in Chaprer 4. In this strain, anaerobic respiration occurs by the respiratory chain formed by NADH: menaquinone oxide reductase (*yieF* and *wrbA*) and fumarate reductase (*frdABCD*), where electrons are transferred from NADH to fumarate to menaquinone pool (Yagi and Matsuno-Yagi 2003), resulting in the formation of succinate (Portnoy et al. 2010). Unlike the case

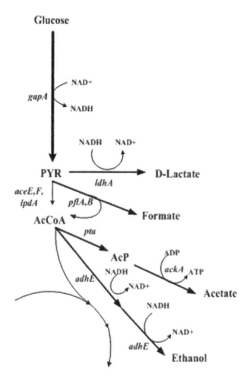

**Figure 8.** Mechanism of D-lactate production and associated gene expression analysis. Pathways of conversion of pyruvate to common organic acids are presented with corresponding enzyme names. Expression of the *ldhA*, *aceF*, and *pflA* genes was measured and is presented by the bar diagrams. Gene expression was measured under oxic (dark gray bars) or anoxic (light gray bars) conditions. *ldhA* showed a significant upregulation, while no upregulation was observed for the *aceF* and *pflA* genes (Portnoy et al. 2008).

of anaerobic conditions, Fnr is not active under oxic conditions, while ArcA is active as stated above, and thus *pfl* operon may not be active without producing formate in this strain, where *pfl* operon is co-regulated by both ArcA and Fnr.

*Corynebacterium glutamicum* has been used for the industrial production of amino acids such as glutamate and lysine, etc. (Herman 2003, Wendisch 2014) as will be explained later in this chapter. This strain produces L-lactate as well as succinate and acetate under oxygen-limiting conditions (Inui et al. 2004, Okino et al. 2005). The glycolytic and anaplerotic pathway fluxes as well as the reductive branch of the TCA cycle increase under oxygen deprivation conditions, where L-lactate is the main product by LDH encoded by *ldhA* with small amount of production of succinate and acetate (Inui et al. 2004). In particular, the glycolytic flux increased by about 1.5 fold under oxygen deprivation conditions as compared to aerobic conditions. By introduction of *ldhA* gene encoding D-lactate dehydrogenase of *Lactobacillus delbrueckii* into a *C. glutamicum* yielded 120 g/*l* of D-lactate with 99.9% optical purity, the yield of 1.73 mol/mol, and the productivity of 4.0 g/*l*.h (Okino et al. 2008). Noting that *ppc* gene knockout of this strain causes the reduction of the glycolytic flux, the simultaneous overexpression of *glk* encoding glucokinase, *gapA* encoding glyceraldehyde 3-phosphate dehydrogenase, *pfk* encoding phosphofructkinase, *tpi* encoding triose phosphate isomerase, and *fba* encoding fructose 1,6-bisphosphate aldorase enables the performance improvement as 195 g/*l* of D-lactate, the yield of 1.80 mol/mol-glucose, and the productivity of 2.4 g/*l*.h (during production phase) (Tsuge et al. 2015).

Saitoh et al. (2005) genetically engineered wine yeast that produced high concentrations of L-lactate with high optical purity. Another candidate may be Enterococcus isolate *Candida* sp. (Christopher et al. 2014, Subramanian et al. 2015).

## 2.3 Acetic acid fermentation

The *Acetobacter* and *Gluconobacter* species are commonly used for industrial vinegar production because of their ability to oxidize ethanol for acetate production and their resistant characteristics to acetic acid (Nakano et al. 2006). One of the mechanisms of acetic acid resistance may be the efflux pump in the cytoplasmic membrane, which pumps out acetic acid and enables the cell to grow in the presence of high acetic acid concentrations (Matsushita et al. 2005). The acetic acid bacteria produce vinegar fron alcohol-containing liquids such as wine, beet, etc. Ethanol is oxidized via alcohol dehydrogenase and acetaldehyde dehydrogenase to yield acetate, yielding the reducing power in the form of pyrroloquinoline quinol ($PQQH_2$) (Gottschalk 1988). The pyrroloquinoline quinone (PQQ) is the preferred $H^+$ acceptor of these hydrogenases and not $NAD^+$. The $PQQH_2$ generates proton-motive force (PMF) to synthesize ATP. The acetic acid bacteria utilize glucose dehydrogenase (GDH) for ATP production by forming PQQ, and then $PQQH_2$. Moreover, Pfk activity is quite low, and thus the glucose is metabolized exclusively through the oxidative PP pathway, and the glycolysis flux is low with low flux of Ppc, resulting in the activation of another anaplerotic pathway such as glyoxylate pathway (Sarkar et al. 2010). When ethanol is present in addition to glucose, the cell also produces $PQQH_2$ at ADH reaction as well as GDH. When glucose and ethanol were used as carbon sources, the glyoxylate pathway is less active, and the TCA cycle is more active than the case of using glucose and acetate as carbon sources. This also causes less flux through OAA decarboxylase (OAADC) for the case of using glucose and ethanol as carbon sources, where OAA is supplied from pyruvate originated from ethanol (Sarkar et al. 2010). The

TCA cycle becomes less as the ethanol concentration increases, where this may be due to the repression of aconitase (Nakano et al. 2004).

## 3. Production of TCA Cycle Intermediates

The US Department of Energy (DOE) has identified malic acid, fumaric acid, and succinic acid as potential building block (bulk) chemicals among 12 most valuable chemicals (Werpy and Petersen 2004). Microbial production of four carbon 1,4-dicarboxylic acids such as malate, fumarate, and succinate have been paid recent attention (Fig. 9), but currently, all of these C4 diacids are mainly produced by the petrochemical processes due to cheap cost. The trends toward green chemicals production and future petroleum shortage require the development of innovative processes that are cost effective, energy saving, and environmentally friendly.

The intermediates of the TCA cycle such as citric, fumaric, and succinic acids are localized in mitochondria in eukaryotes. These organic acids are synthesized and excreted in the medium at relatively high concentrations by filamentous fungi such as *Aspergillus* spp. and *Rhizopus* sp. (Goldberg et al. 2006). Itaconic acid is produced from aconitate in the TCA cycle in *Aspergillus* spp. In filamentous fungi, the citric, fumaric, and malic acids are accumulated under specific stress conditions. The overproduction of fumaric and L-malic acids is made by a separate and unique pathway such as the reductive TCA pathway, localized in the cytosol (Goldberg 1991). Citric acid is, on the other hand, produced via citrate synthase (CS), an enzyme in the mitochondrial oxidative TCA cycle, yet citosolic L-malic acid is an intermediate in citric acid synthesis.

One of the reasons why filamentous fungi accumulate organic acids in the medium is the presence of pyruvate carboxylase (Pyc) in the cytosol, in contrast to eukaryotic and mammarian cells, where it is located in the mitochondria (Bercovitz et al. 1990, Attwood 1995), while the reductive TCA cycle enzymes such as malate dehydrogenase (MDH) and fumarase (Fum) are located in the cytosol in higher eukaryotes, that may be involved in the scavenging of the organic acids in the cytosol (Goldberg et al. 2006).

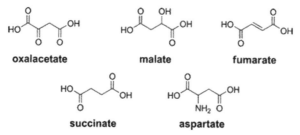

oxalacetate      malate      fumarate

succinate      aspartate

**Figure 9.** Structures of the five kinds of four-carbon 1,4-dicarboxylic acids which are naturally produced by *E. coli* (Cao et al. 2011).

Pyc is a biotin-dependent tetrametric enzyme that catalyzes the carboxylation of pyruvate to yield OAA. In general, Pyc is localized in mitochondria in eukaryotic organisms, while it is located in the cytosol in certain filamentous fungi and in *S. cerevisiae*, where it is important for the ability of such microorganisms to produce organic acids. The fumaric and/or L-malic acids can be over-produced by modulating Pyc activity by such effectors as biotin and avidin (Kenealy et al. 1986, Schwartz and Radler 1988).

MDH catalyzes NADH-dependent reversible conversion of L-malic acid to OAA. There are at least two forms, where a mitochondrial MDH (MDH1) functions in the oxidative TCA cycle, while a cytosolic MDH (MDH2) involves in the gluconeogenesis for pyruvate formation. Another MDH (MDH3) is located in peroxysomes in *S. cerevisiae*. The ability of MDH is related to the L-malic acid production in *Aspergillus* strains (Goldberg et al. 2006).

Fumarase catalyzes the reversible hydration of fumaric acid to L-malic acid, where two classes of fumarases are present depending on the organisms (Moir et al. 1984, Wu et al. 1987, Woods et al. 1988, Suzuki et al. 1989, Acuna et al. 1991, Colombo et al. 1994, Friedberg et al. 1995), where class I fumarases are oxygen-labile homodimeric fumarase hydratases, while class II fumarases are stable homotetrametric enzymes. In *E. coli*, these enzymes are encoded by *fumABC*, where FumA and FumB belong to class II, while FumC belongs to class I (Woods et al. 1988). Members of class I have been characterized in *Euglena gracilis* (Shibata et al. 1985), *Bacillus stearothermophilus* (Reaney et al. 1993), as well as in *E. coli*, while class II fumarases have been characterized in other organisms. The fumarases in *S. cerevisiae*, rat, and human are encoded by a single gene, yet the enzymes are present in both cytosol and mitochondria. The fact that L-malic acid accumulates, whereas fumaric acid does not may be explained by the activity of cytosolic fumarase, which catalyzes the conversion of fumaric acid to L-malic acid, and not the reverse reaction (Pines et al. 1996). Thus, the cytosolic enzyme scavenges fumaric acid to L-malic acid. This seems to occur in yeast and probably also in *A. flavus* (Battat et al. 1991).

On the other hand, fair amount of fumaric acid is produced in *Rhizopus* strains (Goldberg et al. 1985, Kenealy et al. 1986, Foster et al. 1949). There might be several reasons for this. One possibility is that the cytosolic fumarase is kinetically different from the mitochondrial isozymes, due to post-translational modifications, or due to specific conditions in the two compartments in *Rhizopus* spp. Another possibility is that *Rhizopus* spp. harbor two genes encoding two different fumarases, one in mitochondria, which catalyzes the conversion of fumaric acid to L-malic acid, while another one in cytosol, which catalyzes the reverse conversion to form fumaric acid (Goldberg et al. 2006).

In the cultivation of fungi, the cell growth declines, while organic acids such as fumaric acid and/or L-malic acid are formed in response to nitrogen limitation. In addition to nitrogen limitation, $CaCO_3$ is required for the high concentrations of acid salt, where it ensures culture pH to retain around 6.0, provides $CO_2$ for the reaction at Pyc, and also causes precipitation of calcium salts of the acids, thus moving the reaction equilibrium towards the acid products (Goldberg et al. 2006).

On the other hand, the culture condition for citric acid production by *A. niger* is different, such as low manganese concentration, low pH, etc. (Krebs 1970, Demain and Sanchez 2006). Citric acid is produced in acid form, whereas fumaric and L-malic acids are produced as calcium salts. Namely, citric acid production is typically made by *A. niger* using a medium with high nitrogen concentration, low pH without ferrous ions, while L-malic acid is produced by *A. flavus* using a medium with low nitrogen concentration, at relatively higher ferrous ion concentration, containing biotin and $CaCO_3$ (Goldberg et al. 2006).

## 3.1 Citric acid production

Citric acid (2-hydroxypropane-1,2,3-tricarboxylic acid) was first isolated from lemon juice by Carl Wilhelm Scheele in 1784, where it is a symmetric tricarboxylic acid, and its name originates from the Latin word *"citrus"*, the citron tree, the fruit of which resembles a lemon (Mattey and Kristiansen 1999). In view of its three carboxylic acid functional groups, it has three pKa values at pH 3.1, 4.7, and 6.4. Citric acid is a natural constituent of a variety of citrus fruits, pineapple, pear, peach and fig, and is the important food acidulant. It has been typically used in candies, jellies, jams, gelatin desserts, soft drinks, fruits and vegetable juices, wines, ciders, frozen fruits, dairy products, animal fats, oils, pharmaceuticals, cosmetics, and toiletries (Grewal and Karla 1995).

Commercial production of citric acid started around 1826 in England by extracting the acid from Italian lemons (Goldberg et al. 2006). After the first finding of citric acid accumulation in *Citromyces* (now identified as *Penicilium*), many microorganisms such as *A. niger*, *A. awamori*, *A. fluvus*, etc. have been also shown to accumulate citric acid (Papagianni 2007). In 1917, Currie (1917) discovered that some *A. niger* strains grew and excreted large amounts of citric acid with high glucose concentration and at low pH around 2.5–3.5, where the formation of oxalic and gluconic acids was suppressed at low pH. The industrial citric acid fermentation using *A. niger* began in 1919 in Belgium (Mattey and Kristiansen 1999), and the citric acid production is also made by Pfizer in 1923 in USA (Papagianni 2007). The size of the citric acid market is massive, and continuously growing, largely because of the expanding food and beverage markets in developing countries, and increasing use of citric acid salts (e.g., calcium salt) in health related consumer products (Goldberg et al. 2006).

Although fungi such as *Aspergillus* spp. and *Rhizopus* sp. have been commercially used in large scale production, yeast species such as *Yarrowia, Candida, Hansenula, Pichia, Debaromyces, Torula, Torulopsis, Kloekera, Saccharomyces, Zygosaccharomyces* strains have also been used since 1960s (Yamada 1977, Papagianni 2007). The advantages of using yeast as compared to filamentous fungi are (1) more tolerant to high glucose concentrations, (2) higher conversion rates of substrates to acids, (3) less sensitive to metal ions, (4) capable of utilizing a wider range of carbon sources (Zarowska et al. 2001). The citric acid production in yeast starts under nitrogen limitation, where fair amount of isocitric acid (unwanted by-product) is also produced. The production rate of citric acid by yeast is lower, and in particular, *Y. lipolytica* cannot utilize molasses and whey (Zarowska et al. 2001), and therefore, *A. niger* is exclusively used in citric acid fermentation, and remain the main industrial producer (Anastassiadis et al. 2002).

For citric acid formation, two moles of pyruvate is formed from one mole of glucose as well as other hexose sugars, where one mole of pyruvate is converted to AcCoA by decarboxylation at PDH, releasing one mole of $CO_2$, while the other mole of pyruvate is converted to OAA by fixing one mole of $CO_2$ at Pyc reaction. These precursor metabolites such as AcCoA and OAA are condensed to form one mole of citric acid by citrate synthase (CS) (Fig. 10), where aconitase (Acn) must be inhibited by modulating culture conditions such as removal of iron, an activator of aconitase.

The typical citric acid production by yeasts starts only after nitrogen source limitation some time at the end of the growth phase, where the TCA cycle is largely inoperative, and the anaplerotic supply of OAA via Pyc pathway is important (Yalcin et al. 2010). The culture pH also affects citric acid production as well as its transport, where the optimum pH is around 5, and the citric acid production is adversely affected below pH at 5 due to accumulation of some polyalcohols such as erythritol, arabitol, and mannitol (Mattey 1992).

In *A. niger*, the OAA formed via Pyc is not directly converted to citrate by the mitochondrial CS, but is first reduced to L-malic acid by the cytosolic MDH, thereby entering the mitochondria via a malate-citrate antiport (Fig. 10) (Kubicek 1988, Pekset et al. 2002, Kraffa and Kubicek 2003). The yield of citric acid by *A. niger* strains often exceeds 70% of the theorectical yield on the carbon source.

Since citric acid production is an aerobic process, the aeration and agitation also affect the citric acid production (Yalcin et al. 2010). In *A. niger*, the overexpression of the cytosolic fumarases FumR from *Rhizopus oryzae* and fumarate reductase (Frds1) from *S. cerevisiae* gave the maximum yield of 0.9 g/g glucose, and a maximum specific productivity of 0.025 g/gDCW.h in the presence of trace manganese.

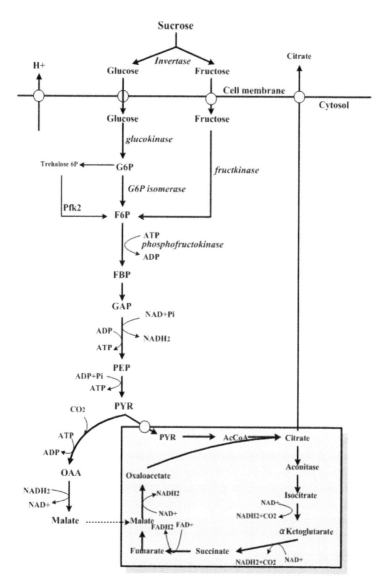

**Figure 10.** Schematic representation of the metabolic reactions involved in citric acid production, the enzymes (italics), the known feedback loops (dashed lines) and their locations within the cellular structure of *A. niger* (Papagianni et al. 2007).

As mentioned in Chapter 4, the glycolytic flux can be increased by modulating the respiratory chain, giving low ATP production, and by modulating NADH/NAD+ ratio by overexpression of NADH oxidase. The increase in the glycolytic flux directly affects the citric acid production. This may be attained by adding antimycin A, an inhibitor of oxidative phosphorylation, which inhibits succinate-cytochrome c reductase in the electron transfer chain to block NADH oxidation and ATP synthesis. This is effective for *Torulopsis glabrata* CCTCC M202019, where ATP concentration decreased to 27.7%, while the glucose consumption rate increased to 240% (Liu et al. 2006).

Another way is to use 2,4-dinitrophenol (DNP), a typical uncoupler of oxidative phosphorylation (Abo-Khatwa et al. 1996), that separates the coupled process of NADH oxidation and ATP synthesis into two independent processes, where DNP does not influence on the electron transfer chain but blocks ATP synthesis, thus the accumulated ADP promotes substrate level phosphorylation and the glycolytic flux is accelerated as shown in *E. coli* (Dietzler et al. 1975).

By the appropriate amount of addition of either antimycin A (0.2 mg/*l*) or DNP (0.1 mg/*l*) at 24 h-time point in the cultivation of *A. niger* CGMCC 5751 and *A. niger* ATCC 1015, the citric acid production could be increased by 19.89% and 7.32%, respectively (Wang et al. 2015).

In the submerged culture of filamentous fungi, the morphological forms such as clumps, filamentous, and pellets change in response to culture condition (Xu et al. 2012). *Aspergillus niger* preferentially synthesizes enzymes during filamentous growth (Driouch et al. 2010), whereas citric acid is accumulated in pellet form. The changes in morphology affect nutrient consumption, oxygen uptake, and also affect broth rheology. Cells with clump morphology suffer from serious internal oxygen limitation, resulting in low growth rate, and thus the clump morphology may be avoided in fermentation processes. Filamentous growth with highly branched mycelia is favorable for the production of enzymes such as amylase and lipase, but causes the broth to be viscous, limiting the oxygen and mass transfer. On the other hand, the broth viscosity can be reduced by changing the cell morphology to pellets, thus improving the mass and oxygen transfer, and reduce the power consumed in mixing and aeration. Therefore, pellets may be the most suitable and favorable morphological form for fungal organic acid production (Xu et al. 2012).

### 3.2  Malic acid production

Malic acid was first discovered by Carl Wilhelm Scheele in 1785 from apple juice (Meek 1975), where this name originates from the Latin word "*malum*" meaning apple (Goldberg et al. 2006). Malic acid (2-hydroxybutandioic acid, 2-hydroxy succinic acid) is a dicarboxylic acid, and it has an asymmetric carbon position. It is abundant in fruits and vegetables, and is currently produced by petrochemical processes at higher temperature and pressure through either the hydration of fumaric or maleic acid (Werpy and Petersen 2004), where a racemic mixture of D- and L-malate is formed. These processes require petroleum derived maleic unhydride of fumaric acid.

Malic acid is typically used in food industry as an acidulant for beverages (Takata et al. 1980, Takata and Tosa 1993, Bressler et al. 2002, Goldberg et al. 2006), metal cleaning, textile finishing, pharmaceuticals (Goldberg et al. 2006, Vrsalovic Presecki et al. 2009), and the synthesis of biodegradable polymer, polymalic acid (PMA) (Gross and Rhano 2002, Tsao et al. 1999, Zou et al. 2013). Malic acid can be produced from renewable biomass, and can be used to produce various products such as polymer resins (Werpy and Petersen 2004).

Malic acid was identified in a product of yeast fermentation in 1924 (Dakin 1924). Since then, a variety of microorganisms have been used for malic acid production. Fermentative production of malic acid started in the early 1960s using the natural malate producer *Aspergillus flavus* (Abe et al. 1962). The performance of such fermentation has been improved to 63% of the maximum theoretical yield (1.28 mol/mol glucose) with relatively higher production rate (0.59 g/l.h) and a titers (113 g/) (Table 2) (Battat et al. 1991). However, this filamentous fungus produces mycotoxin, aflatoxin (Hesseltine et al. 1966, Geiser et al. 1998), which prevented the use of this microorganism for large-scale industrial application (Hedayati et al. 2007). *Aspergillus oryzae* is a very close relative to *A. flavus*, but does not produce mycotoxin, and therefore, this microorganism has been used for malic acid production (Brown et al. 2013, Knuf et al. 2013, 2014).

*A. oryzae* NRRL 3488 produces malic acid through the reductive TCA branch such as a two-step cytosolic pathway of Pyc-MDH from pyruvate via OAA with the theoretical yield of 2 mol/mol glucose. Thus the overexpression of the native cytosolic alleles of *pyc* and *mdh* as well as a native C4-dicarboxylate transporter (C4T318) achieved a malate titer of 154 g/l, a productivity of 0.94 g/l.h with a yield of 1.38 mol/mol glucose (69% of the theoretical maximum yield) (Brown et al. 2013).

The stress condition such as nitrogen starvation may affect the metabolism of *A. oryzae* NRRL 3488 during the transition from the exponential growth phase to the stationary phase, and affect malic acid production. At the stationary phase, cells respond to nitrogen starvation by recycling nitrogen by degradation of proteins, etc., where the ubiquitination of proteins might be responsible for this, which requires the significant up-regulation of *pyc*, and thus Pyc might be the important target of the limiting pathway for malic acid production (Knuf et al. 2013). As nitrogen depletion hampers the cell growth, the ATP generation in the oxidative phosphorylation as well as reducing equivalents are not needed, where the malic acid production from hexose sugars can be made through the reductive cytosolic TCA branch from pyruvate via OAA. This indicates that the malic acid production can be uncoupled from the cell growth, which is the important characteristic in industrial application. Moreover, *A. oryzae* is tolerant to acidic pHs ranging from 3 to 7, where malic acid production is efficient in its acidifying potential with pKa's such as 3.46 and 5.10 (Knuf et al. 2013). The switch from the ATP production for the cell growth to malic acid secretion may be regulated by the transcription factor similar to *S. cerevisiae* Msn2/4 for the stress conditions such as nitrogen starvation (Knuf et al. 2013). The engineered *A. orzae* 2103a-68 strain overexpressing *pyc*, *mdh*, and the putative malic acid transporter C4T318 yielded a maximum specific malate production rate during stationary phase of 1.87 mmol/gDCW.h, with a yield of 1.49 mol/mol glucose (Knuf et al. 2014). Moreover, compared with *S. cerevisiae* (Shen et al. 2012), *C. glutamicum* (Buschke et al. 2011), or *E. coli* (Liu et al. 2012), *A. oryzae* could assimilate other sugars such as xylose without any engineering, where glucose repression may occur (Knuf et al. 2014).

Although *Aspergillus* species have been used for malic acid production as stated above (Abe et al. 1962, Battat et al. 1991, Knuf et al. 2014, West 2011), other microorganisms such as *S. cerevisiae* (Zelle et al. 2008), *E. coli* (Moon et al. 2008, Zhang et al. 2011b), *Zygosaccharomyces rouxii* (Taing and Taing 2007), *Schizophyllum commune* (Kawagoe et al. 1997), and *Aureobasidium pullulans* (Zou et al. 2013) have been also investigated (Table 2).

The wild type *S. cerevisiae* is generally regarded as safe (GRAS), and has a relatively high tolerance to organic acids and low pH. Although the wild type *S. cerevisiae* produces only a little amount of malate, rational design of its metabolism may allow industrial production of food-grade malic acid (Zelle et al. 2008, 2010). There are mainly four metabolic pathways for the production of L-malic acid from glucose (Fig. 11) (Zelle et al. 2008). The first one is to produce oxaloacetate (OAA) from pyruvate by Pyc (*S. cerevisiae* does not have Ppc), followed by the reduction of OAA to malate via MDH. This non-oxidative pathway is ATP neutral with $CO_2$ fixation at Pyc, resulting in a maximum theoretical yield of 2 mols malate/mole glucose, where NADH produced at GAPDH can be reoxidized at MDH reaction. The second pathway is through the TCA cycle via the condensation of OAA and AcCoA at citrate synthase (CS), where the maximum yield by the oxidative pathway reduces to 1 mol of malate/mol glucose, where 3 moles of $CO_2$ are released at PDH, ICDH, and KGDH, while one mole of $CO_2$ is fixed at Pyc. The third pathway is via the glyoxylate pathway, where the maximum yield becomes as 1 mol malate/mol glucose, where $CO_2$ is released at PDH reaction through this alternative oxidative pathway. Fourth pathway is the noncyclic pathway that involves the glyoylate pathway enzymes, and OAA is replenished by Pyc, giving the theoretical maximum yield of 1 plus 1/3 mole malate/mol glucose (Zelle et al. 2008).

Metabolic engineering of *S. cerevisiae* for the production of organic acids requires the elimination of ethanol formation through PDC-ADH pathway, where ethanol is produced even under fully aerobic conditions and at higher glucose concentrations by the so-called Crabtree effect (Postma et al. 1989). Although removal of the three pyruvate decarboxylase genes such as *PDC1*, *PDC5*, and *PDC6* completely shut down the ethanol producing pathway (Hohmann 1991), the resulting engineered strain becomes $C_2$ compound auxotrophy and intolerance to high glucose concentration (Flikweert et al. 1996, 1999). The $C_2$-independent and glucose-tolerant pyruvate decarboxylase-negative (Pdc⁻) strains could be obtained by evolutionary engineering, where pyruvate is substantially produced (van Maris et al. 2004).

The Pdc⁻ *S. cerevisiae* strain can be used for malic acid production by the overexpression of the native Pyc encoded by *PYC2*, high-level expression of an allele of the *MDH3* gene encoding malate dehydrogenase, and functional expression of the *Schizosaccharomyces pombe* malate transporter gene *MAE1* (Zelle et al. 2008). The resulting strain gave the titer up to 59 g/l with a yield of 0.42 mol/mol glucose, where this strain still produced substantial amount of pyruvate (Zelle et al. 2008). The yield could be further increased

(a)

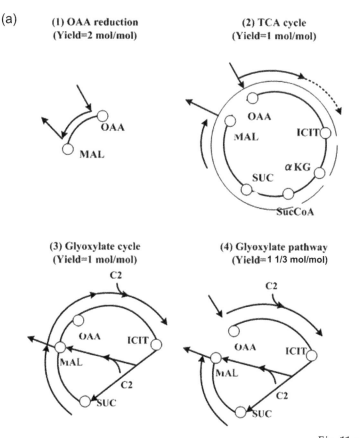

(1) OAA reduction
(Yield=2 mol/mol)

(2) TCA cycle
(Yield=1 mol/mol)

(3) Glyoxylate cycle
(Yield=1 mol/mol)

(4) Glyoxylate pathway
(Yield=1 1/3 mol/mol)

*Fig. 11 contd. ...*

*...Fig. 11 contd.*

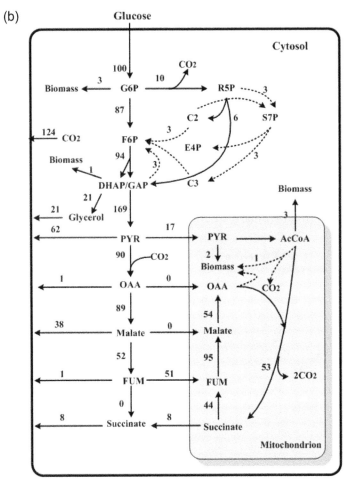

**Figure 11.** (a) Four possible pathways for malate production in *S. cerevisiae*, using oxaloacetate and/or acetyl-CoA as precursors. (I) Direct reduction of oxaloacetate. (II) Oxidation of citrate via the TCA cycle. (III) Formation from acetyl-CoA via the cyclic glyoxylate route. (IV) Formation from acetyl-CoA and oxaloacetate via the noncyclic glyoxylate route. OAA, oxaloacetate; MAL, malate; CIT, citrate; ICI, isocitrate; AKG, α-ketoglutarate; SUCC, succinyl-CoA; SUC, succinate; FUM, fumarate; C2, acetyl-CoA; Yspmax, maximum theoretical yield (in mol malate per mol glucose) (Zelle et al. 2008). (b) Flux profiles for three highlighted situations. Fluxes (in moles) are normalized for a glucose uptake of 100 mol. Arrows indicate the direction of the (net) flux. Some arrows are dashed to improve readability (Zelle et al. 2008).

up to 0.48 mol/mol glucose by using 1-*l* bioreactor with the optimization for pH, concentrations of $CO_2$, calcium, and $O_2$ (Zelle et al. 2010).

*Escherichia coli* has proven to be an excellent biocatalytic platform for metabolic engineering. *E. coli* does not possess Pyc but Ppc, where the amplification of Ppc pathway is critical for malic acid production (Moon et al. 2008). Since Ppc converts PEP to OAA without generating ATP, losing the high-energy phosphate bond of PEP, it is preferred to utilize Pck, where ATP is used to convert OAA to PEP for gluconeogenesis, and therefore, it is better to force the reaction in the reverse reaction instead of utilizing Ppc. This may be attained by introducing the *pckA* gene of *Mannheimia succiniciproducens*, which encodes Pck that converts PEP to OAA (Fig. 12). Under aerobic conditions, fumarase encoded by *fumA,C* and succinate dehydrogenase encoded by *sdhCDAB* are active, and the TCA cycle functions to convert OAA to malic acid via succinic acid, where the malic acid is not converted further to succinic acid because the anaerobic fumarase encoded by *fumB*, and fumarate reductase encoded by *frdABCD* are not expressed under aerobic condition. Under anaerobic conditions, malic acid is converted to succinic acid as the final product due to the activation of *fumB* and *frdABCD*. The *pta* mutant *E. coli* harboring the plasmid containing *M. succiniciproducens pckA* gene gave the relatively higher productivity of 0.74 g/l.h, but with relatively lower titer of 9.25 g/l and a yield of 0.42 g/g glucose under aerobic condition (Moon et al. 2008).

In order to improve the titer of malic acid, *E. coli* may be cultivated under anaerobic conditions, where a combination of gene deletions is necessary such as Δ*ldhA*, Δ*adhE*, Δ*ackA*, Δ*focA*, Δ*pflB* (KJ060) to avoid by-product formation. The resulting strain still produces some lactate, and further mutation such as KJ060 with Δ*mgsA* Δ*poxB* (KJ073) must be made to minimize lactate and acetate production (Jantama et al. 2008). Under anaerobic conditions, the cell growth of such mutants becomes low, giving low productivity of malic acid, and thus the metabolic evolution may be made to improve the cell growth rate, where the above mutants produced both succinic acid and malic acid (Jantama et al. 2008). Among the mutants, KJ073 strain produced 516 mM of malic acid with a yield of 1.4 mol/mol glucose. In order to reduce the succinate production, fumarate reductase gene (*frd*) may be inactivated. Moreover, malic acid consuming pathways such as malic enzymes encoded by *maeB* and *sfc* may be removed. However, this mutant (KJ073) with Δ*frdBC*, Δ*sfcA*, Δ*maeB*, Δ*fumB*, Δ*fumAC* (XZ658) still accumulates D-lactate despite the deletion of *mgsA* and *ldhA* genes (Zhang et al. 2011). As far as PTS is operative, pyruvate accumulates, and may cause lactate production. This may be reduced by pyruvate kinase gene knockout such as Δ*pykF* and/or Δ*pykA* (Zhang et al. 2011) (Fig. 13). However, many gene knockouts as state above affect the cell growth in the minimal medium under anaerobic conditions, and volumetric production

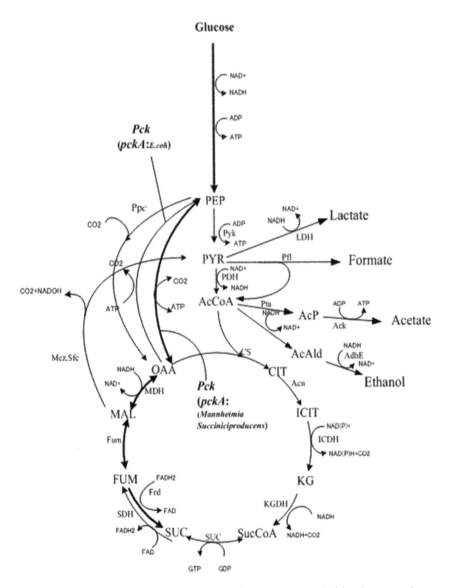

**Figure 12.** Central metabolic pathways in *E. coli*. Enzymes encoded by the genes shown are: aceE/F, pyruvate dehydrogenase multienzyme complex; acnA/B, aconitase; adhE/C, alcohol dehydrogenase; fumA/B/C, fumarase; gltA, citrate synthase; icdA, isocitrate dehydrogenase; ldhA, d-lactate dehydrogenase; mdh, malate dehydrogenase; pckA, phosphoenolpyruvate carboxykinase; ppc, phosphoenolpyruvate carboxylase; pta/ackA, phosphate acetyltransferase/acetate kinase; ptsG, an phosphotransferase system enzyme; pykA, pyruvate kinase; sfcA, malic enzyme; sdhA/B/C/D, succinate dehydrogenase; sucA/B-lpdA, α-ketoglutarate dehydrogenase complex; sucC/D, succinyl-CoA synthetase (Moon et al. 2008).

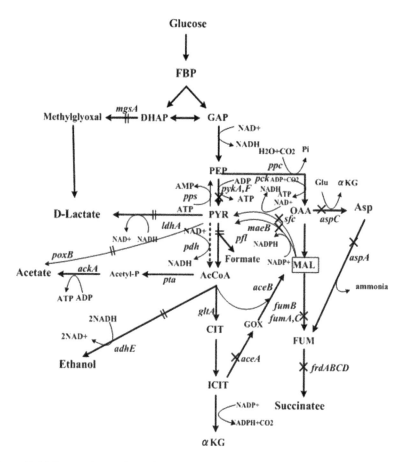

**Figure 13.** Malate production by pathway engineering. Succinate can be produced from OAA through either an aspartate bypass (aspartate aminotransferase and aspartase) or by using the glyoxylate bypass (citrate synthase, aconitate hydratase, and isocitrate lyase). Original gene deletions in the succinate-producing parent (KJ073) are marked by I I. Sites of new deletions described in this paper are also marked (×) (Zhang et al. 2011).

rate remains to be low. The XZ658 strain produced 163 mM of malic acid, with a yield of 1 mol/mol glucose. In order to improve the productivity, a two-stage fermentation may be considered, where the cell growth can be enhance under aerobic condition during the initial 17 h (25 gDCW/l), and then switched to anaerobic conditions to enhance malic acid production until 72 h, where 253 mM of malic acid could be produced with a yield of 1.42 mol/ mol glucose (Table 2) (Zhang et al. 2011).

Although malic acid fermentation by genetically engineered *S. cerevisiae* and *E. coli* may be attractive as stated above, the titer is low less than about 60 g/l (Table 2). This is because malic acid with a low pKa of 3.4 may inhibit the cell growth and hamper the fermentation (Zou et al. 2013). Although metabolic and evolutionary engineering may be useful for malic acid production as mentioned above, the significant performance improvement requires dramatic increases in acid tolerance to improve the titer, which remains a challenge for malic acid fermentation (Cao et al. 2011), where microbial malic acid production has not been industrialized so far yet.

*Aureobacidium pullulans* produces pullulan, poly-malic acid (PMA), xylanase, heavy oil, and lipase (Liu et al. 2008, Manitchotpisit et al. 2011, 2012, Wu et al. 2012). Some *A. pullulans* strains synthesize PMA from monomer L-malic acid and secrete into the culture broth (Gao et al. 2012a, 2012b, Leathers and Manitchotpisit 2012, Liu and Steinbuchel 1997, Manitchotpisit et al. 2012, Zhang et al. 2011a). Unlike malic acid, PMA is not toxic to the cell, and can be easily separated from the fermentation broth by ethanol precipitation (Holler 2010). Although the formation of by-products such as pullulan and organic acids may lower the PMA yield, and also cause burden to the down stream processing (Singh and Saini 2008), a new strain such as *A. pullulans* ZX-10 has been isolated with higher yield and purity together with higher product titer. Separation and purification of PMA may be made by adsorption by anion exchange resins, and acid hydrolysis of PMA to malic acid can be also made (Zou et al. 2013).

### 3.3 Fumaric acid fermentation

Fumaric acid was isolated from plant *Fumaria officinalis*, from which its name originates (Roa Engel et al. 2008). Fumaric acid (2-butenedioic acid *trans*; 1,2-ethylenecarboxylic acid) is a symmetric dicarboxylic acid, and is currently produced chemically from maleic anhydride, which is produced from butane in petrochemical processes. Fumaric acid is 1.5 times more acidic than citric acid, and therefore, it is commonly used as a food acidulant and beverage ingredient (Yang et al. 2011). Fumaric acid has a double bond and two carboxylic groups, and therefore, can be used as starting material for polymerization and esterification reactions in the production of resins. Fumaric acid is non-toxic and provides special properties like hardness in the polymer structure as compared to other carboxylic acids. Other than the application to polymerization, fumaric acid can be also used as a medicine to treat psoriasis, a skin condition (Altmeyer et al. 1994, Mrowietz et al. 1998), where psoriatic individuals are unable to produce fumaric acid in the skin. Thus, they need to take orally fumaric acid in the form of fumaric acid monoethyl or dimethyl ester to treat such disease. Another application is the supplement for the cattle feed in the diet (Mcginn et al. 2004). Compared to the dicarboxylic acids such as malic acid and succinic acid, fumaric acid has a low aqueous solubility (7 g/kg at 25°C, 89 g/kg at 100°C) (Stephen 1965), and less $pK_a$ values (3.03 and 4.44) (Lohbeck et al. 1990), which are preferred from the point of view of product recovery due to low solubility (Roa Engel et al. 2008). In general, the prices of L-malic acid and L-aspartic acid are 1.5–2 times that of fumaric acid, and thus fumarate may be also used as substrates (sodium fumarate or ammonium fumarate) for such production (Goldberg et al. 2006), where L-aspartic acid is an essential raw material for nutritive sweetner known as aspartame (Ager et al. 1998).

Microbial fumaric acid production may be originated in 1911, when Ehrlich found its production by *Rhizopus nigricans* (Xu et al. 2012). Among formate producing species such as *Rhizopus, Mucor, Cunninghamella,* and *Circinella* (Foster and Waksman 1938), *Rhizopus* species have been the best producer of fumaric acid, and have been extensively used (Straathof and Gulik 2012, Wang et al. 2013, Zhou et al. 2014, Gu et al. 2014, Goldberg and Steiglitz 1985, Zhou et al. 2002, Fu et al. 2010, Xu et al. 2010, Ding et al. 2011, Roa et al. 2011, Zhang et al. 2012, Gu et al. 2013).

Several companies such as DuPont attempted to produce fumaric acid using *R. nigrican* and *R. arrhizus* from 1940s to 1970s in industrial scale (Roa Engel et al. 2008). However, such microbial fermentation has been replaced by chemical synthesis from petrochemical feedstock due to cheap price (Xu et al. 2012). Among *Rhizopus* species, *R. nigricans, R. formosa, R. arrhizus,* and *R. oryzae* efficiently produce fumaric acid (Carta et al. 1999, Foster and Davis 1948, Liao et al. 2007, Rhodes et al. 1959). Although *R. arrhizus* gives the high fumarate concentration such as 121 g/$l$, this strain gives low yield (0.37 g/g) (Ling and Ng 1989). Moreover, this strain requires rich nutrients, whereas *R. oryzae* can grow with simple media. Therefore, *R. oryzae* has been considered to be the main producer of fumaric acid after 1990s (Table 2) (Xu et al. 2012). In *R. oryzae*, the pyruvate generated through glycolysis is converted to lactic acid, ethanol, malic acid, fumaric acid, as well as biomass (through TCA cycle) (Fig. 14). The fermentation patterns change depending on the culture conditions, where fumaric, malic, and lactic acids are the main metabolites under aerobic condition (Saito et al. 2004, Abe et al. 2007), while ethanol is the main metabolite under oxygen limitation (Magnuson and Lasure 2004). *Rhizopus* spp. are aerobic microorganisms, where an excess amount of oxygen enhances the cell growth, while the limited oxygen enhances the ethanol formation. DuPont patented a fermentation strategy for the efficient fumaric acid production by limiting the dissolved oxygen (DO) levels between 30% and 80%, achieving a fumaric acid production of 130 g/$l$ in 142 h by *R. arrhizus* NRRL1526, with the production of by-products such as succinic acid, L-malic acid, and α-ketoglutaric acid (Ling and Ng 1989). Noting that higher DO levels enhance the biomass formation, a two-stage fermentation may be considered, where DO concentration was controlled at 80% in the initial 18 h to enhance the cell growth, and then switched to 30% for fumaric acid production, which yielded 56.2 g/$l$ in 80 h using 5-$l$ fermentor, with the yield of 0.54 g/g glucose, and

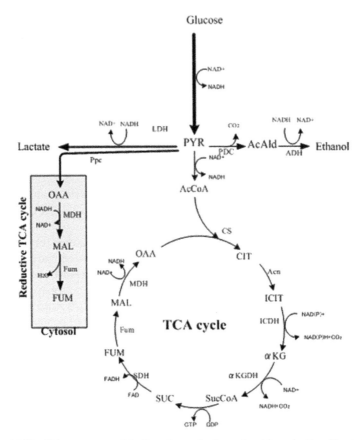

**Figure 14.** The *Rhizopus* spp. metabolic pathway for fumaric acid production (Xu et al. 2012).

the productivity of 1.3 g/l.h, while 7.4 g/l of ethanol was produced (Fu et al. 2010). As for the effect of aeration/agitation on the morphology, the pellet size decreases, while its number increases as the agitation rate was increased (Bai et al. 2003).

Another important culture condition may be the nitrogen concentration, where fumarase activity increased, and fumarate production is enhanced to give a titer of 40.3 g/l (from 14.4 g/l) in 72 h with the yield of 0.51 g/g glucose, while L-malic acid decreased from 2.1 g/l to 0.3 g/l under nitrogen limitation by changing the urea concentration from 2 to 0.1 g/l, whereas the cell growth declined as compared to the case of nitrogen rich conditions (Ding et al. 2011).

Fumarate is produced from pyruvate by the reductive branch of the TCA pathway catalyzed by three cytosolic enzymes such as Pyc, MDH, and Fum in *R. oryzae*. Although this pathway may give the maximal theoretical yield of 200% from glucose (Romano et al. 1967), the experimentally observed yields are lower, which may be explained by the usage of the oxidative TCA cycle to obtain energy and reducing equivalents, since energy is required for the cell maintenance and acid transport even if the cell growth stopped during fermentation (Gangl et al. 1991, Kenealy et al. 1986, Rhodes et al. 1959). In order to increase the yield of fumarate production, it is better to strengthen the reductive branch of the TCA pathway. One attempt has been made to enhance Pyc by overexpressing endogenous *pyc* genes. However, such mutants grew poorly with low fumaric acid yield due to the formation of large pellets that limited oxygen supply, resulting in increased ethanol formation. Part of this may be due to the ATP requirement at Pyc pathway with tight regulation at the pyruvate branch point (Zhang et al. 2012). Therefore, another attempt was made by overexpressing exogenous PEP carboxylase (Ppc) gene from *E. coli* to increase the fumaric acid production, where flask culture with 200 rpm gave the yield of 0.78 g/g glucose, and the titer of about 24 g/*l* in 96 h from initial glucose concentration of about 72 g/*l*, where the glucose consumption rate was 0.32 g/*l*.h with small amount of malic acid (0.1 g/g glucose) and little ethanol (Zhang et al. 2012).

The culture pH and aeration/agitation may affect the metabolism and morphology, where the highest fumaric acid was produced by *R. oryzae* ME-F12 at pH 3 with small pellets (Fu et al. 2009). In the cultivation of *Rhizopus* strains for fumaric acid production, the culture pH decreases without any pH control, and affects the cell growth and fumaric acid production, where the fumaric acid production decreased from 30.21 to 9.39 g/*l*, and more glycerol and ethanol were formed when pH decreased from 5.0 to 3.0. It may be considered to switch off the pH control late at 90 h in the fermentation phase, where pH went down to a final pH of 3.6, giving 20 g/l of fumaric acid in 160 h (Roa Engel et al. 2011). According to the pKa values, the concentration of undissociated fumaric acid increases, and may affect the acid transport through membrane, influencing the acid tolerance. In general, a solid powder of calcium carbonate ($CaCO_3$) is added during fermentation as the neutralizing agent in practice. In lactic acid fermentation, $CaCO_3$ is added during fermentation, while in citric acid fermentation, $CaCO_3$ is added after fermentation. In fumaric acid fermentation, $CaCO_3$ serves as a neutralizer, and also serves as a $CO_2$ source for the reaction at Pyc. Although another agents such as $Na_2CO_3$, $NaHCO_3$, or $(NH_4)_2CO_3$ may be considered, $CaCO_3$ is widely used in fumaric acid fermentation (Xu et al. 2012). In the case of using $(NH_4)_2CO_3$, ammonium fumarate is formed, while sodium fumarate is formed when $Na_2CO_3$ was used.

In the industrial process, the substrate cost may be about 30–40% of the total production costs (Zhang et al. 2007), and therefore, it is desirable to use more cheap renewable raw materials such as starch or lignocellulosic materials. Starch is typically found in potato, cassava, corn, wheat, and rice crop, and several *Ryzopus* strains can produce fumaric acid using an enzymatic hydrolysis of such substrates (Carta et al. 1999, Moresi et al. 1991, 1992).

The bioconversion of starch material to organic acids can be made by the separated two steps of enzymatic hydrolysis and fermentation (SHF) or by a single step of coupling enzymatic hydrolysis of the raw materials and microbial fermentation, known as simultaneous saccharification and fermentation (SSF) (Zhang et al. 2009). In SSF, the product inhibition caused by the mono- or di-saccharide accumulation occur in SHF, can be decreased, and only one reactor is needed, whereas it may be difficult to determine and maintain the culture conditions at optimum in particular pH and temperature, which are different for enzymatic hydrolysis and fermentation, while it can be independently determined in SHF.

In the industrial citric acid fermentation, raw starch material is first liquified into oligo-saccharides, which can be further degraded into mono-saccharides by the glucoamylase generated by *A. niger*, and can be used for citric acid fermentation (Xu et al. 2012). However, in the case of *Rhizopus* strains, raw starch materials cannot be directly used due to lower production of glucoamylase by this fungus. This may be overcome to some extent by ion implantation with higher glucoamylase activity and by SSF, where 39.8 g/l of fumaric acid could be produced from corn starch (Deng et al. 2012).

Although starch is attractive to reduce the cost, its materials are edible. It is highly desirable to utilize lignocelluloses, the most widespread material throughout the world, primary consists of cellulose, hemicellulose, and lignin as mentioned in Chapter 2. After pretreatment, lignocellosic biomass is hydrolyzed to oligosaccharides, primary glucose and xylose. Co-production of fumaric acid and chitin may be attained from dairy manure hydrolysate (rich in nitrogen source) using a pelletized *R. oryzae* in three step processes such as seed culture, cell growth, and fumaric acid fermentation, giving the yield of 31% (Liao et al. 2008). Fumaric acid fermentation may be also produced from *Eucalyptus globules* wood hydrolysates by the ion exchange-treated hydrolysis using *R. arrhizus* DSM5772, giving a yield of 0.44 g/g, and fumaric acid concentration of 9.84 g/l (Julio et al. 2012). Depending on the pretreated method, the release of the reducing sugars contained in the lignocellulosic biomass may be separated to some extent (Charles et al. 2005). Thus a two-stage method may be considered, where corn straw was first pretreated with diluted sulfuric acid to yield a hemicelluloses hydrolysate (30 g/l of xylose) that can be used for fungal cell growth and chitin/chitosan synthesis, followed by the enzymatic hydrolysis to yield glucose-rich liquid (80 g/l of glucose) that can be used for fumaric acid fermentation, yielding 27.79 g/l of fumarate, a yield of 0.35 g/g and a productivity of 0.33 g/l.h (Xu et al. 2010).

Crude glycerol is the primary by-product in biodiesel industry, and it is cheap, but causes a significant environmental problem, which prevents its disposal (Oh et al. 2011). When 80 g/l of crude glycerol was used as a carbon source, 4.37 g/l of fumarate was produced by *R. arrhizus* in 192 h. This can be improved by using a mixture of glucose (40 g/l) and glycerol (40 g/l), where glucose was first consumed and then glycerol was consumed after glucose depletion, giving 22.81 g/l of fumarate production in 144 h (Zhou et al. 2014).

In *S. cerevisiae*, both cytosolic and mitochondrial fumarases catalyze the conversion of fumaric acid to L-malic acid, but not catalyze the reverse direction (Pines et al. 1996). In *R. oryzae*, the cytosolic fumarase converts L-malic acid to fumaric acid, while the mitochondrial fumarase catalyzes the reverse reaction (Goldberg et al. 2006). One of the reasons for the direction of the reaction may be that the cytosolic fumarase (FumR) has the higher affinity to L-malic acid ($K_m$ is 0.46 mM) as compared to fumaric acid ($K_m$ is 3.07 mM) (Song et al. 2011). Another reason may be that the activity of FumR could be inhibited by fumaric acid (Song et al. 2011).

The heterologous expression of the *R. oryzae* genes for MDH (*RoMDH*) and Fum (*RoFUM1*), and the amplification of the expression of endogenous Pyc gene (*PYC2*) in *S. cerevisiae* (FMME-001 strain) allow the fumaric acid production up to 3.18 g/l from 48 g/l of glucose (Xu et al. 2012). Flux balance analysis (FBA) may be made to identify the limiting pathway for fumaric acid production in *S. cerevisiae*, where *FUM1* deletion and the heterologous expression of *R. oryzae* Pyc gene (*RoPYC*) gave 1.675 g/l from 50 g/l of glucose, where *SFC1* gene encoding a succinate-fumarate transporter was introduced (Xu et al. 2012).

*S. cerevisiae* (EN.PK2-1CeTHI2) (FMME003) is capable of accumulating pyruvate, and this may be used for fumaric acid production by *FUM1* mutant (FUMME04) by the oxidative route of the TCA cycle, while the reductive route may be implemented by the heterologous *R. oryzae PYC* gene (*RoPYC*) (FUME004-6), where the native *PYC* gene shows a low degree of control over carbon flow. The cultivation of FMME004-6 strain with addition of 32 µg/l of biotin gave 5.64 g/l of fumaric acid under optimal C/N ratio (Table 2) (Xu et al. 2013).

A multi-vitamin auxotrophic *Torulopsis glabrata* may be engineered for fumaric acid production by modulating arginino-succinate lyase (*ASL*), adenylosuccinate lyase (*ADSL*), fumarylacetoacetase (*FAA*), and fumarase (*FUM1*), where 8.83 g/l of fumarate could be produced from about 60 g/l of glucose by modulating the genes *ASL*, *ADSL*, and *SpMAE1* encoding the $C_4$-dicarboxylic acid transporter (Chen et al. 2014).

Some attempts have been also made to use *E. coli* for fumaric acid production, where the glyoxylate pathway was activated by *iclR* gene knockout, and the fumarate accumulation was made by *fumABC* genes knockout (Song et al. 2013). The cultivation of this engineered *E. coli* under aerobic condition yielded 1.45 g/l of fumarate from glucose, and further improvement could be made by overexpression of *ppc* gene giving 4.09 g/*l* of fumarate. Moreover, further genetic modifications such as *arcA* and *ptsG* mutations to enhance the oxidative TCA cycle, together with *aspA* (encoding aspartate ammonia lyase) to reduce the consumption of fumarate, and replacement of *galP* by a strong *trc* promoter to enhance the glucose uptake rate, where the resulting strain (CWF812) gave the performance such as the titer of 28.2 g/*l*, a yield of 0.389 g/g glucose, in the fed-batch culture of 63 h (Table 2) (Song et al. 2013).

Due to low p*K*a values, the lower solubility of fumaric acid implies that the separation of the product may be relatively easy, indicating that the productivity and yield may be more important than titer (Xu et al. 2012).

### 3.4 Succinic acid production

Succinate has a variety of applications such as a surfactant, detergent extender, foaming agent, ion chelator and food additive, and can be used as a precursor for a variety of chemicals such as tetrahydrofuran and 1, 4 butandiol (Zeikus et al. 1999, Carole et al. 2004).

Under micro aerobic or anaerobic conditions, a mixed-acid fermentation occurs in *E. coli*, forming such metabolites as lactate, acetate, formate, ethanol, $CO_2$, and succinate. Most of the metabolites are produced from pyruvate, while succinate is produced from PEP through OAA, and therefore, one approach to the selective production of succinate is to reduce pyruvate production by disrupting *ptsG* which encodes EIIBC$^{Glc}$ of the PTS, and *pykFA* genes in *E. coli* (Lee et al. 2005). The resulting strain produced 0.35 moles of succinate/mole glucose. Simultaneous overexpression of the *ppc* gene of *Sorghum vulgare* (which encodes Ppc resistant to feedback inhibition by malate) and of the *pyc* gene of *Lactococcus lactis* in *E. coli* increased the succinate yield to 0.91 moles/mole glucose with a concomitant decrease in the lactate formation (Lin et al. 2005a). Additional disruption of such genes as *ldhA*, *ackA* and *pta* increased the succinate yield to 0.3 g/g glucose (Lin et al. 2005a). Further improvement may be made by deletion of the *adhE* and *ldhA* genes, and by overexpression of the *pyc* gene obtained from *L. lactis*, where it increased the yield to 1.3 moles of succinate/mole glucose (Sanchez et al. 2005a). The result indicates that inactivation of *adhE* is effective in increasing the use of reducing equivalents for the formation of succinate. The glyoxylate pathway may be utilized as an alternative route to succinate formation by deleting *iclR* gene, where IclR represses the *aceBAK* operon (Sanchez et al. 2005b). The utilization of the glyoxylate pathway together with the Ppc pathway for succinate production requires less reducing equivalents. The strain lacking the genes *iclR*, *ldhA*, *adhE*, and *ackA, pta* genes with overexpression of the *pyc* gene (from *L. lactis*) gave a succinate yield of 1.6 moles/mole glucose with the production of small amounts of formate and acetate. Further metabolic engineering with metabolic evolution gave higher succinate production by *ldhA, adhE, ackA, focA, pflB* mutant and *ldhA, adhE, ackA, focA, pflB, mgsA, poxB* mutant (Fig. 12) (Jantama et al. 2008).

In order to overcome the disadvantages of anaerobic succinate fermentation, such as low growth rate and the formation of a variety of by-products, aerobic succinate fermentation has been considered using metabolically engineered *E. coli* (Lin et al. 2005b,c). This strain lacks *icd*, *sdhAB*, and *iclR* genes, resulting in inactivation of the oxidative TCA cycle while retaining the glyoxylate pathway. In addition, some of the acetate producing pathway genes such as *pta-ackA* and *poxB* may be deleted. Batch culture of this strain gave 0.7 moles of succinate, 0.2 moles of pyruvate, and 0.25 moles of acetate per mole of glucose (Lin et al. 2005d). The succinate yield could be increased to 0.87 mols/mole of glucose by deletion of the *ptsG* gene. Further improvements in the yield and productivity can be made through overexpression of the *S. vulgare ppc* gene. Specifically, fed-batch cultivation of the strain ($\Delta sdhAB$, $\Delta ackA$-*pta*, $\Delta poxB$, $\Delta iclR$, $\Delta ptsG$, *ppc* overexpression) yielded 494 mM of succinate, with an overall yield of 0.85 moles per mole of glucose, an average volumetric productivity of 1.1 g/l.h, and the specific productivity of 90 mg/gDCW.h (Lin et al. 2005d).

*C. glutamicum* can be also used as an alternative strain to *E. coli*. The deletion of the *ldhA* gene for the NAD$^+$-dependent production of L-LDH and overexpression of the endogenous *ppc* gene in *C. glutamicum* gave a yield of about 1.2 moles of succinate per mole of glucose, and the volumetric productivities can be higher than those of the metabolically engineered *E. coli* as described above (Inui et al. 2004).

## 4. Diol Fermentation

Diols have two hydroxyl groups, and give a wide range of industrial applications for chemicals and fuels. In particular, the microbial production of 1,3-propandiol (1,3-PDO), 1,2-propandiol (1,2-PDO), 2,3-butanediol (2,3-BDO), 1,4-butanediol (1,4-BDO), and 1,3-butanediol (1,3-BDO) has been paid recent attention (Zheng et al. 2011).

### 4.1 Propanediol production

**Propanediol** is a three-carbon diol having its hydroxyl groups at the 1st and 2nd carbon atoms (1,2-propanediol: 1,2-PDO), or at the 1st and the 3rd carbon atoms (1,3-propanediol: 1,3-PDO), where 1,2-PDO is a chiral molecule and present as a rasemic mixture in general. Both 1,2-PDO and 1,3-PDO offer broad range of applications either directly as solvents or as platform chemicals (Sabra et al. 2016).

### 4.1.1  1,3-propanediol production

As mentioned before, glycerol utilization has been paid much attention in relation to biodiesel production. **1,3-Propanediol (1, 3-PDO)** is one of the typical products of fermentative glycerol utilization (Clomburg and Gonzalez 2012, Almeida et al. 2012). 1,3-PDO is used for solvents, adhesives, detergents, resins, and cosmetics, where in particular, the polyester polytrimethylene telephthalate (PTT) has been paid attention due to its application to fiber, textiles, and carpets (Zeng et al. 2011). 1,3-PDO can be produced from glycerol by several microorganisms of such families as *Clostridiacea* and *Enterobacteriaceae*, and such species as *Klebsiella*, *Clostridia*, *Citrobacter*, and *Enterobacter*, where the well-studied species may be *K. pneumoniae* and *Cl. butylicum* (Sabra et al. 2016). Although *Cl. butyricum* is strictly anaerobic microorganism, while *K. pneumoniae* is facultative anaerobic microorganism, and easier to handle, *K. pneumoniae* is regarded as an opportunistic pathogen, and therefore, special safety precautions are required (Sabra et al. 2016). The productive strains must be tolerant to impurities in crude glycerol such as salts, free fatty acids, and methanol (Chatzifragkou et al. 2010). 1,3-PDO is produced by 2-step reactions from glycerol, where glycerol is first dehydrated by glycerol dehydratase to produce 3-hydroxypropionaldehyde (3HPA), which is in turn converted to 1, 3-PDO by the NADH-utilizing 1, 3PDO dehydrogenase (Fig. 15) (Celinska 2010).

### 4.1.2  1,2-propanediol production

**1,2-Propanediol (1,2-PDO)** $(C_3H_8O_2)$ is generally called as propylene glycol, and is a commodity chemical with global demand of about 3 billion lb/y, which can be used for anti-freeze and heat-transfer fluids, plasticizers, thermoplastics, and cosmetics (Shelley 2007). Microbial production of 1,2-PDO has been made from sugars, mainly glucose, by using such microorganisms as *Thermoanaerobacterium thermosacchrolyticum* (Altaras et al. 2001, Sanchez-Riera et al. 1987), *Clostridium sphenoides* (Trandin and Gottschalk 1985), *Bacteroides ruminicola* (Turner and Roberton 1979), *Lactobacillus buchneric* (Oude Elferink et al. 2001), *S. cerevisiae* (Jung et al. 2008, Lee and Da Silva 2006), and *E. coli* (Altaras and Cameron 1999, 2000, Huang et al. 1999, Clomburg and Gonzalez 2011).

Different from the case of 1,3-PDO production, 1,2-PDO is produced from DHAP, where this may be produced either from sugars via EMP pathway or from glycerol (Fig. 16) (Zeng et al. 2011). For the microbial production of 1,2-PDO, two constraints such as the redox balance and ATP production determine the metabolism under anaerobic conditions. The synthesis of 1,2-PDO requires the conversion of the carbon source to DHAP

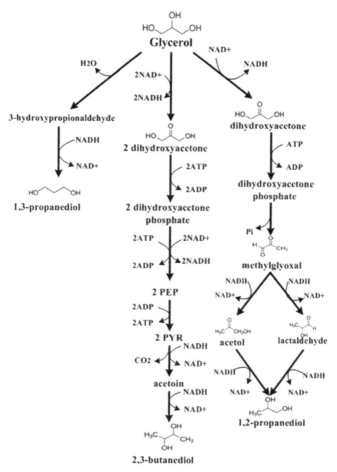

**Figure 15.** 1,3-Propanediol production pathway (Jiang et al. 2014).

together with glycolytic pathways (Fig. 16), and the maximum theoretical yield from glucose is 0.63 g/g by the following equation:

$$C_6H_{12}O_6 \ 2 \text{ and } C_3H_8O_2 + (3/2)CO_2 \tag{1}$$

On the other hand, the maximum theoretical yield from glycerol is higher as 0.72 g/g by the following equation:

$$C_3H_8O_3 \quad (7/8)C_3H_8O_2 + (3/8)CO_2 + (1/2)H_2O \tag{2}$$

There are two possible pathways for 1,2-PDO synthesis. One pathway is to metabolize deoxy sugars such as L-rhamnose and fucose (Bennet and San 2001), where L-rhamnose is first converted to L-rhamnulose 1-phosphate, which is subsequently split into DHAP and S-lactaldehyde by L-rhamnose dehydrogenase (RhaD), while fucose is first isomerized to yield L-fuculose, and then transformed to L-fuculose 1-phosphate by L-fuculose kinase, where L-fuculose 1-phosphate aldorase (FucA) cleaves it into DHAP and L-lactaldehyde (Fig. 16). Depending on the redox conditions, the lactaldehyde can be either reduced to 1,2-PDO or oxidized to lactic acid (Sabra et al. 2016). Anaerobic conditions lead to the conversion to S-1,2-PDO catalyzed by a NAD-oxidoreductase FucO (S)-1,2-propanediol oxidoreductase (1,2-PDOX) (Bennet and San 2001). Although the above synthetic pathways are possible, deoxy sugars are quite expensive, and it is not feasible in practice (Sabra et al. 2016).

In *T. thermosaccharolyticum*, DHAP is converted to methylglyoxal (MG) by methylglyoxal synthase (Mgs), and subsequently MG is reduced to R-1,2-PDO by aldose reductase or glycerol dehydrogenase. If glucose was converted to R-1,2-PDO, acetate, and $CO_2$, a maximum theoretical yield becomes 0.42 g R-1,2-PDO/g glucose.

In *E. coli*, MG is converted to acetol by either NADPH- or NADH-dependent lactaldehyde oxidoreductase, and alcohol or aldehyde dehydrogenase. MG is also converted to R-lactaldehyde by NADH-dependent glycerol dehydrogenase (GlyDH) in *E. coli*. In *S. cerevisiae*, MG is first converted to S-lactaldehyde, which is subsequently converted to S-1,2-PDO by NADH-dependent lactaldehyde reductase, where 1,2-PDO production is low.

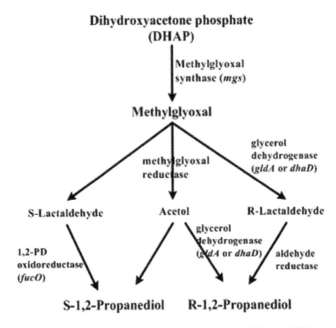

**Figure 16.** 1,2-Propanediol production pathway (Altaras 1999).

The redox balance and ATP production by substrate-level phosphorylation are the key factors to determine the pathways for 1,2-PDO production, where the net consumption of the reducing equivalents is required. This implies the formation of more oxidized co-products such as acetate and lactate to get energy while maintaining the overall redox balance. The stoichiometric equation for the production of 1,2-PDO with acetate becomes as

$$3C_3H_8O_3 \rightarrow 2C_3H_8O_2 + CH_3COOH + HCOOH\ H_2O \tag{3}$$

where the molar ratio of 1,2-PDO to acetate is 2:1 with redox balance. In the case of lactate formation, the stoichiometric equation becomes as:

$$2C_3H_8O_3 \rightarrow C_3H_8O_2 + CH_3CHOHCOOH + H_2O \tag{4}$$

where the molar ratio of 1,2-PDO to lactate is 1:1. The molar ratio of either acetate or lactate to 1,2-PDO required to satisfy redox balance would result in the limited ATP production, where ATP is required for the cell growth and the cell maintenance (Clonburg and Gonzalez 2011). In the case of ethanol formation from glycerol, ATP is generated by satisfying the redox balance (Murarka et al. 2008) as stated in the section of ethanol production.

In *E. coli*, glycerol is converted to DHA by GlyDH encoded by *gldA*, and subsequently to DHAP by DHAK encoded by *dhaKLM*. DHAP is then converted either to MG by MGS encoded by *mgsA*, or GAP by Tpi, where GAP goes down to the lower glycolysis (Fig. 16). The conversion of MG to 1,2-PDO is made through two alternate pathways, where MG is first converted to acetol by aldehyde oxidoreductase (AOR) encoded by *yqhD* (Lee et al. 2010, Soucaille et al. 2008), and then converted to 1,2-PDO by GlyDH in one pathway, while in another pathway, MG is converted to lactaldehyde by GlyDH (Altaras and Cameron 1999, Subedi et al. 2008), and then converted to 1,2-PDO by 1,2-PDO reductase (1,2-PDOR) encoded by *fucO*, where lactaldehyde can be converted to lactate by lactaldehyde dehydrogenase (LALDH) (Fig. 16). Although the accumulation of the starting metabolite, MG is important for 1,2-PDO production, MG is toxic to the cell, and its accumulation at sub-millimolar concentrations inhibits the cell growth, causing even cell death (Booth et al. 2003, MacLean et al. 1998). Because of this, there exist several detoxifying pathways, where one of them is glutathione (GSH)-dependent glyoxalase I (GlxI) encoded by *gloA*, and Glx II encoded by *gloB*, yielding lactate (MacLean et al. 1998), while another one is glyoxalase III (GlxIII) (Misra et al. 1995), yielding lactate, where lactoaldehyde can be also converted to lactate by LALDH as mentioned above, but lactate production through this pathway is low (Clomburg and Gonzalez 2011).

The glycerol uptake rate is significantly low in particular under anaerobic conditions. The glycerol assimilation is coupled with glycolysis, in the sense that the phosphorylation of DHA by DHAK to produce DHAP is made by PEP in the lower glycolysis, which indicates that the glycerol uptake rate depends on the glycolytic flux, which is low when glycerol was used as a carbon source. Therefore, the glycerol uptake rate may be increased by decoupling this relationship by expressing an ATP-dependent DHAK from such microorganism as *Citrobacter freundii (dhaKL)*, which utilizes ATP as the phosphate group donor instead of PEP (Daniel et al. 1995). Then 1,2-PDO production may be improved by overexpressing the 1,2-PDO synthetic pathway genes such as *gldA, mgsA*, and *yqhD* in addition to *dhaKL* from *C. freundii* (Clomburg and Gonzalez 2011). The complete decoupling can be made by the additional disruption of the native *dhaK* gene, and the performance could be further increased, but significant amount of ethanol and other co-metabolies are formed as well (Clomburg and Gonzalez 2011). Therefore, the performance improvement can be made by further disruption of such genes as *ackA-pta* and *ldhA* to suppress acetate and lactate co-production, whereas the disruption of either *frd* or *adhE* gene gives negative effect on the glycerol assimilation due to redox requirement (Clomburg and Gonzalez 2011). The *E. coli* MG1655 ΔackA Δpta ΔldhA ΔdhaK with pTHKLcfgldAmgsAyqhD gave 5.6 g/l of 1,2-PDO with a yield of 21.3% (w/w) but with ethanol production of 33.0% (w/w), formate of 15.1% (w/w), and others (Clomburg and Gonzalez 2011).

### 4.2 Butandiol production

**Butandiol (BDO)** is a four-carbon diol with its hydroxyl groups at the specific positions such as 2,3-BDO, 1,4-BDO, and 1,3-BDO, where only 2,3-BDO is produced naturally by facultative and anaerobic bacteria (Sabra et al. 2016).

#### 4.2.1  2,3-butandiol production

**2,3-Butandiol (2,3-BDO)** is one of the bulk chemicals with many applications to food, fuel, aeronautical and others. 2,3-Butandiol (2,3-BDO) can be used for the production of plastics, anti-freeze solutions, and solvent preparations. Moreover, it can be converted to bulk chemicals such as methyl ethyl ketone (a liquid fuel additives), 1,3-butadiene (used for synthetic rubber), diacetyl (a flavoring agent), or to precursors of polyurethane (Ji et al. 2011, Li et al. 2015). 2,3-BDO has three isomers such as *levo* (2R,3R) [D-(-)-] and *dextro* (2S,3S) [L-(+)-] forms with optical activity, and the *meso*-form with no optical activity (Sabra et al. 2016, Ji et al. 2011).

The typical 2,3-BDO producing bacteria belong to Enterobacteriaceae, where *Klebsiella pneumonia* (Ma et al. 2008, Petrov and Petrova 2009), *K. oxytoca* (Cheng et al. 2010, Ji et al. 2010), and *E. aerogenes* (Zeng et al. 1991) may be the representative species. The 2,3-BDO can be also produced by *Pseudomonas chlororaphis* belonging to the Pseudomonadaceae family, and by *Paenibacillus polymyxa* (Li et al. 2013) belonging to the Paenibacillaceae family, and these have been paid attention due to the formation of optically active stereoisomer (*L*-form) in plant rhizospheres (Sabra et al. 2016). The relatively high concentration of 2,3-BDO could be attained by the fed-batch cultivation of *Bacillus licheniformis* DSMZ 8785 (Jurchescu et al. 2013), and the similar performance could be attained by thermophilic *B. licheniformis* strains from glucose (Li et al. 2013, 2014) and from inulin by SSF (Li et al. 2014). The advantages of using thermophilic strains may be less contamination risk and effectiveness for SSF at higher cultivation temperature (Li et al. 2014). While *B. licheniformis* produces a mixture of *levo* and *meso* (1:1 ratio), *P. polymyxa* is able to form almost exclusively the *levo* isomer (> 98%) under anaerobic conditions (Li et al. 2013, Celinska and Grajek 2009, Nakashimada et al. 1998, 2000, Yu et al. 2011). The reducing equivalent such as NADH may play an important role for the production of chirally pure D-2,3-BDO in *Bacillus subtilis* cultivated under oxygen limitation (Fu et al. 2014).

Under oxygen limitation, a mixed acid fermentation occurs, producing a wide spectrum of metabolites such as acetate, lactate, formate, succinate, ethanol, acetoin, and 2,3-BDO, originating mostly from pyruvate (except succinate that is from PEP), the end product of the glycolysis (Fig. 17). For 2,3-BDO synthesis from pyruvate, three enzymatic reaction steps are involved, where two moles of pyruvate is condensed with a single decarboxylation to form α-acetolactate by α-acetolactate synthase (ALS, EC 4.1.3.18), where ALS is active under slightly acidic and anaerobic conditions, and inactive under aerobic conditions. The genes encoding ALDC, ALS, and BDH are clustered in one operon as *budABC* in *K. ferrigena* and *E. aerogenes* (Blomqvist et al. 1993), where this operon seems to be under control of the putative fumarate nitrate reductase (FNR), which is active under anaerobic conditions (Mayer et al. 1995). The α-acetolactate is then converted to acetoin by α-acetolactate decarboxylase (ALDC, EC 1.1.1.5), and acetoin is converted to 2,3-BDO by NADH dependent 2,3-BD dehydrogenase (BDH, EC 1.1.1.76) (also called as acetoin reductase, EC 1.1.1.4) (Ji et al. 2011), where different isomeric forms of 2,3-BDO is formed by the action of different BDH with different stereospecificities, or by a cyclic pathway (known as butandiol cycle) (Sabra et al. 2016).

One approach to improve the 2,3-BDO production may be to disrupt some by-product-forming pathway genes for *K. oxytoca* (Ji et al. 2008, 2010), and for *Enterobacter cloacae* (Li et al. 2015), where 119 g/*l* of enantiometrically pure (2*R*,3*R*)-2,3-BDO could be attained by the latter strain from lignocelluloses-derived sugars (Li et al. 2015).

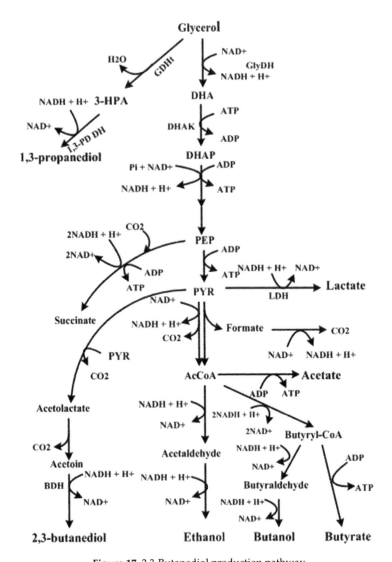

**Figure 17.** 2,3-Butanediol production pathway.

The synthetic pathway can be also constructed in *E. coli* for enantiometrically pure 2,3-BDO by incorporating different stereospecific dehydrogenases (Sabra et al. 2011, Yan et al. 2009). Various 2,3-BDO producing gene clusters have been cloned from native 2,3-BDO producers such as *B. subtilis*, *B. licheniformis*, *K. pneumonia*, *Serratia marcescens*, and *E. cloacae*, and inserted into the expression vector pET28a, where the fed-batch cultivation of the best recombinant *E. coli* gave 74 g/l of 2,3-BDO production within 62 h (Xu et al. 2014). Although no natural producers for the *dextro-* 2,3-BDO (2*S*,3*S*) have been found so far, this can be made by incorporating this diol enantiomer synthesis into engineered *E. coli*, where 26.8 g/*l* of high pure (> 99%) (2*S*,3*S*)-2,3-BDO could be obtained by a fed-batch cultivation of this strain from diacetyl (Wang et al. 2013, Li et al. 2012).

Engineered yeast strains were capable of producing a high titer of 100 g/*l* of enatiometrically pure *levo*-2,3-BDO from a mixture of glucose and galactose with a yield over 70% of its theoretical value (Lian et al. 2014, Nan et al. 2014).

### 4.2.2 1,4-Butandiol production

**1,4-Butandiol (1,4-BDO)** is an important commodity chemical, and mainly used for tetrahydrofuran (THF) and polybutylene telephtalate (PBT), where THF is used for fibers, polymers, resins, solvents, and printing inks for plastics, while PBT is a thermoplastics (Yim et al. 2011). Biodegradable polymers can be also synthesized from 1,4-BDO and dicarboxylic acids (Diaz et al. 2014). Since no natural microorganisms that can produce 1,4-BDO are found so far, unnatural synthetic pathways must be established in such microorganisms as *E. coli* strain (Yim et al. 2011, Burk et al. 2012). One metabolic pathway is the conversion of succinyl-CoA in the TCA cycle to 1,4-BDO via 4-hydroxybutyrate, 4-hydroxybutyryl-CoA, and 4-hydroxybutyraldehyde (Fig. 18) (Yim et al. 2011), where anaerobic operation of the oxidative TCA cycle supplies the reducing equivalents for 1,4-BDO production. The engineered strain has been shown to produce 30–40 g/*l* of 1,4-BDO in a continuous bioreactor (Pharkya et al. 2015).

### 4.2.3 1,3-Butandiol production

**1,3-Butandiol (1,3-BDO)** is typically used as a chemical intermediate in the production of polyester plasticizers, and as a solvent for flavoring, etc., where 1,3-BDO has been synthesized as a racemic mixture of *R*- and *S*-forms, mainly from petroleum-based chemicals (Sabra et al. 2016). (*R*)-1,3-BDO is a non-natural alcohol, and is a valuable building block for the synthesis of various optically active compounds (Matsuyama et al. 2001), and in particular of interest as the key intermediate of menems and carbapenems for industrial synthesis of β-lactam antibiotics, and therefore, bio-based production has been paid attention (Kataoka et al. 2013, 2014). As shown in Fig. 19, excess reducing equivalents and cofactors are required for the production of 1,3-BDO, and therefore, it is critical to control the dissolved oxygen (DO) concentration, where 9 g/*l* of *R*-1,3-BDO could be produced with enanthiometric purity of 98.5% by controlling the DO concentration (Kataoka et al. 2014).

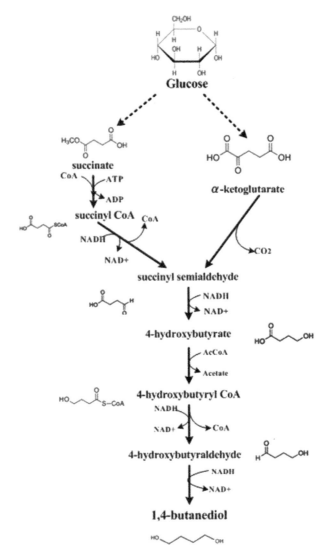

**Figure 18.** 1,4-Butanediol production pathway (Jiang et al. 2014).

## 5. Other Organic Acid Fermentation

Other organic acid fermentations, including acetic acid, ithaconic acid, glucaric acid, muconic acid, and adipic acid may be also considered (Sanchez and Demain 2008, Kramer et al. 2003).

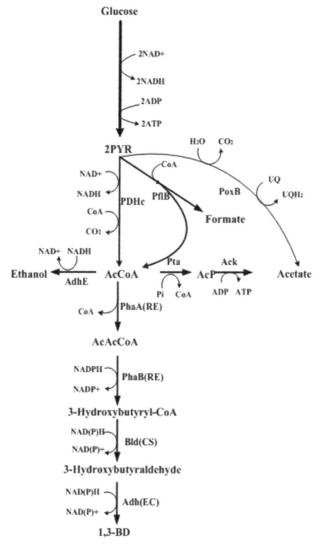

**Figure 19.** 1,3-Butanediol production pathway (Kataoka et al. 2014).

## 5.1 *Itaconic acid fermentation*

**Itaconic acid (IA)** (2methylidenenenebutanedioic acid or methylenesuccinic acid) is an unsaturated dicarboxylic acid, in which one carboxyl group is conjugated to the methylene group with a molecular formula $C_5H_6O_4$ and a molecular weight of 130.1 g/mol. As dicarboxylic acid, it has three different protonation states with pKa values of 3.83 and 5.55, and it can be easily separated using ion exchange chromatography (Klement and Buchs 2013).

IA has gained increased interest as one of the top 12 building block chemicals that can be produced from biomass (Welpy and Peterson 2004). It is commercially used for the synthesis of resins, lattices, fibers, detergents, cleaners, and bioactive compounds (Okabe et al. 2009, Wilke and Vorlop 2001), and can be converted to methacrylic acid (Choi et al. 2015). Although many fungi such as *Ustilago*, *Candida*, and *Rhodotorula* species can produce IA, *Aspergillus terreus* is the dominant production host, which produces more than about 80 g/*l* of IA (Okabe et al. 2009, Wilke and Vorlop 2001). In fact, under optimized fermentation condition, IA titer reached 146 g/*l* with a productivity of 0.48 g/*l*.h and a yield of 0.81 mol/mol glucose (Hevekerl et al. 2014). The productivity may be improved to 1.15 g/*l*.h with the decreased titer of 129 g/*l* by controlling the pH low at around 3.

By considering the higher citric acid production more than about 200 g/*l* by *Aspergillus niger*, some attempts have been made to use *A. niger* by over-expressing the *cadA* gene encoding *cis*-aconitate decarboxylase (EC 4.1.1.6), where IA is produced via the decarboxylation of *cis*-aconitate, an intermediate of the aconitase reaction in the TCA cycle in *A. terreus*. However, IA production by such engineered *A. niger* is very low (Li et al. 2011, 2012, 2013, Blumhoff et al. 2013). One reason for this may be the wrong compartmentalization, where citrate is produced in mitochondria, and anti-port of malate and citrate may cause the efficient citric acid production in *A. niger* (De Jongh and Nielsen 2008, Rohr and Kubicek 1981). In *A. terreus* and *A. niger,* aconitase exists primarily in mitochondria, while CadA is exclusively located in the cytosol (Jaklitsch et al. 1991b). Although overexpression of either the mitochondrial carrier or plasma membrane carrier allows the increased IA production in *A. niger* (Li et al. in press, Jore et al. 2011), the titer is low (Li et al. 2013). Another strategy of co-overexpressing aconitase and *cis*-aconitate decarboxylase in the same compartment may be considered (Bluhoff et al. 2013), but the production rate is still low. In the genome of *A. terreus*, two putative transporters flank the *cadA* gene, where mitochondrial transporter *mttA* is crucial for the efficient IA production in *A. niger* and to a minor extent the plasma membrane transporter *mfsA* has a positive effect for IA production (Li et al. 2013). In addition to IA synthesis cluster (*cadA*, *mttA*, and *mfsA*), the over-expression of putative cytosolic citrate synthase gene *citB* in *A. niger* strain gave the maximum titer of 26.2 g/*l* with a productivity of 0.35 g/*l*.h (Hossain et al. 2016).

A specific truncated version of 6-phospho-1-fructokinase, *pfkA* gene from *A. niger* resistant to citrate inhibition was over expressed in *A. terreus*, and gave increased IA production (Tevz et al. 2010). A modified *pfkA* gene was expressed in *A. niger* in combination with *cis*-aconitate decarboxylase from *A. terreus*, and also in combination with IA biosynthetic cluster from *A. terreus*, such as *cadA*, *mttA*, and *mfsA*, which gave a productivity of 0.15 g/*l*.h (van der Staat et al. 2013).

One of the drawbacks of using *A. terreus* is its low growth rate (Kuenz et al. 2012), which limits the increase in the IA productivity. Another drawback is the dependence on and high sensitivity to oxygen supply (Kuenz et al. 2012). Since *cadA* gene has been identified, cloned, and characterized (Kanamasa et al. 2008), some attempts have been made to express this gene in *E. coli* (Li et al. 2011, Liao and Chang 2010, Okamoto et al. 2014, Vuorist et al. 2015a), *S. cerevisiae* (Blazeck et al. 2014), *C. glutamicum* (Otten et al. 2015), and in *Y. lipolytica* (Blazeck et al. 2015), where the titer ranges from 0.17 g/*l* for *S. cerevisiae* (Blazeck et al. 2014), 4.3 g/*l* for *E. coli* (Okamoto et al. 2014), to 7.8 g/*l* for *C. glutamicum* (Otten et al. 2015). Among these, *E. coli* seems to be promising due to its fast growth and its capability of producing IA even under anaerobic condition (Vuoristo et al. 2015). Model-based metabolic engineering of *E. coli* gave 2.27 g/*l* of IA production with an yield of 0.77 mol/(mol glucose), where *aceA*, *sucCD*, *pykAF*, *pta* genes were knocked out, and *icdA* gene expression was reduced, while *cadCS* gene was over-expressed (Harder et al. 2016). The performance can be further increased to 3.2 g/*l* of IA production with an overall yield of 0.68 mol/(mol glucose), and a peak productivity of 0.45 g/*l*.h by a fed-batch cultivation (Harder et al. 2016).

### *5.2 Glucaric acid production*

D-Glucaric acid is found in fruits, vegetables, and mammals. It may be used as a dietary supplement in the form of calcium D-glucarate, and it has been investigated for a variety of therapeutic purposes such as cholesterol reduction (Walaszek et al. 1996), diabetes treatment (Bhattachaya et al. 2013), and cancer chemotherapy (Gupta and Singh 2004). D-Glucaric acid is also identified as a top value-added chemical from biomass (Werpy and Petersen 2004). D-Glucaric acid is a highly functionalized compound with four chiral carbons, and is currently produced by chemical oxidation of glucose, a nonselective, expensive, and environmentally not friendly process using nitric acid as the solvent and oxidant (Werpy and Petersen 2004). Biological production may be able to avoid costly catalyst and harsh reaction conditions, and offers the potential for a cheaper and more environmentally friendly process.

In mammarian cells, D-glucaric acid and L-ascorbic acid are both the end products of the D-glucuronic acid pathway. Although this pathway may be considered to produce D-glucaric acid, it is complex with 10 conversion steps. Therefore, a biosynthetic pathway from D-glucose to D-glucaric acid, consisting of three heterologous genes has been investigated in recombinant *E. coli* (Moon et al. 2009). This synthetic pathway starts from the uptake of D-glucose by PTS giving G6P. This G6P is then isomerized to *myo*-inositol-1-phosphate by *myo*-inositol-1-phosphate synthase (INO1) from *S. cerevisiae*. An endogenous phosphatase dephosphorylates *myo*-inositol-1-phosphate to yield *myo*-inositol, which is then oxidized to D-glucuronic acid by *myo*-inositol oxygenase (MIOX) from *Mus muscalus* (mouse). D-glucuronic acid is then further oxidized by urinate dehydrogenase (Udh) from *Pseudomonas syringae* to produce D-glucaric acid. This synthetic pathway in *E. coli* gave 1.13 g/*l* of D-glucaric acid from 10 g/*l* of glucose. Noting that the MIOX activity is the limiting, the increase in MIOX activity gave an improved titer of 2.5 g/*l* (Moon et al. 2010). The performance could be further improved by feeding *myo*-inositol rather than D-glucose, where 4.85 g/*l* of D-glucaric acid could be produced from 10.8 g/*l* of *myo*-inositol in recombinant *E. coli* as mentioned above (Shiue and Prather 2014).

### 5.3 Muconic acid production

Muconic acid (MA) or 2,4-hexadienedioic acid is an unsaturated dicarboxylic acid, and is present in three isomers such as *cis,cis*-MA, *cis,trans*-MS, and *trans,trans*-MA (Bui et al. 2013, Burk et al. 2011). MA is a precursor for the commercially important bulk chemicals such as adipic acid, telephthalic acid, and trimellitic acid, where these have been used for the manufacture of nylon-6,6, polytrimethylene terephthalate, polyethylene terephthalate (PET), dimethyl terephthalate, trimellitic anhydride, plastics, resins, polyester polyols, food ingredients, pharmaceuticals, plasticizers, cosmetics, and engineering polymers (Xie et al. 2014). Annual production of adipic acid and terephthalic acid is more than 2.8 million and 71 million tones, respectively (Burridge 2011, Mirasol 2011). Currently, these chemicals are mainly produced from non-renewable petroleum feedstock with toxic intermediates environmentally not friendly.

Aromatic compounds comprise about one-quarter of the Earth's biomass and are second most widely distributed organic compounds in nature (Valderrama et al. 2012). Aromatic compounds such as benzoate, toluene, benzene, phenol, aniline, anthranilate, mandelate, and salycylate can be oxidized by some bacteria with the formation of catecol and the central aromatic intermediate through a variety of ring modification reactions (Harwood and Parales 1996). The aromatic ring of catecol can be cleaved either by the *ortho*-cleavage or the *meta*-cleavage pathways depending on the types of bacteria (Vaillancourt et al. 2006, Wells and Ragauskas 2012). Benzoate can be metabolized by such microorganisms as *Pseudomonas, Anthrobacter, Corynebacterium, Brevibacterium, Mycobacterium*, and *Sphingobacterium* species via catechol branch of the β-ketoadipate pathway to produce MA (Xie et al. 2014).

In order to produce MA from aromatic compounds, the followings may be required: (i) ability to metabolize aromatic compounds via catechol by means of *ortho* cleavage pathway, (ii) lacking or reduced functional muconate cycloisomerase, which degrades MA, thus allowing the accumulation of

MA, (iii) resistant to aromatic compounds that are toxic to microorganisms in general, (iv) excreting MA into the medium, (v) and showing high CatA activity (Xie et al. 2014a, van Furren et al. 2012). Benzoate-tolerant mutant of *Pseudomonas* sp. has been used to produce 7.2 g/*l* of MA from 12 g/*l* of benzoate within 11 h in shake-flask culture (Xie et al. 2014b). Another strain *P. putida* obtained by random mutagenesis has been used to produce MA from benzoate with eight times higher specific production rate (van Durren et al. 2011). Another common aromatic compound such as toluene can be used to produce 45 g/*l* of MA over 4 d by the fed-batch cultivation (Chua and Hsieh 1990).

An alternative pathway for producing MA from D-glucose may be promising (Draths and Frost 1994), where artificial synthetic pathway is required, since it is not known to exist naturally (Xie et al. 2014a). This pathway utilizes the shikimic acid pathway, where it begins with the condensation of PEP and E4P in the central metabolic pathway to form DHAP catalyzed by DHAP synthase (EC 2.5.1.54) in *E. coli*, where DHAP synthase consists of three isozymes encoded by *aroF*, *aroG*, and *aroH*, each regulated by the feedback inhibition by L-tyrosine, L-phnylalanine, and L-triptophane, respectively (Chen et al. 2013, Ikeda et al. 2006). DHAP is then converted to 3-dehydroquinate (DHQ) catalyzed by DHQ synthase (EC 4.2.3.4) encoded by *aroB*. Thereafter, DHQ is converted to 3-dehydroshikimate (DHS) by DHQ dehydratase (EC 4.2.1.19) encoded by *aroD*. DHS is in turn converted to shikimic acid by shikimate dehydrogenase (EC 1.1.1.25) encoded by *aroE* with the cofactor NADPH (Ghosh et al. 2012).

In order to redirect the carbon flux from the shikimic acid pathway to the synthetic MA production pathway, shikimate dehydrogenase must be inactivated, while the heterologous genes such as *aroZ* encoding 3-dehydroshikimate dehydratase (AroZ), *aroY* encoding protocatechuate decarboxylase (AroY), and *catA* must be expressed (Bui et al. 2013, Niu et al. 2002). Such engineered *E. coli* together with overexpression of DHQ synthase and transketorase as well as feedback resistant for *aroF* produced 36.8 g/*l* of MA with a yield of 22% (mol/mol) within 48 h of fed-batch cultivation (Niu et al. 2002). The complete inactivation of shikimate dehydrogenase also blocks the biosynthesis of aromatic amino acids and other aromatic metabolites, and therefore such metabolites must be supplied for the cultivation of such auxotroph in minimal medium (Bui et al. 2013, Niu et al. 2002). Otherwise a leaky *aroE* mutation may be made (Xie et al. 2014a).

In order to enhance the shikimate pathway, Pyk and PTS mutations may be effective to increase PEP, and amplify Tkt and Tal to increase E4P. Moreover, *csrA* gene disruption can increase the gluconeogenesis (in particular Pps) to accumulate PEP, and contribute to the aromatic acid production (Tatark and Romeo 2001).

Another microorganism such as *S. cerevisiae* may be used for MA production, where *S. cerevisiae* prefers a lower pH environment than

*E. coli*. Three synthetic pathway genes such as *aroZ*, *aroY*, and *catA* may be incorporated into *S. cerevisiae* from *Podospora anserine*, *Enterobacter cloacae*, and *Candida albicans*, respectively (Curran et al. 2013). The knockout of *ARO3* gene and overexpression of a feedback-resistant *aro4* (*aro4*[FBR]) reduced feedback inhibition in the shikimic acid pathway. Moreover, *zwf1* deletion and overexpression of *TKL1* increased E4P, and the resultant engineered *S. cerevisiae* gave 0.141 g/*l* of MA in the shake flask culture (Curran et al. 2013).

### 5.4 Adipic acid production

Adipic acid or hexanedioic acid ($C_6H_{10}O_4$) with a molecular weight of 146.14 is the commercial aliphatic dicarboxylic acid (Musser 2005). The primal application of adipic acid is in the chemical production of nylon-6,6 polyamide. Adipic acid is also used to produce polyurethane, as a reactant to form plasticizers, lubricant components and polyester polyols. Others are a food ingredient in gelatins, desserts and other foods that require acidulation (Polen et al. 2013).

Adipic acid can be found in nature in juice of sugar beet and beet red. Adipic acid is an intermediate in the degradation pathways of cyclohexane, cyclohexanol or cyclohexanone, and of ε-caprolactum typically seen in *Pseudomonas* strains (Chen et al. 2002). The degradation of these compounds gives the common intermediate such as adipate semialdehyde, which is further oxidized to adipic acid by 6-oxohexanoate dehydrogenase (EC 1.2.1.63). This enzyme requires NADP$^+$ as coenzyme in *Acinetobacter* sp. (Iwaki et al. 1999). The oxidation can be alternatively catalyzed by NADP$^+$-dependent aldehyde dehydrogenases (EC 1.2.1.4), which catalyze the oxidation of long chain aliphatic aldehydes to the corresponding acids (Ahvazi et al. 2000). Subsequently, adipic acid is activated to adipyl-CoA and degraded to SucCoA and AcCoA via *β*-oxidation (Polen et al. 2013).

The synthetic pathway of the reversal of β-oxidation of dicarboxylic acids may produce adipic acid, where the condensation of AcCoA and SucCoA to form C6 backbone and subsequent reduction, dehydration, dehydrogenation, and release of adipic acid from its thioester in *E. coli* (Yu et al. 2014). In this case, *E. coli* QZ1111 (MG1655 Δ*ptsG*Δ*poxB*Δ*pta*Δ*sdhA* Δ*iclR*) was transformed with the plasmid harboring adipic acid synthetic pathway genes such as *paaJ*, *paaH1*, *ter*, *ptb*, and *buk1*, which resulted in adipic acid production of 639 µg/*l* (Yu et al. 2014).

## 6. Amino Acids and Related Fermentation

The industrial amino acid (AA) production has been attracted much attention during more than 50 years (Razak and Viswanath 2015). Among them, L-glutamic acid may be the largest production scale, followed by L-lysine and DL-methionine (Anastassiadis 2007). The important usages of AAs are food preservative, feed supplements, therapeutic agents and precursors for the production of agrochemicals (Leuchtenberger 2005). Among the AA producing microorganisms, Gram-positive, non-sporulating bacterium, *C. glutamicum* has been extensively used for its GRAS (Generally Regarded As Safe) status. Historically, the isolation of *C. glutamicum* was made for the production of glutamate, and then mutagenesis and screening were made for L-lisine production by Kyowa Hakko Co., Japan (Kinoshita et al. 1957, 1958, Udaka 1960). The biotechnological production of L-lysine has been improved since then (Pfefferle et al. 2003, Wendisch and Bott 2005, Razak and Viswanath 2015). Metabolic engineering techniques have been applied to *C. glutamicum* for the production of other AAs such as L-threonine (Shiio 1990, Kase and Nakayama 1974, Shiio et al. 1991), L-methionine (Nakayama and Araki 1973, Kalinowski et al. 2003), L-serine (Eggeling 2007), L-histidine (Araki et al. 2007), L-valine (Ruklisha et al. 2007), L-tryptophan (Ikeda 2006), L-phenylalanine and L-tyrosine (Ikeda and Katsumata 1992), L-leucine (Patek 2007), and L-isoleucine (Guillout et al. 2002).

The aromatic compounds are used for a variety of industrial applications such as organic solvents, dyes, and precursors for the synthesis of numerous products for food, pharmaceutical, and chemical industry (Gosset 2009). The aromatic amino acids (AAs) such as L-phenylalanine (L-Phe), L-tyrosine (L-Tyr), and L-tryptophan (L-Trp) originate from the common shikimic acid (SA), which is synthesized from PEP and E4P in the central metabolic pathways (Fig. 20).

### 6.1 Shikimic acid fermentation

In *E. coli*, the 1st pathway for SA synthesis is the condensation of PEP and E4P into 3-deoxy-D-arabinoheptulosonate 7-phosphate (DAHP) by the DAHP synthase isozymes AroF, AroG, and AroH, encoded by *aroF*, *aroG*, and *aroH*, respectively (Fig. 20). DAHP is then converted to 3-dehydroquinic acid (DHQ) by DHQ synthase encoded by *aroD*, and then 5-dehydroshikimic acid (DHS) is transformed to SA by shikimate dehydrogenase encoded by *aroE*. SA is then converted to S3P by shikimate kinase I and II encoded by *aroL* and *aroK*, respectively. Finally S3P is converted to 5-enolpyruvil-shikimate-3-phosphate (EPSP) by 3-phosphoshikimate-1-carboxyvinyl transferase encoded by *aroA*. The last step in the SA pathway is the conversion of EPSP to chorismic acid (CHA) by CHA synthase encoded by *aroC*. CHA is the common building block for aromatic amino acids and such compounds as quinone, menaquinone, and enterobactin (Kramer et al. 2003, Ghosh et al. 2012, Tripathi et al. 2013).

SA has gained recent interest as the starting compound for the specific chemicals such as OSP (Oseltamivir phosphate) which serves as the potent inhibitor of the neuraminidase enzyme located on the surface of

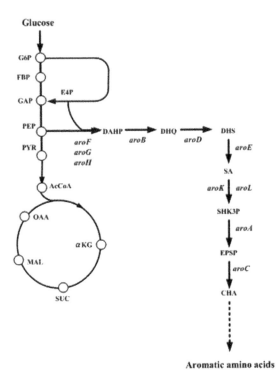

**Figure 20.** Shikimic acid and aromatic acid synthetic pathways.

the influenza virus, commercially known as Tamiflu[R] (Kramer et al. 2003, Ghosh et al. 2012, Tripathi et al. 2013). OSP prevents the release of newly formed virus particles from influenza virus type A and B, avian influenza virus H5N1, and human influenza virus H1N1 (Cortes-Tolalpa et al. 2014). Due to recent threat of possible pandemic influenza outbreak, it has been speculated that antiviral would be insufficient in particular in the developing countries. With this background, microbial SA production has been paid recent attention (Cortes-Tolalpa et al. 2014).

SA production has been made by metabolically engineered *E. coli* by amplifying *aroF*, *aroG*, and *tktA* genes, as well as *aroB* and *aroE* genes together with inactivation of *aroK* and *aroL* genes with the resultant yield ranging from 0.08 to 0.42 mol SA/mol glucose (Cortes-Tolalpa et al. 2014, Chandran et al. 2003, Johansson et al. 2005, Escalante et al. 2010, Chen et al. 2012, Rodriguez et al. 2013, Cui et al. 2014). The basic idea behind such mutations is to increase the precursor metabolites such as PEP and E4P by PTS mutation and amplify TktA pathway. The *pykF* gene knockout also causes the accumulation of PEP together with the activation of the PP pathway (Siddiquee et al. 2004a, 2004b), and SA pathway may be activated

(Kedar et al. 2007). In fact, the inactivation of *ptsHIcrr*, *aroK*, *aroL*, *pykF*, and *lacI* shows higher titer, yield, and productivity up to 43 g/L of SA in 30 h (Rodriguez et al. 2013).

### 6.2 Phenylalanin and biofuels production by modulating Csr

The global regulator such as carbon storage regulator Csr influences a variety of physiological processes such as central carbon metabolism, biofilm formation, motility, peptide uptake, virulence and pathogenesis, quorum sensing, and oxidative stress response (Chapter 4). Csr is controlled by the RNA-binding protein CsrA, a posttranscriptional global regulator that regulates mRNA stability and translation. CsrA is regulated by two sRNAs called CsrB and CsrC in *E. coli*. In principle, CsrA represses glycogen synthetic pathway genes such as *pgm, glgC, glgA*, and *glgB*, as well as gluconeogenic pathway genes such as *fbp, ppsA*, and *pckA* genes, while glycolysis genes such as *pgi, pfkA, tpiA, eno*, and *pykF* genes are activated (Chapter 4).

Manipulation of Csr has been demonstrated to increase the production of phenylalanine (Tatarko and Romeo 2001, Yakandawara et al. 2007). Namely this gene knockout activates gluconeogenesis gene such as *ppsA*, and thus PEP is accumulated, which contributes for the aromatic amino acids production.

### 6.3 Production of L-Tyrosine and its derived compounds

L-Tyrosine is used as a dietary supplemental and as a precursor for the synthesis of drugs, polymers, and phenylpropanoids (Lutke-Eversloh et al. 2007, Leonard et al. 2007). L-Tyrosine can be produced in *E. coli* by over-expressing *aroG* and *tyrA* encoding feed-back-inhibition resistant of 3-deoxy-D-arabino-heptulosonate-7-phosphate synthase (DAHPS) and chorismate-mutase-prephenate dehydrogenase (CM-PDH), respectively. In order to increase the precursor metabolites such as PEP and E4P, *ppsA* and *tktA* genes may be over-expressed, which yields 9.7 g/L of L-Tyr from glucose (Luthke-Eversloh et al. 2007).

## 7. Isoprenoid, Polyketide, and Alkanoid Production

Isoprenoid is synthesized in recombinant *E. coli* from such precursors as pyruvate and GAP (Fig. 21), where lycopene biosynthesis branches from the isoprenoid pathway. The phosphoenol pyruvate synthase (Pps) controls the balance between PYR and GAP, where it may be considered to control the carbon flux toward lycopene production by modulating *glnAp2* by acetyl phosphate (while *glnAp1* is under control of cAMP, where the *glnAp2*

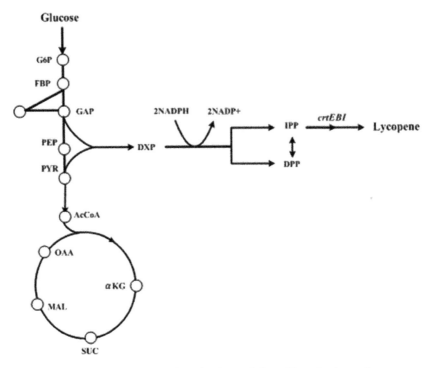

**Figure 21.** Isoprenoid (lycopen) production and alcanoid production pathways.

promoter controls the expression of *idi* (coding for isopentenyl diphosphate isomerase) and *ppsA* (Farmer and Liao 2000).

Polyketides are used as antibiotic, immunosuppressant, antitumor, antifungal and antiparasitic agents. All polyketides are assembled by successive round of decarboxylative condensation between an acyl thioester and α-carboxythioester in a similar way as fatty acid synthesis (Yuzawa et al. 2011). Polyketides are commonly produced from the precursors such as malonyl-CoA and (2S)-methylmalonyl-CoA, where *E. coli* produces only the former metabolite at the sufficient level to promote polyketide synthesis, and thus some metabolic engineering strategy is required to generate (2S)-methylmalonyl-CoA (Yuzawa et al. 2011).

## 8. Biofuel Production

As mentioned in introduction, biofuel production from biomass has been paid much attention from both environmental protection and energy generation points of view. The typical biofuels include hydrogen, ethanol, 1-butanol, isobutanol, 1-propanol, isopropanol, 3-methyl-1-butanol, etc. (Stephanopoulos 2008, Liao et al. 2016).

The elevation of CsrB level changed metabolite and protein levels in the glycolysis, the TCA cycle, and amino acid levels, resulting in the improved production of biofuels through native fatty acid, and heterologous n-butanol and isoprenoid pathways with concomitant decreases in acetate and $CO_2$ productions (Mckee et al. 2012).

## 8.1 Hydrogen production

Hydrogen is a clean energy without producing pollutants such as CO, $CO_2$, and $SO_2$ during combustion. Microbial fermentation is a potential approach for large-scale hydrogen production (Das and Veziroglu 2001). There are two primary means for this such as photosynthesis and fermentation, where the hydrogen production rates by photosynthetic systems are orders of magnitude lower than those of microbial fermentation (Levin et al. 2004).

In *E. coli*, hydrogen is produced from formate through the reaction $HCOO^- + H_2O \Leftrightarrow H_2 + HCO_3^-$, catalyzed by the formate hydrogenlyase (FHL) complex (Wood 1936). The FHL complex is composed of such components as formate dehydrogenase H, encoded by *fdhF* (Axley et al. 1990), which converts formate to $2H^+$, $2e^-$, and $CO_2$; hydrogenase 3 (Hyd-3), encoded by *hycE* (large subunit) and *hycG* (small subunit), a NiFe hydrogenase, that synthesize molecular hydrogen from $2H^+$ and $2e^-$ (Sauter et al. 1992); and the electron transported proteins encoded by *hycBCDE*, whch are shuttle electrons between formate dehydrogenase H and Hyd-3 (Sawer 1994). An active FHL complex also requires the protease HycI, the putative electron carrier HydN, and maturation proteins such as HycH, HypF, and HypABCDE (Sanchez-Torres et al. 2009).

The FHL complex has the regulators such as FhlA and HycA, where FhlA is transcriptionally activates the expression of *fdhF* gene and *hyc, hyp*, and *hydN-hypF* operons, which form the formate regulon (Korsa and Bock 1997). The *fhlA* gene is also regulated by Fnr (Rossmann et al. 1991). The translation of FhlA is inhibited by the RNA regulator OxyS (Argaman and Altuvia 2000). HycA, the product of the first *hyc* operon, represses *hyc* and *hyp* (Sauter et al. 1992).

Several attempts have been made to enhance hydrogen production in *E. coli* by metabolic engineering (Penfold et al. 2003, 2006, Yoshida et al. 2005). In particular, significant performance improvement could be made by the mutations in *hyaB, hybC, hycA*, and *fdoG* together with overexpression of *fhlA* (*fhlA+*), where formate was used as a substrate (Maeda et al. 2008a). Another mutation was considered in *hyaB, hybC, hycA, fdoG, frdC, ldhA*, and *aceE*, where this mutant yielded higher hydrogen production from glucose (Maeda et al. 2007). Further improvement could be made by error-prone PCR (epPCR), DNA shuffling, and saturation mutagenesis of *hycE*, the larger subunit of Hyd-3 (Maeda et al. 2008b). Further protein engineering approach has been made to improve hydrogen production in *E. coli* by epPCR and saturation mutagenesis of *fhlA* (Sanchez-Torres et al. 2009).

## 8.2 Ethanol fermentation

As mentioned in Chapter 2, it is highly desirable to use lignocellulosic biomass for the production of biofuels and biochemicals. The hydrolysates of lignocellulosic biomass contain a monomeric hexoses (glucose, mannose, and galactose) and pentoses (xylose and arabinose) as well as ligin. The most abundant sugar is glucose, followed by xylose (from hardwoods and agricultural residues) or mannose (from softwoods), where other sugars are much lower. While many microorganisms can efficiently assimilate hexoses such as glucose for the cell growth, the assimilation of pentoses such as xylose remain relatively inefficient (Chu and Lee 2007).

Several microorganisms have been investigated for ethanol fermentation from lignocellulosic hydrolysates. Among them, *S. cerevisiae, Pichia stipitis, Zymomonas mobilis,* and *E. coli* have been extensively investigated for ethanol fermentation. For the ethanol production from lignocellulosic hydrolysates, saccharolytic enzyme is needed for the hydrolysis of polysaccharides available from pre-treated biomass, and fermentation must be made using hexose and pentose sugars obtained after saccharification. The conversion of lignocelluloses to ethanol may be made in separate hydrolysis and fermentation (SHF), or in one process such as simultaneous saccharification and co-fermentation (SSCF) or consolidated bioprocessing (CBP).

Traditionally, *S. cerevisiae* (Gray et al. 2006, Lin and Tanaka 2006, Mielenz 2001, Service 2007) and *Z. mobilis* (Deanda et al. 1996, Zhang et al. 1995) have been employed for the ethanol production from hexose sugars, but these organisms cannot assimilate pentose sugars (Gray et al. 2006, Lin and Tanaka 2006, Mielendez 2001, Service 2007). Many metabolic engineering strategies have been considered to develop mutant strains that can assimilate both hexose and pentose sugars (Chen 2011, Chu and Lee 2007, Hahn-Hagerdal et al. 2007, Jeffries 2000, Jeffries and Jin 2004).

*S. cerevisiae* is a **GRAS (Generally Regarded as Safe)** microorganism, which can assimilate hexose sugars such as D-glucose, D-mannose, and D-galactose, as well as disaccharides like sucrose and maltose, and commonly used for ethanol fermentation. Moreover, *S. cerevisiae* is relatively tolerant to lignocelluloses-derived inhibitors and high osmotic pressure (Girio et al. 2010). Although *S. cerevisiae* harbor the xylose assimilation pathway genes (Batt et al. 1986), encoding xylose reductase (XR) (*YHR104w, GRE3*) (Traff et al. 2002), xylitol dehydrogenase (XDH) (*YLR070c,ScXYL2*) (Richard et al. 1999), and xylulokinase (XK) (*XKS1*) (Yang and Jeffries 1997, Richard et al. 2000), their expression levels are too low, and do not support the growth on xylose (Barnett 1976). In fact, *S. cerevisiae* can grow on xylose with very slow growth under aerobic condition (Attfield and Bell 2006), or by overexpression of the above endogenous genes (Toivari et al. 2004).

While *S. cerevisiae* has been extensively used for ethanol fermentation from glucose, xylose-fermenting yeasts such as *Pichia stipitis, Candida shehatae, Pachysolen tannophilus* have been also paid attention, where *P. stipitis* and *C. shehatae* are the native ethanol producers with the theoretical maximum yield of 0.51 g ethanol/g xylose (Table 2). However, such pentose-assimilating yeasts generally require oxygen for efficient xylose assimilation (Neirinck et al. 1984, Ligthelm et al. 1988, Yu et al. 1995). The aerobic condition shifts the metabolism from ethanol formation to ethanol oxidation and activates respiration for energy generation, causing lower ethanol yield (Mahmourides et al. 1985, Maleszka and Schneider 1982). Moreover, these yeasts are less tolerant to low pH, ethanol, and lignocellulosic hydrolysate inhibitors as compared to *S. cerevisiae* (Harn-Hagerdal et al. 1994).

Filamentous fungi can degrade hemicelluloses yielding oligosaccharides and monosaccharides, and convert to ethanol, but its production rate is low and less tolerant to ethanol as compared to *S. cerevisiae* (Singh et al. 1992). Some yeasts such as *Candida guilliermondii* and *Debaryomyces hansenii* can produce ethanol from glucose, while these produce xylitol from xylose (Mussatto and Roberto 2004, Carvalheiro et al. 2005).

On the other hand, most bacteria can assimilate a broad range of substrates, but ethanol is rarely a single metabolite, where other metabolites are produced as well. For example, *E. coli* has the ability to ferment a wide range of hexose and pentose sugars, but the problems of using *E. coli* are the narrow pH range (6–8) for growth (which is susceptible to contamination), low tolerance to lignocelluloses-derived inhibitors, low ethanol tolerance, and mixed metabolite formation (formate, lactate, acetate, succinate, as well as ethanol), resulting in low ethanol yield (Dien et al. 2003).

*Z. mobilis* has higher ethanol production rate than *S. cerevisiae*, and tolerant to higher concentrations up to 120–135 g/l (Osman and Ingram 1985). However, *Z. mobilis* uses exclusively Entner-Doudoroff (ED) pathway under anaerobic condition, and thus only one ATP is generated per mole of glucose in contrast to 2ATP via EMP pathway, resulting in lower cell growth rate. Moreover, *Z. mobilis* lacks the ability to assimilate sugars available from lignocellulosic biomass except D-glucose (and fructose) (Girio et al. 2010).

Other enteric bacteria such as *Klebsiella oxytoca* and *Erwinina ohrysanthemi*, etc. may be also considered for ethanol production, but these have the similar characteristics as *E. coli*. Some bacteria such as *Klebsiella, Bacillus, Clostridium, Thermotoga* are able to produce cellulase and hemicellulases to assimilate lignocellulosic hydrolysates, and therefore, these can be considered to be a candidate as a suitable cell factories for CBP (Girio et al. 2010).

After hydrolysis of hemicellulose, monosaccharides, disaccharides, and oligosaccharides must be transported across the cell membrane.

The transport systems for microbial sugar uptake may be classifies as (1) facilitated diffusion system, which does not require energy, where the concentration gradient across the membrane is the driving force for the sugar transport, and (2) active transport system, which requires energy as (i) membrane potential (e.g., proton symporters), as (ii) ATP (ATP-binding cassette-ABC transporters), or as (iii) PEP (carbohydrate-PEP phosphotransferase system: PTS transport systems) (Girio et al. 2011).

Xylose is taken up by the facilitated diffusion through the hexose transporters encoded by the *HXT* gene family in *S. cerevisiae* (Kruckeberg 1996), where the hexose transporters are classified into two such as *HXT2*, *HXT6*, *HXT7* with high affinity, and *HXT1*, *HXT3*, *HXT4* with low affinity to glucose. The former transporters are active under low glucose concentration, while the latter transporters are active under high glucose concentration (Lee et al. 2002). The high affinity system is repressed by glucose, while the low affinity system is constitutively expressed (Kotter and Ciriacy 1993, Ramos et al. 1988). The $K_m$ values for xylose are high about 137–190 mM and 1.5 M for the high and low affinity transport systems, respectively, and thus xylose uptake through such transporters is significantly low as compared to glucose (Kotter and Ciriacy 1993).

As shown in Fig. 22, the metabolic pathways for the assimilation of D-xylose and L-arabinose are different between bacteria and fungi (Harn-Hagerdal et al. 2007). In bacteria, D-xylose is first isomerized to D-xylulose by xylulose isomerase (XI: EC 5.3.1.5), which is then phosphorylated by xylulokinase (XK: EC 2.7.1.17) to yield D-xylulose 5-phosphate (X5P), where X5P is the entry to the PP pathway. In the case of arabinose assimilation, L-arabinose is first isomerized by L-arabinose isomerase (AI: EC 5.3.1.4) to yield L-ribulose, which is then phosphorylated by L-ribulokinase (RK: EC 2.7.1.16) to yield L-ribulose 5-phosphate, which is then epimerized by L-ribulose-5P 4-epimerase (L-RPE: EC 5.1.3.1) to yield X5P (Harn Hagerdal et al. 2007).

In most fungi and some yeasts, D-xylose is first reduced to xylitol by NAD(P)H-dependent xylose (aldose) reductase (XR: EC 1.1.1.21), and then xylitol is oxidized to D-xylulose by NAD⁺-dependent xylitol dehydrogenase, a $Zn^{2+}$-dependent medium chain alcohol/sorbitol dehydrogenase (XDH: EC 1.1.1.9). D-Xylulose is then phosphorylated at the C5-OH position by xylulokinase (XK: EC 2.7.1.17) by the phosphate donner such as ATP to yield X5P (Matsushika et al. 2009).

The oxygen level significantly affects the cell growth, ethanol formation, and xylitol accumulation in pentose-fermenting yeasts. It is important to control the oxygen level at low for the maximal ethanol production in *P. stipitis* and *C. shehatae* (Delgenes et al. 1986, Skoog and Hahn-Hagerdel 1990). The xylitol formation depends on the activities of XR (for incoming flux to xylitol) and XDH (for outgoing flux from xylitol), where XR is strictly NADPH dependent, and XDH is NAD⁺-linked in such yeasts as *D. hansenii* (Girio et al. 1994) and *C. utilis* (Bruinenberg et al. 1984b), while

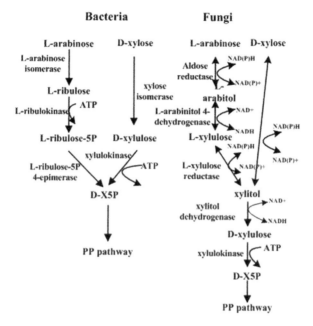

**Figure 22.** D-Xylose and L-arabinose utilization pathways in bacteria and fungi (Hahn-Hagerdal et al. 2007).

XR has dual co-factor specificity for NADPH and NADH, yet the former being the preferred cofactor in such yeasts as *P. stipitis* (Rizzi et al. 1988, Verduyn et al. 1985), *C. shehatae*, and *P. tannophilus* (Bruinenberg et al. 1984a). The different cofactor requirement for XR and XDH reactions (NADPH and NAD⁺, respectively) affect to NADP⁺ and NADH formation.

Unlike bacteria such as *E. coli*, yeast does not possess transhydrogenase, and therefore, the interconversion between NADPH and NADH cannot be made (Bruinenberg et al. 1985). Under aerobic condition, NADH goes into respiratory chain and generates energy, regenerating NAD⁺, while under oxygen limitation, NAD⁺ is not efficiently regenerated, and deactivates XDH, causing xylitol accumulation. In particular, xylitol is produced as a major product from xylose under oxygen limitation in such yeasts as *D. hansenii* and *Candida guilliermondii*, which are strictly NADPH-dependent XR (Girio et al. 1994, Silva et al. 1996), while xylitol is less produced in the yeasts having dual cofactor specificity for XR (Bruinenberg et al. 1983). For the fermentation using the hydrolysates from hemicelluloses under oxygen limitation, the accumulation of NADH may be modulated by induction of the NADH-specific glutamate dehydrogenase in the glutamate decarboxylase bypass as made in *P. stipitis* (Jeffries 2008).

On the other hand, L-arabinose is first reduced to L-arabitol, which undergoes additional oxidation and reduction steps, yielding xylitol (Fig. 22b). L-arabinose is first converted to L-ribulose by L-arabinose isomerase (AI) encoded by *araA*, and to L-ribulose 5-phosphate by L-ribulokinase (RK) encoded by *araB*, and converted to X5P by L-ribulose-5P 4-epimerase (L-RPE) encoded by *araD* in bacteria such as *E. coli* (Englesberg 1961, Englesberg et al. 1969), *B. subtilis* (Sa-Nogueira and de Lencastre 1989), and *Lactobacillus plantarum* (Burma and Horeoker 1958a, Burma and Horecker 1958b, Heath et al. 1958), where *araBAD* form an operon.

In fungi, L-arabinose is first converted to L-arabitol by L-arabinose reductase (AR) (Chiang and Knight 1961), and L-arabitol is oxidized to L-xylulose by L-arabitol 4-dehydrogenase (LADH), and then L-xylulose is converted to xylitol by L-xylulose reductase (LXR), where xylitol is metabolized as the same pathway as mentioned above for the assimilation of D-xylose (Girio et al. 2010).

The fungal L-arabinose catabolic pathway has been identified in *Penicilium chrysogenum* (Chiang and Knight 1961), *A. niger* (Wittevee et al. 1989), and *H. jecorina* (Richard et al. 2002), as well as in yeasts such as *C. arabinofermentans* and *P. guilliermondii* (Fonseca et al. 2007a,b). The reductases such as AR and LXR require NADPH as cofactor, while dehydogenases such as LADH and XDH require $NAD^+$. Under oxygen limitation, NADH is less reoxidized in the respiratory chain, and causes a shortage in $NAD^+$, resulting in L-arabitol accumulation and secretion (Fonseca et al. 2007b). The aldose reductase can use either NADPH or NADH in *P. stipitis* (Verduyen et al. 1985), while LXR is NADH dependent in yeast *Ambrosiozyma monospora* (Verho et al. 2004) and in *C. arabinofermentans* (Fonseca et al. 2007a). These may affect L-arabinose fermentation (Girio et al. 2010). Figure 23 shows various pathways of different strains for ethanol production from hemicelluloses derived monosaccharides.

Unlike fungi, *S. cerevisiae* cannot hydrolyze either cellulosic or hemicellulosic biomass. Several attempts have, therefore, been made to express the heterologous genes encoding cellulolytic and hemicellulolytic enzymes on the cell surface aiming at the ethanol fermentation by consolidated bioprocess (CBP). Xylanase expression on the cell surface of *S. cerevisiae* allows the direct conversion of birchwood xylan to ethanol with a yield of 0.3 g/g (Fujita et al. 2004, Katahira et al. 2004). Several other hemicellulases were also expressed in *S. cerevisiae*, including mannan degrading enzymes (Setati et al. 2001, Stalbrand et al. 1995), xylanases and side chain-splitting enzymes (La Grange et al. 2001, Perez-Gonzalez et al. 1996). The hydrolytic enzymes from *T. reessei*, α-L-arabinofuranosidase and β-xylosidase were co-expressed in *S. cerevisiae* (Margolles-Clark et al. 1996), releasing L-arabinose and D-xylose, respectively. The industrial *S. cerevisiae* has been engineered for the fermentation of cellobiose, which could be achieved by expressing minicellulosome on the cell surface of *S. cerevisiae* (Lilly et al. 2009). However, the expression of so many enzymes on the cell surface may affect the substrate up take as mentioned in Chapter 4, and may represent a metabolic burden, that negatively influences the fermentation capacity (Gorgens et al. 2001).

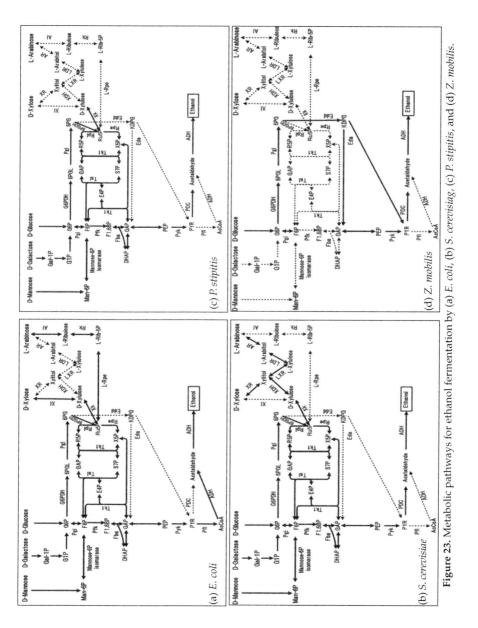

**Figure 23.** Metabolic pathways for ethanol fermentation by (a) *E. coli*, (b) *S. cerevisiae*, (c) *P. stipitis*, and (d) *Z. mobilis*.

In order to improve the assimilation of D-xylose, the heterologous genes such as *XYL1* encoding XR and *XYL2* encoding XDH from *P. stipitis* were expressed in *S. cerevisiae* (Kotter and Ciriacy 1990, Tantirungkij et al. 1993). This strain, however, produces xylitol instead of ethanol from D-xylose. This may be due to the limitation of XK encoded by *XKS1*, and thus the additional overexpression of the endogeneous *XKS1* lowers xylitol accumulation, giving higher ethanol production, but with low growth rate under anaerobic condition in *S. cerevisiae* (Eliasson et al. 2000b, Ho et al. 1998, Johansson et al. 2001, Tovari et al. 2001). There might be an optimal XR:XDH ratio (about 0.06) for higher ethanol production with lower xylitol accumulation (Walfridsson et al. 1997).

As such, the site-directed mutagenesis in XR and XDH modified the cofactor specificity towards NADH or NADP[+], to overcome the redox imbalance, leading to higher xylose consumption rate, with higher ethanol production and lower xylitol production (Bengtsson et al. 2009, Krahulec et al. 2009, Petschacher and Nidetzky 2008). The disruption of *ZWF1* gene encoding glucose 6-phosphate dehydrogenase (G6PDH) reduces xylitol production with lower xylose consumption in xylose-utilizing recombinant *S. cerevisiae* (TMB3255) (Jeppsson et al. 2002). Although the modulation of G6PDH activity at lower level may enhance xylose consumption, keeping less production of xylitol and more ethanol production, low G6PDH-activity strains are sensitive to lignocelluloses-derived inhibitors and $H_2O_2$ together with low growth rate (Jeppsson et al. 2003). The expression of heterologous transhydrogenase from *Acetobacter vinelandii* allows to convert NADPH to NADH, resulted in lower xylitol yield, and enhance glycerol yield during xylose utilization (Jeppsson et al. 2003). On the other hand, the impairment of trehalose synthesis (Hohmann et al. 1996) and the reduction of Pgi activity (Boles et al. 1993) allowed the improved ethanol yield from xylose (Eliasson et al. 2000a). This may be explained by the accumulation of the intracellular F6P and FBP, which are required for the induction of ethanologenic enzymes, where FBP level is significantly low in the case of assimilating D-xylose as compared to the case of glucose (Senac and Hahn-Hagerdal 1990).

Pentose fermentation by *S. cerevisiae* can be made by the combined expression of both *XYL1 / XYL2* for D-xylose assimilation, and the bacterial *E. coli araB, araD* and *B. subtilis araA* for L-arabinose assimilation (Karhumaa et al. 2006, Bettiga et al. 2008, 2009). The overexpression of *Piromyces XylA*, *S. cerevisiae XKS1*, and *L. plantanum araA, araB*, and *araD*, as well as endogenous *RPE1, RKI1, TKL1, TAL1* together with optimized evolution (under anaerobic condition) allowed the relatively higher ethanol production (0.43 g/g of total sugar) from a mixture of glucose (30 g/*l*), xylose (15 g/*l*) and arabinose (15 g/*l*) (Wisselink et al. 2009).

Metabolic engineering of xylose-assimilating pathway allows *S. cerevisiae* to produce 40 g/*l* of ethanol with a yield of 0.46 and a productivity of 0.8 g/*l*.h from a mixture of 57.5 g/*l* of glucose and 28.1 g/*l* of xylose obtained from ammonia fiber expansion (AFEX)-corn stover

(CS)-hydrolysates (Lau and dale 2009). Moreover, engineered *S. cerevisiae* could produce 48 g/*l* of ethanol with a yield of 0.37 g/g, and a productivity of 0.8 g/*l*.h from a mixture of 80 g/*l* of cellobiose, 40 g/*l* of xylose, and 10 g/*l* of glucose (Ha et al. 2011).

Contrary to *S. cerevisiae*, *P. stipitis* is able to naturally assimilate L-arabinose and/or D-xylose, and produces ethanol (Jeffries et al. 2007, Jeffries 2008, Jeffries and van Vleet 2009). The ethanol production by *P. stipitis* has been investigated from a variety of lignocellulosic biomass such as recycled paper sludge (by SHF and SSF) (Marques et al. 2008), hardwood (Nigam et al. 2001a, 2001b), wheat straw (Nigam et al. 2001c), and sugar cane bagasse (van Zyl et al. 1988). Several metabolic engineering strategies have been applied to develop the mutants with higher fermentation capacity and xylose utilization for ethanol production (Jeffries 2008). The *P. stipitis* is, however, unable to grow anaerobically, and is more sensitive to ethanol and inhibitors as compared to *S. cerevisiae*. The heterologous expression of *URA1* gene encoding the dihydroorotate dehydrogenase of *S. cerevisiae* allows *P. stipitis* to grow under anaerobiosis (Shi and Jeffries 1998). Moreover, the disruption of cytochrome c gene increased xylose fermentation for ethanol production (yield) (Shi et al. 1999). The heterologous expression of fungal xylanase on the cell surface of *P. stipitis* allows the direct conversion of xylan to ethanol (CBP) (Den Haan and van Zyl 2003).

*Klebsiella oxytoca* is able to naturally metabolize xylo-oligo-saccharides (XOS) (Qian et al. 2003). *K. oxytoca* M5A1 has been engineered by incorporating *pdc* and *adh* genes from *Z. mobilis* for ethanol production from glucose and xylose, where the maximum productivity from xylose was comparable to glucose, and almost twice that of *E. coli* KO11 (Ohta et al. 1991b). In order to improve the genetic stability, chromosomal intergration of the heterologous genes was made, allowing ethanol fermentation by SSF from hydrolysates by the resulting strain (Doran et al. 1994). This *K. oxytoca* P2 strain can co-ferment glucose, arabinose, and xylose by this order of preference (Bothast et al. 1994).

*Zmomonas mobilis* CP4 has been shown to be the efficient ethanol producer from glucose (Skotnicki et al. 1981). *Z. mobilis* has a facilitated diffusion transport system for glucose (Dimarco and Romano 1985, Parker et al. 1995, Weisser et al. 1995). Since *Z. mobilis* lacks the ability of assimilating hemicellulose-derived monosaccharides except glucose, an attempt has been made to incorporate a plasmid of the *E. coli* genes encoding XI and XK as well as the PP pathway enzymes such as transaldorase (Tal) and transketolase (Tkt) to attain xylose fermentation, giving 86% of the theoretical yield from xylose (Zhang et al. 1995). The similar approach has also been considered in *Z. mobilis* ATCC39676 for arabinose fermentation, where the genes from the *E. coli* operon *araBAD* encoding AI, RK, and L-RPE, as well as Tal and Tkt, giving a high yield of 96% from L-arabinose, but with low production rate (Deanda et al. 1996). The same ATCC 39676 strain expressing xylose assimilating pathway genes has been subjected to long-term (149 d) adaptation in continuous culture of hemicelluloses hydrolysate containing xylose, glucose, and acetic acid (Lowford et al. 1999). Moreover, the DNA integration of the xylose and arabinose assimilating pathway genes (AX101 strain) allows co-fermentation of glucose, xylose, and arabinose by this order of preference, giving a yield of 84% (Mohagheghi et al. 2002). Engineering of Z. *mobilis* allows the production of ethanol from a mixture of 75 g/l of glucose and 75 g/l of xylose, yielding a titer of 62 g/l, a yield of 0.46 g/g, and a productivity of 1.29 g/l.h (Joachimsthal et al. 1999).

In bacteria, glucose, mannose, and galactose are transported by PTS-systems. In E. *coli*, D-xylose is transported by a high-affinity ABC-xylose transporter and a low affinity proton symporter (Lam et al. 1980, Sumiya et al. 1985). Since xylose transport requires energy, xylose uptake is not efficient (Khankal et al. 2008). E. *coli* has the similar transporters for L-arabinose such as high-affinity ABC-arabinose transporter and a low affinity arabinose-symporter (Kolodrubets and Schleif 1981, Novotny and Englesberg 1966).

Unlike S. *cerevisiae* and Z. *mobilis*, E. *coli* has the ability to ferment hexose and pentose sugars, and is able to produce ethanol by various genetic manipulation (Ingram et al. 1999), although this organism cannot tolerate higher ethanol concentration. The native fermentation pathways in E. *coli* may be changed by the introduction of the heterologous ethanol-forming pathway genes for pyruvate decarboxylase (*pdc*) and alcohol dehydrogenase (*adhB*) from Z. *mobilis* into the chromosome of the *pfl* gene (to avoid AcCoA formation from pyruvate). Moreover, the disruption of fumarate reductase gene (*frd*) (to prevent succinate production) allows the ethanol production of 54.4 g/l from 100 g/l of glucose with nearly the theoretical yield and the productivity of 1.2 g/l.h. In the case of using xylose as a carbon source, 41.6 g/l of ethanol could be produced from 80 g/l of xylose, with nearly the theoretical yield and the productivity of 1.3 g/l.h (Ohta et al. 1991a), where nearly or more than the theoretical yield could be attained, which may be due to the usage of rich LB medium. Although *ldhA* gene was not disrupted, lactate production is low (Ohta et al. 1991a), which may be due to low $K_m$ of PDC for pyruvate as compared to LDH.

Since this strain (KO11) cannot grow at high ethanol concentration, subsequent repeated experiments were conducted to select ethanol tolerant strains, and the resultant strain (LY01) was able to produce 61.0 g/l of ethanol with 85.4% of the theoretical yield from 140 g/l of glucose in 96 h, while 63.2 g/l of ethanol could be produced in 96 h from 140 g/l of xylose with 88.5% of the theoretical yield (Yomano et al. 1998). However, this strain requires complex nutrients, and the major problems for KO11 and the related strain are the suboptimal expression from the *pflB* integration site of the Z. *mobilis* genes, absence of a complete set of alcohol producing genes from Z. *mobilis* (*pdc* and *adhB* but lacking *adhA*), and the presence of an antibiotic resistant genes such as *cat* (Yomano et al. 2008). Then the complete Z. *mobilis* ethanol pathway genes such as *pdc*, *adhA*, and *adhB* were integrated into the chromosome, where the resulting strain (LY160) (KO11, Δ*frd*, Δ*adhE*, Δ*ldhA*, Δ*ackA*, *pflB*⁺) produced 40 g/*l* of ethanol from 90 g/*l* of xylose in the minimal salts medium, which is comparable to KO11 in LB medium (Yomano et al. 2008).

*E. coli* KO11 and the related strains were nutritionally starved for biosynthesis precursors, and restricted carbon flow into the oxidative TCA cycle, limiting the cell growth and the rate of ethanol production (Underwood et al. 2002a). Both the cell growth and the ethanol production can be improved by the expression of NADH-insensitive citrate synthase (CS) (encoded by *citZ*) from *Bacillus subtilis* (Underwood et al. 2002b), where unlike *E. coli*, Gram-positive CS is not repressed by the higher level of NADH. Genetic changes to optimize carbon partitioning between ethanol formation and biosynthesis have been considered (Underwood et al. 2002a), and an ethanol-tolerant mutant (LY01) has been developed (Gonzalez et al. 2003).

Inui et al. (2004) considered to express the *pdc* and *adhB* genes under the control of the endogenous *ldhA* promoter. Under oxygen limited conditions, the growth rate of this strain is low, but significant amount of ethanol is produced with high yield, while forming some amount of acetate, lactate and succinate as byproducts. The addition of small amounts of pyruvate (or acetaldehyde) to the culture medium increased the ethanol production rate. Disruption of the *ldhA* gene led to a further increase in the ethanol production rate without any production of lactate.

The strain with simultaneous mutations on *pfl* and *ldh* genes cannot grow under anaerobic condition without regenerating NAD⁺, where the transformation of the strain with plasmid pLOI297, which encodes *pdc* and *adhB* genes of which expression is under the control of the native *lac* promoter, restored the fermentative growth, and allows the ethanol production from arabinose, glucose, or xylose (Dien et al. 2003). The maintenance of pLOI297 is positively selected under anaerobic condition, because the cells lost the plasmid cannot grow (Dien et al. 2003). The resulting strain FBR5 produced 41.5 g/*l* of ethanol with 90% of the theoretical yield, and a productivity of 0.59 g/*l*.h (Dien et al. 2003).

One of the drawbacks of KO11 and its related strains is the phenotypic instability during repeated batch or continuous culture, losing the ability of producing ethanol in the absence of antibiotics. Namely, the transformation of the cell by overexpression of heterologous genes sometimes causes genetic instability, toxicity, containment, etc. in large scale fermentation (Munjal et al. 2012).

Since *E. coli* has ethanol producing pathway gene *adhE*, it is much preferred to utilize this pathway for ethanol production to overcome the problems associated with long-term genetic stability, etc. (Kim et al. 2007, Zhou et al. 2008). In fact, this was attempted by modulating endogenous pathways to enhance the yield and productivity of bioethanol in *E. coli* (Munjal et al. 2012). The wild type *E. coli* produces a mixture of metabolites under anaerobic condition, where lactate, formate (and $H_2$ and $CO_2$), ethanol, acetate, and succinate are the main metabolites to be produced, where NADH reoxidation is essential for the formation of such metabolites as lactate, ethanol, and succinate. Under aerobic condition, NADH ($FADH_2$) produced goes into respiratory chain to produce ATP, and the NADH/$NAD^+$ ratio is low, resulting in little production of such metabolites. As the oxygen level decreases, NADH/$NAD^+$ ratio increases and such metabolites tend to accumulate.

In the case of ethanol fermentation by *E. coli* without introducing heterologous genes such as *pdc* and *adhA* from *Z. mobilis*, *ldhA* and *pfl* genes must be first knocked out to avoid lactate and formate productions (Zhou et al. 2008). Under anaerobic condition, PDHc is repressed by ArcA, pyruvate is accumulated, and the cell growth of *ldhA.pfl* mutant is suppressed under anaerobic condition mainly due to shortage of $NAD^+$, since no pathway for NADH reoxidation is available for the mutant. If *arcA* gene was knocked out, the pyruvate can be converted to AcCoA with the production of $CO_2$ and NADH in PDH pathway. If further *pta* or *ackA* gene was knocked out, AcCoA can be exclusively converted to ethanol. However, acetate production cannot be repressed completely by such gene knockout, since Pox pathway may produce acetate directly from pyruvate (Rahman and Shimizu 2008, Rahman et al. 2008). Moreover, it must be careful about the knockout of *poxB* gene, since Pox pathway is connected to the respiratory chain, and thus this pathway knockout will cause some growth defect (Abdel-Hamid et al. 2001, Li et al. 2007). Since ArcA is a global regulator, and it affects not only the expression of *aceE, F,* and *lpdA* genes, but also the expression of the TCA cycle genes and the respiratory pathway genes (Appendix A), this gene knockout activates the TCA cycle with some improvement of the cell growth rate (Nizam and Shimizu 2008, Toya et al. 2012). It may thus be better to express PDH operon under *gapA* promoter, where *ldhA, frdA, ackA,* and *pflB* genes knockout mutant may produce ethanol exclusively, but this causes significant growth defect (Munjal et al. 2012). In order to overcome this growth defect, Munjal et al. (2012) modulated acetate kinase expression to regain the cell growth. As such, metabolic engineering practice may not necessarily give the expected result without deep understanding of the metabolic regulation mechanism in particular on the roles of transcription factors.

Bioethanol may be converted to ethylene and ethylene glycol (Jang et al. 2012), where ethylene is widely used as a platform chemicals, where its worldwide production is 120 million tons in 2008 (Seddon 2010). Bio-based ethylene and ethylene glycol can be used as building brock chemicals for the polyethylene (PE) and polyethylene terephthalate (PETP) (Jang et al. 2012).

## 8.3 Higher alcohol production

The higher alcohol such as 2-phenylethanol has the advantage of a higher energy density as compared to ethanol. This compound can be produced by recombinant *E. coli* from L-phenylalanine precursor, phenylpyruvate (PPY) by first converting it to 2-phenylacetaldehyde by 2-keto-acid decarboxylase (Kivd) from *Lactococcus lactis* and then to 2-phenylethanol by alcohol dehydrogenase 2 (Adh2) from *S. cerevisiae* (Atsumi et al. 2008a).

Although ethanol has been paid much attention with respect to its fuel potential, it may not be an ideal replacement for gasoline as it has a high water content and a low energy density relative to gasoline. Higher alcohols (C4 and C5) have energy densities similar to gasoline and are less volatile than ethanol, and thus may be considered to be the next generation biofuels (Atsumi et al. 2008b).

1-Butanol is hydrophobic and its energy content is similar to that of gasoline (Atsumi et al. 2008b). Moreover, its vapor pressure is lower than that of ethanol. Thus, 1-butanol may be a substitute for, or supplement to, gasoline as a transportation fuel. Microbial production of 1-butanol has been achieved using *Clostridium acetobutylicum* (Lin et al. 1983). *C. acetobutylicum* is a gram positive anaerobe and produces such metabolites as butylate, acetone, and ethanol (Fig. 24) (Jones et al. 1986). The industrial application is limited to some extent because of its slow growth rate, spore-forming life cycle, and the by-products formation.

Some attempts have been made to produce 1-butanol using recombinant *E. coli,* in which 1-butanol production pathway genes from AcCoA such as *thl, hbd, crt, bcd, etfAB* and *adhE2* were cloned and expressed in *E. coli* (Atsumi et al. 2008b). Moreover, Further mutation in such genes as *adhE, ldhA, frdBC, fnr* and *pta* may be considered to increase the yield of 1-butanol (Fig. 24).

Longer chain alcohols such as 1-propanol, 1-butanol, isopropanol, isobutanol, etc. are of interest from the points of view of high energy densities and their low hygroscopicities, which contribute to compact storage and easier distribution. 1-Propanol can be esterified to produce diesel fuels, and it can be dehydrated to produce propylene, where currently polypropylene is produced from petroleum due to cheap cost. 1-Butanol has been proposed to supplement gasoline as transportation fuel, where it has been mainly produced by using *Clostridium* sp. (Tracy et al. 2012, Jang et al. 2013). Some attempts have been made to produce such higher alcohols by engineered *E. coli* (Hanai et al. 2007, Atsumi et al. 2008b, Shen et al. 2008, Atsumi and Liao 2008), where it takes advantage of the amino acid and biosynthesis capacity to produce various 2-keto acids as well as the heterologous enzymes such as 2-keto acid decarboxylases (KDCs) and alcohol dehydrogenases (ADHs).

1-Propanol and 1-butanol can be formed through 2-keto-butyrate via the KDC and ADH pathway (Atsumi et al. 2008b, Atsumi and Liao

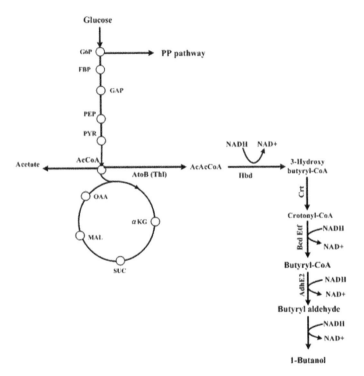

**Figure 24.** Higher alcohol (1-butanol) production pathways. AtoB: acetyl-CoA acetyltransferase, Thl: thiolase, Hbd: 3-hydroxybutyryl-CoA dehydrogenase, Crt: crotonase, Bcd: butyryl-CoA dehydrogenase, Etf: electrol transfer flavoprotein, AdhE2: aldehyde/alcohol dehydrogenase.

2008). 2-Keto butyrate is produced from threonine, and it is a precursor of isoleusine. 2-Ketobutyrate can be converted to 2-ketovalerate and 2-keto-3-methyl-valerate, where 1-butanol and 2-methyl-1-butanol are produced from such intermediates, respectively. In general, 2-ketobutyrate is synthesized via threonine originated from PEP. An alternative route to produce 2-keto-butyrate from PYR and AcCoA may be seen in such organism as *Methanococcus jannaschee* (Howell et al. 1999), and this pathway is shorter as compared to L-threonine synthetic pathway, and does not involve transamination followed by deamination. It may be, therefore, considered to introduce citramalate synthase (CimA) of *M. jannaschii* to produce (R)-citramalate from PYR and AcCoA, and then CimA is converted to 2-ketobutyrate via LeuCD- and LeuB-mediated reactions (Atsumi and Liao 2008), where the increased 1-propanol and 1-butanol can be made by directed evolution based on the requirement of L-isoleusine (Atsumi and Liao 2008).

As shown in Fig. 25, isopropanol can be produced from AcCoA by introducing such heterologous genes as *thl* for thiolase (Thl), *ctfAB* for

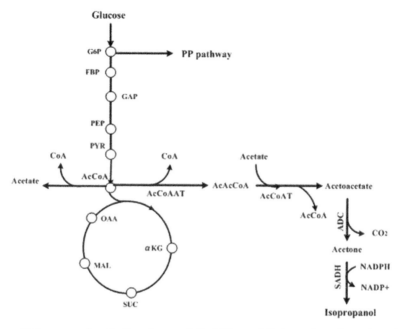

**Figure 25.** Isopropanol synthetic pathways. AcCoAAT: acetyl-CoA acetyltransferase (encoded by *E. coli atoB* or *C. acetobutyricum thl* gene), AcCoAT, acetyl-CoA transferase (encoded by *C. acetobutyricum ctfAB* or *E. coli atoAB* gene), ADC: acetoacetate decarboxylase (encoded by *C. acetobutyricum adc* gene), SADH: secondary alcohol dehydrogenase (encoded by *C. beijerinckii adh* or *T. brockii adh* gene).

CoA-transferase (Ctf), *adc* for acetoacetate decarboxylase (ADC), and *adh* for primary-secondary alcohol dehydrogenase (ADH) from *Clostridium acetobutyricum* into *E. coli* (Hanai et al. 2007, Jojima et al. 2008).

## 8.4 Fatty acid fermentation

Fatty acid metabolism has been paid recent attention for the production of liquid transportation fuels and higher value oleochemicals (Lennen and Pfleger 2013). Fatty acid metabolism belongs to anabolic pathways capable of producing large hydrophobic metabolites similar to those found in petroleum. Microbially produced fatty acid derivatives may be considered to be a supplement for petro-based diesel and jet fuels. Unlike short chain alcohols, the low water solubility of longer carbon chain length hydrocarbons and esters may result in reduced recovery cost and reduced toxicity in the culture broth due to phase separation (Lennen and Pfleger 2013).

The fatty acid synthesis is made from AcCoA in the main metabolism, and produces fatty acyl-ACP (acyl carrier protein), which is directed cellular components such as structural and storage lipids. The fatty acid-derived chemicals are formed from thioesters (in fatty acid synthesis), fatty acyl-CoA thioesters (in fatty acid catabolism), or free fatty acids (from thioesterase) of cleavage of acyl-thioesters (Lennen and Pfleger 2013).

## 9. Tolerance to Solvent Stresses

In the practical application of biofuels production from various resources, another important factor must be taken into consideration. Namely, many biofuels are known to be toxic to microorganisms, and reduce the cell viability through damage to the cell membrane and interference with essential physiological processes (Dunlop 2011), where the higher concentration of biofuels in the culture broth is preferred by considering the down-stream processes. Therefore, the trade-offs between biofuel production and the cell's survical must be considered. Several attempts have been made to improve the tolerance to the biofuels, where biofuel export systems, heat shock proteins, membrane modifications, general stress responses, and the approaches that integrate multiple strategies are considered (Dunlop 2011). The bacterial cell such as *E. coli* has the ability to sense the environmental condition, integrate such information, and regulate the metabolism against perturbations to maintain the cell system with robustness (Shimizu 2013, 2014, 2016).

Several attempts have been made to engineer microbes to cope with the tolerance to the stresses caused by biofuels produced (Dunlop 2011, Nicolaou et al. 2010, Atsumi et al. 2010). Increasing tolerance by engineering microbes may increase the performance such as yields and productivities (Dunlop et al. 2011, Alper et al. 2006, Tomas et al. 2003). The effect of biofuels on the cell is through hydrophobicity of the cytoplasmic membrane, where the accumulation of solvent in the cytoplasmic membrane increases the permeability of the membrane, diminishes energy transduction, interferes with membrane protein function, and increases fluidity (Dunlop 2011, Nicolaou et al. 2010, Isken et al. 1998, Ramos et al. 2002). This may cause the release of ATP, ions, and phospholipids, RNA and proteins, and thus the cell growth is depressed due to reduced ATP production by diminished proton motive force (PMF). Moreover, the increase in fluidity affects the nutrient transport as well as energy transduction.

Toxicity levels vary depending on the microbes and the types of biofuels and biochemicals. In general, longer chain alcohols are more toxic than short chain alcohols. Efflux pumps are membrane transporters that recognize and export toxic compounds from the cell by PMF, where this is important for the cell to survive by exporting bile salts, antimicrobial drugs, and solvents, etc. *Pseudomonas putida* exports hexane, octanol, and several other hydrocarbons (Kieboom et al. 1998). Moreover, *acrAB-tolC* pump in *E. coli* provides tolerance to hexane, heptanes, octane, and nonane (Takatsuka et al. 2010). Efflux pumps are effective for increasing tolerance and production of biofuels, in particular, for long chain alcohols such as alkanes, alkenes, and cyclic hydrocarbons, but those may not be effective for exporting short chain alcohols such as 1-propanol, isobutanol, etc. (Ankarloo et al. 2010).

Heat shock proteins are up-regulated in response to short chain alcohols (Piper 1995), where a sigma factor for heat shock, RpoH is activated (Brynildsen and Liao 2009), and several heat shock and protein refolding genes such as *rpoH, dnaJ, htpG,* and *ibpAB* were up-regulated (Rutherford et al. 2010), and *groESL, dnaKJ, hsp18, hsp90* were up-regulated in *Clostridium acetobutylicum* (Tomas et al. 2004). Heat shock proteins are molecular chaperones that are involved in the synthesis, transport, folding, and degradation of the specific proteins, where under stress condition, these prevent protein aggregation and assist with refolding (Mogk et al. 2003). This implies that over-expression of heat shock proteins may increase tolerance against biofuels (Fiocco et al. 2007, Keyes et al. 2011).

In general, solvents disrupt the cell membrane structure and have a strong impact on physiological function, and eventually leading to the cell death (Sikkema et al. 1995). To overcome this problem, solvent tolerant microbes change the composition of the fatty acids from *cis* to *trans* unsaturated fatty acids catalyzed by the *cis-trans* isomerase (*cti*), thus decreasing membrane fluidity resulting in the block of the entry of solvents into the cell (Holtwick et al. 1997, Kiran et al. 2004). In addition, modifications to phospholipid headgroups or phospholipid chain length have been shown to increase solvent tolerance (Ramos et al. 2002).

In relation to solvent stresses caused by the accumulation of biofuels in the culture broth, the primary role to protect the cell such as *E. coli*, etc. from such stresses is made by outer membrane porin proteins, where the specific molecules can only move across these channels. Porins are the outer membrane proteins that produce large, open but regulated water-filled pores that form substrate specific, ion-selective, or nonspecific channels that allow the influx of small hydrophilic molecules and efflux of waste metabolites (Nikaido and Nakano 1980). The porin genes are under control of two-component system such as EnvZ-OmpR system, where EnvZ is an inner membrane sensor kinase, and OmpR is the cytosolic response regulator. In response to the environmental signals such as osmolarity, pH, temperature, nutrients, and toxins, EnvZ phosphorylates OmpR, where the phosphorylated OmpR controls the membrane protein expressions to protect the cell from the stresses (De la Cruz et al. 2007) as mentioned in Chapter 3.

Since cytosolic membrane is also under stress condition, respiration and membrane proteins as well as general stress response mechanism are affected (Nicolaou et al. 2010). Several transcription factors are affected by isobutanol in *E. coli* (Brynildsen and Liao 2009). The reactive oxygen species (ROS) are highly increased in response to the stress caused by n-butanol in *E. coli* (Rutherford et al. 2010). Moreover, it has been shown that improved tolerance against n-butanol can be made by over-expression of ion transport and metabolism genes such as *entC* and *feoA*, as well as acid resistance-related gene *astE* and inner membrane protein gene *ygiH* (Reyes et al. 2011).

As for the enhancement of *crp* (*crp⁺* or *crp\**), this contributes to co-consumption of multiple sugars as mentioned in Chapter 4. In fact, this effect is more than that. Namely, Crp plays important roles under various stress conditions such as acid shock, heat shock, osmo shock as well as nutrient starvation including carbon and nitrogen limitation (see also chapter 5) (Shimizu 2013, 2014, 2016). Thus the enhancement of *crp* also contributes to the stress resistance.

The general stress global regulator RpoS plays important roles in response to the stresses caused by the accumulation of biofuels (Brynildsen and Liao 2009, Minty et al. 2011). It is also of interest that some of the global regulators such as ArcA, Fur, and PhoB are activated probably indirectly by isobutanol, etc. (Brynildsen and Liao 2009).

Since the mechanism of stress responses are complicated, one approach for strain improvement may be to use mutagenesis and evolutionary engineering during long term cultivation (Parekh et al. 2000, Patnaik et al. 2002). The improvement in stress-tolerance does not necessarily guarantee the performance improvement for biofuels production (Dunlop et al. 2011).

## 10. Concluding Remarks

For the efficient production of biochemicals and biofuels by microbes, the primary concern is the available raw materials or carbon sources and the host microbes to be used, where the enhancement of the substrate uptake rate contributes to the improvement of the target metabolite production. However, as mentioned in Chapter 4, it is not obvious to improve the substrate uptake rate. It is also of interest to co-consume multiple carbon sources by modulating carbon catabolite repression, thus increasing the total carbon source uptake rate. However, simultaneous consumption of multiple carbon sources does not necessarily give the improved performance.

It may be better to employ two-stage fermentation with aerobic cultivation to enhance the cell growth, followed by the anaerobic conditions to promote the productivity of the target metabolite(s).

In order to improve the yield, it may be considered to block the by-product forming pathway genes, but it must be careful about the redox balance as well as ATP production by substrate level phosphorylation under anaerobic conditions.

# References

Abdel-Hamid, A.M., M.M. Attwood and J.R. Guest. 2001. Pyruvate oxidase contributes to the aerobic growth efficiency of *Escherichia coli*. Microbiology. 147: 1483–1498.

Abdel-Rahman, M.A., Y. Tashiro and K. Sonomoto. 2013. Recent advances in lactic acid production by microbial fermentation processes. Biotechnol Adv. 31: 877–902.

Abe, A., Y. Oda, K. Asano and T. Sone. 2007. *Rhizopus delemar* is the proper name for *Rhizopus oryzae* fumaric acid producers. Mycologia. 99(5): 714–722.

Abe, S., A. Furuya, T. Saito and K. Takayama. 1962. Method of producing l-malic acid by fermentation. U.S. Patent 3,063,910.

Abo-Khatwa, A.N., A.A. Al-Robai and D.A. Al-Jawhari. 1996. Lichen acids as uncouplers of oxidative phosphorylation of mouse-liver mitochondria. Nat Toxins. 4: 96–102.

Acuna, G., S. Ebeling and H. Hennecke. 1991. Cloning, sequencing, and mutational analysis of the *Bradyrhizobium japonicum* fum C-like gene: evidence for the existence of two different fumarases. J Gen Microbiol. 137: 991–1000.

Adams, M.A. and Z. Jia. 2005. Structural and biochemical evidence for an enzymatic quinine redox cycle in *Escherichia coli*: identification of a novel quinol monooxygenase. J Biol Chem. 280: 8358–8363.

Ager, D.J., D.P. Pantaleone, S.A. Henderson, A.R. Katritzky, I. Prakash and D.E. Walters. 1998. Commercial, synthetic nonnutritive sweeteners. Angew Chem Int Edit. 37: 1802–1817.

Ahvazi, B., R. Coulombe, M. Delarge, M. Vadadi, L. Zhang, E. Meighen and A. Vrielink. 2000. Crystal structure of the NADP+-dependent aldehyde dehydrogenase from *Vibrio harveyi*: structural implications for cofactor specificity and affinity. Biochem J. 349(Pt 3): 853–861.

Almeida, J.R.M., L.C.L. Favaro and B.F. Quirino. 2012. Biodiesel biorefinery: opportunities and challenges for microbial production of fuels and chemicals from glycerol waste. Biotechnol for Biofuels. 5: 48.

Alper, H., J. Moxley, E. Nevoigt, G.R. Fink and G. Stephanopoulos. 2006. Engineering yeast transcription machinery for improved ethanol tolerance and production. Science. 314: 1565–1568.

Altaras, N., M. Etzel and D.C. Cameron. 2001. Conversion of sugars to 1, 2-propanediol by *Thermoanaerobacterium thermosaccharolyticum* HG-8. Biotechnol Prog. 17: 52–56.

Altaras, N.E. and D.C. Cameron. 1999. Metabolic engineering of a 1, 2-propanediol pathway in *Escherichia coli*. Appl Environ Microbiol. 65(3): 1180–1185.

Altaras, N.E. and D.C. Cameron. 2000. Enhanced production of (R)-1, 2-propanediol by metabolically engineered *Escherichia coli*. Biotechnol Prog. 16(6): 940–946.

Altmeyer, P.J., U. Matthes, F. Pawlak, K. Hoffmann, P.J. Frosch, P. Ruppert, S.W. Wassilew, T. Horn, H.W. Krysel, G. Lutz et al. 1994. Antisoriatic effect of fumaric acid derivatives-results of multicenter double blind study in 100 patients. J Am Acad Dermatol. 30: 977–981.

Anastassiadis, S., A. Thiersch, E. Weissbrodt and U. Stottmeister. 2002. Citric acid production by *Candida* strains under intracellular nitrogen limitation. Appl Microbiol Biotechnol. 60: 81–87.

Anastassiadis, S. 2007. l-Lysine fermentation. Recent Pat Biotechnol. 1(1): 11–24.

Ankarloo, J., S. Wikman and I.A. Nicholls. 2010. *Escherichia coli mar* and *acrAB* mutants display no tolerance to simple alcohols. Int J Mol Sci. 11: 1403–1412.

Araki, K., F. Kato, Y. Arai and K. Nakayama. 1974. Histidine production by auxotrophic histidine analog-resistant mutants of Corynebacterium glutamicum. Agri Biol Chem. 38: 837.

Argaman, L. and S. Altuvia. 2000. fhlA repression by OxyS RNA: kissing complex formation at two sites results in a stable antisense-target RNA complex. J Mol Biol. 300: 1101–1112.

Ashok, S., S.M. Raj, C. Rathnasingh and S. Park. 2011. Development of recombinant Klebsiella pneumoniae ΔdhaT strain for the co-production of 3-hydroxypropionic acid and 1,3-propanediol from glycerol. Appl Microbiol Biotechnol. 90(4): 1253–1265.

Atsumi, S. and J.C. Liao. 2008. Directed evolution of *Methanococcus jannaschii* citramalate synthase for biosynthesis of 1-propanol and 1-butanol by *Escherichia coli*. Appl Environ Microbiol. 74(24): 7802–7808.

Atsumi, S., T. Hanai and J.C. Liao. 2008a. Non-fermentative pathways for synthesis of branched-chain higher alcohols as biofuels. Nature. 451(Jan 3): 86–89.

Atsumi, S., A.F. Cann, M.R. Connor, C.R. Shen, K.M. Smith, M.P. Brynildsen, K.J.Y. Chou, T. Hanai and J.C. Liao. 2008b. Metabolic engineering of *Escherichia coli* for 1-butanol production. Metab Eng. 10(6): 305–11.

Atsumi, S., T.-Y. Wu, I. Machado, I.M. Machado, W.-C. Huang, P.-Y. Chen, M. Pellegrini and J.C. Liao. 2010. Evolution, genomic analysis, and reconstruction of isobutanol tolerance in *Escherichia coli*. Mol Syst Biol. 6: 449.

Attfield, P.V. and P.J. Bell. 2006. Use of population genetics to derive nonrecombinant *Saccharomyces cerevisiae* strains that grow using xylose as a sole carbon source. FEMS Yeast Res. 6: 862–868.

Attwood, P.V. 1995. The structure and mechanism of action of pyruvate carboxylase. Int J Biochem Cell Biol. 27: 231–249.

Axley, M.J., D.A. Grahame and T.C. Stadtman. 1990. *Escherichia coli* formate-hydrogen lyase. Purification and properties of the selenium-dependent formate dehydrogenase component. J Biol Chem. 265: 18213–18218.

Bachler, C., P. Schneider, P. Bahler, A. Lustig and B. Erni. 2005. *Escherichia coli* dihydroxyacetone kinase controls gene expression by binding to transcription factor DhaR. EMBO J. 24: 283–293.

Bai, D.M., M.Z. Jia, X.M. Zhao, R. Ban, F. Shen, X.G. Li et al. 2003. L(+)–lactic acid production by pellet-form *Rhizopus oryzae* R 1021 in a stirred tank fermentor. Chem Eng Sci. 58(3-6): 785–791.

Balderas-Hernández, V.E., A. Sabido-Ramos, P. Silva, N. Cabrera-Valladares, G. Hernández-Chávez, J.L. Báez-Viveros, A. Martínez, F. Bolívar and G. Gosset. 2009. Metabolic engineering for improving anthranilate synthesis from glucose in *Escherichia coli*. Microb Cell Fact. 8: 19.

Balderas-Hernández, V.E., L.G. Treviño-Quintanilla, G. Hernández-Chávez, A. Martinez, F. Bolívar and G. Gosset. 2014. Catechol biosynthesis from glucose in *Escherichia coli* anthranilate-overproducer strains by heterologous expression of anthranilate 1,2-dioxygenase from *Pseudomonas aeruginosa* PAO1. Microb Cell Fact. 13: 136.

Barnett, J.A. 1976. The utilization of sugars by yeasts. Adv Carbohydr Chem Biochem. 32: 125–234.

Batt, C.A., S. Carvallo, D.D. Easson, M. Akedo and A.J. Sinskey. 1986. Direct evidence for a xylose metabolic pathway in *Saccharomyces cerevisiae*. Biotechnol Bioeng. 28: 549–553.

Battat, E., Y. Peleg, A. Bercovitz, J.S. Rokem and I. Goldberg. 1991. Optimization of l-malic acid production by *Aspergillus flavus* in a stirred fermentor. Biotechnol Bioeng. 37: 1108–1116.

Bengtsson, O., B. Hahn-Hagerdal and M.F. Gorwa-Grauslund. 2009. Xylose reductase from *Pichia stipitis* with altered coenzyme preference improves ethanolic xylose fermentation by recombinant *Saccharomyces cerevisiae*. Biotechnol Biofuels. 2: 9.

Bennett, G.N. and K.Y. San. 2001. Microbial formation, biotechnological production and applications of 1, 2-propanediol. Appl Microbiol Biotechnol. 55: 1–9.

Benthin, S. and J. Villadsen. 1995. Production of optically pure d-lactate by *Lactobacillus bulgaricus* and purification by crystallization and liquid–liquid extraction. Appl Microbiol Biotechnol. 42: 826–829.

Bercovitz, A., Y. Peleg, E. Battat, J.S. Rokem and I. Goldberg. 1990. Localization of pyruvate carboxylase in organic acid producing Aspergillus strains. Appl Environ Microbiol. 56: 1594–1597.

Bettiga, M., B. Hahn-Hagerdal and M.F. Gorwa-Grauslund. 2008. Comparing the xylose reductase/xylitol dehydrogenase and xylose isomerase pathways in arabinose and xylose fermenting *Saccharomyces cerevisiae* strains. Biotechnol Biofuels. 1(1): 16.

Bettiga, M., O. Bengtsson, B. Hahn-Hagerdal and M.F. Gorwa-Grauslund. 2009. Arabinose and xylose fermentation by recombinant *Saccharomyces cerevisiae* expressing a fungal pentose utilization pathway. Microb Cell Fact. 8: 40.

Bhattachaya, S., P. Manna, R. Gachhui and P.C. Sil. 2013. D-saccharic acid 1, 4-lactone protects diabetic rat kidney by ameliorating hyperglycemia-mediated oxidative stress and renal inflammatory cytokines via NF-κB and PKC signaling. Toxicol Appl Pharmacol. 267: 16–29.

Blazeck, J., A. Hill, M. Jamoussi, A. Pan, J. Miller and H.S. Alper. 2015. Metabolic engineering of *Yarrowia lipopytica* for itaconic acid production. Metab Eng. 32: 66–73.

Blazeck, J., J. Miller, A. Pan, J. Gengler, C. Holden, M. Jamoussi and H.S. Alper. 2014. Metabolic engineering of *Saccharomyces cerevisiae* for itaconic acid production. Appl Microbiol Biotechnol. 98: 8155–8164.

Blombach, B., T. Riester, S. Wieschalka, C. Ziert, J.W. Youn, V.F. Wendisch and B.J. Eikmanns. 2011. *Corynebacterium glutamicum* tailored for efficient isobutanol production. Appl Microbiol Biotechnol. 77(10): 3300–3310.

Blomqvist, K., M. Nikkola, P. Lehtovaara, M.L. Suihko, U. Airaksinen, K.B. Straby et al. 1993. Characterization of the genes of the 2,3-butanediol operons from *Klebsiella terrigena* and *Enterobacter aerogenes*. J Bacterol. 175: 1392–1404.

Blumhoff, M.L., M.G. Steiger, D. Mattanovich and M. Sauer. 2013. Targeting enzymes to the right compartment: metabolic engineering for itaconic acid production by *Aspergillus niger*. Metab Eng. 19: 26–32.

Boles, E., W. Lehnert and F.K. Zimmermann. 1993. The role of the NAD-dependent glutamate dehydrogenase in restoring growth on glucose of a *Saccharomyces cerevisiae* phosphoglucose isomerase mutant. Eur J Biochem. 217(1): 469–477.

Booth, I.R., G.P. Ferguson, S. Miller, C. Li, B. Gunasekera and S. Kinghorn. 2003. Bacterial production of methylglyoxal: a survival strategy or death by misadventure? Biochem Soc Trans. 31: 1406–1408.

Bothast, R.J., B.C. Saha, A.V. Flosenzier and L.O. Ingram. 1994. Fermentation of L-arabinose, D-xylose and D-glucose by ethanologenic recombinant *Klebsiella oxytoca* strain P2. Biotechnol Lett. 16(4): 401–406.

Bressler, E., O. Pines, I. Goldberg and S. Braun. 2002. Conversion of fumaric acid to L-malic acid by sol-gel immobilized *Saccharomyces cerevisiae* in a supported liquid membrane bioreactor. Biotechnol Prog. 18: 445–450.

Brown, S.H., L. Mashkirova, R. Berka, T. Chandler, T. Doty, K. McCall, M. McCulloch, S. McFarland, S. Tompson, D. Yaver and A. Berry. 2003. Metabolic engineering of *Aspergillus oryzae* NRNL 3488 for increased production of L-malic acid. Appl Microbiol Biotechnol. 97: 8903–8912.

Bruinenberg, P.M., P.H.M. Debot, J.P. van Dijken and W.A. Scheffers. 1984a. NADH linked aldose reductase: the key to anaerobic alcoholic fermentation of xylose by yeasts. Appl Microbiol Biotechnol. 19(4): 256–260.

Bruinenberg, P.M., J.P. van Dijken and W.A. Scheffers. 1984b. Production and consumption of NADPH and NADH during growth of *Candida utilis* on xylose. Antonie Van Leeuwenhoek. 50(1): 81–82.

Bruinenberg, P.M., R. Jonker, J.P. van Dijken and W.A. Scheffers. 1985. Utilization of formate as an additional energy source by glucose-limited chemostat cultures of *Candida utilis* CBS 621 and *Saccharomyces cerevisiae* CBS 8066. Evidence for the absence of transhydrogenase activity in yeasts. Arch Microbiol. 142(3): 302–306.

Brynildsen, M.P. and J.C. Liao. 2009. An integrated network approach identifies the isobutanol response network of *Escherichia coli*. Mol Syst Biol. 5: 277.

Bui, V., M.K. Lau, D. MacRae and D. Schweitzer. 2011. Method for producing isomers of muconic acid and muconic salts. Patent Application Publication Pub. No.: US 8809583 B2.

Bui, V., M.K. Lau, D. MacRae and D. Schweitzer. 2013. Method for producing isomers of muconic acid and muconic salts. Patent Application Publication (10) Pub. No.: US 2013/0030215 A1.

Burk, M.J., A.P. Burgard, R.E. Osterhout and J. Sun. 2012. Microorganisms for the production of 1, 4-Butanediol. US Patent 8, 178, 327, B2. (13/009,813).

Burk, M.J., R.E. Osterhout and J. Sun. 2011. Semi-synthetic telephthalic acid via microorganisms that produce muconic acid. US Patent 20110124911 A1.

Burma, D.P. and B.L. Horecker. 1958. Pentose fermentation by *Lactobacillus plantarum*. IV. l-Ribulose-5-phosphate 4-epimerase. J Biol Chem. 231: 1053–1064.

Burma, D.P. and B.L. Horecker. 1958a. Pentose fermentation by *Lactobacillus plantarum*. III. Ribulokinase. J Biol Chem. 231: 1039–1051.

Burridge, E. 2011. Adipic acid. ICIS Chem Bus. 279: 43–43.

Buschke, N., H. Schroder and C. Wittmann. 2011. Metabolic engineering of *Corynebacterium glutamicum* for production of 1,5-diaminopentane from hemicellulose. Biotechnol J. 6(3): 306–317.

Cann, A.F. and J.C. Liao. 2008. Production of 2-methyl-1-butanol in engineered *Escherichia coli*. Appl Microbiol Biotechnol. 81(1): 89–98.

Cao, W., J. Luo, J. Zhao, C. Qiao, L. Ding, B. Qi, Y. Su and Y. Wan. 2012a. Intensification of β-poly(L-malic acid) production by *Aureobacidium pullulans* ipe-1 in the late exponential growth phase. J Ind Microbiol Biotechnol. 39: 1073–1080.

Cao, Y., Y. Cao and X. Lin. 2011. Metabolically engineered *Escherichia coli* for biotechnological production of four-carbon 1,4-dicarboxylic acids. J Ind Microbiol Biotechnol. 38: 649–656.

Carole, T.M., J. Pellegrino and M.D. Paster. 2004. Opportunities in the industrial biobased products industry. Appl Biochem Biotechnol. 113-116: 871–885.

Carta, F.S., A.R. Soccol, L.P. Ramos and J.D. Fontana. 1999. Production of fumaric acid by fermentation of enzymatic hydrolysis derived from cassava bagasse. Bioresour Technol. 68(1): 23–28.

Carvalheiro, F., L.C. Duarte, S. Lopes, J.C. Parajo, H. Pereira and F.M. Girio. 2005. Evaluation of the detoxification of brewery's spent grain hydrolysate for xylitol production by *Debaryomyces hansenii* CCMI 941. Process Biochem. 40(3-4): 1215–1223.

Causey, T.B., K.T. Shanmugam, L.P. Yomano and L.O. Ingram. 2004. Engineering *Escherichia coli* for efficient conversion of glucose to pyruvate. NAS USA. 101(8): 2235–2240.

Celinska, E. 2010. Debottlenecking the 1, 3-propanediol by metabolic engineering. Biotechnol Adv. 28: 519–530.

Celinska, E. and W. Grajek. 2009. Biotechnological production of 2,3-butanediol-current state and prospects. Biotechnol Adv. 27: 715–725.

Chandran, S.S., J. Yi, K.M. Draths, R. von Daeniken, W. Weber and J.W. Frost. 2003. Phosphoenolpyruvate availability and the biosynthesis of shikimic acid. Biotechnol Prog. 19: 808–814.

Chang, D.-E., H.-C. Jung, J.-S. Rhee and J.-G. Pan. 1999. Homofermentative production of D- or L-lactate in metabolically engineered *Escherichia coli* RR1. Appl Environ Microbiol. 65(4): 1384–1389.

Charles, E.W., E.D. Bruce, T.E. Richard, H. Mark, R.L. Michael and Y.Y. Lee. 2005. Comparative sugar recovery data from laboratory scale application of leading pretreatment technologies to corn stover. Bioresour Technol. 96(18): 2026–2032.

Chatterjee, R., C. Sanville Millard, K. Champion, D.P. Clark and M.I. Donnelly. 2001. Mutation of the ptsG gene results in increased production of succinate in fermentation of glucose by *Escherichia coli*. Appl. Environ. Microbiol. 67: 148–154.

Chatzifragkou, A., D. Dietz, M. Komaitis, A.P. Zeng and S. Papanikolaou. 2010. Effect of biodiesel-derived waste glycerol impurities on biomass and 1, 3-propanediol production of *Clostridium butyricum* VPI 1718. Biotechnol Bioeng. 107: 76–84.

Chen, K., J. Dou, S. Tang, Y. Yang, H. Wang, H. Fang and C. Zhou. 2012. Deletion of the *aroK* gene is essential for high shikimic acid accumulation through the shikimate pathway in *E. coli*. Bioresour Technol. 119: 141–147.

Chen, X., L. Zhou, K. Tian, A. Kumar, S. Singh, B.A. Prior and Z. Wang. 2013. Metabolic engineering of *Escherichia coli*: A sustainable industrial platform for bio-based chemical production. Biotechnol Adv. 31: 1200–1223.

Chen, Y. 2011. Development and application of co-culture for ethanol production by co-fermentation of glucose and xylose: a systematic review. J Ind Microbiol Biotechnol. 38: 581–597.

Chen, X., J. Wu, W. Song, L. Zhang, H. Wang and L. Liu. 2014. Fumaric acid production by *Torulopsis glabrata*: Engineering the urea cycle and the purine nucleotide cycle. Biotechnol Bioeng. 112(1): 156–167.

Cheng, K., Q. Liu, J.A. Zhang, J.P. Li, J.M. Xu and G.H. Wang. 2010. Improved 2, 3-butanediol production from corncob acid hydrolysate by fed-batch fermentation using *Klebsiella oxytoca*. Process Biochem. 45(4): 613–616.

Cheng, Q., S.M. Thomas and P. Rouviere. 2002. Biological conversion of cyclic alkanes and cyclic alcohols into decarboxylic acids: biochemical and molecular basis. Appl Microbiol Biotechnol. 58: 704–711.

Chiang, C. and S.G. Knight. 1961. L-arabinose metabolism by cell-free extracts of *Penicillium chrysogenum*. Biochim Biophys Acta. 46(2): 271–278.

Choi, S., C.W. Song, J.H. Shin and S.Y. Lee. 2015. Biorefineries for the production of top building block chemicals and their derivatives. Metab Eng. 28: 223–239.

Christopher, L.P., V. Kapatral, B. Vaisvil, G. Emel and L.C. DeVeaux. 2014. Draft Genome sequence of a new homofermentative, lactic acid-producing *Enterococcus faecalis* isolate, CBRD01. Genome Announc. 2(2): e00147–14.

Chu, B.C.H. and H. Lee. 2007. Genetic improvement of *Saccharomyces cerevisiae* for xylose fermentation. Biotechnol Adv. 25: 425–441.

Chua, J.W. and J.H. Hsieh. 1990. Oxidative bioconversion of toluene to 1,3-butadiene-1,4-dicarboxylic acid (cis,cis-muconic acid).World J Microbiol Biotechnol. 6: 127–143.

Cirino, P.C., J.W. Chin and L.O. Ingram. 2006. Engineering *Escherichia coli* for xylitol production from glucose-xylose mixtures. Biotechnol Bioeng. 95(6): 1167–76.

Clark, D.P. 1989. The fermentative pathways of *Escherichia coli*. FEMS Microbiol Lett. 63(3): 223–234.

Clomburg, J.M. and R. Gonzalez. 2010. Biofuel production in *Escherichia coli*: the role of metabolic engineering and synthetic biology. Appl Microbiol Biotechnol. 86(2): 419–434.

Clomburg, J.M. and R. Gonzalez. 2011. Metabolic engineering of *Escherichia coli* for the production of 1, 2-propanediol from glycerol. Biotechnol Bioeng. 108: 867–879.

Clomburg, J.M. and R. Gonzalez. 2012. Anaerobic fermentation of glycerol: a platform for renewable fuels and chemicals. Trends Biotechnol. 31(1): 20–28.

Colombo, S., M. Grisa, P. Tortooora and M. Vanoni. 1994. Molecular cloning, nucleotide sequence and expression of a *Sulfolobus solfataricus* gene encoding a class II fumarase. FEBS Lett. 337: 93–98.

Cortés-Tolalpa, L., R.M. Gutiérrez-Ríos, L.M. Martínez, R. de Anda, G. Gosset, F. Bolívar and A. Escalante. 2014. Global transcriptomic analysis of an engineered *Escherichia coli* strain lacking the phosphoenolpyruvate: carbohydrate phosphotransferase system during shikimic acid production in rich culture medium. Microb Cell Fact. 13: 28.

Crolla, A. and K.J. Kennedy. 2004. Fed-batch production of citric acid by *Candida lipolytica* grown on *n*-paraffins. J Biotechnol. 110(1): 73–84.

Cui, Y.-Y., C. Ling, Y.-Y. Zhang, J. Huang, J.-Z. Liu et al. 2014. Production of shikimic acid from *Escherichia coli* through chemically inducible chromosomal evolution and cofactor metabolic engineering. Microb Cell Factories. 13: 21.

Curran, K.A., J.M. Leavitt, A.S. Karim and H.S. Alper. 2013. Metabolic engineering of muconic acid production in *Saccharomyces cerevisiae*. Metab Eng. 15: 55–66.

Currie, J.N. 1917. The citric acid fermentation of *A. niger*. J Biol Chem. 31: 5.

da Silva, G.P., M. Mack and J. Contiero. 2009. Glycerol: A promising and abundant carbon source for industrial microbiology. Biotechnol Adv. 27: 30–39.

Dakin, H.D. 1924. The formation of *l*-malic acid as a product of alcoholic fermentation by yeast. J Biol Chem. 61: 139–145.

Daniel, R., K. Stuertz and G. Gottschalk. 1995. Biochemical and molecular characterization of the oxidative branch of glycerol utilization by Citrobacter freundii. J Bacteriol. 177(15): 4392–4401.

Das, D. and T.N. Vezirog˘lu. 2001. Hydrogen production by biological processes: a survey of literature. Int J Hydrogen Energy. 26: 13–28.

Datta, R., S.P. Tsai, P. Bonsignore, S.H. Moon and J.R. Frank. 1995. Technological and economic potential of poly(lactic acid) and lactic acid derivatives. FEMS Microbiol Rev. 16: 221–231.

De Jongh, W.A. and J. Nielsen. 2008. Enhanced citrate production through gene insertion in *Aspergillus niger*. Metab Eng. 10: 87–96.

De la Cruz, M.A., M. Fernandez-Mora, C. Guadarrama et al. 2007. LeuO antagonizes H-NS and StpA-dependent repression in *Salmonella* enteric *ompS1*. Mol Microbiol. 66(3): 727–743.

Deanda, K., M. Zhang, C. Eddy and S. Picataggio. 1996. Development of an arabinose fermenting *Zymomonas mobilis* strain by metabolic pathway engineering. Appl Environ Microbiol. 62(12): 4465–4470.

Delgenes, J.P., R. Moletta and J.M. Navarro. 1986. The effect of aeration on D-xylose fermentation by Pachysolen tannophilus, *Pichia stipitis, Kluyveromyces marxianus* and *Candida shehatae*. Biotechnol Lett. 8(12): 897–900.

Dellomonaco, C. et al. 2011. Engineered reversal of the β-oxidation cycle for the synthesis of fuels and chemicals. Nature. 476(7360): 355–359.

Demain, A.L. and S. Sanchez. 2006. Microbial synthesis of primary metabolites: current status and future prospects. *In*: E.M.T. El-Mansi (ed.). Fermentation Microbiology and Biotechnology. Taylor and Francis, London.

Den Haan, R. and W.H. van Zyl. 2003. Enhanced xylan degradation and utilisation by *Pichia stipitis* overproducing fungal xylanolytic enzymes. Enzyme Microb Technol. 33(5): 620–628.

Deng, Y.F., S. Li, Q. Xu, M. Gao and H. Huang. 2012. Production of fumaric acid by simultaneous saccharification and fermentation of starchy materials with 2-deoxyglucose-resistant mutant strains of *Rhizopus oryzae*. Bioresour Technol. 107: 363–367.

Diaz, A., R. Katsarava and J. Puiggali. 2014. Synthesis, properties and applications of biodegradable polymers derived from diols and dicarboxylic acids: from polyesters to poly (ester amide)s. Int J Mol Sci. 15: 7064–7123.

Dien, B.S., N.N. Nichols and R.J. Bothast. 2002. Fermentation of sugar mixture using *Escherichia coli* catabolite repression mutants engineered for production of l-lactic acid. J Ind Microbiol Biotechnol. 29: 221–229.

Dien, B.S., M.A. Cotta and T.W. Jeffries. 2003. Bacteria engineered for fuel ethanol production: current status. Appl Microbiol Biotechnol. 63: 258–266.

Dietzler, D.N., M.P. Leckie, J.L. Magnani, M.J. Sughrue and P.E. Bergstein. 1975. Evidence for the coordinate control of glycogen synthesis, glucose utilization, and glycolysis in *Escherichia coli*. II. Quantitative correlation of the inhibition of glycogen synthesis and the stimulation of glucose utilization by 2,4-dinitrophenol with the effects on the cellular levels of glucose 6-phosphate, fructose 1, 6-diphosphate, and total adenylates. J Biol Chem. 250(18): 7195–203.

Dimarco, A.A. and A.H. Romano. 1985. D-glucose transport system of *Zymomonas mobilis*. Appl Environ Microbiol. 49(1): 151–7.

Ding, Y., S. Li, C. Dou et al. 2011. Production of fumaric acid by *Rhizopus oryzae*: role of carbon-nitrogen ratio. Appl Biochem Biotechnol. 164: 1461–1467.

Doran, J.B., H.C. Aldrich and L.O. Ingram. 1994. Saccharification and fermentation of sugar cane bagasse by *Klebsiella oxytoca* P2 containing chromosomally integrated genes encoding the *Zymomonas mobilis* ethanol pathway. Biotechnol Bioeng. 44(2): 240–247.

Driouch, H., B. Sommer and C. Wittmann. 2010. Morphology engineering of *Aspergillus niger* for improved enzyme production. Biotechnol Bioeng. 105(6): 1058–1068.

Dunlop, M.J. 2011. Engineering microbes for tolerance to next-generation biofuels, Biotechnol for Biofuels. 4: 32.

Durnin, G., J. Clomburg, Z. Yeates, P.J.J. Alvarez, K. Zygourakis, P. Campbell and R. Gonzalez. 2009. Understanding and harnessing the microaerobic metabolism of glycerol in *Escherichia coli*. Biotechnol Bioeng. 103: 148–161.

Eitman, M.A., S.A. Lee and E. Altman. 2008. A co-fermentation strategy to consume sugar mixtures effectively. J Biol Eng. 2: 3.

Eitman, M.A., S.A. Lee, R. Altman and E. Altman. 2009. A substrate-selective co-fermentation strategy with *Escherichia coli* produces lactate by simultaneously consuming xylose and glucose. Biotechnol Bioeng. 102(3): 822–827.

Eliasson, A., E. Boles, B. Johansson, M. Osterberg, J.M. Thevelein, I. Spencer-Martins, H. Juhnke and B. Hahn-Hagerdal. 2000a. Xylulose fermentation by mutant and wild-type strains of *Zygosaccharomyces* and *Saccharomyces cerevisiae*. Appl Microbiol Biotechnol. 53(4): 376–382.

Eliasson, A., C. Christensson, C.F. Wahlbom and B. Hahn-Hägerdal. 2000b. Anaerobic xylose fermentation by recombinant *Saccharomyces cerevisiae* carrying XYL1, XYL2, and XKS1 in mineral medium chemostat cultures. Appl Environ Microbiol. 66(8): 3381–6.

Englesberg, E. 1961. Enzymatic characterization of 17 L-arabinose negative mutants of *Escherichia coli*. J Bacteriol. 81: 996–1006.

Englesberg, E., C. Squires and F. Meronk Jr. 1969. The L-arabinose operon in *Escherichia coli* B-r: a genetic demonstration of two functional states of the product of a regulator gene. Proc Natl Acad Sci USA. 62(4): 1100–1107.

Escalante, A., R. Calderón, A. Valdivia, R. de Anda, G. Hernández, O.T. Ramírez, G. Gosset and F. Bolívar. 2010. Metabolic engineering for the production of shikimic acid in an evolved *Escherichia coli* strain lacking the phosphoenolpyruvate: carbohydrate phosphotransferase system. Microb Cell Fact. 9: 21.

Farmer, W.R. and J.C. Liao. 2000. Improving lycopene production in *Escherichia coli* by engineering metabolic control. Nature Biotechnol. 18: 533–537.

Fiocco, D., V. Capozzi, P. Goffin, P. Hols and G. Spano. 2007. Improved adaptation to heat, cold, and solvent tolerance in *Lactobacillus plantarum*. Appl Microbiol Biotechnol. 77: 909–915.

Flikweert, M.T., L. Van der Zanden, W.M.T.M. Janssen, H.Y. Steensma, J.P. van Dijken and J.T. Pronk. 1996. Pyruvate decarboxylase: an indispensable enzyme for growth of *Saccharomyces cerevisiae* on glucose. Yeast. 12: 247–257.

Flikweert, M.T., M. Kuyper, A.J.A. van Maris, P. Kötter, J.P. van Dijken and J.T. Pronk. 1999. Steady-state and transient-state analysis of growth and metabolite production in a *Saccharomyces cerevisiae* strain with reduced pyruvate-decarboxylase activity. Biotechnol Bioeng. 66: 42–50.

Fonseca, C., R. Romao, H. Rodrigues de Sousa, B. Hahn-Hagerdal and I. Spencer-Martins. 2007a. L-arabinose transport and catabolism in yeast. FEBS J. 274(14): 3589–3600.

Fonseca, C., I. Spencer-Martins and B. Hahn-Hagerdal. 2007b. L-arabinose metabolism in *Candida arabinofermentans* PYCC 5603T and *Pichia guilliermondii* PYCC 3012: influence of sugar and oxygen on product formation. Appl Microbiol Biotechnol. 75(2): 303–310.

Forster, A., A. Aurich, S. Mauersberger and G. Barth. 2007. Citric acid production from sucrose using a recombinant strain of the yeast Yarrowia lipolytica. Appl Microbiol Biotechnol. 75: 1409–17.

Foster, I.W. and S.A. Waksman. 1938. The production of fumaric acid by molds belonging to the genus *Rhizopus*. J Am Chem Soc. 61: 127–135.

Foster, J.W., S.F. Carson, D.S. Anthony, J.B. Davis, W.E. Jefferson and M.W. Long. 1949. Aerobic formation of fumaric acid in the mold *Rhizopus nigricans*: synthesis by direct $2C_2$ condensations. PNAS USA. 35: 663–671.

Foster, J.W. and J.B. Davis. 1948. Anaerobic formation of fumaric acid by the model *Rhizopus nigricans*. J Bacteriol. 56(3): 329–338.

Friedberg, D., Y. Peleg, A. Monsonego, S. Maissi, E. Battat, J.S. Rokem et al. 1995. The fumR gene encoding fumarase in the filamentus fungus *Rhizopus oryzae*: cloning, structure and expression. Gene. 163: 139–144.

Fu, J., Z. Wang, T. Chen, W. Liu, T. Shi, G. Wang, Y.J. Tang and X. Zhao. 2014. NADH plays the vital role for chiral pure D-(–)-2,3-butanediol production in *Bacillus subtilis* under limited oxygen conditions. Biotechnol Bioeng. 111: 2126.

Fu, Y.Q. et al. 2010. Enhancement of fumaric acid production by *Rhizopus oryzae* using a two-stage dissolved oxygen control strategy. Appl Biochem Biotechnol. 162(4): 1031–1038.

Fu, Y.Q., Q. Xu, S. Li, H. Huang and Y. Chen. 2009. A novel multi-stage preculture strategy of *Rhizopus oryzae* ME-F12 for fumaric acid production in a stirred-tank reactor. World J Microbiol Biotechnol. 25(10): 1871–1876.

Fu, Y.Q., S. Li, Y. Chen et al. 2010. Enhancement of fumaric acid production by *Rhizopus oryzae* using a two-stage dissolved oxygen control strategy. Appl Biochem Biotechnol. 162: 1031–1038.

Fujita, Y., J. Ito, M. Ueda, H. Fukuda and A. Kondo. 2004. Synergistic saccharification, and direct fermentation to ethanol, of amorphous cellulose by use of an engineered yeast strain codisplaying three types of cellulolytic enzyme. Appl Environ Microbiol. 70(2): 1207–1212.

Gancedo, C. and K. Schwerzmann. 1976. Inactivation by glucose of phosphoenolpyruvate carboxykinase from *Saccharomyces cerevisiae*. Arch Microbiol. 109: 221–225.

Gangl, I.C., W.A. Weigand and F.A. Keller. 1991. Metabolic modeling of fumaric acid production by *Rhizopus arrhizus*. Appl Biochem Biotechnol. 28-29(1): 471–486.

Gao, W., B. Qi, J. Zhao, C. Qiao, Y. Su and Y. Wan. 2012b. Control strategy of pH dissolved oxygen concentration and stirring speed for enhancing β-poly(malic acid) production by *Aureobacidium pullulans* ipe-1. J Chem Technol Biotechnol. DOI: 10.1002/jctb.3905.

Geiser, D.M., J.I. Pitt and J.W. Taylor. 1998. Cryptic speciation and recombination in the aflatoxin-producing fungus *Aspergillus flavus*. Proc Natl Acad Sci USA. 95: 388–393.

Ghosh, S., Y. Chisti and U.C. Banerjee. 2012. Production of shikimic acid. Biotechnol Adv. 30: 1425–1431.

Girio, F.M., C. Fonseca, F. Carvaheiro, L.C. Duarte, S. Marques and R. Bogel-Lukasil. 2010. Hemicelluloses for fuel ethanol: A review. Bioresour Technol. 101: 4775–4800.

Girio, F.M., J.C. Roseiro, P. Sa-Machado, A.R. Duarte-Reis and M.T. Amaral-Collaco. 1994. Effect of oxygen transfer rate on levels of key enzymes of xylose metabolism in *Debaryomyces hansenii*. Enzyme Microb Technol. 16(12): 1074–1078.

Goldberg, I. and B. Steiglitz. 1985. Improved rate of fumaric acid production by Tweens and vegetable oils in *Rhizopus arrhizus*. Biotechnol Bioeng. 27: 1067–1069.

Goldberg, I., J.S. Rokem and O. Pines. 2006. Organic acids: old metabolites, new themes. J Chem Technol Biotechnol. 81: 1601–1611.

Goldberg, I., Y. Peleg and J.S. Rokem. 1991. Citric, fumaric and malic acids. pp. 349–374. In: I. Goldberg and Williams R. Van Nostrand Reinhold (eds.). Biotechnology and Food Ingredients. New York.

Gonzalez, R., H. Tao, J.E. Purvis, S.W. York, K.T. Shanmugam and L.O. Ingram. 2003. Gene array-based identification of changes that contribute to ethanol tolerance in ethanologenic *Escherichia coli*: comparison of KO11 (parent) to LY01 (resistant mutant). Biotechnol Prog. 19: 612–623.

González-Pajuelo, M., I. Meynial-Salles, F. Mendes, J.C. Andrade, I. Vasconcelos and P. Soucaille. 2005. Metabolic engineering of *Clostridium acetobutylicum* for the industrial production of 1,3-propanediol from glycerol. Metab Eng. 7(5-6): 329–336.

Gorgens, J.F., W.H. van Zyl, J.H. Knoetze and B. Hahn-Hagerdal. 2001. The metabolic burden of the PGK1 and ADH2 promoter systems for heterologous xylanase production by *Saccharomyces cerevisiae* in defined medium. Biotechnol Bioeng. 73(3): 238–245.

Gosset, G. 2009. Production of aromatic amino acids in bacteria. Curr Opinion Biotechnol. 20: 651–658.

Grabar, T.B., S. Zhou, K.T. Shanmugam, L.P. Yomano and L.O. Ingram. 2006. Methylglyoxal bypath identified as source of chiral contamination in L(+) and L(–)-lactate fermentations by recombinant *Escherichia coli*. Biotechnol Lett. 28: 1527–1535.

Granstrom, T.B., K. Izumori and M. Leisola. 2007a. A rare sugar xylitol. Part I: the biochemistry and biosynthesis of xylitol. Appl Microbiol Biotechnol. 74: 277–281.

Granstrom, T.B., K. Izumori and M. Leisola. 2007b. A rare sugar xylitol. Part II: biotechnological production and future applications of xylitol. Appl Microbiol Biotechnol. 74: 273–276.

Gray, K.A., L.S. Zhao and M. Emptage. 2006. Bioethanol. Curr Opin Chem Biol. 10(2): 141–146.

Grewall, H.S. and K.L. Kalra. 1995. Fungal production of citric acid. Biotechnol Adv. 13: 209–234.

Gross, R.A. and K. Rhano. 2002. Biodegradable polymers for the environment. Science. 297: 803–807.

Gu, C., Y. Zhou, L. Liu et al. 2013. Production of fumaric acid by immobilized *Rhizopus arrhizus* on net. Bioresour Technol. 131: 303–307.

Gu, S., Q. Xu, H. Huang et al. 2014. Alternative respiration and fumaric acid production of *Rhizopus oryzae*. Appl Microbiol Biotechnol. 98: 5145–5152.

Guettler, M.V., M.K. Jain and B.K. Soni. 1996a. Process for making succinic acid, microorganisms for use in the process and methods of obtaining the microorganisms. U.S. patent 5,504,004.

Guettler, M.V., M.K. Jain and D. Rumler. 1996b. Method for making succinic acid, bacterial variants for use in the process, and methods for obtaining variants. U.S. patent 5,573,931.

Guillout, S., A.A. Rodal, P.A. Lessard and A.J. Sinskey. 2002. Methods for producing l-isoleucine. USA patent (US 6451564).

Gupta, K.P. and J. Singh. 2004. Modulation of carcinogen metabolism and DNA interaction by calcium glucarate in mouse skin. Toxicol Sci. 79: 47–55.

Ha, S.-J., J.M. Galazka, S.R. Kim, J.-H. Choi, X. Yang, J.-H. Seo, N.L. Glass, J.H.D. Cate and Y.-S. Jin. 2011. Engineered *Saccharomyces cerevisiae* capable of simultaneous cellobiose and xylose fermentation. PNAS USA. 108(2): 504–509.

Hahn-Hegerdal, B., K. Karhumaa, C. Fonseca, I. Spencer-Martins and M.F. Gorwa-Grauslund. 2007. Towards industrial pentose-fermenting yeast strains. Appl Microbiol Biotechnol. 74: 937–953.

Hanai, T., S. Atsumi and J.C. Liao. 2007. Engineered synthetic pathway for isopropanol production in *Escherichia coli*. Appl Environ Microbiol. 73: 7814–7818.

Harder, B.-J., K. Bettenbrock and S. Klamt. 2016. Model-based metabolic engineering enables high yield itaconic acid production by *Escherichia coli*. Metab Eng. 38: 29–37.

Harn-Hagerdal, B., H. Jeppsson, K. Skoog and B.A. Prior. 1994. Biochemistry and physiology of xylose fermentation by yeasts. Enzyme Microbiol Technol. 16(11): 933–943.

Harris, L.M., R.P. Desai, N.E. Welker and E.T. Papoutsakis. 2000. Characterization of recombinant strains of the Clostridium acetobutylicum butyrate kinase inactivation mutant: need for new phenomenological models for solventogenesis and butanol inhibition? Biotechnol Bioeng. 67(1): 1–11.

Harwood, C.S. and R.E. Parales. 1996. The β-ketoadipate pathway and the biology of self-identity. Ann Rev Microbiol. 50: 553–590.

Hasunuma, T. and A. Kondo. 2011. Development of yeast cell factories for consolidated bioprocessing of lignocelluloses to bioethanol through cell surface engineering. Biotechnology Adv. 30(6): 1207–18.

Hawkins, K.M. and C.D. Smolke. 2008. Production of benzylisoquinoline alkaloids in *Saccharomyces cerevisiae*. Nat Chem Biol. 4: 564–573.

Heath, E.C., B.L. Horecker, P.Z. Smyrniotis and Y. Takagi. 1958. Pentose fermentation by *Lactobacillus plantarum*. 2. L-arabinose isomerase. J Biol Chem. 231(2): 1031–1037.

Hedayati, M.T., A.C. Pasqualotto, P.A. Warn, P. Bowyer and D.W. Denning. 2007. *Aspergillus flavus*: human pathogen, allergen and mycotoxin producer. Microbiology. 153(Pt 6): 1677–92.

Hermann, T. 2003. Industrial production of amino acids by coryneform bacteria. J Biotechnol. 104: 155–172.

Hesseltine, C.W., O.L. Shotwell, J.J. Ellis and R.D. Stubblefield. 1966. Aflatoxin formation by *Aspergillus flavus*. Bacteriol Rev. 30: 795–805.

Hevekerl, A., A. Kuenz and K.D. Vorlop. 2014. Influence of the pH on the itaconic acid production with *Aspergillus terreus*. Appl Microbiol Biotechnol. 98: 10005–10012.

Ho, N.W., Z. Chen and A.P. Brainard. 1998. Genetically engineered *Saccharomyces* yeast capable of effective cofermentation of glucose and xylose. Appl Environ Microbiol. 64(5): 1852–1859.

Hofvendahl, K. and B. Hahn-Hagerdal. 2000. Factors affecting the fermentative lactic acid production from renewable resources. Enzyme Microbiol Technol. 26: 87–107.

Hohmann, S. 1991. Characterization of PDC6, a third structural gene for pyruvate decarboxylase in *Saccharomyces cerevisiae*. J Bacteriol. 173: 7963–7969.

Hohmann, S., W. Bell, M.J. Neves, D. Valckx and J.M. Thevelein. 1996. Evidence for trehalose-6-phosphate-dependent and -independent mechanisms in the control of sugar influx into yeast glycolysis. Mol Microbiol. 20(5): 981–991.

Holler, E. 2010. Production of long-chain unbranched beta-poly(L-malic acid) by large scale physarum cultivation and high-grade purification of the same. US20100216199.

Holtwick, R., F. Meinhardt and H. Keweloh. 1997. Cis-trans isomerization of unsaturated fatty acids: cloning and sequencing of the cti gene from *Pseudomonas putida* P8. Appl Environ Microb. 63: 4292–4297.

Hossain, A.H., A. Li, A. Brickwedde, L. Wilms, M. Caspers, K. Overkamp and P.J. Punto. 2016. Rewiring a secondary metabolite pathway towards itaconc acid production in *Aspergillus niger*. Microb Cell Fact. 15: 130.

Howell, D.M., H. Xu and R.H. White. 1999. (R)-citramalate synthase in methanogenic archaea. J Bacteriol. 181(1): 331–333.

Huang, K.X., F.B. Rudolph and G.N. Bennet. 1999. Characterization of methylglyoxal synthase from *Clostridium acetobutylicum* ATCC824 and its use in the formation of 1, 2-propanediol. Appl Environ Microbiol. 65(7): 3244–3247.

Huffer, S., C.M. Roche, H.W. Blanch and D.S. Clark. 2012. *Escherichia coli* for biofuel production: bridging the gap from promise to practice. Trends Biotechnol. 30(10): 538–545.

Ikeda, M. 2006. Towards bacterial strains overproducing l-tryptophan and other aromatics by metabolic engineering. Appl Microbiol Biotechnol. 69: 615–626.

Ikeda, M. and R. Katsumata. 1992. Metabolic engineering to produce tyrosine or phenylalanine in a tryptophan producing Corynebacterium glutamicum strain. Appl Env Microbiol. 58: 781–785.

Ikram-ul, H., S. Ali, M.A. Qadeer et al. 2004. Citric acid production by selected mutants of *Aspergillus niger* from cane molasses. Bioresour Technol. 93(2): 125–130.

Ingram, L.O., H.C. Aldrich, A.C.C. Borges, T.B. Causey, A. Martinez, F. Morales, A. Saleh, S.A. Underwood, L.P. Yomano, S.W. York, J. Zaldivar and S. Zhou. 1999. Enteric bacteria catalyst for fuel ethanol production. Biotechnol Prog. 15: 855–866.

Ingram, L.O., T. Conway, D.P. Clark, G.W. Sewell and J.F. Preston. 1987. Appl Environ Microbiol. 53: 2420–2425.

Inui, M., H. Kawaguchi, S. Murakami, A.A. Vertes and H. Yukawa. 2004. Metabolic engineering of *Corynebacterium glutamicum* for fuel ethanol production under oxygen-deprivation conditions. J Mol Microbiol Biotechnol. 8: 243–254.

Inui, M., S. Murakami, S. Okino, H. Kawaguchi, A.A. Vertes and H. Yukawa. 2004. Metabolic analysis of *Corynebacterium glutamicum* during lactate and succinate productions under oxygen deprivation condition. J Mol Microbiol Biotechnol. 7: 182–196.

Isken, S. and J.A.M. de Bont. 1998. Bacteria tolerant to organic solvents. Extremophiles. 2: 229–238.

Iwaki, H., Y. Hasegawa, M. Teraoka, T. Yokuyama, H. Bergeron and P.C. Lau. 1999. Identification of a transcriptional activator (ChnR) and a 6-oxohexanoate dehydrogenase (ChnE) in the cyclohexanol catabolic pathway in *Acinetobacter* sp. strain NCIMB 9871 and localization of the genes that encode them. Appl Environ Microbiol. 65: 5158–5162.

Jaklitsch, W.M., C.P. Kubicek and M.C. Scrutton. 1991. The subcellular organization of itaconic biosynthesis in *Aspergillus terreus*. J Gen Microbiol. 137: 533–539.

Jang, Y.-S., B. Kim, J.H. Shin, Y.J. Choi, S. Choi, C.W. Song, J. Lee, H.G. Park and S.Y. Lee. 2012. Bio-based production of C2–C6 platform chemicals. Biotechnol Bioeng. 109(10): 2437–2459.

Jang, Y.-S., A. Malaviya and S.Y. Lee. 2013. Acetone–butanol–ethanol production with high productivity using *Clostridium acetobutylicum* BKM19. Biotechnol Bioeng. 110(6): 1646–1653.

Jantama, K., M.J. Haupt, S.A. Svoronos, X. Zhang, J.C. Moore, K.T. Shanmugam and L.O. Ingram. 2008. Combining metabolic engineering and metabolic evolution to develop non-recombinant strains of *Escherichia coli* C that produce succinate and malate. Biotechnol Bioeng. 99(8): 1140–1153.

Jeffries, T.W. and Y.S. Jin. 2004. Metabolic engineering for improved fermentation of pentoses by yeasts. Appl Microbiol Biotechnol. 63: 495–509.

Jeffries, T.W. 2008. Engineering the *Pichia stipitis* genome for fermentation of hemicellulose hydrolysates. pp. 37–47. *In*: J.D. Wall, C.S. Hardwood and A. Demain (eds.). Bioenergy. ASM Press. Washington DC.

Jeffries, T.W., I.V. Grigoriev, J. Grimwood, J.M. Laplaza, A. Aerts, A. Salamov, J. Schmutz, E. Lindquist, P. Dehal, H. Shapiro, Y.S. Jin, V. Passoth and P.M. Richardson. 2007. Genome sequence of the lignocellulose-bioconverting and xylose fermenting yeast *Pichia stipitis*. Nat Biotechnol. 25(3): 319–326.

Jeffries, T.W. and Y.S. Jin. 2000. Ethanol and thermotolerance in the bioconversion of xylose by yeasts. Adv Appl Microbiol. 2000: 221–268.

Jeppsson, M., B. Johansson, B. Hahn-Hagerdal and M.F. Gorwa-Grauslund. 2002. Reduced oxidative pentose phosphate pathway flux in recombinant xylose utilizing *Saccharomyces cerevisiae* strains improves the ethanol yield from xylose. Appl Environ Microbiol. 68(4): 1604–1609.

Jeppsson, M., B. Johansson, P.R. Jensen, B. Hahn-Hagerdal and M.F. Gorwa-Grauslund. 2003. The level of glucose-6-phosphate dehydrogenase activity strongly influences xylose fermentation and inhibitor sensitivity in recombinant *Saccharomyces cerevisiae* strains. Yeast. 20(15): 1263–1272.

Ji, X.J., H. Huang, S. Li, J. Du and M. Lian. 2008. Enhanced 2,3-butanediol production by altering the mixed acid fermentation pathway in *Klebsiella oxytoca*. Biotechnol Lett. 30: 731–734.

Ji, X.J., H. Huang, J.G. Zhu, L.J. Ren, Z.K. Nie, J. Du et al. 2010. Engineering *Klebsiella oxytoca* for efficient 2,3-butanediol production through insertional inactivation of acetaldehyde dehydrogenase gene. Appl Microbiol Biotechnol. 85: 1751–1758.

Ji, X.J., H. Huang and P.K. Ouyang. 2011. Microbial 2,3-butanediol production: A state-of-the-art review. Biotechnol Adv. 29: 351–364.

Jiang, L., J. Wang, S. Liang, J. Cai and Z. Xu. 2011. Control and optimization of *Clostridium tyrobutyricum* ATCC 25755 adhesion into fibrous matrix in a fibrous bed bioreactor. Appl Biochem Biotechnol. 165: 98–108.

Joachimsthal, E., K.D. Haggett and P.L. Rogers. 1999. Evaluation of recombinant strains of Zymomonas mobilis for ethanol production from glucose/xylose media. Appl Biochem Biotechnol. 77(1): 147–157.

Johansson, B., C. Christensson, T. Hobley and B. Hahn-Hagerdal. 2001. Xylulokinase overexpression in two strains of *Saccharomyces cerevisiae* also expressing xylose reductase and xylitol dehydrogenase and its effect on fermentation of xylose and lignocellulosic hydrolysate. Appl Environ Microbiol. 67(9): 4249–4255.

Johansson, L., A. Lindskog, G. Silfversparre, C. Cimander, K.F. Nielsen and G. Lidén. 2005. Shikimic acid production by a modified strain of *E. coli* (W3110.shik1). under phosphate-limited and carbon-limited conditions. Biotechnol Bioeng. 92: 541–552.

John, R.P., K.M. Nampoothiri and A. Pandey. 2007. Fermentative production of lactic acid from biomass: an overview on process developments and future perspectives. Appl Microbiol Biotechnol. 74: 524–534.

Jojima, T., M. Inui and H. Yukawa. 2008. Production of isopropanol by metabolically engineered *Escherichia coli*. Appl Microbiol Biotechnol. 77(6): 1219–1224.

Jones, D.T. and D.R. Woods. 1986. Acetone–butanol fermentation revisited. Microbiol Rev. 50: 484–524.

Jore, J.P.M., P.J. Punt and M.J. van der Werf. 2011. Production of itaconic acid. Neder-landse organisatie voor toegepast-naturrweteenschappelijk onderzoek. US patent 201101240066 A1.

Julio, R.L., J.S. Antonio, M.G. Diana, R. Aloja and C.P. Juan. 2012. Fermentative production of fumaric acid from *Eucalyptus globules* wood hydrolysates. J Chem Technol Biotechnol. 87(7): 1036–1040.

Jung, J.Y., E.S. Choi and M.K. Oh. 2008. Enhanced production of 1, 2-propanediol by tpi1 deletion in *Saccharomyces cerevisiae*. J Microbiol Biotechnol. 18(11): 1797–1802.

Jung, Y.K., K.Y. Kim, S.J. Park and S.Y. Lee. 2010. Metabolic engineering of *Escherichia coli* for the production of polylactic acid and its copolymers. Biotechnol Bioeng. 105: 161–171.

Jurchescu, I.M., J. Hamann, X. Zhou, T. Ortmann, A. Kuenz, U. Prusse and S. Lang. 2013. Enhanced 2,3-butanediol production in fed-batch cultures of free and immobilized *Bacillus licheniformis* DSM 8785. Appl Microbiol Biotechnol. 97: 6715–6723.

Kabir, M., P.Y. Ho and K. Shimizu 2005. Effect of *ldh*A gene deletion on the metabolism of *E. coli* based on gene expression, enzyme activities, intracellular metabolite concentrations and metabolic flux distribution. Biochm Eng J. 26: 1–11.

Kalinowski, J., B. Bathe, D. Bartels, N. Bischoff, M. Bott, A. Burkovski, N. Dusch, L. Eggeling, B.J. Eikmanns and L. Gaigalat. 2003. The complete Corynebacterium glutamicum ATCC 13032 genome sequence and its impact on the production of aspartate derived amino acids and vitamins. J Biotechnol. 104: 5–25.

Kanamasa, S., L. Dwiarti, M. Okabe and E.Y. Park. 2008. Cloning and functional characterization of the cis-aconitic acid decarboxylase (CAD) gene from *Aspergillus terreus*. Appl Microbiol Biotechnol. 80: 223–229.

Karaffa, L. and C.P. Kubicek. 2003. *Aspergillus niger* citric acid accumulation: do we understand this well working black box? Appl Microbiol Biotechnol. 61: 189–196.

Karhumaa, K., B. Wiedemann, B. Hahn-Hagerdal, E. Boles and M.F. Gorwa-Grauslund. 2006. Co-utilization of L-arabinose and D-xylose by laboratory and industrial *Saccharomyces cerevisiae* strains. Microb Cell Fact. 5: 18–28.

Kase, H. and K. Nakayama. 1974. Studies on l-threonine fermentation, mechanism of l-threonine and l-lysine production by analog-resistant mutants of Corynebacterium glutamicum. Agric Biol Chem. 38: 993–1000.

Katahira, S., Y. Fujita, A. Mizuike, H. Fukuda and A. Kondo. 2004. Construction of a xylan-fermenting yeast strain through codisplay of xylanolytic enzymes on the surface of xylose-utilizing *Saccharomyces cerevisiae* cells. Appl Environ Microbiol. 70(9): 407–5414.

Kataoka, N., A.S. Vangnai, H. Ueda, T. Tajima, Y. Nakashimada and J. Kato. 2014. Enhancement of (*R*)-1, 3-butanediol production by engineered *Escherichia coli* using a bioreactor system with strict regulation of overall oxygen transfer coefficient and pH. Biosci Biotechnol Biochem. 78: 695–700.

Kataoka, N., A.S. Vangnai, T. Tajima, Y. Nakashimada and J. Kato. 2013. Improvement of (*R*)-1, 3-butanediol production by engineered *Escherichia coli*. J Biosci Bioeng. 115: 475–480.

Kawagoe, M., K. Hyakumura, S.-I. Suye et al. 1997. Application of bubble column fermentors to submerged culture of *Schizophyllum commune* for production of L-malic acid. J Ferment Bioeng. 84: 333–336.

Keasling, J.D. 2010. Manufacturing molecules through metabolic engineering. Science. 330: 1355–1358.

Kedar, P., R. Colah and K. Shimizu. 2007. Proteomic investigation on the *pyk-F* gene knockout *Escherichia coli* for aromatic amino acid production. Enzym Microbial Technol. 41: 455–465.

Kenealy, W., E. Zaady, J.C. du Preez, B. Stieglitz and I. Goldberg. 1986. Biochemical aspects of fumaric acid accumulation by *Rhizopus arrhizus*. Appl Environ Microbiol. 52(1): 128–133.

Khanal, S.K., M. Ramussen, P. Shrestha, H. van Leeuwen, C. Visvanathan and H. Liu. 2008. Bioenergy and biofuel production from wastes/residues of emerging biofuel industries. Water Environ Res. 80: 1625–1647.

Khankal, R., J.W. Chin and P.C. Cirino. 2008. Role of xylose transporters in xylitol production from engineered *Escherichia coli*. J Biotechnol. 134(3-4): 246–252.

Kieboom, J., J.J. Dennis, G.J. Zylstra and J.A.M. de Bont. 1998. Active efflux of organic solvents by *Pseudomonas putida* S12 is induced by solvents. J Bacteriol. 180: 6769–6772.

Kim, Y., L.O. Ingram and K.T. Shanmugam. 2007. Construction of an *Escherichia coli* K-12 mutant for homoethanogenic fermentation of glucose or xylose without foreign genes. Appl Environ Microbiol. 73: 1766–1771.

Kim, S., E. Seol, Y.K. Oh, G.Y. Wang and S. Park. 2009. Hydrogen production and metabolic flux analysis of metabolically engineered *Escherichia coli* strains. Int J Hydogen Energy. 34: 7417–7427.

Kim K., S.K. Kim, Y.C. Park and J.H. Seo. 2014. Enhanced production of 3-hydroxypropionic acid from glycerol by modulation of glycerol metabolism in recombinant *Escherichia coli*. Bioresour Technol. 156: 170–175.

Kinoshita, S., S. Udaka and M. Shimono. 1957. Amino acid fermentation I. Production of l-glutamic acid by various microorganism. J Gen Appl Microbiol. 3: 193–205.

Kinoshita, S., K. Nakayama and S. Akita. 1958. Taxonomical study of glutamic acid accumulating bacteria, Micrococcus glutamicus. Agri Chem Soc Jpn. 22: 176.

Kiran, M., J. Prakash, S. Annapoorni, S. Dube, T. Kusano, H. Okuyama, N. Murata and S. Shivaji. 2004. Psychrophilic *Pseudomonas syringae* requires transmonounsaturated fatty acid for growth at higher temperature. Extremophiles. 8: 401–410.

Klement, T. and J. Buchs. 2013. Itaconic acid-a biotechnological process in change. Bioresour Technol. 135: 422–431.

Knuf, C., I. Nookaew, I. Remmers, S. Khoomrung, S. Brown, A. Berry and J. Nielsen. 2014. Physiological characterization of the high malic acid-producing *Aspergillus oryzae* strain 2103a-68. Appl Microbiol Biotechnol. 98: 3517–3527.

Knuf, C., I. Nookaew, S.H. Brown, M. McCulloch, A. Berry and J. Nielsen. 2013. Investigation of malic acid production in *Aspergillus oryzae* under nitrogen starvation conditions. Appl Environ Microbiol. 79(19): 6050–6058.

Kolodrubetz, D. and R. Schleif. 1981. L-arabinose transport systems in *Escherichia coli* K-12. J Bacteriol. 148(2): 472–479.

Kondo, A., J. Ishii, K.Y. Hara, T. Hasunuma and F. Matsuda. 2012. Development of microbial cell factories for bio-refinery through synthetic bioengineering. J Biotechnol. 163(2): 204–16.

Korsa, I. and A. Böck. 1997. Characterization of fhlA mutations resulting in ligand-independent transcriptional activation and ATP hydrolysis. J Bacteriol. 179: 41–45.

Kotter, P. and M. Ciriacy. 1990. Xylose fermentation by *Saccharomyces cerevisiae*. Appl Microbiol Biotechnol. 38(6): 776–783.

Kotter, P. and M. Ciriacy. 1993. Xylose fermentation by *Saccharomyces cerevisiae*. Appl Microbiol Biotechnol. 38: 776–783.

Krahulec, S., M. Klimacek and B. Nidetzky. 2009. Engineering of a matched pair of xylose reductase and xylitol dehydrogenase for xylose fermentation by *Saccharomyces cerevisiae*. Biotechnol J. 4(5): 684–694.

Kramer, M., J. Bongaerts, R. Bovenberg, S. Kremer, U. Muller, S. Orf, M. Wubbolts and L. Raeven. 2003. Metabolic engineering for microbial production of shikimic acid. Metab Eng. 5: 277–283.

Krebs, H.A. 1970. The history of the tricarboxylic acid cycle. Perspect Biol Med. 14: 154–170.

Kruckeberg, A.L. 1996. The hexose transporter family of *Saccharomyces cerevisiae*. Arch Microbiol. 166: 283–292.

Kubicek, C.P. 1988. The role of the citric acid cycle in fungal organic acid fermentations. Biochem Soc Symp. 54: 113–126.

Kuenz, A., Y. Gallenmuller, T. Willke and K.D. Vorlop. 2012. Microbial production of itaconic acid: developing a stable platform for high product concentratios. Appl Microbiol Biotechnol. 96: 1209–1216.

Kyla-Nikkila, K., M. Hujamen, M. Leisola and A. Palva. 2000. Metabolic engineering of *Lactobacillus helveticus* CNRZ32 for production of l-(+)-lactic acid. Appl Environ Microbiol. 66: 3835–3841.

La Grange, D.C., I.S. Pretorius, M. Claeyssens and W.H. van Zyl. 2001. Degradation of xylan to D-xylose by recombinant *Saccharomyces cerevisiae* coexpressing the *Aspergillus niger* beta-xylosidase (*xlnD*) and the Trichoderma reesei xylanase II (*xyn2*) genes. Appl Environ Microbiol. 67(12): 5512–5519.

Lam, V.M., K.R. Daruwalla, P.J. Henderson and M.C. Jones-Mortimer. 1980. Proton-linked D-xylose transport in *Escherichia coli*. J Bacteriol. 143(1): 396–402.

Lau, M.W. and B.E. Dale. 2009. Cellulosic ethanol production from AFEX-treated corn stover using *Saccharomyces cerevisiae* 424A (LNH-ST). PNAS USA. 106: 1368–1373.

Lawford, H.G., J.D. Rousseau, A. Mohagheghi and J.D. McMillan. 1999. Fermentation performance characteristics of a prehydrolyzate-adapted xylose-fermenting recombinant *Zymomonas* in batch and continuous fermentations. Appl Biochem Biotechnol. 77-9: 191–204.

Leathers, T.D. and P. Manitchotpisit. 2012. Production of poly(beta-L-malic acid) (PMA) from agricultural biomass substrates by *Aureobasidium pullulans*. Biotechnol Lett. 35: 83–89.

Lee, C., I. Kim, J. Lee, K.L. Lee, B. Min and C. Park. 2010. Transcriptional activation of the aldehyde reductase YqhD by YqhC and its implication in glyoxal metabolism in *Escherichia coli* K-12. J Bacteriol. 192(16): 4205–4214.

Lee, S.J., D.Y. Lee, T.Y. Kim, B.H. Kim, J. Lee and Y.S. Lee. 2005. Metabolic engineering of *Escherichia coli* for enhanced production of succinic acid, based on genome comparison and *in silico* gene knockout simulation. Appl Environ Microbiol. 71: 7880–7887.

Lee, W. and N.A. da Silva. 2006. Application of sequential integration for metabolic engineering of 1, 2-propanediol production in yeast. Metab Eng. 8(1): 58–65.

Lee, W.J., M.D. Kim, Y.W. Ryu, L.F. Bisson and J.H. Seo. 2002. Kinetic studies on glucose and xylose transport in *Saccharomyces cerevisiae*. Appl Microbiol Biotechnol. 60: 186–191.

Lee, J., Y.-S. Jang, S.J. Choi, J.A. Im, H. Song, J.H. Cho, D.Y. Seung, E.T. Papoutsakis, G.N. Bennett and S.Y. Lee. 2012. Metabolic Engineering of *Clostridium acetobutylicum* ATCC 824 for Isopropanol-Butanol-Ethanol Fermentation. 78(5): 1416–1423.

Lennen, R.M. and B.F. Pfleger. 2013. Microbial production of fatty acid-derived fuels and chemicals. Curr Opin in Biotechnol. 24(6): 1044–1053.

Leonard, E., K.H. Lim, P.N. Saw and M.A. Koffas. 2007. Engineering central metabolic pathways for high-level flavonoid production in *Escherichia coli*. Appl Microbiol Biotechnol. 73: 3877–3886.

Leonardo, M.R., P.R. Cunningham and D.P. Clark. 1993. Anaerobic regulation of the *adhE* gene, encoding the fermentative alcohol dehydrogenase of *Escherichia coli*. J Bacteriol. 175: 870–878.

Leonardo, M.R., Y.P. Dailly and D.P. Clark. 1996. Role of NAD in regulating the *adhE* gene of *Escherichia coli*. J Bacteriol. 178: 6013–6018.

Leuchtenberger, W., K. Huthmacher and K. Drauz. 2005. Biotechnological production of amino acids and derivatives: current status and prospects. Appl Microbiol Biotechnol. 69(1): 1–8.

Levin, D.B., L. Pitt and M. Love. 2004. Biohydrogen production: prospects and limitations to practical application. Int J Hydrogen Energy. 29: 173–185.

Li, H., T. Qiu, G. Huang and Y. Cao. 2010. Production of gamma-aminobutyric acid by *Lactobacillus brevis* NCL912 using fed-batch fermentation. Microb Cell Fact. 9: 85.

Li, A., N. van Luijk, M. ter Beek, M. Caspers, P. Punt and M. van der Werf. 2011. A clone-based transcriptomics approach for the identification of genes relevant for itaconic acid production in *Aspergillus*. Fungal Genet Biol. 48: 602–611.

Li, A., N. Pfelzer, R. Zuijderwijk, A. Brickwedde, C. van Zeijl and P. Punt. 2013. Reduced by-product formation and modified oxygen availability improve itaconic acid production in *Aspergillus niger*. Appl Microbiol Biotechnol. 97: 3901–3911.

Li, A., N. Pfelzer, R. Zuijderwijk and P. Punt. 2012. Enhanced itaconic acid production in *Aspergillus niger* using genetic modification and medium optimization. BMC Biotechnol. 12: 57.

Li, J., W. Wang, Y. Ma and A.P. Zeng. 2013. Medium optimization and proteome analysis of (R, R)-2,3-butanediol production by *Paenibacillus polymyxa* ATCC 12321. Appl Microbiol Biotechnol. 97: 585–597.

Li, L., C. Chen, K. Li, Y. Wang, C. Gao, C. Ma and P. Xu. 2014. Efficient simultaneous saccharification and fermentation of inulin to 2,3-butanediol by thermophilic *Bacillus licheniformis* ATCC 14580. Appl Environ Microbiol. 80: 6458–6464.

Li, L., K. Li, Y. Wang, C. Chen, Y. Xu, L. Zhang, B. Han, C. Gao, F. Tao and C. Ma. 2015. Metabolic engineering of *Enterobacter cloacae* for high-yield production of enantiopure (2R, 3R)-2,3-butanediol from lignocellulose-derived sugars. Metab Eng. 2015(28): 19–27.

Li, L., L. Zhang, K. Li, Y. Wang, C. Gao, B. Han, C. Ma and P. Xu. 2013. A newly isolated *Bacillus licheniformis* strain thermophilically produces 2,3-butanediol, a platform and fuel biochemical. Biotechnol Biofuels. 6: 123.

Li, L., Y. Wang, L. Zhang, C. Ma, A. Wang, F. Tao and P. Xu. 2012. Biocatalytic production of (2S, 3S)-2,3-butanediol from diacetyl using whole cells of engineered *Escherichia coli*. Bioresour Technol. 115: 111–116.

Li, M., S. Yao and K. Shimizu. 2007. Effect of *poxB* gene knockout on metabolism in *Escherichia coli* based on growth characteristics and enzyme activities. World J Microbiol Biotechnol. 23(4): 573–580.

Li, Y., J. Chen and S.-Y. Lun. 2001. Biotechnological production of pyruvic acid. Appl Microbiol Biotechnol. 57: 451–459.

Lian, J., R. Chao and H. Zhao. 2014. Metabolic engineering of a *Saccharomyces cerevisiae* strain capable of simultaneously utilizing glucose and galactose to produce enantiopure (2R, 3R)-butanediol. Metab Eng. 23: 92–99.

Liao, J.C. and P.-C. Chang (eds.). 2010. Genetically modified microorganisms that produce itaconic acid at high yields and uses thereof. I.T.R. Institute.

Liao, W., Y. Liu and S.L. Chen. 2007. Studying pellet formation of a filamentous fungus *Rhizopus oryzae* to enhance organic acid production. Appl Biochem Biotechnol. 136-141(1-12): 689–701.

Liao, W., Y. Liu, C. Frear and S.L. Chen. 2008. Co-production of fumaric acid and chitin from a nitrogen-rich lignocellulosic material-dairy manune-using a pelletized filamentous fungus *Rhizopus oryzae* ATCC20344. Bioresour Technol. 99: 5859–5866.

Liao, J.C., L. Mi, S. Pontrelli and S. Luo. 2016. Fueling the future: microbial engineering for the production of sustainable biofuels. Nature Rev Microbiol. 14(5): 288–304.

Ligthelm, M.E., B.A. Prior and J.C. du Preez. 1988. The oxygen requirements of yeasts for the fermentation of D-xylose and D-glucose to ethanol. Appl Microbiol Biotechnol. 28(1): 63–68.

Lilly, M., H.P. Fierobe, W.H. van Zyl and H. Volschenk. 2009. Heterologous expression of a Clostridium minicellulosome in *Saccharomyces cerevisiae*. FEMS Yeast Res. 9(8): 1236–1249.

Lin, H., K.Y. San and G.N. Bennett. 2005a. Effect of Sorgham vulgare phosphoenol pyruvate carboxylase and *Lactococcus lactis* pyruvate carboxylase coexpression on succinate production in mutant strains of *Escherichia coli*. Appl Microbiol Biotechnol. 67: 515–523.

Lin, H., G.N. Bennett and K.Y. San. 2005b. Genetic reconstruction of the aerobic central metabolism in *Escherichia coli* for the absolute aerobic production of succinate. Biotechnol Prog. 89: 148–156.

Lin, H., G.N. Bennett and K.Y. San. 2005c. Metabolic engineering of aerobic succinate production systems in *Escherichia coli* to improve process productivity and achieve the maximum theoretical succinate yield. Metab Eng. 7: 116–127.

Lin, H., G.N. Bennett and K.Y. San. 2005d. Fed-batch culture of a metabolically engineered *Escherichia coli* strain designed for high-level succinate production and yield under aerobic conditions. Biotechnol Bioeng. 90: 775–779.

Lin, Y. and H.P. Blaschek. 1983. Butanol production bu butanol-tolerant strain of *Clostridium aceto-butylicum* in extruded corn broth. Appl Environ Microbiol. 45: 966–973.

Lin, Y. and S. Tanaka. 2006. Ethanol fermentation from biomass resources: current state and products. Appl Microbiol Biotechnol. 69: 627–642.

Ling, L.B. and T.K. Ng. 1989. Fermentation process for carboxylic acids. United States patent. US4877731 (Oct 31).

Litsanov, B., M. Brocker and M. Bott. 2012. Toward homosuccinate fermentation: Metabolic engineering of *Corynebacterium glutamicum* for anaerobic production of succinate from glucose and formate. Appl Environ Microbiol. 83(4): 3325–3337.

Liu, L.M., Y. Li, H.Z. Li and J. Chen. 2006. Significant increase of glycolytic flux in *Torulopsis glabrata* by inhibition of oxidative phosphorylation. FEMS Yeast Res. 6(8): 1117–1129.

Liu, X., Y. Zhu and S.T. Yang. 2006. Construction and characterization of ack deleted mutant of *Clostridium tyrobutyricum* for enhanced butyric acid and hydrogen production. Biotechnol Prog. 22: 1265–1275.

Liu, X.Y., Z. Chi, G.L. Liu, C. Madzak and Z.M. Chi. 2013. Both decrease in ACL1 gene expression and increase in ICL1 gene expression in marine-derived yeast Yarrowia lipolytica expressing INU1 gene enhance citric acid production from inulin. Mar Biotechnol. 15: 26–36.

Liu, P., X. Zhu, Z. Tan, X. Zhang and Y. Ma. 2016. Construction of *Escherichia coli* cell factories for production of organic acids and alcohols. Adv Biochem Eng Biotechnol. 155: 107–140.

Liu, R., L. Liang, K. Chen, J. Ma, M. Jiang, P. Wei and P. Ouyang. 2012. Fermentation of xylose to succinate by enhancement of ATP supply in metabolically engineered *Escherichia coli*. Appl Microbiol Biotechnol. 94(4): 954–968.

Liu, S.J. and A. Steinbuchel. 1997. Production of poly(malic acid) from different carbon sources and its regulation in *Aureobacidium pullulans*. Biotechnol Lett. 19: 11–14.

Liu, Z., Z. Chi, L. Wang and J. Li. 2008. Production, purification, and characterization of an extracellular lipase from *Aureobasidium pullulans* HN2.3 with potential application for the hydrolysis of edible oils. Biochem Eng J. 40: 445–451.

Lohbeck, K., H. Haferkorn, W. Fuhrmann and N. Fedtke. 1990. Maleic and fumaric acids. Ullmann's Encyclopedia of Industrial Chemistry, Vol. A16. VCH Weinheim, Germany. pp. 53–62.

Lotfy, W.A., K.M. Ghanem and E.R. El-Helow. 2007. Citric acid production by a novel *Aspergillus niger* isolate: II. Optimization of process parameters through statistical experimental designs. Bioresour Technol. 98(18): 3470–3477.

Lutke-Eversloh, T., C.N.S. Santos and G. Stephanopoulos. 2007. Perspectives of biotechnological production of L-tyrosine and its applications. Appl Microbiol Biotechnol. 77: 751–762.

Ma, C., A. Wang, J. Qin, L. Li, X. Ai, T. Jiang, H. Tang and P. Xu. 2009. Enhanced 2,3-butanediol production by *Klebsiella pneumoniae* SDM. Appl Microbiol Biotechnol. 82: 49–57.

MacLean, M.J., L.S. Ness, G.P. Ferguson and I.R. Booth. 1998. The role of glyoxalase I in the detoxification of methylglyoxal and in the activation of the KefB K+ efflux system in *Escherichia coli*. Mol Microbiol. 27(3): 563–571.

Maeda, T., V. Sanchez-Torres and T.K. Wood. 2007. Enhanced hydrogen production from glucose by metabolically-engineered *Escherichia coli*. Appl Microbiol Biotechnol. 77: 879–890.

Maeda, T., V. Sanchez-Torres and T.K. Wood. 2008a. Metabolic engineering to enhance bacterial hydrogen production. Microb Biotechnol. 1: 30–39.

Maeda, T., V. Sanchez-Torres and T.K. Wood. 2008b. Protein engineering of hydrogenase 3 to enhance hydrogen production. Appl Microbiol Biotechnol. 79: 77–86.

Magnuson, J.K. and L.L. Lasure. 2004. Organic acid production by filamentous fungi. pp. 307–340. *In*: S. Jan and Lange Lene (eds.). Advanced in Fungal Biotechnology for Industry, Agriculture, and Medicine. New York, Kluwer Academic/Plenum Publishers.

Mahmourides, G., H. Lee, N. Maki and H. Schneider. 1985. Ethanol accumulation in cultures of *Pachysolen tannophilus* on D-xylose is associated with a transition to a state of low oxygen consumption. Biotechnol. 3: 59–62.

Maleszka, R. and H. Schneider. 1982. Concurrent production and consumption of ethanol by cultures pf *Pachysolen tannophilus* growing on D-xylose. Appl Environ Microbiol. 44: 909–912.

Manitchotpisit, P., C.D. Skory, S.W. Peterson, N.P.J. Price, K.E. Vermillion and T.D. Leathers. 2012. Poly(beta-L-malic acid) production by diverse phylogenetic clades of *Aureobacidium pullulans*. J Ind Microbiol Biotechnol. 39: 125–132.

Manitchotpisit, P., N.P.J. Price, T.D. Leathers and H. Punnapayak. 2011. Heavy oils produced by *Aureobacidium pullulans*. Biotechnol Lett. 33: 1151–1157.

Margolles-Clark, E., M. Tenkanen, T. Nakari-Setala and M. Penttila. 1996. Cloning of genes encoding alpha-L-arabinofuranosidase and beta-xylosidase from *Trichoderma reesei* by expression in *Saccharomyces cerevisiae*. Appl Environ Microbiol. 62(10): 3840–3846.

Marques, S., L. Alves, J.C. Roseiro and F.M. Girio. 2008. Conversion of recycled paper sludge to ethanol by SHF and SSF using *Pichia stipitis*. Biomass Bioenerg. 32(5): 400–406.

Martinez, F.A.C., E.M. Balciunas, J.M. Salgado, J.M.D. Gonzalez, A. Converti and R.P.S. Oliveira. 2012. Lactic acid properties and production: A review. Food Sci Technol. 30: 70–83.

Matsumoto, K. and S. Taguchi. 2013. Enzyme and metabolic engineering for the production of novel biopolymers: crossover of biotechnological and chemical processes. Curr Opin Biotechnol. 24: 1054–1060.

Matsuoka, Y. and K. Shimizu. 2015. Metabolic flux analysis for *Escherichia coli* by flux balance analysis. pp. 237–260. *In*: J.O. Kromer, L.K. Nielsen and L.M. Blank (eds.). Metabolic Flux Analysis. Humana press.

Matsushika, A., H. Inoue, T. Kodaki and S. Sawayama. 2009. Ethanol production from xylose in engineered *Saccharomyces cerevisiae* strains: current state and perspectives. Appl Microbiol Biotechnol. 84: 37–53.

Matsuyama, A., H. Yamamoto, N. Kawada and Y. Kobayashi. 2001. Industrial production of (*R*)-1, 3-butanediol by new biocatalysts. J Mol Catal B Enzym. 11: 513–521.

Mattey, M. 1992. The production of organic acids. Critical Rev in Biotechnol. 12(1/2): 87–132.

Mattey, M. 1999. Biochemistry of citric acid production by yeasts. pp. 33–54. *In*: B. Kristiansen, M. Mattey and J. Linden (eds.). Citric Acid Biotechnology. London: Taylor and Francis.

Mayer, D., V. Schlensog and A. Bock. 1995. Identification of the transcriptional activator controlling the butandiol fermentation pathway in *Klebsiella terrigena*. J Bacteriol. 177: 5261–5269.

Mazumdar, S., J.M. Clomburg and R. Gonzalez. 2010. *Escherichia coli* strains engineered for homofermentative production of D-lactic acid from glycerol. Appl Environ Microbiol. 76(13): 4327–4336.

Mazumdar, S., M.D. Blankschien, J.M. Clomburg and R. Gonzalez. 2013. Efficient synthesis of L-lactic acid from glycerol by metabolically engineered *Escherichia coli*. Microb Cell Fact. 12: 7.

McGinn, S.M., K.A. Beauchemin, T. Coates and D. Colombatto. 2004. Methane emissions from beef cattle: Effects of monensin, sunflower oil, enzymes, yeast and fumaric acid. J Anim Sci. 82: 3346–3356.

Mckee, A.E., B.J. Rutherford, D.C. Chivian, E.K. Baidoo, D. Juminaga, D. Kuo, P.I. Benke, J.A. Dietrich, S.M. Ma, A.P. Arkin, C. Petzold, P.D. Adams, J.D. Keasling and S.R. Chhabra. 2012. Manipulation of the carbon storage regulator system for metabolic remodeling and biofuel production in *Escherichia coli*. Microb Cell Fact. 11: 79.

Meek, J.S. 1975. The determination of a mechanism of isomerization of maleic acid to fumaric acid. J Chem Educ. 52(8): 541.

Mielenz, J.R. 2001. Ethanol production from biomass: technology and commercialization status. Curr Opin Microbiol. 4: 324–329.

Minty, J.J., A.A. Lesnefsky, F. Lin, Y. Chen, T.A. Zaroff, A.B. Velose, B. Xie, C.A. McConnell, R.J. Ward, D.R. Schwartz, J.M. Rouillard, Y. Gao, E. Gulari and X.N. Lin. 2011. Evolution combined with genomic study elucidates genetic bases of isobutanol tolerance in *Escherichia coli*. Microb Cell Fact. 10: 18.

Mirasol, F. 2011. PTA. ICIS Chem Bus. 279: 43–43.

Misra, K., A.B. Banerjee, S. Ray and M. Ray. 1995. Glyoxalase III from *Escherichia coli*: a single novel enzyme for the conversion of methyglyoxal into D-lactate with reduced glutathione. Biochem J. 305: 999–1003.

Miyata, R. and T. Yonehara. 1996. Improvement of fermentative production of pyruvate from glucose by *Torulopsis glabrata* IFO 0005. J Ferment Bioeng. 82: 475–479.

Mogk, A.C., Schlieker, K.L. Friedrich, H.-J. Schönfeld, E. Vierling, B. Bukau. 2003. Refolding of substrates bound to small Hsps relies on a disaggregation reaction mediated most efficiently by ClpB/DnaK. The J of Biol Chem. 278(33): 31033–31042.

Mohagheghi, A., K. Evans, Y.C. Chou and M. Zhang. 2002. Cofermentation of glucose, xylose, and arabinose by genomic DNA-integrated xylose/arabinose fermenting strain of *Zymomonas mobilis* AX101. Appl Biochem Biotechnol. 98-100: 885–898.

Moir, A., I.M. Feavers and J.R. Guest. 1984. Characterization of the fumarase gene of *Bacillus subtilis* 168 cloned and expressed in *Escherichia coli* K12. J Gen Microbiol. 130: 3009–3017.

Moon, S.Y., S.H. Hong, T.Y. Kim and S.Y. Lee. 2008. Metabolic engineering of *Escherichia coli* for the production of malic acid. Biochem Eng J. 40: 312–320.

Moon, T.S., J.E. Dueber, E. Shiue and K.L.J. Prather. 2010. Use of modular, synthetic scaffolds for improved production of glucaric acid in engineered *E. coli*. Metab Eng. 12: 298–305.

Moon, T.S., S.-H. Yoon, A.M. Lanza, J.D. Roy-Mayhew and K.L.J. Prather. 2009. Production of glucaric acid from a synthetic pathway in recombinant *Escherichia coli*. Appl Environ Microbiol. 75(3): 589–595.

Moresi, M., E. Parente, M. Petruccioli and F. Federici. 1991. Optimization of fumaric acid production from potato flour by *Rhizopus arrhizus*. Appl Microbiol Biotechnol. 36: 35–39.

Moresi, M., E. Parente, M. Petruccioli and F. Federici. 1992. Fumaric acid production from hydrolysates of starch-based substrates. J Chem Technol Biotechnol. 54: 283–290.

Mrowietz, U., E. Christophers and P. Altmeyer. 1998. Treatment of psoriasis with fumaric acid esters: results of a prospective multicenter study. Brit J Dermatol. 138: 456–460.

Munjal, N., A.J. Mattam, D. Pramanik, P.S. Srivastava and S.S. Yazdani. 2012. Modulation of endogenous pathways enhances bioethanol yield and productivity in *Escherichia coli*. Microb Cell Fact. 11: 145.

Murarka, A., Y. Dharmadi, S.S. Yazdani and R. Gonzalez. 2008. Fermentative utilization of glycerol in *Escherichia coli* and its implications for the production of fuels and chemicals. Appl Environ Microbiol. 74(4): 1124–1135.

Mussatto, S.I. and I.C. Roberto. 2004. Kinetic behavior of *Candida guilliermondii* yeast during xylitol production from highly concentrated hydrolysate. Proc Biochem. 39(11): 1433–1439.

Musser, M.T. 2005. Sdipic acid. ULLMANN'S Encyclopedia of Industrial Chemistry. Vol. 1. Wiley-VCH Verlag GmbH & Co. KGaA, Weinheim. pp. 537–548.

Nakamura, C.E. and G.M. Whitedy. 2003. Metabolic engineering for the microbial production of 1,3-propanediol. Cuur Opin Biotechnol. 14: 454–459.

Nakashimada, Y., B. Marwoto, T. Kashiwamura, T. Kakizono and N. Nishio. 2000. Enhanced 2,3-butanediol production by addition of acetic acid in *Paenibacillus polymyxa*. J Biosci Bioeng. 90: 661–664.

Nakashimada, Y., K. Kanai and N. Nishio. 1998. Optimization of dilution rate, pH and oxygen supply on optical purity of 2,3 butanediol produced by *Paenibacillus polymyxa* in chemostat culture. Biotechnol Lett. 20: 1133–1138.

Nakayama, H.V. and K. Araki. 1973. Process for producing l-lysine, US patent (3708395).

Nakano, S., M. Fukaya and S. Horinouchi. 2004. Enhanced expression of aconitase raises acetic acid resistance in *Acetobacter aceti*. FEMS Microbiol Lett. 235: 315–322.

Nakano, S., M. Fukaya and S. Horinouchi. 2006. Putative ABC Transporter Responsible for Acetic Acid Resistance in *Acetobacter aceti*. Appl Environ Microbiol. 72(1): 497–505.

Nan, H., S.O. Seo, E.J. Oh, J.H. Seo, J.H. Cate and Y.S. Jin. 2014. 2,3-Butanediol production from cellobiose by engineered *Saccharomyces cerevisiae*. Appl Microbiol Biotechnol. 98: 5757–5764.

Narayanan, N., R.K. Royehoundhury and A. Srivastava. 2004. L(+) lactic acid fermentation and its product polymerization. J Biotechnol. 7: 167–178.

Neirinck, L.G., R. Maleszka and H. Schneider. 1984. The requirement of oxygen for incorporation of carbon from D-xylose and D-glucose by *Pachysolen tannophilus*. Arch Biochem Biophys. 228: 13–21.

Nicolaou, S.A., S.M. Gaida and E.T. Papoutsakis. 2010. A comparative view of metabolite and substrate stress and tolerance in microbial processing: from biofuels and chemicals, to biocatalysis and bioremediation. Metabolic Eng. 12: 307–331.

Nigam, J.N. 2001a. Development of xylose-fermenting yeast *Pichia stipitis* for ethanol production through adaptation on hardwood hemicellulose acid prehydrolysate. J Appl Microbiol. 90(2): 208–215.

Nigam, J.N. 2001b. Ethanol production from hardwood spent sulfite liquor using an adapted strain of *Pichia stipitis*. J Ind Microbiol Biotechnol. 26(3): 145–150.

Nigam, J.N. 2001c. Ethanol production from wheat straw hemicellulose hydrolysate by *Pichia stipitis*. J Biotechnol. 87(1): 17–27.

Nikaido, H. and T. Nakano. 1980. The outer membrane of gram-negative bacteria. Adv Microbiol Physiol. 20: 163–250.

Niu, W., K.M. Draths and J.W. Frost. 2002. Benzene-Free Synthesis of Adipic Acid. Biotechnol Prog. 18: 201–211.

Niu, D., K. Tian, B.A. Prior, M. Wang, Z. Wang, F. Lu and S. Singh. 2014. Highly efficient L-lactate production using engineered *Escherichia coli* with dissimilar temperature optima for L-lactate formation and cell growth. Microb Cell Fact. 13: 78.

Nizam, S.A. and K. Shimizu. 2008. Effects of *arc* A and *arc* B genes knockout on the metabolism in *Escherichia coli* under anaerobic and microaerobic conditions. Biochem Eng J. 42: 229–236.

Novotny, C.P. and E. Englesberg. 1966. The L-arabinose permease system in *Escherichia coli* B/r. Biochim Biophys Acta. 117(1): 217–230.

Oh, B.R., J.W. Seo, S.Y. Heo, W.K. Hong, L.H. Luo, M.H. Joe and C.H. Kim. 2011. Efficient production of ethanol from crude glycerol by a *Klebsiella pneumonia* mutant strain. Bioresour Technol. 102(4): 3918–3922.

Ohta, K., D.S. Beall, J.P. Mejia, K.T. Shanmugam and L.O. Ingram. 1991. Genetic improvement of *Escherichia coli* for ethanol production: chromosomal integration of *Zymomonanmobilis* genes encoding pyruvate decarboxylase and alcohol dehydrogenase II. Appl Environ Microbiol. 57: 893–900.

Ohta, K., D.S. Beall, J.P. Mejia, K.T. Shanmugam and L.O. Ingram. 1991b. Metabolic engineering of *Klebsiella oxytoca* M5A1 for ethanol production from xylose and glucose. Appl Environ Microbiol. 57(10): 2810–2815.

Okabe, M., D. Lies, S. Kanamasa and E.Y. Park. 2009. Biotechnological production of itaconic acid and its biosynthesis in *Aspergillus terreus*. Appl Microbiol Biotechnol. 84: 597–606.

Okamoto, S., T. Chin, K. Hiratsuka, Y. Aso, Y. Tanaka, T. Takahashi and H. Ohara. 2014. Production of itaconic acid using metabolically engineered *Escherichia coli*. J General Appl Micribiol. 60: 191–197.

Okano, K., T. Tanaka, C. Ogino, H. Fukuda and A. Kondo. 2010. Biotechnological production of enantiomeric pure lactic acid from renewable resources: recent achievements, perspectives, and limits. Appl Microbiol Biotechnol. 85: 413–423.

Okino, S., M. Inui and H. Yukawa. 2005. Production of organic acids by *Corynebacterium glutamicum* under oxygen deprivation. Appl Microbiol Biotechnol. 68: 475–480.

Okino, S., M. Suda, K. Fujikura, M. Inui and H. Yukawa. 2008. Production of D-lactic acid by *Corynebacterium glutamicum* under oxygen deprivation. Appl Microbiol Biotechnol. 78: 449–454.

Osman, Y.A. and L.O. Ingram. 1985. Mechanism of ethanol inhibition of fermentation in *Zymomonas mobilis* CP4. J Bacteriol. 164(1): 173–180.

Otten, A., M. Brocker and M. Bott. 2015. Metabolic engineering of *Corynebacterium glutamicum* for the production of itaconate. Metab Eng. 30: 156–165.

Oude Elferink, S.J., J. Krooneman, J.C. Gottschal, S.F. Spoelstra, F. Faber and F. Driehuis. 2001. Anaerobic conversion of lactic acid to acetic acid and 1, 2-propanediol by *Lactobacillus buchneri*. Appl Environ Microbiol. 67: 125–132.

Pandey, R.P., P. Parajuli, M.A.G. Koffas and J.K. Sohng. 2016. Microbial production of natural and non-natural flavonoids: Pathway engineering, directed evolution and systems/synthetic biology Biotechnol Adv. 34(5): 634–662.

Papanikolaou, S., M. Galiotou-Panayotou, I. Chevalot, M. Komaitis, I. Marc and G. Aggelis. 2006. Influence of glucose and saturated free-fatty acid mixtures on citric acid and lipid production by *Yarrowia lipolytica*. Curr Microbiol. 52(2): 134–142.

Pang, X., X. Zhuang, Z. Tang and X. Chen. 2010. Polylactic acid (PLA): research development and industrialization. Biotechnol J. 5: 1125–1136.

Papagianni, M. 2007. Advances in citric acid fermentation by *Aspergillus niger*. Biochemical aspects, membrane transport and modeling. Biotechnol Adv. 25: 244–263.

Papagianni, M. 2012. Recent advances in engineering the central carbon metabolism of industrially important bacteria. Microb Cell Fact. 11: 50.

Parekh, S., V.A. Vinci and R.J. Strobel. 2000. Improvement of microbial strains and fermentation processes. Appl Microbiol Biotechnol. 54: 287–301.

Parker, C., W.O. Barnell, J.L. Snoep, L.O. Ingram and T. Conway. 1995. Characterization of the *Zymomonas mobilis* glucose facilitator gene product (*glf*) in recombinant *Escherichia coli*: examination of transport mechanism, kinetics and the role of glucokinase in glucose transport. Mol Microbiol. 15(5): 795–802.

Patek, M. 2007. Branched chain amino acids biosynthesis pathways, regulation and metabolic engineering. Springer, Heidelberg 778.

Patnaik, R., S. Louie, V. Gavrilovic, K. Perry, W. Stemmer, C. Ryan and S. del Cardayre. 2002. Genome shuffling of Lactobacillus for improved acid tolerance. Nat Biotechnol. 20: 707–712.

Peksel, A., N.V. Torres, J. Liu and G. Juneau. 2002. 13C-NMR analysis of glucose metabolism during acid production by *Aspergillus niger*. Appl Microbiol Biotechnol. 58: 157–163.

Penfold, D.W., C.F. Forster and L.E. Macaskie. 2003. Increased hydrogen production by *Escherichia coli* strain HD701 in comparison with the wild-type parent strain MC4100. Enzyme Microb Technol. 33: 185–189.

Penfold, D.W., F. Sargent and L.E. Macaskie. 2006. Inactivation of the *Escherichia coli* K-12 twin-arginine translocation system promotes increased hydrogen production. FEMS Microbiol Lett. 262: 135–137.

Peralta-Yahya, P.P., F. Zhang, S.B. del Cardayre and J.D. Keasling. 2012. Microbial engineering for the production of advanced biofuels. Nature. 488: 320–328.

Perez-Gonzalez, J.A., L.H. De Graaff, J. Visser and D. Ramon. 1996. Molecular cloning and expression in *Saccharomyces cerevisiae* of two *Aspergillus nidulans* xylanase genes. Appl Environ Microbiol. 62(6): 2179–2182.

Petrov, K. and P. Petrova. 2009. High production of 2,3-butanediol from glycerol by *Klebsiella pneumoniae* G31. Appl Microbiol Biotechnol. 84: 659–665.

Petschacher, B. and B. Nidetzky. 2008. Altering the coenzyme preference of xylose reductase to favor utilization of NADH enhances ethanol yield from xylose in a metabolically engineered strain of *Saccharomyces cerevisiae*. Microb Cell Fact. 7: 9.

Pfeifer, B.A., S.J. Admiraal, H. Gramajo, D.E. Cane and C. Khosla. 2001. Biosynthesis of complex polyketides in a metabolically engineered strain of *E. coli*. Science. 291: 1790–1792.

Pfefferle, W., B. Moeckel, B. Bathe and A. Marx. 2003. Biotechnological manufacture of l-Lysine. *In*: Scheper, T. (ed.). Advances in Biochemical Engineering/Biotechnology. Vol 79. Springer, Berlin, p. 59.

Pharkya, P., A.P. Burgard, S.J. Van Dien, R.E. Osterhout, M.J. Burk, J.D. Trawick, M.P. Kuchinskas and B. Steer. 2015. Microorganisms and methods for production of 4-Hydroxybutyrate, 1, 4-Butanediol and related compounds. US patent 20150148513.

Pines, O., S. Even Ram, N. Elnathan, E. Battat, O. Aharonov, D. Gibson et al. 1996. The cytosolic pathway of L-malic acid synthesis in *Saccharomyces cerevisiae*: the role of fumalase. Appl Microbiol Biotechnol. 46: 393–399.

Piper, P. 1995. The heat-shock and ethanol stress responses of yeast exhibit extensive similarity and functional overlap. FEMS Microbiol Lett. 134: 121–127.

Polen, T., M. Spelberg and M. Bott. 2013. Toward biotechnological production of adipic acid and precursors from biorenewables. J Biotechnol. 167: 75–84.

Portnoy, V.A., D.A. Scott, N.E. Lewis, Y. Tarasova, A.L. Osterman and B.Ø. Palsson. 2010. Deletion of genes encoding cytochrome oxidases and quinol monooxygenase blocks the aerobic-anaerobic shift in *Escherichia coli* K-12 MG1655. Appl Environ Microbiol. 76(19): 6529–6540.

Portnoy, V.A., M.J. Herrgard and B.O. Palsson. 2008. Aerobic fermentation of D-glucose by an evolved cytochrome oxidase-defficient *Escherichia coli* strain. Appl Environ Microbiol. 74(24): 7561–7569.

Postma, E., C. Verduyn, W.A. Scheffers and J.P. Vandijken. 1989. Enzymic analysis of the Crabtree effect in glucose-limited chemostat cultures of *Saccharomyces cerevisiae*. Appl Environ Microbiol. 55: 468–477.

Pratt, E.A., L.W. Fung, J.A. Flowers and C. Ho. 1979. Membrane-bound D-lactate dehydrogenase from *Escherichia coli*: purification and properties. Biochemistry. 18: 312–316.

Qian, Y., L.P. Yomano, J.F. Preston, H.C. Aldrich and L.O. Ingram. 2003. Cloning, characterization, and functional expression of the *Klebsiella oxytoca* xylodextrin utilization operon (*xynTB*) in *Escherichia coli*. Appl Environ Microbiol. 69(10): 5957–5967.

Qian, Z.-G., X.-X. Xia and S.Y. Lee. 2009. Metabolic engineering of *Escherichia coli* for the production of putrescine: A four carbon diamine. Biotechnol Bioeng. 104: 651–662.

Qian, Z.G., X.X. Xia and S.Y. Lee. 2011. Metabolic engineering of *Escherichia coli* for the production of cadaverine: a five carbon diamine. Biotechnol Bioeng. 108(1): 93–103.

Rahman, M. and K. Shimizu. 2008a. Altered acetate metabolism and biomass production in several *Escherichia coli* mutants lacking *rpoS*–dependent metabolic pathway genes. Mol Biosys. DOI:10.1039/b712023k. 1–17.

Rahman, M., M.M. Hasan and K. Shimizu. 2008b. Growth phase-dependent changes in the expression of global regulatory genes and associated metabolic pathways in *Esherichia coli*. Biotechnol Lett. 30: 853–60.

Ramos, J.L., E. Duque, M.-T. Gallegos, P. Godoy, M.I. Ramos-Gonzalez, A. Rojas, W. Teran and A. Segura. 2002. Godoy P, Ramos-Gonzalez MI, Rojas A, Teran W, Segura A, Mechanisms of solvent tolerance in Gram-negative bacteria. Annu Rev Microbiol. 56: 743–768.

Ramos, J., K. Szkutnicka and V.P. Cirillo. 1988. Relationship between low and high-affinity glucose transport systems of *Saccharomyces cerevisiae*. J Bacteriol. 170: 5375–5377.

Rathnasingh, C., S.M. Raj, Y. Lee, C. Catherine, S. Ashok and S. Park. 2012. Production of 3-hydroxypropionic acid via malonyl-CoA pathway using recombinant *Escherichia coli* strains. J Biotechnol. 157(4): 633–640.

Rathnasingh, C., S.M. Raj, J.E. Jo and S. Park. 2009. Development and evaluation of efficient recombinant *Escherichia coli* strains for the production of 3-hydroxypropionic acid from glycerol. Biotechnol Bioeng. 104(4): 729–739.

Rausch, K.D. and R.L. Belyea. 2006. The future of co-products from corn processing. Appl Biochem Biotechnol. 128: 47–86.

Razak, M.A. and B. Viswanath. 2015. Comparative studies for the biotechnological production of L-lysine by immobilized cells of wild type Corynebacterium glutamicum ATCC 13032 and mutant MH 20-22 B. 3 Biotech. 5: 765–774.

Reaney, S.K., S.J. Bungard and J.R. Guest. 1993. Molecular and enzymological evidence for two classes of fumarases in *Bacillus stearothermophillus* (var non-diastaticus). J Gen Microbiol. 139: 403–416.

Reyes, L.H., M.P. Almario and K.C. Kao. 2011. Genomic library screens for genes involved in n-butanol tolerance in *Escherichia coli*. PLoS One. 6: e17678.

Rhodes, R.A., A.J. Moyer and M.L. Smith. 1959. Production of fumaric acid by *Rhizopus arrhizus*. Appl Microbiol. 7: 74–80.

Richard, P., M.H. Toivari and M. Penttilä. 2000. The role of xylulokinase in *Saccharomyces cerevisiae* xylulose catabolism. FEMS Microbiol Lett. 190: 39–43.

Richard, P., M. Putkonen, R. Vaananen, J. Londesborough and M. Penttila. 2002. The missing link in the fungal L-arabinose catabolic pathway, identification of the Lxylulose reductase gene. Biochemistry. 41(20): 6432–6437.

Richard, P., M.H. Toivari and M. Penttila. 1999. Evidence that the gene *YLR070c* of *Saccharomyces cerevisiae* encodes a xylitol dehydrogenase. FEBS Lett. 457(1): 135–138.

Rizzi, M., P. Erlemann, N.A. Buithanh and H. Dellweg. 1988. Xylose fermentation by yeasts. 4. Purification and kinetic studies of xylose reductase from *Pichia stipitis*. Appl Microbiol Biotechnol. 29(2–3): 148–154.

Roa Engel, C.A., A.J.J. Straathof, T.W. Zilmans, W.M. van Gulik and L.A.M. van der Wielen. 2008. Fumaric acid production by fermentation. Appl Microbiol Biotechnol. 78(3): 379–389.

Roa Engel, C.A., W.M. van Gulik, L. Marang et al. 2011. Development of a low pH fermentation strategy for fumaric acid production by *Rhizopus oryzae*. Enzyme Microb Technol. 48: 39–47.

Rodriguez, A., J.A. Martínez, J.L. Báez-Viveros, N. Flores, G. Hernández-Chávez, O.T. Ramírez, G. Gosset, F. Bolivar et al. 2013. Constitutive expression of selected genes from the pentose phosphate and aromatic pathways increases the shikimic acid yield in high-glucose batch cultures of an *Escherichia coli* strain lacking PTS and *pykF*. Microb Cell Factories. 12: 86.

Rohr, M. and C.P. Kubicek. 1981. Regulatory aspects of citric acid fermentation by *Aspergillus niger*. Process Biochem. 16: 34–37.

Romano, A.H., M.M. Bright and W.E. Scott. 1967. Mechanism of fumaric acid accumulation in *Rhizopus nigricans*. J Biotechnol. 93(2): 600–604.

Rossmann, R., G. Sawers and A. Bock. 1991. Mechanism of regulation of the formate-hydrogenlyase pathway by oxygen, nitrate, and pH: definition of the formate regulon. Mol Microbiol. 5: 2807–2814.

Ruijter, G.J., H. Panneman, D.B. Xu and J. Visser. 2000. Properties of *Aspergillus niger* citrate synthase and effects of *citA* overexpression on citric acid production. FEMS Microbiol Lett. 184: 35–40.

Ruklisha, M., L. Paegle and I. Denina. 2007. L-Valine biosynthesis during batch and fed batch cultivations of Corynebacterium glutamicum: relationship between changes and bacterial growth rate and intracellular metabolism. Prog Biochem. 42: 634–640.

Rutherford, B.J., R.H. Dahl, R.E. Price, H.L. Szmidt, P.I. Benke, A. Mukhopadhyay and J.D. Keasling. 2010. Functional genomic study of exogenous n-butanol stress in *Escherichia coli*. Appl Environ Microbiol. 76: 1935–1945.

Sabra, W., C. Groeger and A.P. Zeng. 2016. Microbial cell factories for diol production. Adv Biochem Eng Biotechnol. 155: 165–197.

Sabra, W., J.Y. Dai, H. Quitmann, A.P. Zeng and Z.L. Xiu. 2011. Microbial production of 2,3-butanediol. Industrial biotechnology and commodity products. Compr Biotechnol. 3: 87–97.

Saito, K., A. Saito, M. Ohnishi and Y. Oda. 2004. Genetic diversity in *Rhizopus oryzae* strains as revealed by the sequence of lactate dehydrogenase genes. Arch Microbiol. 182(1): 30–6.

Saito, S., N. Ishida, T. Onishi, K. Tokuhiro, E. Nagamori, K. Kitano and H. Takahashi. 2005. Genetic engineered wine yeast produces a high concentration of l-lactic acid of extremely high optical purity. Appl Environ Microbiol. 71: 2789–2792.

Sanchez, A.M., G.N. Bennett and K.Y. San. 2005a. Efficient succinic acid production from glucose through overexpression of pyruvate carboxylase in an *Escherichia coli* alcohol dehydrogenase and lactate dehydrogenase mutant. Biotechnol Prog. 21: 358–365.

Sanchez, A.M., G.N. Bennett and K.Y. San. 2005b. Novel pathway engineering design of the anaerobic central metabolic pathway in *Escherichia coli* to increase succinate yield and productivity. Metab Eng. 7: 229–239.

Sanchez, S. and A.L. Demain. 2008. Metabolic regulation and overproduction of primary metabolites. Microb Biotechnol. 1(4): 283–319.

Sanchez-Riera, F., D.C. Cameron and C.L. Cooney. 1987. Influence of environmental factors in the production of R(-)-1,2-propanediol by *Clostridium thermosaccharolyticum*. Biotechnol Lett. 9(7): 449–454.

Sanchez-Torrez, V., T. Maeda and T.K. Wood. 2009. Protein engineering of the transcriptional activator FhlA to enhance hydrogen production in *Escherichia coli*. Appl Environ Microbiol. 75(17): 5639–5646.

Santos, E., H. Kung, I.G. Young and H.R. Kaback. 1982. *In vitro* synthesis of the membrane-bound D-lactate dehydrogenase of *Escherichia coli*. Biochemistry. 21: 2085–2091.

Sarkar, D., M. Yabusaki, Y. Hasebe, P.Y. Ho, S. Kohmoto, T. Kaga and K. Shimizu. 2010. Fermentation and metabolic characteristics of *Gluconoacetobacter oboediens* for different carbon sources. Appl Microbiol Biotechnol. 87: 127–136.

Sauer, M., D. Porro, D. Mattanovich and P. Branduardi. 2008. Microbial production of organic acids: expanding the markets. Trends Biotechnol. 26(2): 100–108.

Sauter, M., R. Bo¨hm and A. Bo¨ck. 1992. Mutational analysis of the operon (*hyc*) determining hydrogenase 3 formation in *Escherichia coli*. Mol Microbiol. 6: 1523–1532.

Sawer, G. 1994. The hydrogenases and formate dehydrogenases of *Escherichia coli*. Antonie Van Leeuwenhoek. 66: 57–88.

Schwartz, H. and F. Radler. 1988. Formation of L-malate by *Saccharomyces cerevisiae* during fermentation. Appl Microbiol Biotechnol. 27: 553–560.

Schweizer, H. and T.J. Larson. 1987. Cloning and characterization of the aerobic sn-glycerol-3-phosphate dehydrogenase structural gene *glpD* of *Escherichia coli* K-12. J Bacteriol. 169: 507–513.

Seddon, D. 2010. Petrochemical Economics: Technology Section in a Carbon Constrained World; Imperial College Press: London, UK. 8: 1–19.

Senac, T. and B. Hahn-Hagerdal. 1990. Intermediary metabolite concentrations in xylulose- and glucose-fermenting *Saccharomyces cerevisiae* cells. Appl Environ Microbiol. 56(1): 120–126.

Seol, E., A. Manimaran, Y. Jang, S. Kim, Y.K. Oh and S. Park. 2011. Sustained hydrogen production from formate using immobilized recombinant *Escherichia coli* SH5. Int. J. Hydrogen Energy. 36: 8681–8686.

Service, R.F. 2007. Cellulosic ethanol: biofuel researchers prepare to reap a new harvest. Science. 315: 1488–1491.

Setati, M.E., P. Ademark, W.H. van Zyl, B. Hahn-Hagerdal and H. Stalbrand. 2001. Expression of the *Aspergillus aculeatus* endo-beta-1, 4-mannanase encoding gene (*man1*) in *Saccharomyces cerevisiae* and characterization of the recombinant enzyme. Protein Expr Purif. 21(1): 105–114.

Shelley, S. 2007. A nenewable route to propylene glycerol. Chem Eng Prog. 103: 6–9.

Shen, C.R. and J.C. Liao. 2008. Metabolic engineering of *Escherichia coli* for 1-butanol and 1-propanol production via the keto-acid pathways. Metab Eng. 10(6): 312–320.

Shen, C.R., E.I. Lan, Y. Dekishima, A. Baez, K.M. Cho and J.C. Liao. 2011. Driving forces enable high-titer anaerobic 1-butanol synthesis in *Escherichia coli*. Appl Environ Microbiol. 77(9): 2905–2915.

Shen, Y., X. Chen, B. Peng, L. Chen, J. Hou and X. Bao. 2012. An efficient xylose-fermenting recombinant *Saccharomyces cerevisiae* strain obtained through adaptive evolution and its global transcription profile. Appl Microbiol Biotechnol. 96(4): 1079–1091.

Shi, N.Q. and T.W. Jeffries. 1998. Anaerobic growth and improved fermentation of *Pichia stipitis* bearing a *URA1* gene from *Saccharomyces cerevisiae*. Appl Microbiol Biotechnol. 50(3): 339–345.

Shi, F. and Y. Li. 2011. Synthesis of γ-aminobutyric acid by expressing *Lactobacillus brevis*-derived glutamate decarboxylase in the *Corynebacterium glutamicum* strain ATCC 13032. Biotechnol Lett. 33: 2469

Shibata, H., W.E. Gardiner and S.D. Schwartzbach. 1985. Purification, characterization, and immunological properties of fumarase from *Euglena gracilis* var *bacillus*. J Bacteriol. 164: 762–768.

Shiio, I. 1990. Threonine production by dihydrodipicolinate synthase-defective mutants of Brevibacterium flavum. Biotechnol Adv. 8: 97–103.

Shiio, I., Y. Toride, A. Yokota, S. Sugimoto and K. Kawamura. 1991. Process for the production of l-threonine by fermentation, USA Patent (5077207).

Shimizu, K. 2013. Metabolic regulation of a bacterial cell system with emphasis on *Escherichia coli* metabolism. ISRN Biochemistry. Article ID 645983: doi: 10.1155/2013/645983.

Shimizu, K. 2014. Metabolic regulation of *Escherichia coli* in response to nutrient limitation and environmental stress conditions. Metabolites. 4: 1–35.

Shimizu, K. 2016. Metabolic regulation and coordination of the metabolism in bacteria in response to a variety of growth conditions. Adv Biochem Eng/Biotechnol. 155: 1–54.

Shiue, E. and K.L.J. Prather. 2014. Improving D-glucaric acid production fom *myo*-inositol in *E. coli* by increasing MIOX stability and myo-inositol transport. Metab Eng. 22: 22–31.

Shukla, V.B., S. Zhou, L.P. Yomano, K.T. Shanmugam, J.F. Preston and L.O. Ingram. 2004. Production of d-(−)-lactate from sucrose and molasses. Biotechnol Lett. 26: 689–693.

Siddiquee, K.A.Z., M. Arauzo-Bravo and K. Shimizu. 2004. Effect of pyruvate kinase (*pykF* gene) knockout mutation on the control of gene expression and metabolic fluxes in *Escherichia coli*. FEMS Microbiol Lett. 235: 25–33.

Siddiquee, K.A.Z., M. Arauzo-Bravo and K. Shimizu. 2004. Metabolic flux analysis of *pykF* gene knockout *Escherichia coli* based on [13]C-labeled experiment together with measurements of enzyme activities and intracellular metabolite concentrations. Appl Microbiol Biotechnol. 63: 407–417.

Sikkema, J., J.A.M. de Bont and B. Poolman. 1995. Mechanisms of membrane toxicity of hydrocarbons. Microbiol Rev. 59: 201–222.

Silva, S.S., M. Vitolo, A. Pessoa and M.G.A. Felipe. 1996. Xylose reductase and xylitol dehydrogenase activities of D-xylose-xylitol-fermenting *Candida guilliermondii*. J Basic Microbiol. 36(3): 187–191.

Singh, R.S. and G.K. Saini. 2008. Pullulan-hyperproducing color variant strain of *Aureobacidium pullulans* FB-1 newly isolated from phylloplane of *Ficus* sp. Bioresour Technol. 99: 3896–3899.

Singh, A., P.K.R. Kumar and K. Schugerl. 1992. Bioconversion of cellulosic materials to ethanol by filamentous fungi. Adv Biochem Eng Biotechnol. 45: 29–55.

Skoog, K. and B. Hahn-Hagerdal. 1990. Effect of oxygenation on xylose fermentation by *Pichia stipitis*. Appl Environ Microbiol. 56(11): 3389–3394.

Skotnicki, M.L., K.J. Lee, D.E. Tribe and P.L. Rogers. 1981. Comparison of ethanol production by different *Zymomonas strains*. Appl Environ Microbiol. 41(4): 889–893.

Song, C.W., D.I. Kim, S. Choi et al. 2013. Metabolic engineering of *Escherichia coli* for the production of fumaric acid. Biotechnol Bioeng. 110: 2025–2034.

Song, P., S. Li, Y.Y. Ding, Q. Xu and H. Huang. 2011. Expression and characterization of fumarase (FUMR) from *Rhizopus oryzae*. Fungal Biol. 115: 49–53.

Soucaille, P., I. Meynial-Salles, F. Voelker and R. Figge. 2008. Microorganisms and methods for production of 1, 2-propanediol and acetol. WO. 2008/116853.

Stalbrand, H., A. Saloheimo, J. Vehmaanpera, B. Henrissat and M. Penttila. 1995. Cloning and expression in *Saccharomyces cerevisiae* of a Trichoderma reesei betamannanase gene containing a cellulose binding domain. Appl Environ Microbiol. 61(3): 1090–1097.

Stephanopoulos, G. 2008. Metabolic engineering: enabling technology for biofuels production. Metab Eng. 10: 293–294.

Stephen, W.I. 1965. Solubility and pH calculations. Ana Chim Acta. 33: 227.

Straathof, A.J. and W.M. van Gulik. 2012. Production of fumaric acid by fermentation. Subcell Biochem. 64: 225–240.

Subedi, K.P., I. Kim, J. Kim, B. Min and C. Park. 2008. Role of GldA in dihydroxyacetone and methylglyoxal metabolism of *Escherichia coli* K12. FEMS Microbiol Lett. 279(2): 180–187.

Subramanian, M.R., S. Talluri and L.P. Christopher. 2015. Production of lactic acid using a new homofermentative *Enterococcus faecalis* isolate. Microb Biotechnol. 8(2): 221–229.

Sumiya, M., E.O. Davis, L.C. Packman, T.P. McDonald and P.J. Henderson. 1995. Molecular genetics of a receptor protein for D-xylose encoded by the gene *xylF* in *Escherichia coli*. Recept Channel. 3(2): 117–128.

Suzuki, T., M. Sato, T. Yoshida and S. Tuboi. 1989. Rat river mitochondrial and cytosolic fumarases with identical amino acid sequences are encoded from a single gene. J Biol Chem. 264: 2581–2586.

Taing, O. and K. Taing. 2007. Production of malic and succinic acids by sugar-tolerant yeast *Zygosaccharomyces rouxii*. Eur Food Res Technol. 224: 343–347.

Takata, I., K. Yamamoto, T. Tosa and I. Chibata. 1980. Immobilization of *Brevibacterium flavum* with carrageenan and its application for continuous production of L-malic acid. Enzyme Microbial Technol. 2: 30–36.

Takata, I. and T. Tosa. 1993. Production of L-malic acid. pp. 53–65. *In*: A. Tanaka, T. Tosa and T. Kobayashi (eds.). Industrial Application of Immobilized Biocatalysts. Marcel Dekker Inc., New York.

Takatsuka, Y., C. Chen and H. Nikaido. 2010. Mechanism of recognition of compounds of diverse structures by the multidrug efflux pump AcrB of *Escherichia coli*. PNAS USA. 107: 6559–6565.

Tantirungkij, M., N. Nakashima, T. Seki and T. Yoshida. 1993. Construction of xylose assimilating *Saccharomyces cerevisiae*. J Ferment Bioeng. 75(2): 83–88.

Tatarko, M. and T. Romeo. 2001. Disruption of a global regulatory gene to enhance central carbon flux into phenylalanine synthesis in *Escherichia coli*. Curr Microbiol. 43: 26–32.

Tevz, G., M. Bencina and M. Legisa. 2010. Enhancing itaconic acid production by *Aspergillus terreus*. Appl Microbiol Biotechnol. 87: 1657–1664.

Thakker, C., I. Martínez, K.Y. San and G.N. Bennett. 2012. Succinate production in *Escherichia coli*. Biotechnol J. 7(2): 213–224.

Toivari, M.H., A. Aristidou, L. Ruohonen and M. Penttila. 2001. Conversion of xylose to ethanol by recombinant *Saccharomyces cerevisiae*: importance of xylulokinase (XKS1) and oxygen availability. Metab Eng. 3(3): 236–249.

Toivari, M.H., L. Salusjarvi, L. Ruohonen and M. Penttila. 2004. Endogenous xylose pathway in *Saccharomyces cervisiae*. Appl Environ Microbiol. 70: 3681–3686.

Tomar, A., M.A. Eiteman and E. Altman. 2003. The effect of acetate pathway mutations on the production of pyruvate in *Escherichia coli*. Appl Microbiol Biotechnol. 62: 76–82.

Tomas, C., J. Beamish and E. Papoutsakis. 2004. Transcriptional analysis of butanol stress and tolerance in *Clostridium acetobutylicum*. J Bacteriol. 186: 2006–2018.

Tomas, C., N. Welker and E. Papoutsakis. 2003. Overexpression of groESL in *Clostridium acetobutylicum* results in increased solvent production and tolerance, prolonged metabolism, and changes in the cell's transcriptional program. Appl Environ Microbiol. 69: 4951–4965.

Toya, Y., K. Nakahigashi, M. Tomita and K. Shimizu. 2012. Metabolic regulation analysis of wild-type and *arcA* mutant *Escherichia coli* under nitrate conditions using different levels of omics data. Affiliation Information Mol BioSyst. 8: 2593–2604.

Tracy, B.P., S.W. Jones, A.G. Fast, D.C. Indurthi and E.T. Papoutsakis. 2012. Clostridia: the importance of their exceptional substrate and metabolite diversity for biofuel and biorefinery applications. Curr Opin Biotechnol. 23(3): 364–381.

Traff, K.L., L.J. Jonsson and B. Harn-Hagerdal. 2002. Putative xylose and arabinose reductases in *Saccharomyces cerevisiae*. Yeast. 19(14): 1233–1241.

Tran-Din, K. and G. Gottschalk. 1985. Formation of D (–)-1, 2-propanediol and D (–)-lactate from glucose by *Clostridium sphenoides* under phosphate limitation. Arch Microbiol. 142: 87–92.

Tripathi, P., G. Rawat, S. Yadav and R.K. Saxena. 2013. Fermentative production of shikimic acid: a paradigm shift of production concept from plant route to microbial route. Bioprocess Biosyst Eng. 11: 1665–1673.

Tsao, G.T., N.J. Cao, J. Du and C.S. Gong. 1999. Production of multifunctional organic acids from renewable resources. Adv Biochem Eng Biotechnol. 65: 243–280.

Tseng, H.-C., C.L. Harwell, C.H. Martin and K.L.J. Prather. 2010. Biosynthesis of chiral 3-hydroxyvalerate from single propionate-unrelated carbon sources in metabolically engineered *E. coli*. Microb Cell Fact. 9: 96.

Tsuge, Y., H. Kawaguchi, K. Sasaki and A. Kondo. 2016. Engineering cell factories for producing building block chemicals for bio-polymer synthesis. Microb Cell Fact. 15: 19.

Tsuge, Y., S. Yamamoto, N. Kato, M. Suda, A.A. Vertes and H. Yukawa. 2015. Overexpression of the phosphofructokinase encoding gene is crucial for achieving high production of

D-lactate in *Corynebacterium glutamicum* under oxygen deprivation. Appl Microbiol Biotechnol. 99: 79–89.

Turner, K.W. and A.M. Roberton. 1979. Xylose, arabinose, and rhamnose fermentation by *Bacteroides ruminicola*. Appl Environ Microbiol. 38: 7–12.

Udaka, S. 1960. Screening method for microorganisms accumulating metabolites and its use in the isolation of Micrococcus glutamicus. J Bacteriol. 79: 754–755.

Underwood, S.A., M.L. Buszko, K.T. Shanmugam and L.O. Ingram. 2002a. Flux through citrate synthase limits the growth of ethanogenic *Escherichia coli* KO11 during xylose fermentation. Appl Environ Microbiol. 68: 1071–1081.

Underwood, S.A., S. Zhou, T.B. Causey, L.P. Yomano, K.T. Shanmugam and L.O. Ingram. 2002b. Genetic changes to optimize carbon partitioning between ethanol and biosynthesis in ethanogenic *Escherichia coli*. Appl Environ Microbiol. 68: 6263–6272.

Upadyaya, B.P., L.C. DeVeaux and L.P. Christopher. 2014. Metabolic engineering as a tool for enhanced lactic acid production. Trends Biotechnol. 32(12): 637–644.

Vaillancourt, F.H., J.T. Bolin and L.D. Eltis. 2006. The ins and outs of ring-cleaving dioxigenases. Crit Rev Biochem Mol Biol. 41: 241–267.

van der Staat, L., M. Vernooij, M. Lammers, W. van der Berg, T. Schonewille, J. Cordewener, I. van der Meer, A. Koops and L.H. Graaff. 2013. Expression of the *Aspergillus terreus* itaconic acid biosynthesis cluster in *Aspergillus niger*. Microb Cell Fact. 13: 11.

van Duuren, J.B.J.H., D. Wijte, B. Karge, V.A.P. Martins dos Santos, Y. Yang, A.E. Mars and G. Eggink. 2012. pH-stat fed-batch process to enhance the production of *cis*, *cis*-muconate from benzoate by *Pseudomonas putida* KT2440-JD1. Biotechnol Prog. 28: 85–92.

van Maris, A.J.A., J.M.A. Geertman, A. Vermeulen, M.K. Groothuizen, A.A. Winkler, M.D.W. Piper, J.P. van Dijken and J.T. Pronk. 2004. Directed evolution of pyruvate decarboxylase-negative *Saccharomyces cerevisiae*, yielding a $C_2$-independent, glucose-tolerant, and pyruvate-hyperproducing yeast. Appl Environ Microbiol. 70: 159–166.

van Zyl, C., B. Prior and J. Du Preez. 1988. Production of ethanol from sugar cane bagasse hemicellulose hydrolyzate by *Pichia stipitis*. Appl Biochem Biotechnol. 17(1): 357–369.

Vemuri, G.M. Eiteman and E. Altman. 2002. Succinate production in dual-phase *Escherichia coli* fermentations depends on the time of transition from aerobic to anaerobic conditions. J Ind Microbiol Biotechnol. 28: 325–332.

Verduyn, C., R. Van Kleef, J. Frank, H. Schreuder, J.P. van Dijken and W.A. Scheffers. 1985. Properties of the NAD(P)H-dependent xylose reductase from the xylose fermenting yeast *Pichia stipitis*. Biochem J. 226(3): 669–677.

Verho, R., M. Putkonen, J. Londesborough, M. Penttila and P. Richard. 2004. A novel NADH-linked L-xylulose reductase in the L-arabinose catabolic pathway of yeast. J Biol Chem. 279(15): 14746–14751.

Vickers, C.E., D. Klein-Marcuschamer and J.O. Krömer. 2012. Examining the feasibility of bulk commodity production in *Escherichia coli*. Biotechnol Lett. 34(4): 585–596.

Vrsalović Presečki, A., Z. Findrik and D. Vasić-Rački. 2009. Starch hydrolysis by the synergistic action of amylase and glucoamylase. New Biotechnol. 25(1): 170.

Vuoristo, K.S., A.E. Mars, J.V. Sangra, J. Springer, G. Eggink, J.P. Sanders and R.A. Weusthuis. 2015a. Metabolic engineering of itaconate production in *Escherichia coli*. Appl Microbiol Biotechnol. 99: 221–228.

Vuoristo, K.S., A.E. Mars, J.V. Sangra, J. Springer, G. Eggink, J.P. Sanders and R.A. Weusthuis. 2015b. Metabolic engineering of the mixed-acid fermentation pathway of *Escherichia coli* for anaerobic production of glutamate and itaconate. AMB Express. 5: 61.

Walaszek, Z., J. Szemraj, M. Hanausek, A.K. Adams and U. Sherman. 1996. Glucaric acid content of various fruits and vegetables and choresterol-lowering effects of dietary glucarate in the rat. Nutr Res. 16: 673–681.

Walfridsson, M., M. Anderlund, X. Bao and B. Hahn-Hagerdal. 1997. Expression of different levels of enzymes from the *Pichia stipitis* XYL1 and XYL2 genes in *Saccharomyces cerevisiae* and its effects on product formation during xylose utilisation. Appl Environ Microbiol. 48(2): 218–224.

Walz, A.C., R.A. Demel, B. de Kruijff and R. Mutzel. 2002. Aerobic sn-glycerol-3-phosphate dehydrogenase from *Escherichia coli* binds to the cytoplasmic membrane through an amphipathic alpha-helix. Biochem J. 365: 471–479.

Wang, C.L., S. Takenaka, S. Murakami and K. Aoki. 2001. Isolation of a benzoate-utilizing Pseudomonas strain from soil and production of catechol from benzoate by transpositional mutants. Microbiol Res 156: 151–158.

Wang L., B. Zhao, F. Li, K. Xu, C. Ma, F. Tao, Q. Li and P. Xu. 2011. Highly efficient production of D-lactate by *Sporolactobacillus* sp. CASD with simultaneous enzymatic hydrolysis of peanut meal. Appl Microbiol Biotechnol. 89(4): 1009–1017.

Wang, X., E.N. Miller, L.P. Yomano, X. Zhang, K.T. Shanmugam and L.O. Ingram. 2011. Increased furfural tolerance due to overexpression of NADH-dependent oxidoreductase FucO in *Escherichia coli* strains engineered for the production of ethanol and lactate. Appl Environ Microbiol. 77(15): 5132–5140.

Wang, G., D. Huang, H. Qi et al. 2013. Rational medium optimization based on comparative metabolic profiling analysis to improve fumaric acid production. Bioresour Technol. 137: 1–8.

Wang, L., J. Zhang, Z. Cao, Y. Wang, Q. Gao, J. Zhang and D. Wang. 2015. Inhibition of oxidative phosphorylation for enhancing citric acid production by *Aspergillus niger*. Microb Cell Fact. 14: 7.

Wang, Y., L. Li, C. Ma, C. Gao, F. Tao and P. Xu. 2013. Engineering of cofactor regeneration enhances (2S, 3S)-2,3-butanediol production from diacetyl. Sci Rep. 2013(3): 2643.

Wang, Y., T. Tian, J. Zhao, J. Wang, T. Yan, L. Xu, Z. Liu, E. Garza, A. Iverson, R. Manow, C. Finan and S. Zhou. 2012. Homofermentative producton of D-lactic acid from sucrose by a metabolically engineered *Escherichia coli*. Biotechnol Lett. 34: 2069–2075.

Wang, Y., Y. Tashiro and K. Sonomoto. 2015. Fermentative production of lactic acid from renewable materials: Recent achievements, prospects, and limits. J Biosci Bioeng. 119(1): 10–18.

Wee, Y., J. Kim and H. Ryu. 2006. Biotechnological production of lactic acid and its recent applications. Food Technol Biotechnol. 44: 163–172.

Weisser, P., R. Kramer, H. Sahm and G.A. Sprenger. 1995. Functional expression of the glucose transporter of *Zymomonas mobilis* leads to restoration of glucose and fructose uptake in *Escherichia coli* mutants and provides evidence for its facilitator action. J Bacteriol. 177(11): 3351–3354.

Wells, Jr, T. and A.J. Ragauskas. 2012. Biotechnological opportunities with the β-ketoadipate pathway. Trends Biotechnol. 30: 627–637.

Wendisch, V.F. and M. Bott. 2005. In Handbook of Corynebacterium glutamicum, CRC Press. Taylor & Francis, Boca Raton, pp. 377–396.

Wendisch, V.F. 2014. Microbial production of amino acids and derived chemicals: synthetic biology approaches to strain development. Curr Opin Biotechnol. 30: 51–58.

Wendisch, V.F., M. Bott and B.J. Eikmanns. 2006. Metabolic engineering of *Escherichia coli* and Corynebacterium for biotechnological production of organic acids and amino acids. Curr Opin Microbiol. 9: 268–274.

Werpy, T. and G. Petersen. 2004. Top value added chemicals from biomass: I. Results of screening for potential candidates from sugars and synthesis gas. U.S. Department of Energy, Washington, DC.

West, T.P. 2011. Malic acid production from thin stillage by *Aspergillus* species. Biotechnol Lett. 33(12): 2463–7.

Wierckx, N.J.P., H. Ballerstedt, J.A.M. de Bont and J. Wery. 2005. Engineering of Solvent-Tolerant *Pseudomonas putida* S12 for Bioproduction of Phenol from Glucose. Appl Environ Microbiol. 71(12): 8221–8227.

Wilke, T. and K.D. Vorlop. 2001. Biotechnological production of itaconic acid. Appl Microbiol Biotechnol. 56: 289–295.

Wisselink, H.W., M.J. Toirkens, Q. Wu, J.T. Pronk and A.J.A. van Maris. 2009. Novel evolutionary engineering approach for accelerated utilization of glucose, xylose, and arabinose mixtures by engineered *Saccharomyces cerevisiae* strains. Appl Environ Microbiol. 75(4): 907–914.

Witteveen, C.F.B., R. Busink, P. van de Vondervoort, C. Dijkema, K. Swart and J. Visser. 1989. L-arabinose and D-xylose catabolism in *Aspergillus niger*. J Gen Microbiol. 135: 2163–2171.

Woods, D.D. 1936. Hydrogenlyases. The synthesis of formic acid by bacteria. Biochem J. 30: 515–527.

Woods, S.A., S.D. Schwartzbach and J.R. Guest. 1988. Two biochemically different distinct classes of fumarase in *Escherichia coli*. Biochim Biophys Acta. 954: 14–26.

Wu, M. and A. Tzagoloff. 1987. Mitochondrial and cytoplasmic fumarases in *Saccharomyces cerevisiae* are encoded by a single nuclear gene FUM1. J Biol Chem. 262: 12275–12282.

Wu, S., J. Chen and S. Pan. 2012. Optimization of fermentation conditions for the production of pullulan by a new strain of *Aureobacidium pullulans* isolated from sea mud and its characterization. Carbohyd Polym. 87: 1696–1700.

Xie, N.-Z., H. Liang, R.-B. Huang and P. Xu. 2014a. Biotechnological production of muconic acid: current status and future prospects. Biotechnol Adv. 32: 615–622.

Xie, N.-Z., Q.Y. Wang, Q.X. Zhu, Y. Qin, F. Tao, R.B. Huang et al. 2014b. Optimization of medium composition for cis, cis-muconic acid production by a *Pseudomonas* sp. mutant using statistical methods. Prep Biochem Biotechnol. 44: 342–354.

Xu, Q., S. Li, H. Huang and J. Wen. 2012. Key technologies for the industrial production of fumaric acid by fermentation. Biotechnol Adv. 30: 1685–1696.

Xu, Q., S. Li, Y. Fu et al. 2010. Two-stage utilization of corn straw by *Rhizopus oryzae* for fumaric acid production. Bioresour Technol. 101: 6262–6264.

Xu, Y., H. Chu, C. Gao, F. Tao, Z. Zhou, K. Li, L. Li, C. Ma and P. Xu. 2014. Systematic metabolic engineering of *Escherichia coli* for high-yield production of fuel biochemical 2,3-butanediol. Metab Eng. 23: 22–33.

Xu, G., W. Zou, X. Chen, N. Xu, L. Liu and J. Chen. 2012. Fumaric acid production in *Saccharomyces cerevisiae* by *in silico* aided metabolic engineering. PLoS One. 7(12): e52086.

Xu, G., X. Chen, L. Liu and L. Jiang. 2013. Fumaric acid production in *Saccharomyces cerevisiae* by simultaneous use of oxidative and reductive routes. Bioresour Technol. 148: 91–96.

Xu, Q., L. Liu and J. Chen. 2012. Reconstruction of cytosolic fumaric acid biosynthetic pathways in *Saccharomyces cerevisiae*. Microb Cell Fact. 11: 24.

Xu, Q., S. Li, Y.Q. Fu, C. Tai and H. Huang. 2010. Two-stage utilization of corn straw by *Rhizopus oryzae* for fumaric acid production. Bioresour Technol. 101(15): 6262–6264.

Yadav, V.G., M.D. Mey, C.G. Lim, P.K. Ajikumar and G. Stephanopoulos. 2012. The future of metabolic engineering and synthetic biology: Towards a systematic practice. Metab Eng. 14: 233–241.

Yagi, T. and A. Matsuno-Yagi. 2003. The proton-translocating NADH-quinone oxide reductase in the respiratory chain: the secret unlocked. Biochemistry. 42: 2266–2274.

Yahiro, K., T. Takahama, Y.S. Park and M. Okabe. 1995. Breeding of *Aspergillus terreus* mutant TN-484 for itaconic acid production with high yield. J Biosci Bioeng. 79(5): 506–508.

Yakandawara, N., T. Romeo, A.D. Friesen and S. Madhyastha. 2007. Metabolic engineering of *Escherichia coli* to enhance phenylalanine production. Appl Microbiol Biotechnol. 78: 283–291.

Yalcin, S.K., M.T. Bozdemir and Y. Ozbas. 2010. Citric acid production by yeasts: Fermentation conditions, process optimization and strain improvement in current research. *In*: A. Mendez-Vilas (ed.). Technology and Education Topics in Applied Microbiology and Microbial Biotechnology. Formatex.

Yamada, K. 1977. Recent advances in industrial fermentation in Japan. Biotechol Bioeng. 19: 1563–1622.

Yan, Y., C.C. Lee and J.C. Liao. 2009. Enantioselective synthesis of pure (R, R)-2,3-butanediol in *Escherichia coli* with stereospecific secondary alcohol dehydrogenases. Org Biomol Chem. 7: 3914–3917.

Yang, S.T., K. Zhang, B. Zhang and H. Huang. 2011. Fumaric acid. pp. 163–177. *In*: Moo-Young and Bulter Michel (eds.). Comprehensive Biotechnology. 2nd ed. 3.

Yang, V.W. and T.W. Jeffries. 1997. Regulation of phosphotransferases in glucose- and xylose-fermenting yeasts. Appl Biochem Biotechnol. 63-65: 97–108.

Yao, R. and K. Shimizu. 2013. Recent progress in metabolic engineering for the production of biofuels and biochemicals from renewable sources with particular emphasis on catabolite regulation and its modulation. Process Biochem. 48: 1409–1417.

Yoshida, A., T. Nishimura, H. Kawaguchi, M. Inui and H. Yukawa. 2005. Enhanced hydrogen production from formic acid by formate hydrogen lyase-overexpressing *Escherichia coli* strains. Appl Environ Microbiol. 71: 6762–6768.

Yazdani, S.S. and R. Gonzalez. 2007. Anaerobic fermentation of glycerol: a path to economic viability for the biofuel industry. Curr Opin Biotechnol. 18: 213–219.

Yim, H., R. Haselbeck, W. Niu, C. Pujol-Baxley, A. Burgard, J. Boldt, J. Khandurina, J.D. Trawick, R.E. Osterhout, R. Stephen, J. Estadilla, S. Teisan, H.B. Schreyer, S. Andrae, T.H. Yang, S.Y. Lee, M.J. Burk and D.S. Van. 2011. Metabolic engineering of *Escherichia coli* for direct production of 1, 4-butanediol. Nat Chem Biol. 7: 445–452.

Yokota, A., Y. Shimizu, Y. Terasawa, N. Takaoka and F. Tomita. 1994. Pyruvic acid production by a lipoic acid auxotroph of *Escherichia coli* W1485. Appl Microbiol Biotechnol. 41: 638–643.

Yomano, L.P., S.W. York and L.O. Ingram. 1998. Isolation and characterization of ethanol-tolerant mutants of *Escherichia coli* KO11 for fuel ethanol production. J Ind Microbiol Biotechnol. 20(2): 132–138.

Yomano, L.P., S.W. York, S. Zhou, K.T. Shanmugam and L.O. Ingram. 2008. Reengineering *Escherichia coli* for ethanol production. Biotechnol Lett. 30(12): 2097–2103.

Yonehara, T. and R. Miyata. 1995. Fermentative production of pyruvate from glucose by *Torulopsis glabrata*. J Ferment Bioeng. 78: 155–159.

Yoshida, A., T. Nishimura, H. Kawaguchi, M. Inui and H. Yukawa. 2005. Enhanced hydrogen production from formic acid by formate hydrogen lyase-overexpressing *Escherichia coli* strains. Appl Environ Microbiol. 71: 6762–6768.

Yu, B., J.-B. Sun, R. Bommareddy, L.F. Song and A.P. Zeng. 2011. A novel (2R, 3R)-2,3-butanediol dehydrogenase from an industrially potential strain *Paenibacillus polymyxa* ATCC12321. Appl Environ Microbiol. 77: 4230.

Yu, C., Y. Cao, H. Zou and M. Xian. 2011. Metabolic engineering of *Escherichia coli* for biotechnological production of high-value organic acids and alcohols. Appl Microbiol Biotechnol. 89: 573–583.

Yu, J.-L., X.-X. Xia, J.-J. Zhong and Z.-G. Qian. 2014. Direct biosynthesis of adipic acid from a synthetic pathway in recombinant *Escherichia coli*. Biotechnol Bioeng. 111(12): 2580–2586.

Yu, S., H. Jeppsson and B. Hahn-Hagerdal. 1995. Xylose fermentation by *Saccharomyces cerevisiae* and xylose-fermenting yeast strains. Appl Microbiol Biotechnol. 44: 314–320.

Yun, N.R., K.Y. San and G.N. Bennett. 2005. Enhancement of lactate and succinate formation in *adhE* or *pta–ackA* mutants of NADH dehydrogenase-deficient *Escherichia coli*. J Appl Microbiol. 99: 1404–1412.

Yuzawa, S., W. Kim, L. Katz and J.D. Keasling. 2011. Heterologous production of polyketides by modular type I polyketide synthases in *Escherichia coli*. Curr Opin Biotechnol. 23: 1–9.

Zarowska, B., M. Wojtatowicz, W. Rymowicz and M. Robak. 2001. Production of citric acid on sugar beet molasses by single and mixed cultures of *Yarrowia lipolytica*. Biotechnol. 4: 1–8.

Zeikus, J., M.K. Jain and P. Elankovan. 1999. Biotechnology of succinic acid production and markets for derived industrial products. Appl Microbiol Biotechnol. 51: 545–552.

Zelic, B., S. Gostovic, K. Vuorilehto, D. Vasic-Racki and R. Takours. 2004. Process strategies to enhance pyruvate production with recombinant *Escherichia coli*: from repetitive fed-batch to *in situ* product recovery with fully integrated electrodyalysis. Biotechnol Bioeng. 85: 638–646.

Zelle, R.M., E. de Hulster, W.A. van Winden, P. de Waard, C. Dijkema, A.A. Winkler, J.A. Geertman, J.P. van Dijken, J.T. Pronk and A.J.A. van Maris. 2008. Malic acid production by *Saccharomyces cerevisiae*: engineering of pyruvate carboxylation, oxaloacetate reduction, and malate export. Appl Environ Microbiol. 74: 2766–2777.

Zelle, R.M., E. de Hulster, W. Kloezen, J.T. Pronk and A.J.A. van Maris. 2010. Key process conditions for production of C$_4$ dicarboxylic acids in bioreactor batch cultures of an engineered *Saccharomyces cerevisiae* strain. Appl Environ Microbiol. 76(3): 744–750.

Zeng, A.P., H. Biebl and W.D. Deckwer. 1991. Production of 2,3-butanediol in a membrane bioreactor with cell recycle. Appl Microbiol Biotechnol. 34: 463–468.

Zeng, A.P. and W. Sabra. 2011. Microbial production of diols as platform chemicals: recent progress. Curr Opin Biotechnol. 22: 749–757.

Zhang, M., C. Eddy, K. Deanda, M. Finkelstein and S. Picataggio. 1995. Metabolic engineering of a pentose metabolism pathway in Ethanogenic *Zymomonas mobilis*. Science. 267: 240–243.

Zhang, B., C.D. Skory and S.T. Yang. 2012. Metabolic engineering of *Rhizopus oryzae*: effects of overexpressing *pyc* and *pepc* genes on fumaric acid biosynthesis from glucose. Metab Eng. 14: 512–520.

Zhang, M., C. Eddy, K. Deanda, M. Finkelstein and S. Picataggio. 1995. Metabolic engineering of a pentose metabolism pathway in ethanologenic *Zymomonas mobilis*. Science. 267(5195): 240–243.

Zhang, K., M.R. Sawaya, D.S. Eisenberg and J.C. Liao. 2008. Expanding metabolism for biosynthesis of nonnatural alcohols. PNAS USA. 105(52): 20653–8.

Zhang, M., R. Su, W. Qi and Z. He. 2009. Enzymatic conversion of lignocellulose into sugars. Prog Chem. 21: 1070–1074.

Zhang A. and S.T. Yang. 2009. Propionic acid production from glycerol by metabolically engineered *Propionibacterium acidipropionici*. Process Biochem. 44: 1346–1351.

Zhang, L., J.A. Sun, Y. Hao, J. Zhu, J. Chu, D. Wei and Y. Shen. 2010. Microbial production of 2,3-butanediol by a surfactant (serrawettin)-deficient mutant of *Serratia marcescens* H30. J Ind Microbiol Biotechnol. 37: 857–862.

Zhang, X., X. Wang, K.T. Shanmugam et al. 2011. L-malate production by metabolically engineered *Escherichia coli*. Appl Environ Microbiol. 77: 427–434.

Zhang, X., X. Wang, K.T. Shanmugam and L.O. Ingram. 2011a. L-Malate production by metabolically engineered *Escherichia coli*. Appl Envirn Microbiol. 77(2): 427–434.

Zhang, Z.H., B. Jin and J.M. Kelly. 2007. Production of lactic acid from renewable materials by *Rhizopus* fungi. Biochem Eng J. 35(3): 251–263.

Zhang, Y., P.V. Vadlani, A. Kumar, P.R. Hardwidge, R. Covind, T. Tanaka and A. Kondo. 2016. Enhanced D-lactic acid production from renewable resources using engineered *Lactobacillus plantarum*. Appl Microbiol Biotechnol. 100: 279–288.

Zhou, L., D.-D. Niu, K.-M. Tian, X.-Z. Chen, B.A. Prior, W. Shen, G.-Y. Shi, S. Singh and Z.-X. Wang. 2012a. Genetically switched D-lactate production in *Escherichia coli*. Metab Eng. 14: 560–568.

Zhou, L., K.-M. Tian, D.-D. Niu, W. Shen, G.-Y. Shi, S. Singh and Z.-X. Wang. 2012b. Improvement of D-lactate productivity in recombinant *Escherichia coli* by coupling production with growth. Biotechnol Lett. 34: 1123–1130.

Zhou, L., Z.-R. Zuo, X.-Z. Chen, D.-D. Niu, K.-M. Tian, B.A. Prior, W. Shen, G.-Y. Shi, S. Singh and Z.-X. Wang. 2011. Evaluation of genetic manipulation strategies on D-lactate production by *Escherichia coli*. Curr Microbiol. 62: 981–989.

Zhou, S., A.G. Iverson and W.S. Grayburn. 2008. Engineering a native homo ethanol pathway in *Escherichia coli* B for ethanol production. Biotechnol Lett. 30: 335–342.

Zhou, S., T.B. Causey, A. Hasona, K.T. Shanmugam and L.O. Ingram. 2003. Production of optically pure d-lactic acid in mineral salts medium by metabolically engineered *Escherichia coli* W3110. Appl Environ Microbiol. 69: 399–407.

Zhou, S., K.T. Shanmugam and L.O. Ingram. 2003. Functional replacement of the *Escherichia coli* d-(−)-lactate dehydrogenase gene (*ldhA*). with the l-(+)-lactate dehydrogenase gene (*ldhL*). From *Pediococcus acidilactici*. Appl Environ Microbiol. 69: 2237–2244.

Zhou, S., L.P. Yomano, K.T. Shanmugam and L.O. Ingram. 2005. Fermentation of 10% (w/v). sugar to d-(−)-lactate by engineered *Escherichia coli* B. Biotechnol Lett. 27: 1891–1896.

Zhou, Y., J. Du and G.T. Tsao. 2002. Comparison of fumaric acid production by *Rhizopus oryzae* using different neutralizing agents. Bioprocess Biosyst Eng. 25: 179–181.

Zhou, Y., K. Nie, X. Zhang, S. Liu, M. Wang, L. Deng, F. Wang and T. Tan. 2014. Production of fumaric acid from biodiesel-derived crude glycerol by *Rhizopus arrhizus*. Bioresour Technol. 163: 48–53.

Zhou, S., K.T. Shanmugam, L.P. Yomano, T.G. Grabar and L.O. Ingram. 2006. Fermentation of 12% (W/v) glucose to 1.2 M lactate by *Escherichia coli* strain SZ194 using mineral salts medium. Biotechnol Lett. 28: 663–670.

Zhu, Y. and S.T. Yang. 2003. Adaptation of *Clostridium tyrobutyricum* for enhanced tolerance to butyric acid in a fibrous-bed bioreactor. Biotechnol Prog. 19: 365–372.

Zhu, J. and K. Shimizu. 2004. The effect of *pfl* genes knockout on the metabolism for optically pure d-lactate production by *Escherichia coli*. Applied Microbiol Biotechnol. 64: 367–75.

Zhu, J. and K. Shimizu. 2005. Effect of a single-gene knockout on the metabolic regulation in *E. coli* for d-lactate production under microaerobic conditions. Metabolic Eng. 7: 104–15.

Zhu, Y., M.A. Eitman, R. Altman and E. Altman. 2008. High glycolytic flux improves pyruvate production by a metabolically engineered *Escherichia coli* strain. Appl Environ Microbiol. 74: 6649–6655.

Zhu, N., H. Xia, J. Yang, X. Zhao and T. Chen. 2014. Improved succinate production in *Corynebacterium glutamicum* by engineering glyoxylate pathway and succinate export system. Biotechnol Lett. 36(3): 553–560.

Zhu, Y., M.A. Eitman, K. DeWitt and E. Altman. 2007. Homolactate fermentation by metabolically engineered *Escherichia coli* strains. Appl Environ Microbiol. 73(2): 456–464.

Zou, X., Y. Zhou and S.T. Yang. 2013. Production of polymalic acid and malic acid by *Aureobasidium pullulans* fermentation and acid hydrolysis. Biotechnol Bioeng. 110(8): 2105–2113.

**7**

# Biofuel and Biochemical Production by Photosynthetic Organisms

## ABSTRACT

Biofuel and biochemical production by photosynthetic microorganisms such as cyanobacteria and algae is attractive from the points of view of energy security and the reduction of the atmospheric $CO_2$, thus contributing to the environmental problems such as global warming. Although the biofuels produced by photosynthetic microorganisms is called as the 3rd generation biofuels, and significant innovation is necessary for the feasibility in practice, these fuels are attractive due to renewable and potentially carbon neutral resources. Moreover, photosynthetic microorganisms are attractive since they can grow on non-arable land, and utilize saline and wastewater streams. A highly versatile and genetically tractable photosynthetic microorganism can capture solar energy and directly convert atmospheric and waste $CO_2$ to high-energy chemical products as consolidated bioprocesses (CBPs). Understanding of the metabolism and the efficient metabolic engineering of the photosynthetic organisms together with cultivation and separation processes as well as increased $CO_2$ assimilation enable the feasibility of biofuel and biochemical production by photosynthetic organisms in practice. Here, the potential of photosynthetic microorganisms such as algae and cyanobacteria for the production of biofuels and biochemicals is explained.

**Keywords**

Microalgae, cyanobacteria, biofuel, biochemical, metabolic engineering, metabolic regulation, $CO_2$ fixation, photosynthesis

## 1. Introduction

The biofuel and biochemical production by photosynthetic organisms is highly attractive due to $CO_2$ fixation with sunlight (and water) from environmental protection point of view, and therefore, it can contribute the global warming problem. A variety of host organisms such as bacteria, fungi, and microalgae may be considered for the production of biofuels and biochemicals from $CO_2$ with sunlight. Although photosynthetic organisms offer the ability to produce biofuels and biochemicals directly from $CO_2$ and sunlight, significant innovation is inevitable for the metabolic engineering and the process development as well as cultivation, harvesting, and product separation, since the production rate is significantly low.

The commonly used photosynthetic organisms for biofuel and biochemical production are algae and cyanobacteria (Savakis and Hellingwert 2015, Yu 2014). Microalgae are photosynthetic eukaryotic organisms with size ranging from 1 to 100 µm, while cyanobacteria are prokaryotic organisms with size ranging from 1 to 10 µm. Cyanobacteria gave rise to the chloroplasts of eukaryotic algae and also land plants, where they share many features such as the ability to drive photosynthetic water photolysis, and thereby contribute to the production of both atmospheric oxygen and reduced organic carbon (Larkum et al. 2012).

Microalgae are unicellular photosynthetic microorganisms that can convert solar energy to chemical energy with efficiency of 10–50 times greater than terrestrial plants (Khan et al. 2009). Algae have far higher cell growth rates than plants, and therefore, have much smaller footprints for land required for producing energy (Chisti 2007, Mata et al. 2010, Georgianna and Mayfield 2012). Many microalgae accumulate oil under stress conditions such as nitrogen limiting conditions, which can be converted to biodiesel using existing technology. These photosynthetic microorganisms can convert atmospheric $CO_2$ into carbon-rich lipids, only a step or two steps away from biodiesel, of which conversion rate significantly exceeds that of agricultural oleaginous crops, without competing for arable land (Wijffels and Barbosa 2010). They require aquatic environments that may vary from fresh-water to sea-water. Not only do these organisms fix $CO_2$, but they also have the potential to be used for the production of inexpensive bulk chemicals, because the major inputs to the system (sun light and $CO_2$) are essentially free (Boyle and Morgan 2009). Microalgae cells contain approximately 50% of carbon, in which 1.8 kg of $CO_2$ is fixed by producing 1 kg of microalgae biomass (Chisti 2007).

The typical microalgae strains may be *Chlorella* sp., *Chlamydomonas* sp., *Scenedesmus* sp., and *Botryococcus braunii* that have shown promising result to bio-mitigate $CO_2$ emission with typical $CO_2$ consumption rate of 200–1300 mg/$l$.day (Chiu et al. 2008, Rosenberg et al. 2011, Sydney et al. 2010, Yoo et al. 2010, Zhao et al. 2011). Successful commercial utilization of microalgae has been established in low-volume, high-value derivatives such as nutritional supplements, antioxidants, cosmetics, natural dyes, and polyunsaturated fatty acids (PUFA) (Rosenberg et al. 2008).

Cyanobacteria are the photosynthetic Gram-negative prokaryotes, which can also directly fix atmospheric $CO_2$ and convert it to various organic compounds using solar energy by photosynthesis, while generating oxygen on earth. About 20–30% of organic carbon on earth may be originated by the photosynthetic carbon fixation by cyanobacteria (Waterbury et al. 1979), where the photosynthetic efficiency of cyanobacteria is about ten-fold higher than the higher plants (Li et al. 2008). In addition to fast growth, the gene manipulation of cyanobacteria is much easier than the higher plants and algae, and therefore, engineering cyanobacteria into **cyanofactories** is an attractive approach to use solar energy and fix $CO_2$, contributing to the global energy and environmental issues (Zhou et al. 2016, Ono and Cuello 2007, Oliver and Atsumi 2014).

Although photoautotrophic microorganisms are attractive as mentioned above, one of the major drawbacks is its slow growth rate, and slow metabolite production rate as compared to the heterotrophic microorganisms such as *E. coli* and *S. cerevisiae*. In general, the time frames for the cultivation of plants are on the order of months, and those of photosynthetic microorganisms are on the order of weeks, while those of heterotrophic microorganisms are on the order of days or even on the order of hours. Therefore, most metabolites are typically formed in mg/*l* in the photosynthetic microorganisms, while some cyanobacterial strains show in g/*l*. The metabolite production rates are, therefore, one or two orders lower in photosynthetic microorganisms as compared to heterotrophic microorganisms (Angermayr et al. 2015). The performance improvement may be attained by modulating $CO_2$ fixation, metabolic engineering to improve conversion efficiency, and redox balance etc.

In this Chapter, attention is focused on the typical photosynthetic microorganisms such as algae and cyanobacteria, and their metabolic regulation, metabolic engineering, and process development with efficient operation for the production of biofuels and chemicals (Sarkar and Shimizu 2015).

## 2. Candidate Photosynthetic Micro-organisms for Biochemical and Biofuel Production

The most comprehensive evaluation of algal species has been made by the US Department of Energy's Aquatic Species Program (ASP) to develop microalgae as a source of biodiesel (Rosenberg et al. 2008). Over 3000 strains of microalgae have been isolated from ponds and seas. Cellular oil content varies depending on the growth phases (Hu et al. 2008). The chlorophyte microalga *Parietochloris incise* synthesize almost twice as many triacylglycerols (TAGs) in the stationary phase than in the exponential growth phase (Bigogno et al. 2002). Although microalgae have a high level of biodiversity, only a few species can be subjected to genetic manipulation (Radakovits et al. 2010). The alga with the best developed genetic toolbox is the unicellular green microalgae *Chlamydomonas reinhardtii*. It is a well-established model organism for the study of various cellular processes such as photosynthesis, flagella, starch metabolism, and photobiological production of hydrogen (Harris 2009). Like many other algal species, *C. reinhardtii* can accumulate significant amount of oil under stress conditions (Wang et al. 2009, Work et al. 2010, Siaut et al. 2011, Moellering and Benning 2010, Li et al. 2010). *C. reinhardtii* has proven to be a useful model organism to study the improvement of biodiesel production (Merchant et al. 2012, Liu and Benning 2013). It is unicellular and stays as haploid during most of its life cycle (Harris 2009), thus it is particularly useful in the context of a forward genetic approach, because the mutant phenotype can be observed during the first generation, and does not need to reach a diploid homozygous stage (Cagnon et al. 2013). The green microalgae *Parietochloris incise* enhance not only its TAG production under nitrogen starvation, but also the production of arachidonic acid, a valuable nutraceutical (Solovchenko et al. 2008). Green algae including *Spirogyra* species and *Chlorococum* sp. accumulate high levels of polysaccharides both in their complex cell walls and as starch (Jones and Mayfield 2012). This starch can be used for biofuel production such as bioethanol production, where bioethanol production from algae shows significant potential due to their low percentage of lignin and hemicelluloses as compared to other lignocellulosic plants (Harun et al. 2010). Microalgae and cyanobacteria are also able to directly produce hydrogen through photofermentation in an anaerobic process involving oxidation of ferredoxin by the hydrogenase enzyme (Yacoby et al. 2011).

Many species of macroalgae are known to have high levels of carbohydrate, where in many cases these carbohydrates consist of galactose (Park et al. 2011). The red algae *Gelidium amansii* and the brown algae *Laminaria japonica* are both potential biomass source for hydrogen production through anaerobic fermentation (Park et al. 2011, Shi et al. 2011). Microalgae have also been paid attention from the point of view of biogas production in the anaerobic fermentation (Jones and Mayfield 2012).

Cyanobacteria are commonly used as model systems for the metabolism of higher plants. Cyanobacteria possess certain promising properties such as (1) large amounts of lipids, commonly present in thylakoid membranes, (2) higher photosynthetic levels and the cell growth rates compared to algae and higher plants, (3) easy growth with basic nutritional requirements such as air ($CO_2 + N_2$), water, and mineral salts with light (Quintana et al. 2011).

Most metabolic engineering investigations have been made using cyanobacteria such as *Synechocystis* sp. and *Synechococcus* sp. as well as *Aanabaena* sp., whereas much more complex genetic engineering is required for algae (Savakis and Hellingwerf 2015). In particular, *Synechocystis* sp. PCC6803, *Synechococcus* sp. PCC7942, and *Synechococcus* sp. PCC7002 have been extensively used for metabolic engineering (Yu et al. 2015).

## 3. Metabolism of Photosynthetic Microorganisms

### 3.1 Brief overview of the metabolism of photosynthetic organisms

In the case of plants, the metabolite pools exist in more than one location, and therefore, the subcellular location of one or more reactions is uncertain (O'Grady et al. 2012). Entire sections of metabolic pathways like glycolysis are duplicated between organellas; particularly the plastid and cytosol, with both being potentially active and carrying flux (Dennis and Blakely 2000). The simplest way is to examine the metabolites which are formed in only one of the compartments (Sriram et al. 2004, 2007, Schwender et al. 2006, Allen et al. 2007, Lonien and Schwender 2009). Another method involves the fraction of cellular material prior to metabolite analysis (Gerhardt et al. 1983). Unfortunately, even with the supplemental information provided by analyzing compartment-specific metabolites, it may still be difficult to statistically distinguish different configurations of the metabolic map (Masakapallis et al. 2010).

Oxygenic photosynthesis is the process by which plants, algae, and cyanobacteria convert sunlight and $CO_2$ into chemical energy and biomass. The algal photosynthesis is at least able to convert approximately 5–7% of incident light energy to biomass, where a systems-based approach to understand the stresses and efficiencies associated with light energy harvesting, $CO_2$ fixation, and carbon partitioning is necessary to make headway toward improving photosynthetic yields (Peers 2014).

The cell growth conditions are roughly classified as **autotrophic** condition for the case of using only $CO_2$ under light condition, **mixotrophic** condition for the case of using both carbohydrate and $CO_2$ under light condition, and **heterotrophic** condition for the case of using carbohydrate under dark condition. Although autotrophic condition is preferred from the environmental protection point of view using only $CO_2$ as a carbon source, the cell growth rate is significantly low, and thus the productivities of the metabolic products are low.

The atmospheric $CO_2$ is fixed either by $C_3$ photosynthesis where the three carbon molecule such as 3-phosphoglycerate (3PG) is used as the product of ribulose 1,5-bisphosphate carboxylase **(RubisCO)** reaction, or by $C_4$ photosynthesis where four carbon molecule such as oxaloacetate (OAA) is used as the product of phosphoenol pyruvate carboxylase

(Ppc) reaction followed by the decarboxylation at malic enzyme (Mez) from malate yielding pyruvate. The $C_4$ photosynthesis may be created by evolution from ancestral $C_3$ photosynthesis during a global decline in the atmospheric $CO_2$ level (Sage et al. 2012). The $C_4$ pathway will have higher efficiency than $C_3$ pathway in $CO_2$ fixing with which they consume water and nitrogen (Wang et al. 2014).

The green organisms such as plants and algae obtain energy via aerobic respiration, and the metabolism changes depending on the oxygen availability. The green organisms are exposed to a variety of oxygen availability in the environment that may vary from fully aerobic state **(normoxia)** to oxygen deficiency **(hypoxia)** or the anaerobic condition **(anoxia)**. In the context of recent climate change, excess rainfall and frequent flooding may cause the green cells subjected to hypoxia or anoxia condition (Banti et al. 2013). Oxygen is the final electron acceptor in the mitochondrial oxidative phosphorylation to generate ATP, while under hypoxia and anoxia conditions, ATP is generated through glycolysis by the substrate level phosphorylation, and NAD(P)H must be reoxidized by the fermentative pathways. Under oxygen limiting condition, the glycolysis flux is accelerated by the so-called "Pasteur effect", and the plant metabolism uses pyruvate to direct towards ethanolic and lactic fermentations (Geigenberger 2003). In almost all plants, a rapid activation of lactate dehydrogenase (LDH) has been observed under oxygen limitation. However, lactate production causes damage to the cell by lowering cytoplasmic pH, and therefore, lactate production is transient eventually replaced by ethanolic fermentation. The $\alpha$-ketoglutaric acid ($\alpha$KG) in the TCA cycle can be oxidized with the incorporation of $NH_4^+$ and NADH to form glutamate (Glu), which is then decarboxylated to ɤ-amino butyric acid (GABA) by glutamate decarboxylase (GDC), where some protons are utilized in GDC reaction and stabilizes the cytosolic pH (Banti et al. 2013).

### 3.2 Metabolism of algae

Under unfavorable growth condition such as nitrogen starvation, TAG is typically produced in microalgae, where its fraction ranges from 20% to 60% (weight/dry weight) (Griffiths et al. 2009). The efficient production of TAG in microalgae requires a thorough understanding of lipid metabolism and TAG accumulation (Klok et al. 2014). For this, it is important to analyze different levels of information to uncover the molecular mechanism underlying the increased TAG accumulation for microalgae such as *C. reinhardtii* and its starch-less and cell wall-deficient mutant strains (Blaby et al. 2013).

The light energy is incorporated into the cell, where light quanta absorbed by pigments drive the photosynthetic electron transport, where

NADPH instead of NADH is used to generate ATP at the respiratory chain (Yang et al. 2000). The primal pathway for $CO_2$ fixation is the **Calvin-Benson-Bassham (CBB) cycle** or simply called as **Calvin cycle**, where the first step is catalyzed by RubisCO (Fig. 1). This enzyme is also an oxygenase, which can react with $O_2$ and lead to a different pathway called **photorespiration**. Algae have the photorespiration pathway, and photosynthesis is inhibited by high $O_2$ concentration. Photosynthesis reactions such as light reactions, CBB cycle, and starch synthesis occur in chloroplasts. Algae and plant cells have sub-cellular compartments such as chloroplast, mitochondria, and cytoplasm. After the export of gleceraldehyde 3-phosphate (GAP) from chloroplast to cytoplasm, the carbon flow is divided into the sugar synthesis pathway or the glycolytic pathway to form pyruvate. Sucrose is the major storage product in the cytoplasm of plant cells. In plant cells, replenishment of carbon to maintain the operation of the TCA cycle is achieved by anaplerotic reactions involving $CO_2$ fixation by Ppc. The pentose phosphate (PP) pathway operates in the cytoplasm, where CBB cycle is functioning in the chloroplast.

Of all the pigments, chlorophyll takes a major fraction. δ-Aminolevulinic acid (δ-ALA) is the key chlorophyll precursor molecule. The classical succinate-glycine pathway is the condensation of glycine and succinyl-CoA catalyzed by δ-ALA synthetase. In addition, glutamate and α-KG

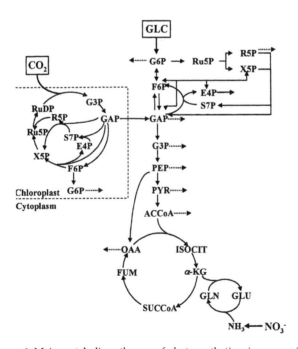

**Figure 1.** Main metabolic pathways of photosynthetic micro-organism.

are incorporated into δ-ALA much more efficiently than are glycine and succinate in many green cells. Although most of the fatty acid synthesis occurs in the chloroplast, the source of AcCoA derives from its synthesis in the mitochondria. The fatty acid composition of the lipids of *Chlorella* cells varies considerably, particularly for the α-linolenic acid (C18:3) content (Yang et al. 2000).

Under **autotrophic condition**, significant ATP is formed from mitochondrial oxidative phosphorylation. The CBB cycle is the main ATP sink in the autotrophic culture. The ATP yield decreases in the following order: heterotroph > mixotroph > autotroph (Yang et al. 2000).

### 3.3 Metabolism of cyanobacteria

The central metabolic network in *Synechocystis* is shown in Fig. 2, which includes those of the glycolysis, PP pathway, CBB cycle, part of TCA cycle, and the C1 metabolism. It has been considered that cyanobacteria have an incomplete TCA cycle lacking αketoglutarate dehydrogenase (KGDH) (Pearce et al. 1969, Vazquez-Bermudex et al. 2000). One of the reasons of the lower cell growth rate in cyanobacteria may be due to this incomplete TCA cycle lacking KGDH and succinyl CoA synthetase (SCS). However,

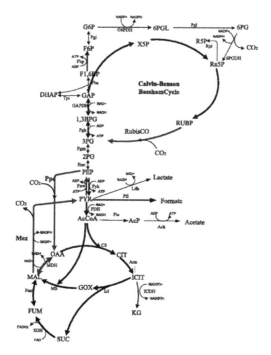

**Figure 2.** Main metabolic pathways of cyanobacteria.

recent investigation revealed that the TCA cycle is closed by the alternative pathways (Steinhouser et al. 2012), where the genes encoding αKG decarboxylase (or 2-oxoglutarate decarboxylase) and succinic semialdehyde dehydrogenase are present in *Synecococcus* sp. PCC7002, where NADPH instead of NADH is produced without producing guanosine triphosphate (GTP) by substrate level phosphorylation (Fig. 3) (Zhang and Bryant 2011). It is important to elucidate the nature of such TCA cycles in cyanobacteria and plants from the point of view of functional significance of the metabolic feature in a broader evolutionary context (Zhang and Bryant 2011).

The enzymes, isocitrate lyase (Icl) and malate synthase (MS), which form the glyoxylate pathway, function in Cyanobacteria (Pearce and Carr 1967). The malic enzyme (Mez) and PEP synthase (Pps) are responsible for the gluconeogenetic steps, where the PEP carboxy kinase (Pck) is absent in

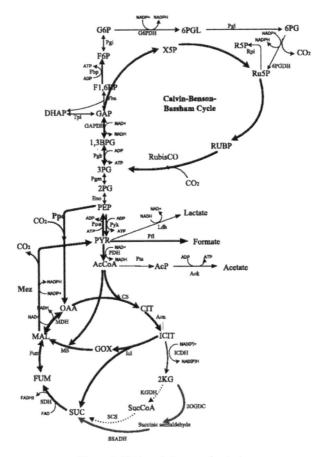

**Figure 3.** TCA cycle in cyanobacteria.

*Synechocystis*. The PP pathway operates for glucose catabolism mainly in the heterotrophic conditions, while the CBB cycle is active under mixotrophic and autotrophic conditions.

In the heterotrophic cultivation of *Synechocystis*, more than 90% of glucose is channeled through the PP pathway (Fig. 4a) (Yang et al. 2002a, 2002b). The high flux through the oxidative PP pathway yields a large amount of NADPH, as well as biosynthetic precursors such as ribose 5-phosphate (R5P) and erythrose 4-phosphate (E4P). In the mixotrophic culture, $CO_2$ is fixed through the CBB cycle. The conventional [13]C-metabolic flux analysis ([13]C-MFA) is based on the steady state and thus limited to heterotrophic and mixotrophic conditions (Yang et al. 2002a, 2002b, Schwender 2008), while [13]C-MFA for autotrophic condition can be also made by the isotopically nonstationary MFA (Wiechert and Nor 2005) with transient measurements of isotope incorporation following a step change from unlabeled to labeled $CO_2$ (Fig. 4c) (Young et al. 2011). It is important to understand the metabolic regulation mechanisms based on different levels of information from gene expression to metabolic fluxes (Yang et al. 2002c).

Since cyanobacteria have negligible photorespiration and produce little or no glycolate during photosynthesis, it is unlikely that serine is synthesized, as in higher plants, from glycine by the glycolate pathway. Serine is synthesized directly from 3-phospho glycerate (3PG) through a phosphorylated route in cyanobacteria (Colman and Norman 1997).

Under both heterotrophic and mixotrophic conditions, the relative flux through Ppc is high (Fig. 4a, b). The reaction catalyzed by Ppc contributes to about 25% of the assimilated $CO_2$ under mixotrophic condition (Yang et al. 2002a, 2002b), indicating that Ppc is important for the fixation of $CO_2$ in cyanobacteria (Owittrim and Colman 1988), where cyanobacterial cells fix significant amount of carbon as C4 acids under light conditions. Considering that Mez in cyanobacteria is NADP-linked (Bricker et al. 2004), it is more likely that Ppc and Mez serve as a device to fix a large amount of $CO_2$ as C4 acids, and then release $CO_2$ and produce NADPH by the decarboxylation of malate. This Ppc and Mez pathways can effectively bypass the Pyk reaction, where its activity is repressed under light condition (Young et al. 2011, Bricker et al. 2004, Knowles et al. 2001). The Pyk activity of *Synechococcus* sp. PCC 6301 is modulated mainly by energy charge, feed-forward activation by hexose monophosphate, ribose 5-phosphate (R5P), and feedback inhibition by several TCA cycle intermediates, where this may be phylogenetically related to *Bacillus* and green algae Pyk (Knowles et al. 2001). This is similar to the carbon metabolism in C4 plants, for which $CO_2$ and NADPH generated by Mez are utilized in the CBB cycle. In fact, although the major pathway of $CO_2$ fixation in the CBB cycle is similar to that in C3 plants, cyanobacteria have many physiological characteristics of C4 plants.

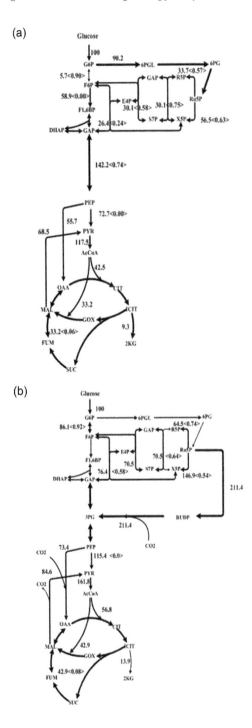

*Fig. 4 contd. ...*

*...Fig. 4 contd.*

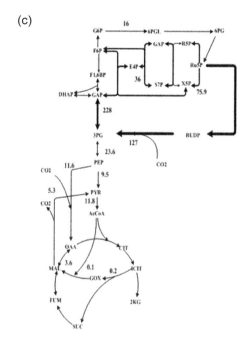

(c)

**Figure 4.** ¹³C-Metabolic flux distribution of *Synecocystis* sp. PCC6803 cultivated under heterotrophic (a), mixotrophic (b) (Yang et al. 2002b), and autotrophic conditions (c) (Young et al. 2011).

*Synechocystis* sp. PCC6803 contains a respiratory electron transport chain (ETC) on both cytoplasmic and thylakoid membranes. The cytoplasmic membrane contains the proteins associated with respiratory electron transport such as NAD(P)H dehydrogenase, cytochrome $b_6f$, and terminal oxidases. In cyanobacteria, two types of NAD(P)H dehydrogenase are present, where NDH-1 encoded by *ndh* prefers NADPH, and contributing to a proton gradient across the membrane, giving ATP production, while NADH-oxidizing type II dehydrogenase (NDH-2) (encoded by *ndbABC*) consisting of a single subunit and presumably not contributing to a proton gradient across the membrane (Cooley and Vermaas 2001). The thylakoid membrane contains a photosynthetic ETC that includes photosystem I (PSI) and PSII and a respiratory ETC containing NDH-1, succinate dehydrogenase (SDH), and a cytochrome $aa_3$-type terminal oxidase (CtaI). The respiratory and photosynthetic ETC in the thylakoids have electron carriers in common, such as cytochrome $b_6f$ complex, the plastoquinone (PQ) pool, and soluble redox active proteins (Cooley and Vermaas 2001).

In the **mixotrophic culture**, large amount of reducing power is required in the CBB cycle to fix $CO_2$ to carbohydrates. Hence, large amounts of

NADPH must be supplied to fulfill the biosynthetic demands. The PP pathway in the heterotrophic culture and the photosynthetic electron transport in the mixotrophic culture accounts for a major fraction of NADPH production. Moreover, cyanobacteria utilize NADPH as an electron donor of the respiratory electron transport chain. Therefore, the excess NADPH is reoxidized in the respiration to gain energy. The CBB cycle is the main ATP sink in the mixotrophic culture.

### 3.4 Nitrogen regulation and carbon storage regulation in cyanobacteria

Like many bacteria, photosynthetic organisms are found in diverse ecological habitats, where the organisms are often exposed to the severe nutrient starvation. In particular, cyanobacteria are found in a wide range of ecological habitats including ocean and lakes (Stomp et al. 2007). They also survive in deserts, polar regions and hot springs, where the nutrient starvation is much more severe.

Cyanobacteria have sophisticated mechanisms to cope with nitrogen limitation, where the primary step is the capture of nitrogen-containing compounds with high affinity, where ammonia, nitrate, and nitrite are the typical nitrogen sources with preference for ammonium (Flores and Herrero 2005). Some strains can fix dinitrogen gas, and may use also urea, cyanate, and amino acids as additional nitrogen sources (Flores and Herrero 2005, Valladares et al. 2002, Garcia-Fernandez et al. 2004). Nitrogen compounds are eventually converted to ammonium and utilized for biosynthesis via the glutamin synthetase (GS)-glutamine oxoglutarate aminotransferase or glutamate synthase (GOGAT) cycle, where glutamate dehydrogenase (GDH) pathway does not function, probably due to low affinity to ammonium.

For survival under nitrogen starvation, cyanobacteria accumulate reserve materials in the form of inclusions and granules, where the induction for such accumulation is made upon high light or $CO_2$, nutrient starvation, as well as addition of arginine or chloramphenicol (Allen 1984). **Cyanophycin** (multi-L-arginyl-poly-[L-aspartic acid]) is a nitrogen reserve, and is a non-ribosomally synthesized peptide consisting of equi-molar quantities of alginine (Alg) and aspatic acid (Asp), where cyanobacteria may consume internal storage compounds such as cyanophycin as nitrogen source upon nitrogen starvation (Allen 1984, Li et al. 2001).

After cyanophycin is exhausted, cells degrade the **phycobilisomes** that are large protein-rich light-harvesting antennae attached to the outside of the thylakoid membranes and support the light-dependent reactions of photosynthesis (Grossman et al. 1993), where it is composed of rod and core proteins to provide nitrogen, which leads to a color change of cells from blue-green to yellow green, known as **bleaching** (Grossman et al. 1993).

Upon availability of nitrogen source again, cyanophycin is immediately synthesized (Li et al. 2001).

Nutrient balance is important for the cell growth, since proteins, nucleic acids, carbohydrates, lipids, and pigments are needed in a suitable ratio for the balanced growth. In eukaryotic microalgae, autophagy is induced by nitrogen starvation to degrade cytoplasmic components including plastids in the large vacuoles (Dong et al. 2013). In cyanobacteria, a unique Nb1A-dependent mechanism is induced to degrade certain **phycobiliproteins,** where the non-bleaching phenotype gene, *nblA* plays an important role for the degradation of phycobiliprotein (Li and Sherman 2002, Baier et al. 2001). The phycobilisome has a role in nitrogen storage as well as photosynthetic antenna (Grossman et al. 1993). Moreover, NblA1/A2-dependent protein turnover contributes to the maintenance of many amino acids (AAs) in NblA1/A2-dependent way, while Lys pool markedly increases under sulfur starvation in cyanobacteria (Kiyota et al. 2014).

The internal C/N ratio is sensed by the P II protein, GlnB, in particular under N-limitation (Schwarz et al. 2014). The global nitrogen regulator NtcA plays important roles for nitrogen-regulation, where it senses $aKG$ levels and regulates the genes involved in nitrogen assimilation. NtcA directly regulates the expression of *nrrA* gene which encodes the nitrogen-regulated response regulator of the OmpR family. NrrA is involved in induction of sugar catabolic genes as well under nitrogen starvation (Azuma et al. 2011). NrrA also regulates glycogen catabolism in *Anabaena* sp. by directly regulating the expression of *glgP* gene encoding glycogen phospholylase and *sigE* gene encoding a group 2σ factor of RNA polymerase (Ehira and Ohmori 2006). Nrr controls cyanophycin accumulation and glycogen catabolism in cyanobacteria (Liu and Yang 2014), where glycogen is accumulated, whereas the expression of sugar catabolic genes is widely up-regulated under nitrogen starvation (Krasikov et al. 2012).

Microalgae produce certain biomass compounds under nutrient limitation (Markou and Nerautzis 2013). In cyanobacteria such as *Synechocystis* sp., polyhydoxy butyric acid (PHB), one of the polyhydroxy alcanoate (PHA), is accumulated under nitrogen or phosphate starvation, where PHB is formed from AcCoA via $\beta$-keto-thiolase (PhaA), acetoacetyl-CoA reductase (PhaB), and PHA synthase (PhaC), where PHA synthase is activated by acetyl phosphate (AcP) (Miyake et al. 1997). Since acetoacetyl-CoA reductase requires NADPH, the pathway modification that may give excess NADPH yields higher PHB production (Hauf et al. 2013).

Cyanobacteria have 9 sigma factors such as SigA–I, where RNA polymerase sigma factor SigE plays important roles under nitrogen starvation (Muro-Pastor et al. 2001). SigE activates the expression of the genes associated with degradation of glycogen and catabolic genes of glycolysis and PP pathway (Osanai et al. 2005). Moreover, SigE activates

PHB synthetic pathway gene expression (Osanai et al. 2011), and thus overexpression of *sigE* allows higher PHB production under nitrogen limitation (Osanai et al. 2013a). Moreover, SigE also activates the expression of hydrogenase gene, and thus the overexpression of *sigE* also allows higher hydrogen production under anaerobic condition (Osanai et al. 2013b).

Cyanobacteria have two types of sunscreen pigments such as scytonemin and mycosporine-like amino acids, where these secondary metabolites play roles against environmental stresses such as UV radiation and desiccation (Wada et al. 2013).

## 4. Metabolic Engineering of Photosynthetic Microorganisms

### 4.1 Metabolic engineering of algae

Algae have the potential for the genetic modification of their lipid pathways by up-regulation of fatty acid biosynthesis or by downregulation of β-oxidation. By disruption or modifying enzymes responsible for the synthesis of polyunsaturated lipids in the cell, it may dramatically increase the production of mono-unsaturated lipids (Schenk et al. 2008). Under optimal growth condition, the wild-type *Chlamydomonas* strains accumulate very low amount of oil (< 1 μg per $10^6$ cells) (Siaut et al. 2011), while when the cells are subjected to nitrogen starvation, oil content can be increased more than 10 fold (up to 10 μg per $10^6$ cells) (Wang et al. 2009, Work et al. 2010, Siaut et al. 2011). Intracellular TAG amounts also fluctuate during the diurnal cycle because TAGs produced during the day provide a carbon and energy source during the night (Lacour et al. 2012). This is the major factor to yield loss in open pond microalgae cultivation.

It is not easy to achieve efficient homologous recombination in the nuclear genome of the commonly transformed laboratory algal strain *C. reinhardtii*, but this is not the case for the marine alga and *Nannochloropsis* (Kilian et al. 2011). As a result of genetic engineering, some obligate photoautotrophs, formerly unable to partake in a sweet diet, have been given a taste of heterotrophy through the introduction of hexose transporters (Singh et al. 2011). In the starchless mutant of *Chlamydomonas*, the flux through starch production redirects to lipid accumulation under nitrogen starved condition. The strain should be designed to switch off completely for starch accumulation under nitrogen starved condition. Blocking oil turnover processes might help increase the level of oil accumulation, as observed in *Arabidopsis* leaves where the oil content was increased 10-fold by disrupting a lipase gene (James et al. 2010).

Eukaryotic algae have been considered for fatty acid production, where they can accumulate lipids up to about 70% of dry biomass (Sivakumar et al. 2012, Radakovits et al. 2010). The limitations of using algae are the complexity of the eukaryotic system and less available genetic tools

(Radakovits et al. 2010, Rasala et al. 2014). Some efforts are being made to identify stress conditions and key enzymes for fatty acid synthesis in the green algae *Haematococcus pluvialis* (Lei et al. 2012). Effects of light conditions on fatty acid production were also investigated for *Nannochloropsis* (Anandarajah et al. 2012).

Methane can be produced by co-culture of *C. reinhardtii* and methanogenic bacteria, where glyconate is produced from the former, while methane is produced from the latter by assimilating glyconate (Gunther et al. 2012). *Rhodobacter* are non-sulfur photosynthetic bacteria that produce hydrogen ($H_2$) from acetate, etc., where hydrogen production may be enhanced by introducing aldehyde dehydrogenase (ALDH) gene from *Rhodospirillum rubrum* into *R. sphaeroides* (Kobayashi et al. 2012). Hydrogen production could be enhanced by introducing exogenous hydrogenase into nitrogen fixing cyanobacteria, where hydrogen production could in fact be improved by introducing hydrogenase from *Clostridium thermocellum* into *Rhodopseudomonas palustris* CGA009 and cultivated at 38°C by considering the outdoor bioreactors (Lo et al. 2012).

Another important aspect of utilizing photosynthetic organisms is their ability of producing pharmaceuticals due to the reduction of ketones (Havel and Weuster-Botz 2007, Nakamura et al. 2000, Yang et al. 2012). Microalgae are attractive for their production of antioxidants, where *Fischerella ambigua* and *Chlorella vulgaris* show higher antioxidant activities (Hajimahmoodi et al. 2010). Phenolic compounds have antioxidant properties, where their production can be enhanced in *Spirulina platensis* by manipulating light intensities (Kepekci and Saygideger 2012). Phycocyanin is also attractive, where its production by *Arthospira* (*Spirulina*) *platensis* was investigated in marine environment (Leema et al. 2010). Lycopene is important food additives and pigment, and can be produced by a purple non-sulfur bacterium, *Rhodospirillum rubrum*, by deletion of downstream phytoene desaturase gene *crtC* and *crtD* (Wang et al. 2012). Moreover, PHB can be produced up to 10.6% of algal dry weight by introducing bacterial PHB forming pathway genes of *R. eutropha* H16 into the diatom *Phaeodactylum tricomutum* (Hempel et al. 2011).

## 4.2 Metabolic engineering of cyanobacteria

Some cyanobacterial strains can accumulate large amounts of lipids that may be converted to biofuels (Quintama et al. 2011). Cyanobacteria produce a wide variety of metabolites such as amino acids, fatty acids, macrolides, lipopeptides, and amides (Burja et al. 2001). The engineered cyanobacteria with the aid of synthetic biology tools have shown the potentials for the industrial production of a variety of chemicals and fuels such as hydrogen, ethanol, ethylene, acetone, isopropanol, diols, isoprene, squalene, *n*-alkanes, and free fatty acids (Englund et al. 2014, Xie et al. 2017).

### 4.2.1 Hydrogen production

Hydrogen is an attractive energy source due to clean gas without evolving greenhouse gas such as $CO_2$ as well as high energy intensity per unit weight in combustion, and thus easily converted to electricity. Biological hydrogen production has several advantages over other conventional hydrogen producing processes, where the former is low energy requirements and it is cost effective, whereas electrochemical hydrogen production via solar battery-based water splitting requires the use of solar batteries with high energy requirements (Dutta et al. 2005, Block and Melody 1992).

Microbial hydrogen production can be made by several cyanobacteria genera under a wide range of culture conditions (Lopes Pinto et al. 2002). Unicellular non-diazotrophic cyanobacteria *Gloeocapsa alpicola* shows increased hydrogen production under sulfur starvation (Antal and Lindblad 2005). *Spirulina (Arthropira) platensis* can produce hydrogen under complete anaerobic and dark conditions (Aoyama et al. 1997). A nitrogen-fixing cyanobacterium, *Anabaena cylindrical* produces hydrogen and oxygen gas simultaneously in an argon atmosphere under light limited condition (Jeffries et al. 1978). *Anabaena cylindrical* produces high amount of hydrogen under nitrogen starved conditions. Hydrogenase-deficient cyanobacteria *Nostoc punctiforme* shows increased hydrogen production under high light condition (Lindberg et al. 2004).

In cyanobacteria, nitrogenase and hydrogenase are the hydrogen producing/metabolizing enzymes (Dutta et al. 2005). **Nitrogenase** is found in heterocysts of filamentous cyanobacteria during nitrogen limiting conditions, where hydrogen is produced as a by-product of nitrogen fixation into ammonia by consuming ATP such as (Rao and Hall 1996)

$$16ATP + 16H_2O + N_2 + 10H^+ + 8e^- \xrightarrow{Nitrogenase} 16ADP + 16P_i + 2NH_4 + +H_2$$

There are two types of **hydrogenases** in different cyanobacterial species. One type encoded by *hupSL* (Tamagnini et al. 2002) has the ability to oxydize hydrogen, while other type encoded by *hoxFUYH* is reversible or bidirectional such that (Dutta et al. 2005)

$$H_2 \xrightarrow{Hydrogenase} 2H^+ + 2e^-$$

Hydrogen photo evolution catalyzed either by nitrogenases or hydrogenases can only function under anaerobic conditions due to their extreme sensitivity to oxygen (Dutta et al. 2005). In hydrogen production by cyanobacteria, culture conditions such as light, temperature, salinity, nutrient availability affect its performance.

Although most cyanobacteria absorb red light near 680 nm (Pinzon-Gamez and Sundaram 2005), *Spirulina platensis* produces hydrogen under anaerobic conditions in the dark (Aoyama et al. 1997), while several other species produce hydrogen only in the light (Stal and Krumbein 1985). *Synechococcus* sp. PCC 7942 produces hydrogen by native hydrogenases under anaerobic condition in the dark (Asada et al. 1999). *Anabaena variabilis* ATCC 29413 and its mutant PK85 produce high amount of hydrogen (Table 1) (Tsygankov et al. 1999). Nitrogen sources such as ammonia, nitrate, and nitrite inhibit nitrogenase in *Anabaena* species (Lambert et al. 1979, Datta et al. 2000, Rawson 1985).

Molecular oxygen inhibits hydrogenase and nitrogenase, and therefore, hydrogen production by nitrogenases or hydrogenases can be made under anaerobic conditions (Fay 1992). Since oxygen is a by-product of photosynthesis, nitrogenase-containing organisms furnish spacial and temporal separation/compartmentation systems (Tamagnini et al. 2002).

Sulfur is an important component in the photosystem II repair cycle, and the biosynthesis is impaired lacking the ability to produce either cysteine or methionine without sulfur. The lack of sulfure inhibits the oxygenic photosynthesis and enhances hydrogen production (Antal and Lindblad 2005).

Hydrogen could be produced by introducing a hydrogenase HydA from *C. acetobutylicum* into *S. elongatus,* and non-nitrogen fixing cyanobacteria (Ducat et al. 2011), and its production can be significantly improved in *Arthosporia* by continuously removing it from the culture broth (Ananyev et al. 2012).

### 4.2.2 Ethylene production

Ethylene production is attractive for the photosynthesis of volatile hydrocarbon fuel with non-invasive monitoring of the *in vivo* ethylene formation. There are three metabolic pathways for ethylene synthesis (Wang et al. 2002, Fukuda et al. 1993). In higher plants, ethylene is synthesized from methionine via 1-aminocyclopropane 1-carboxylic acid (ACC) catalyzed by ACC synthase and ACC oxidase (Lin et al. 2009). In most prokaryotes, ethylene is formed from methionine via 2-keto-4-methyl-thiobutyric acid (KMBA) catalyzed by an NADH: Fe(III) EDTA oxidoreductase (Ogawa et al. 1990). In a few plant pathogens, ethylene is synthesized by an ethylene forming enzyme (EFE) in a complex multi-step reaction using $\alpha$KG, arginine, and dioxygen as substrates (Nagahama et al. 1991). This $\alpha$KG-dependent EFE pathway gene *efe* of *Pseudomonas syringae* pv. *phaseolicola* PK2 was cloned in *E. coli* (Fukuda et al. 1992), and heterologous expression in *Synechococcus* (Fukuda et al. 1994, Sakai et al. 1997) and *Synechocystis* (Ungerer et al. 2012) allows ethylene production (Fig. 5) (Table 1). A *Lac* promoter variant can induce EFE expression in *Synechocystis* at high level and allows a fine level of IPTG-dependent regulation (Guerrero et al. 2012). Ethylene synthesis flux can be increased about 10% by improving TCA cycle activity (Xiong et al. 2015).

**Table 1.** Biofuel and biochemical production by engineered photosynthetic microorganisms.

| Product | Strains | Titer or productivity | Overexpressed or knockout gene(s) | Cultivation | Reference |
|---|---|---|---|---|---|
| Ethanol | *Synechococcus* | 230 mg/L in 28 days | *pdc, adh* | Shake flask | Deng et al. 1995 |
| | *Synechocystis* | 552 mg/L in 6 days | *pdc, adh* | Photobioreactor | Dexter et al. 2009 |
| | *Synechocystis* | 608 mg/L in 18 days | *pdc, adh* | Photobioreactor | Dienst et al. 2014 |
| Isobutyraldehyde | *Synechococcus* | 1,100 mg/L in 8 days | *alsS, ilvC,D, kivd, rbcls* | Bottle with NaHCO3 | Atsumi et al. 2009 |
| Isobutanol | *Synechococcus* | 18 mg/L | *kivd, yqhD* | Shake flask with NaHCO3 | Atsumi et al. 2009 |
| | *synechococcus* | 450 mg/L in 6 days | *alsS, ilvC,D, kivd, yqhD* | Shake flask with NaHCO3 | Atsumi et al. 2009 |
| 2Methyl-1-butanol | *Synechococcus* | 2 mg/L | *kivd, yqhD, cims* | Shake flask with NaHCO3 | Atsumi et al. 2009, Shen et al. 2012 |
| 1-Butanol | *Synechococcus* | 14.5 mg/L in 7 days | *hbd, crt, adhE2, ter, atoB* | Bottle under anoxic cond. | Lan et al. 2011 |
| | *Synechococcus* | 30 mg/L in 18 days | *ter, nphT7, bldh, yqhD, phaJ,B* | Shake flask | Lan et al. 2012 |
| Fatty alcohol | *Synechocystis* | 0.2 mg/L in 18 days | *far* | Photobioreactor with 5% CO2 | Tan et al. 2011 |
| | *Synechocystis* | 0.02 mg/L/OD | *far, aas* | Shake flask | Gao et al. 2012 |
| | *Synechocystis* | 2.87 mg/gDCW | *Δsll0208, Δsll0209* | Flask | |
| Fatty acids | *Synechocystis* | 197 mg/L in 17 days | *tesA, accBCDA, fatB1,B2, Δaas(slr1609), Δpta, ΔPHB, Acyanophycin* | 1% CO2 bubbling | Liu et al. 2011b |
| Alka(e)nes | *Synechocystis* | 0.162 mg/L/OD | *accBCDA* | Bubble column | Tan et al. 2011 |
| | *Synechocystis* | 26 mg/l | *sll0208, sll0209* | Photobioreactor | Wang et al. 2013 |

| Product | Organism | Production | Genes | Condition | Reference |
|---|---|---|---|---|---|
| Hydrogen | *Synechococcus* | 2.8 µmol/h/mg Chlorophyll-a | *hydEF, hydG, hydA* | Anaerobic condition | Ducat et al. 2011 |
| | *Synechococcus* | 54 mmol/1017 cells in 4 days | Δ*ldh* | Anoxic condition | McNeely et al. 2010 |
| L-Lactate | *Synechocystis* | 0.0178 mmol/gDCW/h | *ldh, sth* | Shaking incubator | Angermayr et al. 2012 |
| | *Synechocystis* | 0.2512 mmol/gDCW/h | *pyk, ldh* | Shaking incubator | Angermayr et al. 2014 |
| D-Lactate | *Synechocystis* | 2.17 g/L in 24 days | *gldA, sth* | Photoautotropic with acetate | Varman et al. 2013 |
| 1,2-propanediol | *Synechococcus* | ~150 mg/L | *mgsA, gldA, yqhD* | Shake flask | Li et al. 2013 |
| 1,3-propanedion | | 288 mg/l | | | Horikawa et al. 2016 |
| 2,3-butanediol | | 2.38 g/l | *alsS, alsD, adh* | | Oliver et al. 2013 |
| Isoprene | *Synechocystis* | 50 µg/gDCW/d | *IspS* | Sealed culture | Lindberg et al. 2010 |
| Ethylene | *Synechocystis* | 26 µmol/gDCW/h | *efe(RS1010)* | Rotary shaker | Guerrero et al. 2012 |
| | *Synechocystis* | 111.6 µmol/gDCW/h | *efe(slr068)* | Rotary shaker | Guerrero et al. 2012 |
| | *Synechocystis* | 9.7 µl/l.h | *efe, KgtP* | Semicontinuous | Zhu et al. 2015 |
| | *Synechocystis* | 718 µl/1.h.OD730 | *efe,RBSo4* | | Xiong et al. 2015 |
| | *Synechococcus* | 84.8 µmol/gDCW/h | *efe(pUC303)* | Flask | Sakai et al. 1997 |
| | *Synechococcus* | 80.5 µmol/gDCW/h | *efe(psbAI)* | Flask | Takahama et al. 2003 |
| Acetone | *Synechocystis* | 36.0 mg/L in 4 days | *ctfAB, adc, ΔphaCE, Δpta* | Flask | Wang et al. 2013 |
| PHAs | *Synechocystis* | 1.4 mg/100 mgDCW | *sigE* | Bubbled with 1% CO2 in the air | Osanai et al. 2013a |
| | *Synechocystis* | 533.4 mg/L in 21 days | Δ*slr1829*, Δ*slr1830* | Flask | |

## 4.2.3 Ethanol production

Bio-ethanol production by cyanobacteria has originally been made by the heterologous expression of pyruvate decarboxylase (*pdc*) and alcohol dehydrogenase II (*adh2*) from Z. *mobilis* under the control of *rbcLS* promoter in *Synechococcus* sp. PCC7942, where 5 mM (0.23 g/*l*) of ethanol could be produced (Deng and Coleman 1999). Bio-ethanol could also be produced in *Synechocystis* sp. PCC6803 with a titer of about 10 mM (0.46 g/*l*) (Dexter and Fu 2009). The performance improvement could be made by *Synechocystis* sp. PCC6803 with the integration of *pdc* from Z. *mobilis* and endogenous alcohol dehydrogenase *adhA* (*slr1192*) under the control of different promoters with a titer of 3.6 g/*l* for 38 days (Duhring et al. 2010). Further performance improvement could be made by introducing exogenous *pdc* gene from Z. *mobilis* and by over-expressing endogenous *slr1192* from *Synechocystis* sp. PCC6803 through homologous recombination at two different sites of chromosome, and by disrupting the PHB producing pathway gene *phaAB*, where the titer was 5.50 g/*l*, and the productivity was 0.212 g/*l*.d (Gao et al. 2012). Although the activity of ADH encoded by *adh2* was 94-fold higher than that of *adhA* (*slr1192*), the former requires NADH ($K_m$ = 2.73 mM, while $K_m$ = 9.56 mM for the latter), while the latter requires NADPH ($K_m$ = 1.56 mM, while $K_m$ = 38 mM for the former) (Gao et al. 2012), where the intracellular NADP(H) concentration is about 10-fold that of NAD(H) in *Synechocystis* sp. PCC6803 (Cooley and Vermaas 2001). The similar performance could be obtained under normal oxygen concentration (5% $CO_2$ and 95% air) and anoxic conditions (5% $CO_2$ and 95% $N_2$) (Gao et al. 2012).

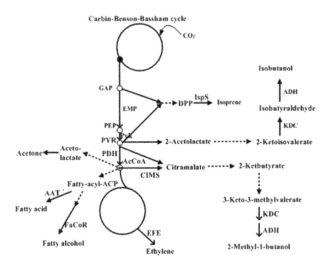

**Figure 5.** A variety of metabolic pathways for the production of biofuels and biochemicals.

The prolonged production of ethanol could be made by *Synechocystis* sp. PCC6803 with *pdc* from *Z. mobilis* and overexpression of *adhA* (*slr1192*) by photo-bioreactor, with a titer 0.608 g/*l* in 18 days, where the transcriptional response with respect to time indicates severe reduction in the expression of picocyanin subunits *cpbB* and *cpbA* causing the bleaching phenotype (Dienst et al. 2014).

### 4.2.4 Isopropanol production

Isopropanol is a secondary alcohol and valuable for the production of propylene by dehydration, and polypropylene, a popular industrial material and currently produced from petroleum. The *Clostridium* species could produce isopropanol up to about 1.8 g/*l* together with acetone and butanol (Chen and Hiu 1986). In *Clostridium beijerinckii*, isopropanol is produced from AcCoA via acetone, where two moles of AcCoA are first condensed to acetoacetylCoA (AcAcCoA) by AcCoA acetyltransferase encoded by *thl*. Then acetoacetyl-CoA transferase encoded by *ctfAB* transfers the CoA of AcAcCoA to acetate or butyrate, forming acetoacetate. Acetoacetate is irreversely converted to acetone and $CO_2$ by an acetoacetate decarboxylase encoded by *adc*. Finally, acetone is converted to isopropanol by NADPH-dependent primary-secondary alcohol dehydrogenase (SADH) encoded by *adh* (Fig. 6) (Kusakabe et al. 2013). The engineered *Synechococcus elongatus* sp. PCC7942 overexpressing heterologous genes such as *thl* from *C. acetobutyricum*, *atoAD* from *E. coli*, *adc* from *C. acetobutyricum*, *adh* from *C. beijerinckii* produced 26.5 mg/*l* of isopropanol in 9 days (Kusakabe et al. 2013).

On the other hand, the engineered *E. coli* overexpressing *thl* gene from *C. acetobutyricum*, and *atoAD* encoding acetoacetyl-CoA transferase from *E. coli*, *adc* from *C. acetobutyricum*, *adh* from *C. beijerinckii* produced 4.9 g/*l* of isopropanol (Hanai et al. 2007). The performance could be improved by the fed-batch culture by removing and recovering isopropanol from the culture broth *in situ*, yielding 143 g/*l* of isopropanol production in 10 days (Inokuma et al. 2010) (see also chapter 6).

### 4.2.5 Lactic acid production

Lactic acid has been used in the food and pharmaceutical industries and for biodegradable polymers, and this can be produced in *Synechosystis* after heterologous expression of a lactate dehydrogenase (LDH), where 50 mg/*l* of L-lactate could be produced by the engineered *S. elongatus* PCC7942 overexpressing LDH and a transporter for exporting hydrophilic products (Niederholtmeyer et al. 2010), and 180 mg/*l* of L-lactate could be produced by the engineered *Synechocystis* sp. PCC6803 over-expressing

L-LDH and soluble transhydrogenase (Fig. 7) (Angermayr et al. 2012, 2014). Since Pyk is inhibited under light condition in *Synechosystis*, heterologous expression of Pyk can enhance the pyruvate production, and in turn enhance the lactate production, where Pyk is allosterically activated by fructose 1,6-bisphosphate (FBP) in the case of PykF transferred form *E. coli*, while the original Pyk does not show such characteristics (Angermayr et al. 2014). Moreover, Ppc may be disrupted to direct the carbon flow from PEP towards lactate production via PYR, but the cell growth is repressed, since Ppc is also important pathway for $CO_2$ fixation (Angermayr et al. 2014). In most bacteria, LDH requires cofactor NADH, whereas NADPH is abundant in *Synechosystis*, and thus NADPH dependent LDH may increase the lactate production, where this may be partly attained by introducing the LDH from *Bacillus subtilis*, of which LDH co-utilizes NADH and NADPH, with preference for NADH (Angermayr et al. 2014, Richter et al. 2011).

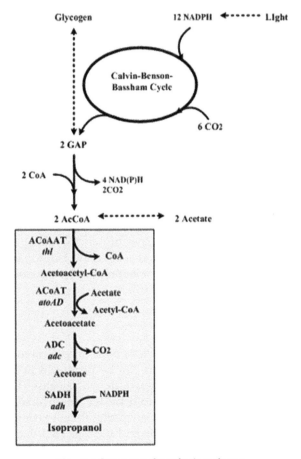

**Figure 6.** Isopropanol synthetic pathway.

Lactate has two optical isomers such as L- and D-lactate, where D-lactate is an essential monomer for the production of thermo-stable polylactide (PLA). PLA is a renewable, biodegradable, and therefore environmentally friendly polymer with many applications such as biodegradable plastics and the artificial biological tissue materials (Okano et al. 2010). By considering the difficulty in separating optical isomers, it is critical to produce optically pure lactate. *Synechocystis* sp. PCC6803 is one of the few microorganisms containing only D-LDH. 1.14 g/l of D-lactate could be produced by the engineered *Synechocystis* sp. PCC6803 overexpressing a D-LDH and a soluble transhydrogenase, where D-lactate production could be further increased to 2.17 g/l when 15 mM of acetate was added (Varman et al. 2013), indicating the limitation of AcCoA. Cyanobacteria convert $CO_2$ into carbohydrates during photosynthesis for the formation of biomass or storage of carbohydrate as glycogen (Stal and Moezelaar 1997). The AcCoA formed from pyruvate is channeled to TCA cycle, and others such as fatty acid-, acetate-, and poly 3-hydroxybutyrate (PHB) pathways, where TCA

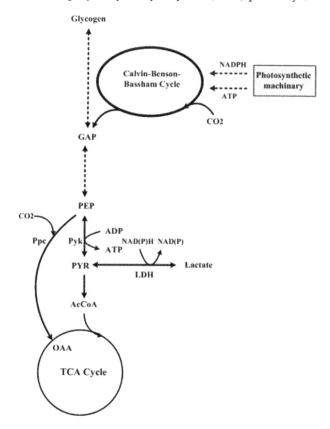

**Figure 7.** D-lactate production by cyanobacteria.

cycle and fatty acid synthesis are essential for the cell growth (Liu et al. 2011a, Stanier and Cohenbazier 1977). PHB can be accumulated up to 40% of its dry cell weight under stress conditions (Asada et al. 1999, Schlebusch and Forchhammer 2010). With this in mind, 1.06 g/*l* of D-lactate could be produced from $CO_2$ by the engineered *Synechocystis* sp. PCC6803, where two competitive pathways from AcCoA such as the native PHB and acetate forming pathways were disrupted, and a more efficient D-LDH was introduced (Fig. 8) (Zhou et al. 2014).

Glycogen allows cyanobacteria to cope with transient starvation as in heterotrophic microbes, where glycogen accumulates when enough carbon source is available but the cell growth is depressed by the lack of other nutrients such as nitrogen or phosphate limitations. Glycogen is used as a substrate during darkness in cyanobacteria. Deletion of the glycogen synthetic pathway gene *glgC* encoding glucose 1-phosphate adenylyl transferase loses the ability to synthesize glycogen in *Synechocystis*. The cyanobacterial cells incapable of synthesizing glycogen are encountered to the dilemma that excess carbon cannot flow into glycogen synthesis, and instead show overflow or energy spilling metabolism by producing pyruvate and *a*KG (Grundel et al. 2012). This pyruvate accumulation can be utilized for the metabolite production originated from pyruvate such as lactate, where *Synechocystis* sp. PCC6803 mutant lacking *glgC* (*ΔglgC*) containing a codon-optimized *ldh* of *Lactococcus lactis* could produce lactate under nitrogen limitation (van der Woude 2014). On the other hand, the deletion of PHB synthetic pathway gene (*ΔphaA*) and introduction of the above *ldh* gene do not allow the cell to produce lactate, where AcCoA accumulates but not contribute to the formation of pyruvate (van der Woude 2014).

### 4.2.6 Acetone production

Ketones are the compounds having a carbonyl group, which play important roles in synthetic chemistry. Acetone is the simplest ketone, and is widely used as a solvent and the precursor of isopropanol. Acetone is typically produced by *Clostridia*, where multiple products such as acetate, ethanol, and butanol are also produced along with acetone. In the case of using hexose such as glucose as a carbon source, the yield is limited, and the redox imbalance becomes a problem. Acetone can be also produced in *Synechocystis* PCC6803 by introducing CoA transferase (CtfAB) and acetoacetate decarboxylase (Adc) from *C. acetobutylicum* for converting AcCoA to acetone (Fig. 5) (Zhou et al. 2012), where the PHB forming pathway genes *phaCE* and the acetate forming pathway gene *pta* were disrupted. Acetone production could be increased six-fold by blocking acetate synthetic pathway in *Synechocystis* PCC 6803, while D-lactate

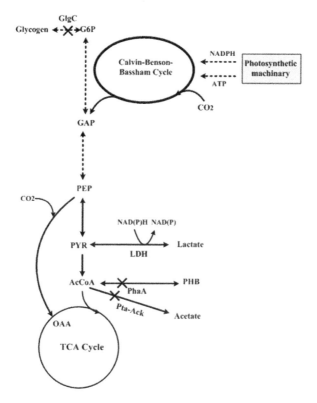

**Figure 8.** Lactate production by cyanobacteria by disrupting glycogen, acetate, and PHB synthetic pathways.

production could be increased two-fold by blocking acetate synthetic and PHB synthetic pathways in *Synechocystis* (Zhou et al. 2012).

### 4.2.7 Butanol production

The *n*-butanol production pathway of *Clostridia* from AcCoA has been incorporated into several other host organisms such as bacteria (Atsumi et al. 2008, Nielsen et al. 2009) and yeast (Steen et al. 2008, Krivoruchko et al. 2013). Since *Synechocystis* possesses PHB synthetic pathway composed of β-ketothiolase encoded by *phaA*, acetoacetyl-CoA reductase encoded by *phaB* (and PHB synthase encoded by *phaE*), additional heterologous enzymes such as enoyl-CoA hydratase (*phaJ*), trans-enoyl-CoA reductase (*ter*), and bifunctional aldehyde/alcohol dehydrogenase (*adhE2*) are required for *n*-butanol production. This may be attained by disrupting *phaE* for the deficient PHB synthesis, but use native *phaA* and *phaB* genes, and by incorporating *phaJ* from *A. caviae*, *ter* from *T. denticola*, *adhE2* from

*C. acetobutylicum* (Anfelt et al. 2015) (Fig. 9). Moreover, AcCoA pool size could be increased under nitrogen limitation, giving the specific butanol productivity to be increased, but without cell growth. The introduction of phosphoketolase increased AcCoA level six fold under nitrogen limitation, giving butanol titers from 27 to 37 mg/$l$ during 8 days, where the flux balance analysis indicates that a CBB cycle with phosphoketolase pathway showed higher theoretical butanol productivity than CBB-EMP pathway (Anfelt et al. 2015).

1-Butanol can be produced from AcCoA by engineered *S. elongatus* cultivated under anaerobic dark condition, where the metabolic engineering was made by introducing acetyl transferase (AtoB) from *E. coli*, 3-hydroxybutyryl-CoA dehydrogenase (Hbd) from *Clostridium acetobutylicum*, trans-2-enoyl-CoA reductase (Ter) from *Treponema denticola*, crotonase (Crt) from *C. acetobutylicum*, and bifunctional aldehyde/ alcohol dehydrogenase (AdhE2) from *C. acetobutylicum* (Lan and Liao 2011). 1-Butanol can be also produced by aerobic culture of *S. elongates* PCC7942, where condensation of AcCoA is made by consuming ATP with $CO_2$ evolution (Lan and Liao 2012), where ATP-dependent malonyl-CoA synthesis enzyme NphT7 was introduced, and NADH-dependent enzymes were replaced by NADPH dependent enzymes in this strain. Butanol tolerance of *Synechocystis* can be improved 1.5 times by evolution by gradually increasing butanol concentration from 0.2 to 0.5% (v/v) (Wang et al. 2014).

2-Methyl-1-butanol (2MB) can be produced in *S. elongatus* by expressing heterologous enzymes for citramalate pathway (Atsumi and Liao 2008) to direct pyruvate toward isoleucine synthesis pathway, where 2-keto-3-methyvalerate is converted to 2MB by Kivd and YqhD (Shen and Liao 2012).

As shown in Fig. 5, 2-ketoisovalerate can be converted to isobutyraldehyde by introducing ketoacid decarboxylase gene *kivd* from *Lactococcus lactis* into the genome of *Synechococcus elongatus* PCC7942 (Atsumi et al. 2009). The flux to 2-ketoisovalerate can be improved by introducing *alsS* gene from *Bacillus subtilis* and the *ilvC* and *ilvD* genes from *E. coli* into the chromosome of *S. elongatus*. Carbon fixation can be improved by over-expression of RubisCO genes to increase the productivity of isobutylaldehyde. Iso-butanol can then be produced by introducing alcohol dehydrogenase YqhD from *E. coli* (Atsumi et al. 2009).

### 4.2.8 Fatty acid and fatty alcohol production

The ideal biofuels may have the same or similar properties to those of petroleum fuels, which may be called as bio-petroleum. Namely, the ideal biofuels should be chemically similar to petroleum, such as fatty acid-based molecules including alcohols and fatty alka(e)nes (Keasling and Chou

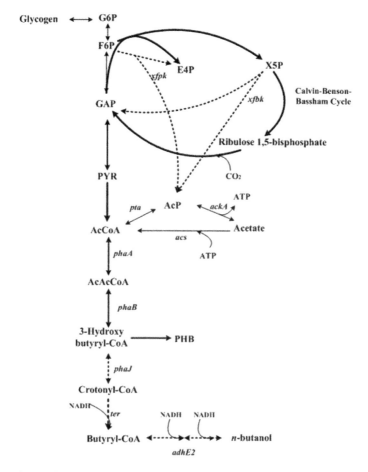

**Figure 9.** Butanol producing pathways in cyanobacteria (Anfelt et al. 2015).

2008). Alkanes with C4–C23 carbon chain length possess higher energy density, hydrophobic property and compatibility with existing liquid fuel infrastructure for gasoline, diesel, and jet fuels (Peralta-Yahya et al. 2012).

As shown in Fig. 10a, AcCoA is converted to malonyl-CoA by a multi-subunit AcCoA carboxylase consisting of AccA, AccB, AccC and AccD encoded by *accBCDA*, the rate-limiting step of the fatty acid synthesis (Davis et al. 2000). Fatty acid substrates as acyl chains of membrane lipids are synthesized by fatty acid synthase (FAS). Acyl-ACPs (acyl-acyl carrier proteins) synthesized by FAS can be incorporated to membrane lipids. Free fatty acids (FFAs) generated by lypolytic enzymes during degradation of membrane lipids can be also activated to acyl-ACPs by acyl-ACO synthetase (AAS, EC 6.2.1.20) (Kaczmarzky and Fulda 2010). The only AAS gene in *Synechocystis* sp. PCC 6803 is *slr1609* (Kaczmarzky and Fulda 2010). Therefore, *slr1609*-knockout mutant is incapable of importing exogeneous fatty acids and secreted fatty acids released from membrane lipids into the culture broth, which indicates the remarkable role of AAS in recycling the released fatty acids (Kaczmarzky and Fulda 2010). AAS also plays an essential role in alka(e)ne production, because *slr1609* deletion mutant of *Synechocystis* sp. PCC6803 showed significantly low production of alka(e) nes (Gao et al. 2012).

Fatty alcohol production can be significantly improved in *Synechocystis* sp. PCC6803 by over-expression of endogenous fatty acyl-ACP synthase gene *slr1609* (Fig. 10b) (Gao et al. 2012). Fatty alcohols and hydrocarbons can be produced from $CO_2$ in cyanobacteria such as *Synechocystis* sp. PCC6803, *Synechococcus* sp. PCC7942, *Anabaena* sp. PCC7120 by amplifying four *acc* genes encoding the subunits of AcCoA carboxylase, where *Synechocystis* gave the highest performance yielding the fatty alcohol production of 0.2 mg/*l* (including hexadecanol and octadecanol) and the productivity of hydrocarbons (including heptadecane and oentadecane, etc.) of 0.162 µg/OD.*l* (Tan et al. 2011). Heptadecane and pentadecane are the major constituents of alka(e)nes in *Synechocystis* PCC6803, and overexpression of alka(e)ne synthetic pathway genes such as *aar* (*sll0209*) encoding acyl-acyl carrier protein reductase and *ado* (*sll0208*) encoding aldehyde-deformylating oxygenase allows the alka(e)ne production of 26 mg/*l* (1.1% of DCW) in column photo-bioreactor (Wang et al. 2013).

Fatty alcohol production can be enhanced by heterologous expression of fatty acyl-CoA reductase gene *maqu 2220* from the marine bacterium *Marinobacter aquaeolei* VT8, and by deleting *sll0208* and *sll0209* genes encoding an acyl-ACP reductase, giving the fatty alcohol production of 2.87 mg/g DCW (Yao et al. 2014).

*E. coli* acyl-acyl carrier protein thioesterase (TE) (encoded by *tesA* gene) is normally a periplasmic enzyme, but *tesA* mutant redirects FAS to FFA secretion to the culture medium. Cyanobacterial FAS type II provides fatty acid substrates for membrane lipids. In FASII, long chain acyl-ACP molecules are important feedback inhibitors for FAS enzymes, such as ACC that catalyzes the conversion of AcCoA to malonyl-CoA, FabH which catalyzes the first step of FASII, and FabI which catalyzes the completion of acyl-ACP elongation (Heath and Rock 1996). Overexpression of TE reduces the acyl-ACP concentrations, thus stimulating the FAS flow by decreasing feedback inhibition (Magnuson et al. 1993). The *Synechocytis* sp. PCC6803 mutant strain lacking *pta*, *aas* (*slr1609*), PHB synthesis genes, and cyanophicin genes, and overexpressing *tesA*, *accBCDA*, *fadB1,B2* produced 197 mg/l of fatty acid (Liu et al. 2011).

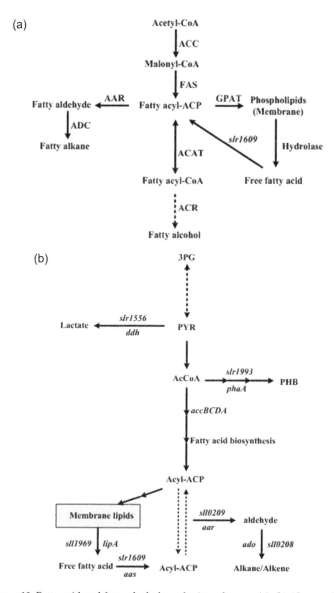

**Figure 10.** Fatty acid and fatty alcohol synthetic pathways: (a), (b) (Gao et al. 2012).

Alcohols such as hexadecanol and octadecanol can be produced by introducing fatty acyl-CoA reductase (*acr*) genes from jojoba, which catalyze a fatty-acyl-ACP (Tan et al. 2011).

## 4.2.9 Isoprene production

Isoprene ($C_5H_8$) is a volatile compound and utilized in the synthesis of rubber, and it is also a repeating unit of many natural products (isoprenoids) such as vitamin A and steroid hormons (Ruzicka 1953). Isoprene is naturally produced by plants (Sharkey and Yeh 2001), and by heterotrophic bacteria such as *Bacillus cereus, Pseudomonas aeruginosa,* and *E. coli* (Kuzma et al. 1995, Fischer et al. 2008, Stephanopoulos 2008). Isoprene is easily evaporated from the culture broth, and thus the toxicity to the cell can be relaxed by evaporation, where it can be trapped in the gas phase.

There are two pathways for isoprene production such as the mevalonic acid (MVA) pathway and the 2-C-methyl-D-erythritol 4-phosphate (MEP) pathway. The MVA pathway is active in archaea and in the cytosol of animals, while the MEP pathway is used in bacteria, algae, and plants (Lange et al. 2000, Lichtenthaler 2000). MEP pathway enzymes are mainly involved in the synthesis of photosynthetic pigments, while MVA pathway is not present in cyanobacteria (Pade et al. 2016). The isoprene synthesis starts from GAP and pyruvate in the glycolysis by 1-deoxy-D-xylulose 5-phosphate synthase (DXS) (Fig. 11), where DXS is the regulatory enzyme in the MEP pathway, and may become rate limiting, and thus it is the primary target for metabolic engineering (Davies et al. 2015). The MEP pathway produces two final products such as isopentenyl diphosphate and dimethylalyl diphosphate (DMAPP), where DMAPP serves as a precursor of carotenoid, the phytol of chlorophyll, and quinones (Nowicka and Kruk 2010). DMAPP also serves as a precursor for isoprene synthesis by isoprene synthase (IspS) in plants (Sharkey and Yeh 2001). Isoprene can be synthesized by *Synechocystis* sp. PCC6803 by introducing *ispS* gene from vine *Pueraria montana* with utilization of the naturally occurring MEP pathway (Fig. 11) (Lindberg et al. 2010).

The photosynthetic carbon is mainly converted to sugars (80–85%), while much less amount is converted to fatty acids (~ 10%) and terpenoids (3–5%) in *Synechocystis* (Lindberg et al. 2010), and thus the carbon partition from sugar to isoprene production is important. In particular, the pool sizes of precursors such as GAP and pyruvate are important for the efficient isoprene production instead of overexpression of IspS and DXS (Pade et al. 2016). Although expression cassette can be localized on chromosome (Lindberg et al. 2010, Bentley and Melis 2012, Bentley et al. 2014, Chaves et al. 2014), a plasmid-based expression may be also considered, where a codon-optimized plant *lspS* was expressed under the control of different *Synechocystis* promoters that give constitutive or light-regulated *ispS* expression (Pade et al. 2016). The MEP pathway enzymes may be inhibited by the intracellular isoprene, and therefore, isoprene production is higher in the open system than in the closed cultivation system. One way of circumventing the negative effect on MEP-pathway activity is to express

MVA pathway, where isoprene production could be increased by 2.5 fold higher in *Synechocystis* strains carrying *ispS* gene with MVA pathway as compared to the strains with only *ispS* gene (Bentley et al. 2014) (Table 1).

It is preferred to cultivate cyanobacteria in seawater to avoid competition with fresh water resources (Chisti 2013). Although *Synechocystis* is an euryhaline strain and can resist even up to two-fold seawater

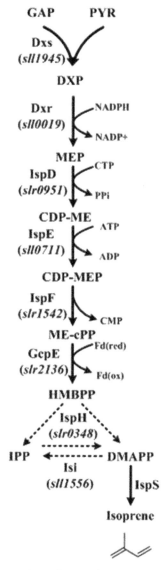

**Figure 11.** Isoprene producing pathway in cyanobacteria (Davies et al. 2015 or Lindberg et al. 2010).

concentration (Pade and Hagemann 2014), isoprene production declines with increased NaCl concentrations despite increased *ispS* expression, where glucosylglycerol allows *Synechocystis* cells to survive at enhanced salinities, accumulated with the increase in NaCl concentration (Pade et al. 2016).

### 4.2.10 Diols production

#### (a) 1,2-propanediol

The racemic 1,2-propanediol (1,2-PDO) is currently produced by petrochemical processes from petroleum-derived propylene oxide, and is used in antifreeze and heat transfer fluids, plasticizers and thermoset plastics, and cosmetics (Clomburg and Gonzalez 2011). Since petrochemical processes are energy intensive and not environmentally friendly, microbial production of 1,2-PDO from glycerol and sugars has been paid attention, where it is naturally produced by *Thermoanaerobacterium thermosaccharolyticum* (Cameron and Cooney 1986), and also by heterologous hosts such as *E. coli* (Altaras and Cameron 1999, Clomburg and Gonzalez 2011, Altaras and Cameron 2000), *Corynebacterium glutamicum* (Niimi et al. 2011), *S. cerevisiae* (Jeon et al. 2009, Jung et al. 2011), and *Pichia pastoris* (Barbier et al. 2011), where the titer of 4.5–6.5 g/*l* could be attained under anaerobic conditions. It is also attractive to produce 1,2-PDO directly from $CO_2$ by photosynthetic organisms. Cyanobacteria produces oxygen in the light reaction of photosynthesis, and do not perform fermentation under light conditions, where $CO_2$ is a more oxidized substrate than sugars or glycerol, and therefore, requires more energy and the reducing equivalent for the production of 1,2-PDO (Li and Liao 2013).

Under light conditions, cyanobacteria fix $CO_2$ via CBB cycle powered by ATP and NADPH generated by photosystems. Fructose 6-phosphate (F6P) in the CBB cycle is branched off to form glycogen for the carbon and energy storage, while glyceraldehydes 3-phosphate (GAP) is the branch point leaving CBB to glycolysis and then TCA cycle, where some of the glycolysis and TCA cycle intermediates are the precursors for the building blocks of the biomass. The synthesis of 1,2-PDO starts from dihydroxy acetone phosphate (DHAP), where DHAP is first converted to methylglyoxal by methylglyoxal synthase (encoded by *mgsA* in *E. coli*). Methyglyoxal is very toxic to the cell, and must be quickly flowed through to the downstream pathways. The 1,2-PDO can be synthesized by two different routes from methylglyoxal, where one route envolves the reduction of methylglyoxal by glycerol dehydrogenase (encoded by *gldA* in *E. coli*) to form lactaldehyde, which is further reduced to 1,2-PDO by 1,2-propanediol reductase (encoded by *fucO* in *E. coli*) (Fig. 12). The second route includes an alcohol dehydrogenase (encoded by *yqhD* in *E. coli*) to produce acetol, which is then converted to 1,2-PDO by *gldA* (secondary alcohol dehydrogenases (sADHs) from

*C. beijerinckii* and *T. brockii*) (Fig. 12) (Li and Liao 2013). The second route is preferred for cyanobacteria due to NADPH dependent pathway. The genetically engineered cyanobacterium *Synechococcus elongatus* PCC 7942 with heterologous genes such as *mgsA*, *gldA*, and *yqhD* from *E. coli* produced about 22 mg/*l* of 1,2-PDO, where the intermediate such as acetol was also accumulated in the stationary phase (Li and Liao 2013). The introduction of NADPH-specific secondary alcohol dehydrogenase from *C. beijerinckii* and *T. brockii* resulted in about 150 mg/*l* of 1,2-PDO, by reducing the acetol accumulation due to NADPH availability instead of NADH dependent (Li and Liao 2013).

## (b) 1,3-Propanediol

As mentioned above, F6P in the CBB cycle is used for the glycogen synthesis for carbon and energy storage, where the glycogen forming rate reaches 0.29 g/*l*.d in *Arthrospira plantensis* (Aikawa et al. 2012) and 0.5 g/*l*.d in *Synechococcus* sp. PCC 7002 (Aikawa et al. 2014). These values are

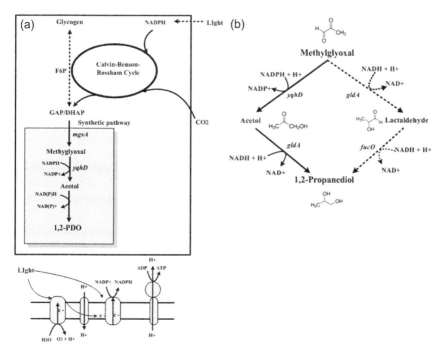

**Figure 12.** 1,2 Propanediol producing pathways in cyanobacteria (Li and Liao 2013).

comparable to the production rate of chemicals by engineered cyanobacteria (Savakis and Hellingwerf 2015).

The **1,3-propanediol (1,3-PDO)** is used for the synthesis of polymethylene telephthalate (PTT) having fine elasticity and anti-fouling properties, and having been used commercially in the textile and carpet industries (Saxena et al. 2009, Liu et al. 2010). The 1,3-PDO is naturally produced by *Klebsiella pneumonia* (Seo et al. 2009), *Citrobacter freundii* (Ainala et al. 2013), and *Clostridium acetobutyricum* (Gungormusler et al. 2010). About 129 g/l (1.7 M) of 1,3-PDO could be produced by engineered *E. coli* by introducing the 1,3-PDO synthetic pathway genes (Emptage et al. 2003). For 1,3-PDO production in cyanobacteria, DHAP in the CBB cycle is first converted to glycerol via glycerol 3-phosphate, and the glycerol is subsequently converted to 1,3-PDO via 3-hydroxypropionaldehyde (3-HPA), where the conversion of glycerol to 3-HPA is catalyzed by the coenzyme $B_{12}$-dependent glycerol dehydratase (Fig. 13). This glycerol hydratase works in *S. elongatus* PCC 7942 without addition of vitamin $B_{12}$, and the highest titer of 1,3-PDO was 288 mg/l (3.79 mM) with 1.16 g/l (12.62 mM) of glycerol produced in 14 days (Horikawa et al. 2016).

## (c) 2,3-butanediol

**2,3-butanediol (2,3-BDO)** has been used in the manufacturing of plasticizers, inks, fumigants, and explosives (Syu 2001). 2,3-BDO can be converted by dehydration to methyl ethyl keton, which is a liquefied fuel additive and useful industrial solvent (Tran and Chambers 1987). 2,3-BDO can be also catalytically converted to 1,3-butadien, which is a precursor for polymer and co-polymer materials (van Haveren et al. 2008).

The synthesis of 2,3-BDO starts from pyruvate, where it is naturally produced through the fixation of three $CO_2$ molecules in the CBB cycle in cyanobacteria. Two pyruvate molecules are converted to 2-acetolactate by acetolactate synthase (ALS) encoded by *alsS*. 2-Acetolactate is then decarboxylated to yield acetoin by 2-acetolactate decarboxylase (ALDC) encoded by *alsD*, where ALDC must be $O_2$ insensitive. Acetoin inhibits the cell growth of *S. elongatus*, where 1 g/l of acetoin prevents the cell growth (Oliver et al. 2013). Acetoin can be reduced to 2,3-BDO by an NADPH-dependent sADH, where the reduction of acetoin by sADH is a diastereoselective reaction allowing to choose enzymes for either *R* or *S* stereocenter (Fig. 14). The engineered *Synecococcus elongatus* sp. PCC 7942 containing *alsS* from *B. subtilis*, *alsD* from *Aeromonas hydrophila*, and *adh* from *C. beijerinckii* produced 2.38 g/l (*R*/*R*)-2,3-BDO in 21 days with the productivity of 9.847 mg/l.h (Oliver et al. 2013).

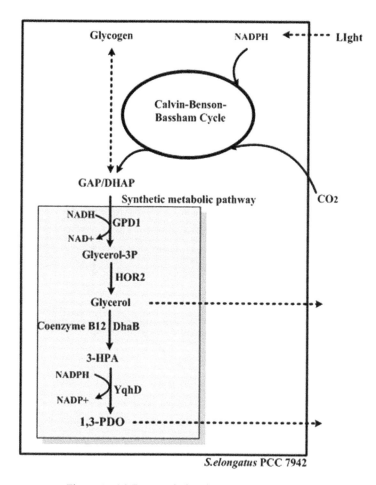

**Figure 13.** 1,3-Propanediol pathway in cyanobacteria.

### 4.2.11 PHB production

In addition to glycogen, poly-3-hydroxyalkanoate (PHA) is another storage carbon source in cyanobacteria (Panda and Mallick 2007, Tsang et al. 2013, Miyake et al. 2000, Wu et al. 2002), where poly-3-hydroxybutyrate (PHB) is the most common PHA that can be used for renewable plastic production. It is naturally synthesized as energy and carbon reserves in many bacteria. PHB can be accumulated up to 40% of the dry cell weight of *Synechocystis* sp. PCC6803 under stress conditions such as nitrogen and phosphate limiting conditions (Asada et al. 1999, Schlebusch and Forchhammer 2010).

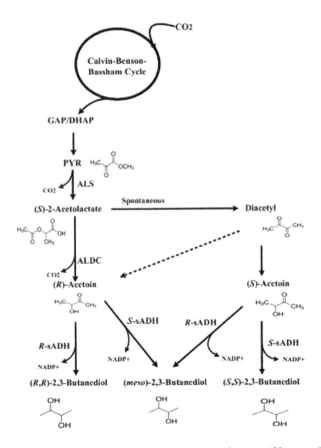

**Figure 14.** 2,3-Butanediol producing pathways in cyanobacteria (Oliver et al. 2013).

Therefore, by blocking PHB synthetic pathway of *Synechocystis* sp. PCC6803, and by cultivation under stress conditions, carbon flux can be rerouted to the metabolites such as acetone (Zhou et al. 2012), 3-hydroxybutyrate (Wang et al. 2013), D-lactate (Zhou et al. 2014), and butanol (Anfelt et al. 2015), whereas L-lactate production could not be increased by this approach (van der Woude et al. 2014).

The cell growth is depressed, while the carbon storage is enhanced under stress conditions such as dark, salt stress, nitrogen and phosphate limiting conditions (Ducat et al. 2012, Asada et al. 1999, Schlebusch and Forchhammer 2010). Since the cell growth is not inhibited by blocking the glycogen synthetic pathway under normal growth conditions (Zhou et al. 2014, Grundel et al. 2012), a two-stage fermentation strategy may be considered, where the cell growth can be enhanced by blocking the carbon storage at normal condition in the first phase, and then switched to the stress condition to increase the target metabolites as mentioned above (Zhou et al. 2016).

The PHB can be produced by *Rhodovulum sulfidophilum* P5 using inexpensive nitrogen and carbon sources (Haase et al. 2012). PHB can be also produced in the filamentous cyanobacterium *Nostoc muscorum* under phosphate limitation by recombinant *Synechocystis* sp. PCC6803 (Tyo et al. 2009).

The PHB synthesis in *Synechocystis* sp. PCC6803 is made by three steps, where AcCoA is first converted to acetoacetyl-CoA (AcAcCoA) by $\beta$-ketothiolase (*slr1993*), and then AcAcCoA is reduced to D-3-hydroxybutyryl-CoA by acetoacethyl-CoA reductase (*slr1994*) (Taroncher-Oldenburg et al. 2000). PHB is then formed by polymerization catalyzed by PHB synthase, which consists of a heterodimer of *phaC* (*slr1829*) and *phaE* (*slr1830*) (Hein et al. 1998). The genes for the two subunits of PHB synthase are organized into one operon.

Under nitrogen limitation or starvation, the transcription factor NtcA and P II-signaling protein play important roles for adaptation to nitrogen starvation. The *slr0873* gene is highly induced under nitrogen limitation. Most of the heterotrophic bacteria harboring this gene cluster can fix nitrogen to produce PHB, where only *Synechocyctis* sp. PCC 6803 can accumulate PHB among cyanobacteria (Schlebusch and Forchhammer 2010). This indicates that *slr0873* gene mutant cannot produce PHB (Schlebusch and Forchhammer 2010).

The direct application of PHB is limited in practice due to its amorphous, elastic, and viscous properties. The incorporation of 4-hydroxybutyrate (4HB) into PHB generates a co-polymer such as poly-3-hydroxybutyrate-co-4-hydroxybutyrate (P3HB-co-4HB), which has improved thermal and mechanical properties (Li et al. 2010). Depending on the co-polymer composition, the physical properties such as crystalline, plastic, and elastic nature can be changed.

As mentioned previously, succinic semialdehyde is an intermediate of the TCA cycle in cyanobacteria (Zhang and Bryant 2011), where succinic semialdehyde can be converted to 4 hydroxybutyryl-CoA (4HB-CoA) by 4 hydroxybutyrate dehydrogenase and 4 hydroxybutyryl-CoA transferase, which may be obtained from *Porphyromonas gingivalis* (Yim et al. 2011). Namely, the genes encoding 2-oxoglutarate decarboxylase (*ogdA*), 4-hydroxybutyrate dehydrogenase (*gbd1*), and 4-hydroxybutyryl-CoA transferase (*cat2*) allow the synthesis of 4HB-CoA (Xu et al. 2011), where poly (3-HB-co-4HB) can be obtained by the coordinated expression of *phaA*, *phaB*, and *phaEC* (Fig. 15). Moreover, LysR-type transcription factor (TF), CcmR controls the genes encoding the carbon concentration mechanism by connecting central metabolism to inorganic carbon aquisition for photosynthesis (Daley et al. 2012). Therefore, the deletion of *ccmR* gene

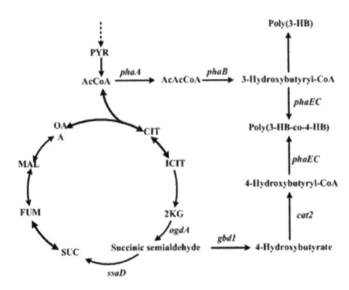

**Figure 15.** Poly (3HB) and poly(3HB-co-4HB) producing pathways in cyanobacteria (Zhang et al. 2015).

improves carbon fixation and thus increases the AcCoA pool, resulting in higher PHB and poly (3HB-co-4HB) production (Zhang et al. 2015).

### 4.2.12 3-hydroxybutyrate production

Intensive research has been conducted on microbial production of PHB, while other attempts have been made to produce PHAs with improved properties such as poly (3-hydroxybutyrate-co-4-hydroxybutyrate) (Doi et al. 1990, Li et al. 2010), poly (3-hydroxybutyrate-co-3-hydroxyvalerate) (Aldor et al. 2002, Chen et al. 2011, Slater et al. 1992), and poly (3-hydroxybutyrate-co-lactate) (Jung et al. 2010). It is, however, difficult to control the monomer composition of the co-polymers, and the carbohydrates used for energy and carbon sources are relatively expensive (Byrom 1987). Moreover, it is energy intensive and costly to crack the cells for PHA extraction (Chisti 2007, Liu and Curtiss 2009).

In order to avoid the cells to accumulate PHAs inside the cells as insoluble granule, it may be considered to block PHA synthesis pathway, but produce hydroxyalkanoate (HAs) (Chen and Wu 2005, Ren et al. 2010), where HAs are the monomers and can be secreted into the culture broth without damage to the cells, which may reduce the cost for product recovery. Moreover, HAs thus obtained can be chemo-catalytically polymerized with other monomers to synthesize PHAs (Tokiwa and Ugwu 2007). (*R*)- or (*S*)-

3HA can also serve as a precursor for stereo-specific fine chemicals (Chen and Wu 2005, Ren et al. 2010, Tokiwa and Ugwu 2007, Tseng et al. 2009).

3-Hydroxybutyrate (3HB) is a common HA and can be produced by engineered *E. coli* (Gao et al. 2002, Lee and Lee 2003, Lee et al. 2008, Liu et al. 2007, Tseng et al. 2009). 3HB is formed from AcCoA by three steps, where two molecules of AcCoA are condensed to produce acetoacetyl-CoA (AcAcCoA) by a thiolase in the first step. AcAcCoA is then reduced to (R)- or (S)-3-hydroxybutyryl-CoA (3HB-CoA) by an acetoacetyl-CoA reductase (Liu et al. 2007, Tseng et al. 2009), where this reaction requires NAD(P)H. There are two types of acetoacetyl-CoA reductase, where the enzyme encoded by *phaB* from *Ralstonia eutropha* reduces AcAcCoA to (R)-3-hydroxybutyryl-CoA, which is eventually converted to (R)-3HB, while the enzyme encoded by *hbd* from *Clostridium acetobutyricul* ATCC 824 reduces AcAcCoA to (S)-3-hydroxybutyryl-CoA, which is eventually converted to (S)-3HB (Fig. 16). *E. coli* thioesterase II encoded by *tesB* can cleave off CoA from both (R)- and (S)-3-hydroxybutyryl-CoA, giving (R)- and (S)-3HB, respectively (Liu et al. 2007, Tseng et al. 2009).

In *Synechocystis* PCC 6803, the native *slr1993* (*phaA2*)-*slr1994* (*phaB2*) operon (Taroncher-Oldenberg et al. 2000) is responsible for the production of (R)-3-hydroxybutyryl-CoA, which is in turn converted to PHB by PHB polymerase encoded by *phaEC* (Hein et al. 1998). As mentioned previously, the efficient PHB production may be made by two-step cultivation, where the culture condition may be switched to nitrogen or phosphate limitation from normal condition (Panda and Mallick 2007, Takahashi et al. 1998, Tyo et al. 2009, Wu et al. 2001). Moreover, the PHB synthase is under complex regulations (Evaggelos et al. 2010). (S)- and (R)-3HB can be photosynthetically produced directly from $CO_2$ and sunlight by *Synechocystis* PCC 6803 by inactivating *slr1829* and *slr1803* encoding PHB polymerase, where 533.4 mg/l of 3HB could be produced in 21 days (Wang et al. 2013).

### 4.2.13 Other metabolite production

Alpha-olefin production can be also made by *Synechococcus* sp. PCC7008 (Mendez-Perez et al. 2011).

The pigment sesquiterpene *β*-caryophyllene can be produced by *Synechocystis* sp. PCC6803 with the aid of a *β*-caryophyllene synthase gene from *Artemisia annua* (Leema et al. 2010), where this compound is used in the fragrance and cosmetic industry and in natural remedies for its anti-inflammatory and anti-microbial properties.

Cyanobacteria can excrete fructose, lactate, and glucose by introducing transport genes from *E. coli* (Datta and Henlry 2006). Many cyanobacteria naturally produce sucrose as an osmotic response to their saline habitat together with manipulation of transport and secretion genes (Ducat et

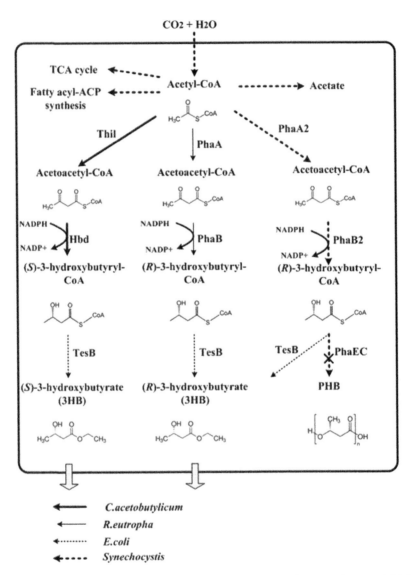

**Figure 16.** 3-Hydroxybutyrate (3HB) synthetic pathways in cyanobacteria (Wang et al. 2013).

al. 2012). The *agp* gene encoding the ADP-glucose pyrophosphorylase is involved in glycogen synthesis and glucosylglycerol formation in cyanobacteria. Partial deletion of *agp* gene prevents the synthesis of glycogen or osmoprotective glucosylglycerol, and allows the production of sucrose with tolerance up to 0.9 M of salt (Miao et al. 2003).

Glycogen can be produced in halophilic bacterium *Arthosporia palentensis* by manipulating growth condition (Aikawa et al. 2012).

Even ammonia can be produced by nitrogen-fixing cyanobacteria *Anabaena* sp. ATCC 33047 (Razon et al. 2012).

### 4.3 Efficient $CO_2$ fixation

Since $CO_2$ fixation is attractive from the environmental protection point of view, several strategies have been considered for the efficient carbon fixation for the cell synthesis (Ducat and Silver 2012), where carbon fixation can be enhanced by amplifying carboxysomes expression (Savage et al. 2010) or heterologous expression of Rubisco gene *rbcLS* (Atsumi et al. 2009). The $CO_2$ fixation can be also enhanced by the hybrid RubisCO, which contains both plant and microalgae subunits (Genkov et al. 2010).

Unlike heterotrophic microorganisms, the major problem toward cyanofactories is the low $CO_2$ fixation rate, where RubisCO catalyzes the carboxylation reaction of ribulose 1,5-bisphosphate (RuBP) using $CO_2$, and initiates Carvin (CBB) cycle in the photosynthetic organisms (Zhou et al. 2016). Many attempts have been made to engineer RubisCO to increase $CO_2$-fixation process by structure-function relationships (Zhou et al. 2016). The direct evolution of RubisCO has been made using phosphoribulokinase (PRK)-expressing *E. coli* (Mueller-Cajar et al. 2007, Parikh et al. 2006, Greene et al. 2007), and carboxylation efficiency of RubisCO could be improved by the appropriate selection system (Cai et al. 2014, Durao et al. 2015).

RubisCO is subject to low carboxylation efficiency and competitive inhibition of oxygen. The carboxylation reaction of RubisCO is confined in the carboxysome of cyanobacteria to avoid the competing oxygenase reaction (Zarzycki et al. 2013, Burnap et al. 2015). As for carboxylation, $CO_2$ is first transported to plasma membrane in the form of bicarbonate by bicarbonate transporter, and then converted to $CO_2$ by carboxylic anhydrase (CA), where $CO_2$ is finally carboxylated by RubisCO in the carboxysome (Zarzycki et al. 2013). This indicates that overexpression or enhancement of both RubisCO and CA in carboxysome may contribute to the increase in $CO_2$ fixation in cyanobacteria (Zhou et al. 2016). In fact, introduction of an extra bicarbonate transporter resulted in the increased cell growth (Kamennaya et al. 2015), and also the overexpression of CA resulted in the increase in the heterotrophic $CO_2$ fixation in *E. coli* (Gong et al. 2015).

Once again, one of the drawbacks of using autotrophic microorganisms is their slow growth rate, irrespective of the advantage of assimilating $CO_2$ using energy obtained from light, hydrogen, and/or sulfur (Gong et al. 2015). Therefore, one idea is to incorporate $CO_2$-fixing enzymes of the cyanobacterial Calvin (CBB) cycle into heterotrophic microorganisms, where two sequential enzymes of cyanobacterial Calvin cycle were incorporated into *E. coli*, giving 13% increase in the $CO_2$ fixation. Moreover, this value was further increased to 17% by incorporating CA, indicating that low intracellular $CO_2$ concentration is one of the limiting factors for $CO_2$ fixation in *E. coli* (Gong et al. 2015).

Another important factor for $CO_2$ fixation is the photorespiration, which gives about 25% loss of the fixed carbon (Sharkey 1988). Although reduction or elimination of photorespiration might contribute to improve the photosynthetic carbon fixation, it is not so easy, because the photorespiration plays the important role to protect the photosynthetic organisms from photoinhibition (Kozaki and Takabe 1996). Nonetheless, introduction of glycolate catabolic pathway of *E. coli* into chloroplasts of *Arabidopsis thaliana* reduced the photorespiration, and resulted in the improvement of biomass formation (Kebeish et al. 2007).

The photosynthetically fixed carbon is usually used either for biomass formation or stored as glycogen (Stal and Moezelaar 1997), where the impaired glycogen synthesis does not affect the growth of *Synechocystis* sp. PCC6803 under light condition (Grundel et al. 2012). The impaired glycogen synthesis causes overflow metabolism, producing pyruvate and $\alpha$KG under nitrogen limitation (Grundel et al. 2012). This implies that the production of pyruvate-dependent and $\alpha$KG-dependent metabolites can be improved, where lactate production could in fact be increased by about two fold in glycogen storage mutant *Synechocystis* sp. PCC6803 cultivated under nitrogen limitation (van der Woude et al. 2014).

Moreover, sucrose secretion could be increased by blocking glycogen synthesis or enhancing glycogen breakdown under salt stress conditions (Ducat et al. 2012).

## 5. Systems Biology Approach and Modeling of the Metabolism

It is critical to properly understand the metabolism in response to culture environment, where the systems biology approaches including metabolite profiling (Veyel et al. 2014), and integration of different levels of information such as transcript and protein abundances, metabolites, and fluxes are useful (Mettler et al. 2014, Yang et al. 2002c).

Although algae and cyanobacteria have been paid recent attention for the potential application to the sustainable biosynthesis, unknown and uncharacterized gene and protein functions hamper the progress toward the future era of algae industrialization and cyanofactories. The systems biology approach plays crucial role for functional analysis of the metabolic systems based on the database with proper metabolic modeling (Reijnders et al. 2014, Dersch et al. 2016). Initial attempts have been made for the modeling of photosynthetic organisms (Poolman et al. 2000, Poolman et al. 2001), while mechanistic model of photosynthesis in microalgae has also been developed (Garcia-Camacho et al. 2012, Camacho-Rubio et al. 2003). The sequential statistical analysis based on experimental design coupled with least squares multiple regression has been made to analyze the dependence of respiratory and photosynthetic responses upon concomitant modulation of light intensity as well as acetate, $CO_2$, nitrate and ammonia concentrations for the culture of *C. reinhardtii* (Gerin et al. 2014).

Metabolic flux analysis (MFA) may be considered to gain insight into the metabolism, where the optimal light intensity can be identified for the biomass yield of *C. reinhardtii* by considering the cell maintenance and biomass formation (Kliphuis et al. 2012). A mixed integer linear programming method was used to find the optimal flux distributions of *C. reinhardtii* cultivated under photoautotrophic conditions in photo-bioreactors functioning in physical light limitation based on the constraint-based model, which includes thermodynamic and energetic constraints on the functioning metabolism, highlighting the existence of a light-driven respiration depending on the incident photon flux density (Cogne et al. 2011).

Flux balance analysis (FBA) based on the network consisting of 484 metabolic reactions and 458 intracellular metabolites for *C. reinhardtii* indicates that aerobic heterotrophic growth on acetate has a low yield on carbon, while mixotrophically and autotrophically grown cells are significantly more efficient (Boyle and Morgan 2009). A genome-scale extension for *C. reinhardtii* has been made with the network consisting of 1080 genes, associated with 2190 reactions and 1068 metabolites (named *i*RC1080), that enables quantitative growth prediction for a given light source, resolving wavelength and photon flux. This offers insight into algae metabolism and potential for genetic engineering and efficient light source design (Chang et al. 2011). Another comprehensive literature-based genome-scale model with the network of 866 ORFs, 1862 metabolites, 2249 gene-enzyme-reaction-association entries, and 1725 reactions has been developed (named AlgaGEM), where it can predict the observable metabolic effects under autotrophic, heterotrophic, and mixotrophic conditions, and predicts increased hydrogen production when cyclic electron flow is disrupted, and the physiological pathway for $H_2$ production, which identified new targets for further improvement of $H_2$ yield (de Oliveira Dal'Molin et al. 2011).

FBA approach has also been employed for cyanobacteria with the emphasis on the alleged glyoxylate shunt and the role of photo-respiration in cellular growth, and analyzed the diurnal light/dark cycles of the metabolism (Knoop et al. 2013). Geneome-scale metabolic model of *Synechococcus elongates* PCC7942 (named *i*Syf715) has also been developed with the network of 851 reactions and 838 metabolites, and the applicability has been demonstrated for autotrophic growth conditions (Triana et al. 2014).

## 6. Cultivation Methods

Microalgae can use sunlight more efficiently than other crop plants to produce oil (Carlozzi et al. 2010). The oil production capacity is almost one or two times higher than any other crop (Chen et al. 2011). The open pond system is better for large scale production. However, the main disadvantages of open pond systems are that the water is lost by evaporation at a rate similar to land crops, and such systems are susceptible to contamination due to open to the atmosphere (Schenk et al. 2008, Lam and Lee 2012). Some protozoa may contaminate the system, and hamper the growth of microalgae.

The effective culture system may have to satisfy the following criteria: (1) effective illumination area, (2) optimal gas-liquid transfer, (3) easy to operate, (4) low contamination, (5) low capital and production cost, and (6) minimal land area requirement (Xu et al. 2009). The system can be made of paddle wheel to avoid microalgae biomass sedimentation, and $CO_2$ may be sparged at the bottom of the raceway as carbon source (Stephenson et al. 2010).

Closed photobioreactor may be considered to overcome the limitations encountered in raceway pond (Lam and Lee 2012). There are several closed photobioreactor systems such as air-lift tubular, flat plate and vertical-column reactors, where the design parameters may be nutrient levels, temperature, amount of inlet $CO_2$, etc. (Rosenberg et al. 2008). The reactor permits selective culture strain, in which optimum growth condition can be maintained to give high biomass and lipid productivities. The tubuler photobioreactor may be one of the most typical cultivation apparatus, where a vertical tubuler photobioreactor can increase the residence time of sparged gas, giving higher $CO_2$ utilization efficiency (Ono and Cuello 2004). The higher intensity of light cannot reach to most of the cell in the large-scale photobioreactor. As a consequence, the metabolism also changes from light to dark condition. This phenomenon is undesirable for large scale production.

## 7. Harvesting of Algal Biomass

The microalgae need to be separated from water to recover their biomass for downstream processing. There are several methods for microalgae harvest: (1) bulk harvesting to separate microalgae from suspension such as natural gravity sedimentation, flocculation, and floatation, and (2) thickening to concentrate the microalgae slurry after bulk harvesting such as centrifugation and filtration (Chen et al. 2011).

Flocculation, the aggregation and sedimentation of algal biomass, is also a very common harvesting method used to concentrate algae. Algal strains can be also engineered such that the addition of a polymer or a change in the environmental variables triggers flocculation (Christenson and Sims 2011, Wu et al. 2012). Conventional flocculation method, however, poses several disadvantages: (1) high dosage of multivalent slat is required to achieve satisfactory result, (2) it produces large quantity of sludge that increases the difficulty to dehydrate the biomass, and (3) flocculation efficiency is highly dependent on pH level (Chen et al. 2011, Renault et al. 2009). By introduction of coagulant that is positively charged into the culture medium, the negative charge surrounding the microalgae cells will be neutralized. At the same time, flocculant can be added to promote agglomeration by creating bridges between the neutralized cells to become dense flocs and settle down due to natural gravity (de Godos et al. 2011). Another possible method to harvest microalgae is through immobilization, in which microalgae are embedded in an entrapment matrix, and continuously grow within the matrix. Only alginate gel entrapment method is feasible to immobilize microalgae so far (Moreno-Garrido 2008). Some of the advantages of using alginate gel are the requirement of only mild condition during immobilization process with negligible toxicity and high transparency (Moreno-Garrido 2008, Moreira et al. 2006). Immobilized microalgae beads can be applied in diverse research areas such as for high-value product synthesis, organic pollutant removal, heavy metal removal, and toxicity measurement (biosensor) (Moreira et al. 2006, Ertugrul et al. 2008, Guedri et al. 2008, Mallick 2006, Perez-Martinez et al. 2010, Ruiz-Marin and Mendoza-Espinosa 2008, Ruiz-Marin et al. 2010). A few issues need to be addressed in immobilization of microalgae before the process can be upgraded to commercialization stage such as (1) stability, (2) leakage of microalgae cells, and (3) mass transfer limitation.

## 8. Downstream Processing

Effective lipid extraction is required particularly for microalgae with low lipid content, since the loss of the lipid during extraction process brings a serious problem for the production cost of microalgae biofuels (Ranjan et al. 2010). The energy consumed in lipid extraction from dried microalgae biomass is a relatively small portion to the overall energy (Stephenson et al. 2010, Sander and Murthy 2010). As for cell disruption, microwave application, sonication, bead beating, autoclaving (Rawat et al. 2011, Amaro et al. 2011), grinding, osmotic shock, homogenization, freeze drying (Rawat et al. 2011), and 10% (w/v) NaCl addition (Amaro et al. 2011, Lee et al. 2010) may be considered.

The solvent must be inexpensive, non-toxic, volatile, non-polar, and it must selectively extract the lipid of the cell (Rawat et al. 2011). The potential of using co-solvent mixtures of ionic liquids and polar covalent molecules has been shown for lipid extraction (Young et al. 2010). The Soxhlet extraction method uses hexane, while the Bligh and Dyer's method uses mixture of chloroform and methanol as solvents to extract lipids (Kim et al. 2012). The other solvents include benzene and ether, but hexane has gained more popularity as a chemical for solvent extraction, and it is also relatively inexpensive (Pragya et al. 2013). Although *n*-hexane is widely used to extract oil from various seed crops, it is inefficient to extract microalgae lipid. This is because microalgae lipid contains high concentration of unsaturated fatty acid, while *n*-hexane is a nonpolar solvent. Thus the selectivity of lipid towards the solvent is not high (Ranjan et al. 2010). Methanol and *n*-hexane are not sustainable, since both solvents are conventionally derived from non-renewable fossil fuels. On the other hand, ethanol is a greener solvent, since it has low toxicity level and can be derived from renewable source such as sugar-based plant and lignocellulosic material (Lam and Lee 2012). The ethanol, however, gives low extraction efficiency. Ultrasonication and microwave can be also used for cell disruption. The cell wall-less mutant is better for oil extraction.

Several super critical fluids are $CO_2$, ethane, methanol, ethanol, benzene, toluene and water (Mendes et al. 2003, Sawangkeaw et al. 2010). The basic principle of this technology is by achieving a certain phase (supercritical) that is beyond the critical point of a fluid, in which meniscus separating the liquid and vapor phase disappears, leaving only a single homogeneous phase (Sawangkeaw et al. 2010). Supercritical-$CO_2$ has received much interest typically in extraction of pharmaceutical and health related products from microalgae (Jaime et al. 2007, Kitada et al. 2009, Macias-Sanchez et al. 2008, Ota et al. 2009). In fact, supercritical-$CO_2$ offers several advantages in comparison with chemical solvent extraction such as (1) non-toxic and provide non-oxidizing environment to avoid degradation of extracts, (2) low critical temperature (around 31°C) which prevents thermal degradation of products, (3) high diffusivity and low surface tension which allow penetration of pores smaller than those accessible by chemical solvents, and (4) easy separation of $CO_2$ at ambient temperature after extraction (Mendes et al. 2003, Jaime et al. 2007, Ota et al. 2009).

The most suitable catalyst for transesterification of oils with low free fatty acids (FFA) content is necessary. The presence of high free fatty acid content in microalgae lipid (more than 0.5% w/w) prevents the use of homogeneous base catalyst for transesterification reaction (Ehimen et al. 2010, Zhu et al. 2008, Ziino et al. 1999). Alkaline metal alkoxides, even at low concentration of 0.5 mol%, are highly active catalysts (Pragya et al. 2013). Metal alkoxides (e.g., potassium methoxide) in methanol are better options than metal hydroxides (NaOH, KOH). In short reaction time of about 30 min, they give high yield of about 98% (Pragya et al. 2013). They performed better in the absence of water, which makes them inappropriate for industrial processes (Schuchardt et al. 1998). FFA will react with base catalyst to form soap leading to lower biodiesel yield, and increase the difficulty to separate biodiesel from glycerol.

Acid catalyst (e.g., $H_2SO_4$) is not sensitive towards FFA level in oil, and thus esterificaiton (FFA is converted to alkyl ester) and transesterification can occur simultaneously. Chemically catalyzed transesterification process requires high amount of energy and separation of catalyst from the product is cost effective. Glycerol produced as a byproduct of alcoholysis, readily adheres to the surface of immobilized lipase and decrease its enzyme activity, where glycerol removal is a complex process, which may hinder the large scale operations (Rawat et al. 2011).

Free fatty acids contained in extracted oil have to be removed prior to reaction in order to maintain activity of the alkaline catalysts. In the superheated method, free fatty acid, and triglycerides are converted to fatty acid methyl ester directly. Academic and commercial enterprises have realized the importance of such research on this topic, and a direct transesterification method for total microalgal lipid content has produced appreciable levels of biodiesel even when there are undetectable levels of neutral lipids (Wahlen et al. 2011, Du et al. 2004). The advantages of non-catalytic alcoholysis reaction for the production of biodiesel are as follows: (1) the purification process to remove catalyst after reaction is not required, (2) the by-product (glycerol) can be directly utilized in other industry, (3) not only the triglycerides but also the free fatty acid might be converted into fatty acid methyl ester, (4) neutralization process for removal of free fatty acid is not required prior to the reaction process, and (5) the yield of the total system will be improved, and the cost required for the process will be reduced.

In large scale operation, the presence of microorganisms, medium composition, and process condition may cause emulsion formation, which lowers the product recovery efficiency. A better understanding on the mechanism of emulsion formation is necessary for the performance improvement based on nanotechnology as well as electrochemical properties such as ion, charge, viscosity, interface stabilization, etc. (Heeres et al. 2014).

## 9. Concluding Remarks

Photosynthetic microorganisms have emerged as one of the most promising sources for biodiesel production. Although algae and cyanobacteria have been paid recent attention from the point of view of sustainable biosynthesis as well as biofuel and biochemical production, the cell growth rate is significantly lower as compared to the typical biofuels-producing microorganisms such as *E. coli* and yeast. It is, therefore, strongly desirable to design microbial cell factories by means of synthetic biology approach with in-depth understanding of the metabolic regulation mechanism with the aid of systems biology approach such as modeling. Several attempts are being made for improving the efficiency of capturing light energy and $CO_2$ fixation as mentioned before.

Even though the production rate by photosynthetic organisms is low as compared to heterotrophic microorganisms, it is attractive that $CO_2$ can be directly fixed with sunlight energy as consolidated bioprocess (CBP).

## References

Aikawa, S., A. Nishida, S.H. Ho, J.S. Chang, T. Hasunuma and A. Kondo. 2014. Glycogen production for biofuels by the euryhaline cyanobacteria *Synechococcus* sp. strain PCC 7002 from an oceanic environment. Biotechnol Biofuels. 7: 88.

Aikawa, S., Y. Izumi, F. Matsuda, T. Hasunuma, J.S. Chang and A. Kondo. 2012. Synergistic enhancement of glycogen production in Arthrospira platensis by optimization of light intensity and nitrate supply. Bioresour Technol. 108: 211–215.

Ainala, S.K., S. Ashok, Y. Ko and S. Park. 2013. Glycerol assimilation and production of 1, 3-propanediol by *Citrobacter amalonaticus* Y19. Appl Microbiol Biotechnol. 97: 5001–5011.

Aldor, I.S., S.W. Kim, K.L. Prather and J.D. Keasling. 2002. Metabolic engineering of a novel propionate-independent pathway for the production of poly (3-hydroxybutyrate-co-3-hydroxyvalerate) in recombinant *Salmonella enterica* serovar Typhimurium. Appl Environ Microbiol. 68: 3848–3854.

Allen, D.K., Y. Shachar-Hill and J.B. Ohlrogge. 2007. Compartment-specific labeling information in [13]C metabolic flux analysis of plants. Phytochem. 68(16): 2197–2210.

Allen, M.M. 1984. Cyanobacterial cell inclusions. Annu Rev Microbiol. 38: 1–25.

Altaras, N.M. and D.C. Cameron. 1999. Metabolic engineering of a 1, 2-propanediol pathway in *Escherichia coli*. Appl Environ Microbiol. 65(3): 1180–1185.

Altaras, N.M. and D.C. Cameron. 2000. Enhanced production of (R)-1, 2-propanediol by metabolically engineered *Escherichia coli*. Biotechol Prog. 16(6): 940–946.

Amaro, H.M., A. Guedes and F.X. Malcata. 2011. Advances and perspectives in using microalgae to produce biodiesel. Appl Energy. 88(10) : 3402–3410.

Anandarajah, K., G. Mahendraperumal, M. Sommerfeld and Q. Hu. 2012. Characterization of microalga *Nannochloropsis* sp. mutants for improved production of biofuels. Appl Energy. 96: 371–377.

Ananyev, G.M., N.J. Skizim and G.C. Dismukes. 2012. Enhancing biological hydrogen production from cyanobacteria by removal of excreted products. J Biotechnol. 162: 97–104.

Anfelt, J., D. Kaczmarzyk, K. Shabestary, B. Renberg, J. Rockberg, J. Nielsen, M. Uhlén and E.P. Hudson. 2015. Genetic and nutrient modulation of acetyl-CoA levels in Synechocystis for n-butanol production. Microb Cell Fact. 14: 167.

Angelmyr, S.A., M. Paszota and K.J. Hellingwerf. 2012. Engineering a cyanobacterial cell factory for production of lactic acid. Appl Environ Microbiol. 78: 7098–7106.

Angermayr, S.A., A.D. van der Woude, D. Correddu, A. Vreugdenhil, V. Verrone and K.J. Hellingwerf. 2014. Exploring metabolic engineering design principles for the photosynthetic production of lactic acid by *Synechosystis* sp. PCC6803. Biotechnol for Biofuels. 7: 99.

Angermayr, S.A., R.A. Gorchs and K.J. Hellingwerf. 2015. Metabolic engineering of cyanobacteria for the synthesis of commodity products. Trends Biotechnol. 33: 352–361.

Antal, T.K. and P. Lindblad. 2005. Production of $H_2$ by sulphur-deprived cells of the unicellular cyanobacteria *Gloeocapsa alpicola* and *Synechocystis* sp. PCC 6803 during dark incubation with methane or at various extracellular pH. J Appl Microbiol. 98: 114–120.

Aoyama, K., I. Uemura, J. Miyake and Y. Asada. 1997. Fermentative metabolism to produce hydrogen gas and organic compounds in a cyanobacterium, *Spirulina platensis*. J Ferment Bioeng. 83: 17–20.

Asada, Y., M. Miyake, J. Miyake, R. Kurane and Y. Tokiwa. 1999. Photosynthetic accumulation of poly-(hydroxybutyrate) by cyanobacteria—the metabolism and potential for CO2 recycling. Int J Biol Macromol. 25: 37–42.

Atsumi, S. and J.C. Liao. 2008. Directed evolution of *Methanococcus jannaschii* citramalate synthase for biosynthesis of 1-propanol and 1-butanol by *Escherichia coli*. Appl Environ Microbiol. 74: 7802–7808.

Atsumi, S., T. Hanai and J.C. Liao. 2008. Non-fermentative pathways for synthesis of branched-chain higher alcohols as biofuels. Nature. 451: 86–89.

Atsumi, S., W. Higashide and J.C. Liao. 2009. Direct photosynthetic recycling of carbon dioxide to isobutyraldehyde. Nat Biotechnol. 27: 1177–1180.

Azuma, M., T. Osanai, M.Y. Hirai and K. Tanaka. 2011. A response regulator Rre37 and an RNA polymerase sigma factor SigE represent two parallel pathways to activate sugar catabolism in a cyanobacterium *Synechocystis* sp. PCC 6803. Plant Cell Physiol. 52: 404–412.

Baier, K., S. Nicklisch, C. Grundner, J. Reinecke and W. Lockau. 2001. Expression of two *nblA*-homologous genes required for phycobilisome degradation in nitrogen-starved *Synechocystis* sp. PCC6803. FEMS Microbiol Lett. 195: 35–39.

Banti, V., B. Giuntoli, S. Gonzali, E. Loreti, L. Magneschi, G. Novi, E. Paparelli, S. Parlanti, C. Pucciariello, A. Santaniello and P. Perata. 2013. Low oxygen response mechanisms in green organisms. Int J Mol Sci. 14: 1–30.

Barbier, G.G., J.L. Ladd, E.R. Campbell and W.H. Campbell. 2011. Genetic modification of *Pichia Pastoris* for production of propylene glycol from glycerol. Int J Genet Eng. 1(1): 6–13.

Bentley, F.K. and A. Melis. 2012. Diffusion-based process for carbon dioxide uptake and isoprene emission in gaseous/aqueous two-phase photobioreactors by photosynthetic microorganisms. Biotechnol Bioeng. 109: 100–109.

Bentley, F.K., A. Zurbriggen and A. Melis. 2014. Heterologous expression of the mevalonic acid pathway in cyanobacteria enhances endogenous carbon partitioning to isoprene. Mol Plant. 7: 71–86.

Bigogno, C., I. Khozin-Goldberg, S. Boussiba, A. Vonshak and Z. Cohen. 2002. Lipid and fatty acid composition of the green oleaginous alga *Parietochloris incisa*, the richest plant source of arachidonic acid. Phytochem. 60(5): 497–503.

Blaby, I.K., A.G. Glaesener, T. Mettler, S.T. Fitz-Gibbon, S.D. Gallaher, B. Liu and S.S. Merchant. 2013. Systems-level analysis of nitrogen starvation-induced modifications of carbon metabolism in a *Chlamydomonas reinhardtii* starchless mutant. The Plant Cell. 25: 4305–4323.

Block, D.L. and I. Melody. 1992. Efficiency and cost goals for photoenhanced hydrogen production processes. Int J Hydrogen Energy. 17: 853–861.

Boyle, N.R. and J.A. Morgan. 2009. Flux balance analysis of primary metabolism in *Chlamydomonas reinhardtii*. BMC Sys Biol. 3(1): 4.

Bricker, T.M., S. Zhang, S.M. Laborde, P.R. Mayer II, L.K. Frankel and J.V. Moroney. 2004. The malic enzyme is required for optimal photoautotrophic growth of *Synechocystis* sp. strain PCC6803. Under continuous light but not under a diurnal light regimen. J Bacteriol. 186(23): 8144–8148.

Burja, A.M., B. Banaigs, E. Abou-Mansour, J.G. Burgess and P.C. Wright. 2001. Marine cyanobacteria—a prolific source of natural products. Tetrahedron. 57: 9347–9377.

Burnap, R., M. Hagemann and A. Kaplan. 2015. Regulation of $CO_2$ concentrating mechanism in cyanobacteria. Life. 5: 348–371.

Byrom, D. 1987. Polymer synthesis by microorganisms. Technology and economics. Trends Biotechnol. 5: 246–250.

Cagnon, C., B. Mirabella, H.M. Nguyen, A. Beyly-Adriano, S. Bouvet, S. Cuiné and Y. Li-Beisson. 2013. Development of a forward genetic screen to isolate oil mutants in the green microalga *Chlamydomonas reinhardtii*. Biotechnol For Biofuels. 6(1): 178.

Cai, Z., G. Liu, J. Zhang and Y. Li. 2014. Development of an activity-directed selection system enabled significant improvement of the carboxylation efficiency of Rubisco. Protein Cell. 5: 552–562.

Camacho-Rubio, F., F.G. Camacho, J.M. Sevilla, Y. Chisti and E.M. Grima. 2003. A mechanistic model of photosynthesis in microalgae. Biotechnol Bioeng. 81: 459–473.

Cameron, D.C. and C.L. Cooney. 1986. A novel fermentation: the production of R(–)-1, 2-propanediol and acetol by *Clostridium thermosaccharolyticum*. Nat Biotechnol. 4(7): 651–654.

Carlozzi, P., A. Buccioni, S. Minieri, B. Pushparaj, R. Piccardi, A. Ena and C. Pintucci. 2010. Production of bio-fuels (hydrogen and lipids) through a photofermentation process. Bioresource Technology. 101(9): 3115–3120.

Chang, R.L., L. Ghamsari, A. Manichaikul, E.F.Y. Hom, S. Balaji, W. Fu, Y. Shen, T. Hao, P.O. Palsson, K. Salehi-Ashtiani and J.A. Papin. 2011. Metabolic network reconstruction of *Chlamydomonas* offers insight into light-driven algal metabolism. Mol Sys Biol. 7: 518.

Chaves, J.E., H. Kirst and A. Melis. 2014. Isoprene production in *Synechocystis* under alkaline and saline growth conditions. J Appl Phycol. doi: 10.1007/s10811-014-0395-2.

Chen, C.Y., K.L. Yeh, R. Aisyah, D.J. Lee and J.S. Chang. 2011. Cultivation, photobioreactor design and harvesting of microalgae for biodiesel production: a critical review. Bioresour Technol. 102(1): 71–81.

Chen, G. and Q. Wu. 2005. Microbial production and application of chiral hydroxyalkanoates. Appl Microbiol Biotechnol. 67: 592–599.

Chen, J.S. and S.F. Hiu. 1986. Acetone-butanol-isopropanol production by *Clostridium beijerinckii* (synonym, *Clostridium butylicum*). Biotechnol Lett. 8: 371–376.

Chen, Q., Q. Wang, G. Wei, Q. Liang and Q. Qi. 2011. Production in *Escherichia coli* of poly(3-hydroxybutyrate-co-3-hydroxyvalerate) with differing monomer compositions from unrelated carbon sources. Appl Environ Microbiol. 77: 4886–4893.

Chisti, Y. 2007. Biodiesel from microalgae. Biotechnol Adv. 25: 294–306.

Chisti, Y. 2013. Constraints to commercialization of algal fuels. J Biotechnol. 167: 201–14.

Chiu, S.Y., C.Y. Kao, C.H. Chen, T.C. Kuan, S.C. Ong and C.S. Lin. 2008. Reduction of $CO_2$ by a high-density culture of *Chlorella* sp. in a semicontinuous photobioreactor. Bioresour Technol. 99(9): 3389–3396.

Christenson, L. and R. Sims. 2011. Production and harvesting of microalgae for wastewater treatment, biofuels, and bioproducts. Biotechnol Adv. 29(6): 686–702.

Clomburg, J.M. and R. Gonzalez. 2011. Metabolic engineering of *Escherichia coli* for the production of 1, 2-propanediol from glycerol. Biotechnol Bioeng. 108(4): 867–879.

Cogne, G., M. Rugen, A. Bockmayr, M. Titica, C.-G. Dussap, J.-F. Cornet and J. Legrand. 2011. A model-based method for investigating bioenergetics processes in autotrophically growing eukaryotic microalgae: Application to the green algal *Chlamydomonas reinhardtii*. Biotechnol Prog. 27(3): 631–640.

Colman, B. and E.G. Norman. 1997. Serine synthesis in cyanobacteria by a nonphotorespiratory pathway. Physiol Plant. 100: 133–136.

Cooley, J.W. and W.F. Vermaas. 2001. Succinate dehydrogenase and other respiratory pathways in thylakoid membranes of *Synechocystis* sp. strain PCC 6803: Capacity comparisons and physiological function. J Bacteriol. 183: 4251–4258.

Daley, S.M., A.D. Kappell, M.J. Carrick and R.L. Burnap. 2012. Regulation of the cyanobacterial CO2-concentrating mechanism involves internal sensing of NADP+ and alpha-ketoglutarate levels by transcription factor CcmR. PLoS One. 7: e41286.

Datta, M., G. Nikki and V. Shah. 2000. Cyanobacterial hydrogen production. World J Microbiol Biotechnol. 16: 8–9.

Datta, R. and M. Henry. 2006. Recent advances in products, processes and technologies-a review. J Chem Technol Biotechnol. 81: 1119–1129.

Davies, F.K., R.E. Jinkerson and M.C. Posewitz. 2015. Toward a photosynthetic microbial platform for terpenoid engineering. Photosynth Res. 123: 265–84.

Davis, M.S., J. Solbiati and J.E. Cronan. 2000. Overproduction of acethyl-CoA carboxylase activity increases the rate of fatty acid biosynthesis in *Escherichia coli*. J Biol Chem. 275: 28593–28598.

de Godos, I., H.O. Guzman, R. Soto, P.A. García-Encina, E. Becares, R. Muñoz and V.A. Vargas. 2011. Coagulation/flocculation-based removal of algal–bacterial biomass from piggery waste water treatment. Bioresour Technol. 102(2): 923–927.

de Oliveira Dal'Molin, C.G., L.-E. Quek, R.W. Palfreyman and L.K. Nielsen. 2011. AlgaGEM a genome-scale metabolic reconstruction of algae based on the *Chalamydomonas reinhardtii* genome. BMC Genomics. 12: (Suppl 4) S5.

Deng, M.D. and J.R. Coleman. 1999. Ethanol synthesis by genetic engineering in cyanobacteria. Appl Environ Microbiol. 65: 523–528.

Dennis, D.T. and S.D. Blakeley. 2000. Carbohydrate metabolism. Biochem Mol Biol of Plants. 630–675.

Dersch, L.M., V. Beckers and C. Wittmann. 2016. Green pathways: Metabolic network analysis of plant systems. Metab Eng. 34: 1–24.

Dexter, J. and P.C. Fu. 2009. Metabolic engineering of cyanobacteria for ethanol production. Energy Environ Sci. 2: 857–864.

Dienst, D., J. Georg, T. Abts, L. Jakorew, E. Kuchmina, T. Borner, A. Wilde, U. Duhring, H. Enke and W.R. Hess. 2014. Transcriptomic response to prolonged ethanol production in the cyanobacterium *Synechosystis* sp. PCC6803. Biotechnol for Biofuels. 7: 21.

Doi, Y., A. Segawa and M. Kunioka. 1990. Biosynthesis and characterization of poly (3-hydroxybutyrate-co-4-hydroxybutyrate) in *Alcaligenes eutrophus*. Int J Biol Macromol. 12: 106–111.

Dong, H.P., E. Williams, D.Z. Wang, Z.X. Xie, R.C. Hsia, A. Jenck, R. Halden, J. Li, F. Chen and A.R. Place. 2013. Responses of *Nanochloropsis oceanica* IMET1 to long-term nitrogen starvation and recovery. Plant Physiol. 162: 1110–1126.

Du, W., Y.Y. Xu, J. Zeng and D.H. Liu. 2004. Novozym 435-catalysed transesterification of crude soya bean oils for biodiesel production in a solvent-free medium. Biotechnol Appl Biochem. 40(2): 187–190.

Ducat, D.C., G. Sachdeva and P.A. Silver. 2011. Rewiring hydrogenase-dependent redox circuits in cyanobacteria. PNAS USA. 108: 3941–3946.

Ducat, D.C., J.A. Avelar-Rivas, J.C. Way and P.A. Silver. 2012. Rerouting carbon flux to enhance photosynthetic productivity. Appl Environ Microbiol. 78: 2660–2668.

Ducat, D.C. and P.A. Silver. 2012. Improving carbon fixation pathways. Curr Opin Chem Bol. 16: 337–344.

Duhring, U., K. Ziegler and D. Kramer. 2010. US Patent Pub. No.:US2010/0003739 A1.

Durão, P., H. Aigner, P. Nagy, O. Mueller-Cajar, F.U. Hartl and M. Hayer-Hartl. 2015. Opposing effects of folding and assembly chaperones on evolvability of Rubisco. Nat Chem Biol. 11: 148–155.

Dutta, D., D. De, S. Chaudhuri and S.K. Bhattacharya. 2005. Hydrogen production by Cyanobacteria. Microb Cell Fact. 4: 36.

Ehimen, E.A., Z.F. Sun and C.G. Carrington. 2010. Variables affecting the *in situ* transesterification of microalgae lipids. Fuel. 89(3): 677–684.

Ehira, S. and M. Ohmori. 2006. NrrA directly regulates expression of hetR during heterocyst differentiation in the cyanobacterium *Anabaena* sp. strain PCC 7120. J Bacteriol. 188: 8520–8525.

Emptage, M., S.L. Haynie, L.A. Laffend, J.P. Pucci and G. Whited. 2003. Process for the biological production of 1, 3-propanediol with high titer. USPatent. No. 6,514,733.

Englund, E., B. Pattanaik, S.J. Ubhayasekera, K. Stensjo, J. Bergquist and P. Lindberg. 2014. Production of squalene in *Synechocystis* sp. PCC 6803. PLoS One. 9: e90270.

Ertuğrul, S., M. Bakır and G. Dönmez. 2008. Treatment of dye-rich wastewater by an immobilized thermophilic cyanobacterial strain: *Phormidium* sp. Ecol Eng. 32(3): 244–248.

Evaggelos, C.T., C.T. Marina and A.K. Dimitrios. 2010. Involvement of the AtoSCDAEB regulon in the high molecular weight poly-(*R*)-3-hydroxybutyrate biosynthesis in *phaCAB*+ *Escherichia coli*. Metab Eng. 14: 354–365.

Fay, P. 1992. Oxygen relations of nitrogen fixation in cyanobacteria. Microbiol Rev. 56: 340–373.

Fischer, C.R., D. Klein-Marcuschamer and G. Stephanopoulos. 2008. Selection and optimization of microbial hosts for biofuels production. Metab Eng. 10: 295–304.

Flores, E. and A. Herrero. 2005. Nitrogen assimilation and nitrogen control in cyanobacteria. Biochem Soc Trans. 33: 164–167.

Fukuda, H., M. Sakai, K. Nagahama, T. Fujii, M. Matsuoka et al. 1994. Heterologous expression of the gene for the ethylene-forming enzyme from *Pseudomonas syringae* in the cyanobacterium *Synechococcus*. Biotechnol Lett. 16: 1–6.

Fukuda, H., T. Ogawa, K. Ishihara, T. Fujii, K. Nagahama et al. 1992. Molecular cloning in *Escherichia coli*, expression, and nucleotide sequence of the gene for the ethylene-forming enzyme of *Pseudomonas syringae* pv. *phaseolicola* PK2. Biochem Biophys Res Commun. 188: 826–832.

Fukuda, H., T. Ogawa and S. Tanase. 1993. Ethylene production by micro-organisms. Adv Microb Physiol. 35: 275–306.

Gao, H.J., Q. Wu and G.Q. Chen. 2002. Enhanced production of D-(–)-3-hydroxybutyric acid by recombinant *Escherichia coli*. FEMS Microbiol Lett. 213: 59–65.

Gao, Q., W. Wang, H. Zhao and X. Lu. 2012. Effects of fatty acid activation on photosynthetic production of fatty acid-based biofuels in *Synechocystis* sp. PCC6803. Biotechnol for Biofuels. 5: 17.

Gao, Z., H. Zhao, Z. Li, X. Tan and X. Lu. 2012. Photosynthetic production of ethanol from carbon dioxide in genetically engineered cyanobacteria. Energy Environ Sci. 5: 9857–9865.

Garcia-Camacho, F., A. Sanchez-Miron, E. Morina-Grima, F. Camacho-Rubio and J.C. Merchuck. 2012. A mechanistic model of photosynthesis in microalgae including photoacclimation dynamics. J Theor Biol. 304: 1–15.

Garcia-Fernandez, J.M., N.T. de Marsac and J. Diez. 2004. Streamlined regulation and gene loss as adaptive mechanisms in *Prochlorococcus* for optimized nitrogen utilization in oligotrophic environments. Microbiol Mol Biol Rev. 68: 630–638.

Geigenberger, P. 2003. Response of plant metabolism to too little oxygen. Curr Opin Plant Biol. 6: 247–256.

Genkov, T., M. Meyer, H. Griffiths and R.J. Spreitzer. 2010. Functional hybrid rubisco enzymes with plant small subunits and algal large subunits: engineered *rbsC* cDNA for expression in *Chlamidomonas*. J Biol Chem. 285: 19833–19841.

Georgianna, D.R. and S.P. Mayfield. 2012. Exploiting diversity and synthetic biology for the production of algal biofuels. Nature. 488: 329–335.

Gerhardt, R., M. Stitt and H.W. Heldt. 1983. Subcellular metabolite determination in spinach leaves through non-aqueous fractionation. Physiol Chem. 364: 1130–1131.

Gerin, S., G. Mathy and F. Franck. 2014. Modeling the dependence of respiration and photosynthesis upon light, acetate, carbon dioxide, nitrate and ammonium in *Chlamydomonas reinhardtii* using design of experiments and multiple regression. BMC Sys Biol. 8: 96.

Gong, F., G. Liu, X. Zhai, J. Zhou, Z. Cai and Y. Li. 2015. Quantitative analysis of an engineered $CO_2$-fixing *Escherichia coli* reveals great potential of heterotrophic $CO_2$ fixation. Biotechnol Biofuels. 8: 86.

Greene, D.N., S.M. Whitney and I. Matsumura. 2007. Artificially evolved *Synechococcus* PCC6301 Rubisco variants exhibit improvements in folding and catalytic efficiency. Biochem J. 404: 517–524.

Griffiths, M.J. et al. 2009. Lipid productivity as a key characteristic for choosing algal species for biodiesel production. J Appl Phycol. 21: 493–507.

Grossman, A.R., M.R. Schaefer, G.G. Chiang and J.L. Collier. 1993. The phycobilisome, a light-harvesting complex responsive to environmental conditions. Microbiol Rev. 57: 725–749.

Gründel, M., R. Scheunemann, W. Lockau and Y. Zilliges. 2012. Impaired glycogen synthesis causes metabolic overflow reactions and affects stress responses in the cyanobacterium *Synechocystis* sp. PCC 6803. Microbiol. 158(Pt 12): 3032–43.

Gründel, M., R. Scheunemann, W. Lockau and Y. Zilliges. 2012. Impaired glycogen synthesis causes metabolic overflow reactions and affects stress responses in the cyanobacterium *Synechocystis* sp. PCC 6803. Microbiol. 158: 3032–3043.

Guedri, H. and C. Durrieu. 2008. A self-assembled monolayers based conductometric algal whole cell biosensor for water monitoring. Microchimica Acta. 163(3-4): 179–184.

Guerrero, F., V. Carbonell, M. Cossu, D. Correddu and P.R. Jones. 2012. Ethylene synthesis and regulated expression of recombinant protein in *Synechocystis* sp. PCC 6803. PLoS One. 7(11): e50470.

Gungormusler, M., C. Gonen, G. Ozdemir and N. Azbar. 2010. 1, 3-Propanediol production potential of *Clostridium saccharobutyricum* NRRL B-643. New Biotechnol. 27: 782–788.

Gunther, A., T. Jakob, R. Goss, S. Konig, D. Spindler, N. Rabiger, S. John, S. Heithoff, M. Fresewinkel and C. Posten. 2012. Methane production from glycolate excreting algae as a new concept in the production of biofuels. Bioresour Technol. 121: 454–457.

Haase, S.M., B. Huchzermeyer and T. Rath. 2012. PHB accumulation in *Nostoc muscorum* under different carbon stress situations. J Appl Phycol. 24: 157–162.

Hajimahmoodi, M., M.A. Faramarzi, N. Mohammadi, N. Soltani, M.R. Oveisi and N. Nafissi-Varcheh. 2010. Evaluation of antioxidant properties and total phenolic contents of some strains of microalgae. J Appl Phycol. 22: 43–50.

Harris, E.H. 2009. The *Chlamydomonas* sourcebook: introduction to *Chlamydomonas* and its laboratory use. Vol. 1. Acad Press.

Harun, R., M.K. Danquah and G.M. Forde. 2010. Microalgal biomass as a fermentation feedstock for bioethanol production. J Chem Technol Biotechnol. 85(2): 199–203.

Hauf, W., M. Schlebusch, J. Huge, J. Kopka, M. Hagemann and K. Forchhammer. 2013. Metabolic changes in *Synechocystis* PCC6803 upon nitrogen-starvation: Excess NADPH sustains polyhydroxybutyrate accumulation. Metabolites. 3: 101–118.

Havel, J. and D. Weuster-Botz. 2007. Cofactor regeneration in phototrophic cyanobacteria applied for asymmetric reduction of ketones. Appl Microbiol Biotechnol. 75: 1031–1037.

Heath, R.J. and C.O. Rock. 1996. Regulation of fatty acid elongation and initiation by acyl-acyl carrier protein in *Escherichia coli*. J Biol Chem. 271: 1833–1836.

Heeres, A.S., C.S.F. Picone, L.A.M. van der Wielen, R.L. Cunha and M.C. Cuellar. 2014. Microbial advanced biofuels production: overcoming emulsification challenges for large-scale operation. Trends in Biotechnol. 32(4): 221–229.

Hein, S., H. Tran and A. Steinbuchel. 1998. *Synechocystis* sp. PCC6803 possesses a two-component polyhydroxyalkanoic acid synthase similar to that of anoxygenic purple sulfur bacteria. Arch Microbiol. 170: 162–170.

Hempel, F., A.S. Bozarth, N. Lindenkamp, A. Klingl, S. Zauner, U. Linne, A. Steinbüchel and U.G. Maier. 2011. Microalgae as bioreactors for bioplastic production. Microb Cell Fact. 10: 81.

Horikawa, Y., Y. Maki, T. Tatsuke and T. Hanai. 2016. Cyanobacterial production of 1, 3-propanediol directly from carbon dioxide using a synthetic metabolic pathway. Metab Eng. 34: 97–103.

Hu, Q., M. Sommerfeld, E. Jarvis, M. Ghirardi, M. Posewitz, M. Seibert and A. Darzins. 2008. Microalgal triacylglycerols as feedstocks for biofuel production: perspectives and advances. Plant. 54(4): 621–639.

Jaime, L., J.A. Mendiola, E. Ibáñez, P.J. Martin-Álvarez, A. Cifuentes, G. Reglero and F.J. Señoráns. 2007. β-Carotene isomer composition of sub-and supercritical carbon dioxide extracts. Antioxidant activity measurement. J Agricultural Food Chem. 55(26): 10585–10590.

James, C.N., P.J. Horn, C.R. Case, S.K. Gidda, D. Zhang, R.T. Mullen and K.D. Chapman. 2010. Disruption of the Arabidopsis CGI-58 homologue produces Chanarin–Dorfman-like lipid droplet accumulation in plants. PNAS USA. 107(41): 17833–17838.

Jeffries, T.W., H. Timourien and R.L. Ward. 1978. Hydrogen production by *Anabaena cylindrica*: Effect of varying ammonium and ferric ions, pH and light. Appl Environ Microbiol. 35: 704–710.

Jeon, E., S. Lee, D. Kim, H. Yoon, M. Oh, C. Park and J. Lee. 2009. Development of a *Saccharomyces cerevisiae* strain for the production of 1, 2-propanediol by gene manipulation. Enzyme Microb Technol. 45(1): 42–47.

Jones, C.S. and S.P. Mayfield. 2012. Algae biofuels: versatility for the future of bioenergy. Curr Opin Biotechnol. 23(3): 346–351.

Jung, J.Y., H.S. Yun, J. Lee and M.K. Oh. 2011. Production of 1, 2-propanediol from glycerol in *Saccharomyces cerevisiae*. J Microbiol Biotechnol. 21(8): 846–853.

Jung, Y.K., T.Y. Kim, S.J. Park and S.Y. Lee. 2010. Metabolic engineering of *Escherichia coli* for the production of polylactic acid and its copolymers. Biotechnol Bioeng. 105: 161–171.

Kaczmarzky, D. and M. Fulda. 2010. Fatty acid activation in cyanobacteria mediated by acyl-acyl carrier protein synthetase enables fatty acid recycling. Plant Physiol. 152: 1598–1610.

Kamennaya, N.A., S. Ahn, H. Park, R. Bartal, K.A. Sasaki, H.-Y. Holman and C. Jansson. 2015. Installing extra bicarbonate transporters in the cyanobacterium *Synechocystis* sp. PCC6803 enhances biomass production. Metab Eng. 29: 76–85.

Keasling, J.D. and H. Chou. 2008. Metabolic engineering delivers next-generation biofuels. Nat Biotechnol. 26: 298–299.

Kebeish, R., M. Niessen, K. Thiruveedhi, R. Bari, H.-J. Hirsch, R. Rosenkranz, N. Stabler, B. Schonfeld, F. Kreuzaler and C. Peterhansel. 2007. Chloroplastic photorespiratory by pass increases photosynthesis and biomass production in *Arabidopsis thaliana*. Nat Biotech. 25: 593–599.

Kepekci, R.A. and S.D. Saygideger. 2012. Enhancement of phenolic compound production in Spirulina platensis by two-step batch mode cultivation. J Appl Phycol. 24: 897–905.

Khan, S.A., M.Z. Hussain, S. Prasad and U.C. Banerjee. 2009. Prospects of biodiesel production from microalgae in India. Renewable and Sustainable Energy Reviews. 13(9): 2361–2372.

Kilian, O., C.S. Benemann, K.K. Niyogi and B. Vick. 2011. High-efficiency homologous recombination in the oil-producing alga *Nannochloropsis* sp. PNAS USA. 108(52): 21265–21269.

Kim, Y.H., Y.K. Choi, J. Park, S. Lee, Y.H. Yang, H.J. Kim and S.H. Lee. 2012. Ionic liquid-mediated extraction of lipids from algal biomass. Bioresour Technol. 109: 312–315.

Kitada, K., S. Machmudah, M. Sasaki, M. Goto, Y. Nakashima, S. Kumamoto and T. Hasegawa. 2009. Supercritical $CO_2$ extraction of pigment components with pharmaceutical importance from *Chlorella vulgaris*. J Chem Technol Biotechnol. 84(5): 657–661.

Kiyota, H., M. Yokota-Hirai and M. Ikeuchi. 2014. NblA1/A2-dependent homeostasis of amino acid pools during nitrogen starvation in *Synechocystis* sp. PCC6803. Metabolites. 4: 517–531.

Kliphuis, A.M.J., A.J. Klok, D.E. Martens, P.P. Lamers, M. Janssen and R.H. Wijffels. 2012. Metabolic modeling of *Chlamydomonas reinhardtii*: Energy requirements for autotrophic growth and maintenance. J Appl Phycol. 24: 253–266.

Klok, A.J., P.P. Lamers, D.E. Martens, R.B. Draaisma and R.H. Wijffels. 2014. Edible oils from microalgae: insights in TAG accumulation. Trends in Biotechnol. 32(10): 521–528.

Knoop, H., M. Grundel, Y. Zilliges, R. Lehmann, S. Hoffmann, W. Lockau and R. Steuer. 2013. Flux balance analysis of cyanobacterial metabolism: The metabolic network of *Synechocystis* sp. PCC6803. Plos Comp Biol. 9(6): 21003081.

Knowles, V.L., C.S. Smith, C.R. Smith and W.C. Plaxton. 2001. Structural and regulatory properties of pyruvate kinase from the cyanobacterium *Synechococcus* PCC 6301. J Biol Chem. 276(24): 20966–20972.

Kobayashi, J., S. Hasegawa, K. Ito, K. Yoshimune, T. Komoriya, A. Asada and H. Kohno. 2012. Expression of aldehyde dehydrogenase gene increases hydrogen production from low concentration of acetate by *Rhodobacter sphaeroides*. Int J Hydrogen Energy. 37: 9602–9609.

Kozaki, A. and G. Takeba. 1996. Photorespiration protects C3 plants from photooxidation. Nature. 384: 557–560.

Krasikov, V., A.E. von Wobeser, H.L. Dekker, J. Huisman and H.C. Matthijs. 2012. Time-series resolution of gradual nitrogen starvation and its impact on photosynthesis in the cyanobacterium *Synechocystis* PCC 6803. Physiol Plant. 145: 426–439.

Krivoruchko, A., C. Serrano-Amatriain, Y. Chen, V. Siewers and J. Nielsen. 2013. Improving biobutanol production in engineered *Saccharomyces cerevisiae* by manipulation of acetyl-CoA metabolism. J Ind Microbiol Biotechnol. 40: 1051–1056.

Kusakabe, T., T. Tatsuke, K. Tsuruno, Y. Hirokawa, S. Atsumi, J.C. Liao and T. Hanai. 2013. Engineering a synthetic pathway in cyanobacteria for isopropanol production directly from carbon dioxide and light. Metab Eng. 20: 101–108.

Kuzma, J., N. Nemecek-Marshall, W.H. Pollock and R. Fall. 1995. Bacteria produce the volatile hydrocarbon isoprene. Curr Microbiol. 2: 97–103.

Lacour, T., A. Sciandra, A. Talec, P. Mayzaud and O. Bernard. 2012. Disel variations of carbohydrates and neutral lipids in nitrogen-sufficient and nitrogen-starved cyclostat cultures of *Isochrysis* sp. 1. J Phycology. 48(4): 966–975.

Lam, M.K. and K.T. Lee. 2012. Microalgae biofuels: a critical review of issues, problems and the way forward. Biotechnol Adv. 30 (3): 673–690.

Lambert, G.R., A. Daday and G.D. Smith. 1979. Hydrogen evolution from immobilized cultures of cyanobacterium Anabena cylindrica. FEBS Letters. 101: 125–128.

Lan, E.I. and J.C. Liao. 2011. Metabolic engineering of cyanobacteria for 1-butanol production from carbon dioxide. Metab Eng. 13: 353–363.

Lan, E.I. and J.C. Liao. 2012. ATP drives direct photosynthetic production of 1-butanol in cyanobacteria. PNAS USA. 109: 6018–6023.

Lange, B.M., T. Rujan, W. Martin and R. Croteau. 2000. Isoprenoid biosynthesis: the evolution of two ancient and distinct pathways across genomes. PNAS USA. 97: 13172–7.

Larkum, A.W.D., I.L. Ross, O. Kruse and B. Hankamer. 2012. Selection, breeding and engineering of microalgae for bioenergy and biofuels production. Trends in Biotechnol. 30(4): 198–205.

Lee, J.-Y., C. Yoo, S.-Y. Jun, C.-Y. Ahn and H.-M. Oh. 2010. Comparison of several methods for effective lipid extraction from microalgae. Biores Technol. 101: 575–577.

Lee, S.H., S.J. Park, S.Y. Lee and S.H. Hong. 2008. Biosynthesis of enantiopure (S)-3-hydroxybutyric acid in metabolically engineered *Escherichia coli*. Appl Microbiol Biotechnol. 79: 633–641.

Lee, S.Y. and Y. Lee. 2003. Metabolic engineering of *Escherichia coli* for production of enantiometrically pure (R)-(-)-hydroxycarboxylic acids. Appl Environ Microbiol. 69: 3421–3426.

Leema, J.T., R. Kirubagaran, N.V. Vinithkumar, P.S. Dheenan and S. Karthikayulu. 2010. High value pigment production from Arthrospira (Spirulina) platensis cultured in seawater. Bioresour Technol. 101: 9221–9227.

Lei, A., H. Chen, G. Shen, Z. Hu, L. Chen and J. Wang. 2012. Expression of fatty acid synthesis genes and fatty acid accumulation in *Haematococcus pluvialis* under different stressors. Biotechnol for Biofuels. 5: 18.

Li, H., D.M. Sherman, S. Bao and L.A. Sherman. 2001. Pattern of cyanophycin accumulation in nitrogen-fixing and non-nitrogen-fixing cyanobacteria anobacteria. Arch Microbiol. 176: 9–18.

Li, H. and J.C. Liao. 2013. Engineering a cyanobacterium as the catalyst for the photosynthetic conversion of $CO_2$ to 1, 2-propanediol. Microb Cell Fact. 12: 4.

Li, H. and L.A. Sherman. 2002. Characterization of *Synechocystis* sp. strain PCC6803 and deltanbl mutants under nitrogen-deficient conditions. Arch Microbiol. 178: 256–266.

Li, Y., D. Han, G. Hu, D. Dauvillee, M. Sommerfeld, S. Ball and Q. Hu. 2010. *Chlamydomonas* starchless mutant defective in ADP-glucose pyrophosphorylase hyper-accumulates triacylglycerol. Metab Eng. 12(4): 387–391.

Li, Y., M. Horsman, N. Wu, C.Q. Lan and N. Dubois-Calero. 2008. Biofuels from microalgae. Biotechnol Prog. 24: 815–20.

Li, Z.J., Y. Shi, J. Jian, Y.Y. Guo, Q. Wu and G.Q. Chen. 2010. Production of poly(3-hydroxybutyrate-co-4-hydroxybutyrate) from unrelated carbon sources by metabolically engineered *Escherichia coli*. Metab Eng. 12: 352–359.

Lichtenthaler, H.K. 2000. Sterols and isoprenoids. Biochem Soc Trans. 28: 785–9.

Lin, Z., S. Zhong and D. Grierson. 2009. Recent advances in ethylene research. J Exp Bot. 60: 3311–3336.

Lindberg, P., P. Lindblad and L. Cournac. 2004. Gas exchange in the filamentous cyanobacterium *Nostoc punctiforme* strain ATCC 29133 and its hydrogenase-deficient mutant strain NHM5. Appl Environ Microbiol. 70: 2137–2145.

Lindberg, P., S. Park and A. Melis. 2010. Engineering a platform for photosynthetic isoprene production in cyanobacteria, using *Synechocystis* as the model organism. Metab Eng. 12: 70–79.

Liu, B. and C. Benning. 2013. Lipid metabolism in microalgae distinguishes itself. Curr Opin Biotechnol. 24(2): 300–309.

Liu, D. and C. Yang. 2014. The nitrogen-regulated response regulator NrrA controls cyanophicin synthesis and glycogen catabolism in the cyanobacterium *Synechocystis* sp. PCC 6803. J Biol Chem. 289(4): 2055–2071.

Liu, H., Y. Xu, Z. Zheng and D. Liu. 2010. 1, 3-Propanediol and its copolymer: research, development and industrialization. Biotechnol J. 5: 1137–1148.

Liu, Q., S. Ouyang, A. Chung, Q. Wu and G. Chen. 2007. Microbial production of R-3-hydroxybutyric acid by recombinant *E. coli* harboring genes of *phbA*, *phbB*, and *tesB*. Appl Microbiol Biotechnol. 76: 811–818.

Liu, X. and R. Curtiss. 2009. Nickel-inducible lysis system in *Synechocystis* sp. PCC 6803. PNAS USA. 106: 21550–21554.

Liu, X., S. Fallon, J. Sheng and R. Curtiss 3rd. 2011a. $CO_2$-limitation-inducible Green Recovery of fatty acids from cyanobacterial biomass. PNAS USA. 108: 6905–6908.

Liu, X., J. Sheng and R. Curtiss 3rd. 2011b. Fatty acid production in genetically modified cyanobacteria. PNAS USA. 108: 6899–6904.

Lo, S.C., S.H. Shih, J.J. Chang, C.Y. Wang and C.C. Huang. 2012. Enhancement of photoheterotrophic biohydrogen production at elevated temperatures by the expression of a thermophilic *Clostridial hydrogenase*. Appl Microbiol Biotechnol. 95: 969–977.

Lonien, J. and J. Schwender. 2009. Analysis of metabolic flux phenotypes for two Arabidopsis mutants with severe impairment in seed storage lipid synthesis. Plant physiol. 151(3): 1617–1634.

Lopes Pinto, F.A., O. Troshina and P. Lindblad. 2002. A brief look at three decades of research on cyanobacterial hydrogen evolution. Int J Hydrogen Energy. 27: 1209–1215.

Macías-Sánchez, M.D., C. Mantell Serrano, M. Rodríguez Rodríguez, E. Martínez de la Ossa, L.M. Lubián and O. Montero. 2008. Extraction of carotenoids and chlorophyll from microalgae with supercritical carbon dioxide and ethanol as cosolvent. J Separation Sci. 31(8): 1352–1362.

Magnuson, K., S. Jackowski, C.O. Rock and J.E. Cronan Jr. 1993. Regulation of fatty acid biosynthesis in *Escherichia coli*. Microbiol Rev. 57: 522–542.

Mallick, N. 2006. Immobilization of microalgae. pp. 373–391. *In*: Immobilization of Enzymes and Cells. Humana Press.

Markou, G. and E. Nerantzis. 2013. Microalgae for high-value compounds and biofuels production: A review with focus on cultivation under stress conditions. Biotechnol Adv. 31: 1532–1542.

Masakapalli, S.K., P. Le Lay, J.E. Huddleston, N.L. Pollock, N.J. Kruger and R.G. Ratcliffe. 2010. Subcellular flux analysis of central metabolism in a heterotrophic Arabidopsis cell suspension using steady-state stable isotope labeling. Plant Physiol. 152(2): 602–619.

Mata, T.M., A.A. Martins and N.S. Caetano. 2010. Microalgae for biodiesel production and other applications: a review. Renew Sustain Energy Re. 14: 217–232.

McNeely, K., Y. Xu, N. Bennette, D.A. Bryant and G.C. Dismukes. 2010. Redirecting reductant flux into hydrogen production via metabolice Engineering of fermentative carbon metabolism in a cyanobacterium. Appl Environ Microbiol. 76(15): 5032–5038.

Mendes, R.L., B.P. Nobre, M.T. Cardoso, A.P. Pereira and A.F. Palavra. 2003. Supercritical carbon dioxide extraction of compounds with pharmaceutical importance from microalgae. Inorganica Chimica Acta. 356: 328–334.

Mendez-Perez, D., M.B. Begemann and B.F. Pfleger. 2011. Modular synthase-encoding gene involved in alpha-olefin biosynthesis in *Synechococcus* sp. strain PCC 7002. Appl Environ Microbiol. 77: 4264–4267.

Merchant, S.S., J. Kropat, B. Liu, J. Shaw and J. Warakanont. 2012. TAG, You're it! *Chlamydomonas* as a reference organism for understanding algal triacylglycerol accumulation. Curr Opin Biotechnol. 23(3): 352–363.

Mettler, T., T. Mühlhaus, D. Hemme, M.A. Schöttler, J. Rupprecht, A. Idoine and M. Stitt. 2014. Systems analysis of the response of photosynthesis, metabolism and growth to an increase in irradiance in the photosynthetic model organism *Chlamydomonas reinhardtii*. The Plant Cell. 26: 2310–2350.

Miao, X., Q. Wu, G. Wu and N. Zhao. 2003. Sucrose accumulation in salt-stressed cells of agp gene deletion-mutant in cyanobacterium *Synechocystis* sp. PCC 6803. FEMS Microbiol Lett. 218: 71–7.

Miyake, M., K. Kataoka, M. Shirai and Y. Asada. 1997. Control of poly-β-hydroxybutyrate synthase mediated by acetyl phosphate in cyanobacteria. J Bacteriol. 179: 5009–503.

Miyake, M., K. Takase, M. Narato, E. Khatipov, J. Schnackenberg, M. Shirai, R. Kurane and Y. Asada. 2000. Polyhydroxybutyrate production from carbon dioxide by cyanobacteria. Appl Biochem Biotechnol. 84-86: 991–1002.

Moellering, E.R. and C. Benning. 2010. RNA interference silencing of a major lipid droplet protein affects lipid droplet size in *Chlamydomonas reinhardtii*. Eukaryotic Cell. 9(1): 97–106.

Moreira, S.M., M. Moreira-Santos, L. Guilhermino and R. Ribeiro. 2006. Immobilization of the marine microalga *Phaeodactylum tricornutum* in alginate for *in situ* experiments: Bead stability and suitability. Enz Microbiol Technol. 38(1): 135–141.

Moreno-Garrido, I. 2008. Microalgae immobilization: current techniques and uses. Bioresour Technol. 99(10): 3949–3964.

Mueller-Cajar, O., M. Morell and S.M. Whitney. 2007. Directed evolution of Rubisco in *Escherichia coli* reveals a specificity-determining hydrogen bond in the form II enzyme. Biochem. 46: 14067–14074.

Muro-Pastor, A.M., A. Herrero and E. Flores. 2001. Nitrogen-regulated group 2 sigma factor from *Synechocystis* sp. strain PCC6803 involved in survival under nitrogen stress. J Bacteriol. 183: 1090–1095.

Nagahama, K., T. Ogawa, T. Fujii, M. Tazaki, S. Tanase et al. 1991. Purification and properties of an ethylene-forming enzyme from *Pseudomonas syringae* pv. *phaseolicola* PK2. J Gen Microbiol. 137: 2281–2286.

Nakamura, K., Y.R. Tohi and K.H. Hamada. 2000. Cyanobacterium-catalyzed asymmetric reduction of ketones. Tetrahedron Lett. 41: 6799–6802.

Niederholtmeyer, H., B.T. Wolfstadter, D.F. Savage, P.A. Silver and J.C. Way. 2010. Engineering cyanobacteria to synthesize and export hydrophilic products. Appl Environ Microbiol. 76: 3462–3466.

Nielsen, D.R., E. Leonard, S.-H. Yoon, H.-C. Tseng, C. Yuan and K.L.J. Prather. 2009. Engineering alternative butanol production platforms in heterologous bacteria. Metab Eng. 11: 262–73.

Niimi, S., N. Suzuki, M. Inui and H. Yukawa. 2011. Metabolic engineering of 1, 2-propanediol pathways in *Corynebacterium glutamicum*. Appl Microbiol Biotechnol. 90(5): 1721–1729.

Nowicka, B. and J. Kruk. 2010. Occurrence; biosynthesis and function of isoprenoid quinines. Biochem Biophys Acta. 1797: 1587–605.

O'Grady, J., J. Schwender, Y. Shachar-Hill and J.A. Morgan. 2012. Metabolic cartography: experimental quantification of metabolic fluxes from isotopic labelling studies. J Exp Bot. 63(6): 2293–2308.

Ogawa, T., M. Takahashi, T. Fujii, M. Tazaki and H. Fukuda. 1990. The role of NADH:Fe(III) EDTA oxidoreductase in ethylene formation from 2-Keto-4-methylthiobutyrate. J Ferment Bioeng. 69: 287–291.

Okano, K. et al. 2010. Biotechnological production of enantiomeric pure lactic acid from renewable resources: recent achievements, perspectives, and limits. Appl Microbiol Biotechnol. 85: 413–423.

Oliver, J.W. and S. Atsumi. 2014. Metabolic design for cyanobacterial chemical synthesis. Photosynth Res. 120: 249–261.

Oliver, J.W.K., I.M.P. Machado, H. Yoneda and S. Atsumi. 2013. Cyanobacterial conversion of carbon dioxide to 2, 3-butanediol. PNAS USA. 110(4): 1249–1254.

Ono, E. and J.L. Cuello. 2004. Design parameters of solar concentration systems for $CO_2$ mitigating algal photobioreactors. Energy. 29: 1651–1657.

Ono, E. and J.L. Cuello. 2007. Carbon dioxide mitigation using thermophilic cyanobacteria. Biosyst Eng. 96: 129–134.

Osanai, T. et al. 2011 Genetic engineering of group 2 sigma factor SigE widely activates expressions of sugar catabolic genes in *Synechocystis* sp. PCC 6803. J Biol Chem. 286: 30962–30971.

Osanai, T., K. Numata, A. Oikawa, A. Kuwahara, H. Iijima, Y. Doi, K. Tanaka, K. Saito and M. Yokota-Hirai. 2013a. Increased bioplastic production with an RNA polymerase sigma factor SigE during nitrogen starvation in *Synechocystis* sp. PCC6803. DNA Res. 20: 525–535.

Osanai, T., A. Kuwahara, H. Iijima, K. Toyooka, M. Sato, K. Tanaka, M. Ikeuchi, K. Saito and M. Yokota-Hirai. 2013b. Pleiotropic effect of *sigE* over-expression on cell morphorogy, photosynthesis, and hydrogen production in *Synechocystis* sp. PCC6803. The Plant J. 76: 456–465.

Osanai, T., Y. Kanesaki, T. Nakano, H. Takahashi, M. Asayama, M. Shirai, M. Kanehisa, I. Suzuki, N. Murata and K. Tanaka. 2005. Positive regulation of sugar catabolic pathways in the cyanobacterium *Synechocystis* sp. PCC 6803 by the group 2 sigma factor *sig*E. J Biol Chem. 280: 30653–30659.

Ota, M., H. Watanabe, Y. Kato, M. Watanabe, Y. Sato, R.L. Smith and H. Inomata. 2009. Carotenoid production from Chlorococcum littorale in photoautotrophic cultures with downstream supercritical fluid processing. J Separation Sci. 32(13): 2327–2335.

Owittrim, G.W. and B. Colman. 1988. Phosphoenolpyruvate carboxylase mediated carbon flow in a cyanobacterium. Biochem Cell Biol. 66: 93–99.

Pade, N. and M. Hagemann. 2014. Salt acclimation of cyanobacteria and their application in biotechnology. Life. 5: 25–49.

Pade, N., S. Erdmann, H. Enke, U. Duhring, J. Georg, J. Wambutt, J. Kopka, W.R. Hess, R. Zimmermann, D. Kramer and M. Hagemann. 2016. Insights into isoprene production using the cyanobacterium *Synechocystis* sp. PCC 6803. Biotechnol Biofuels. 9: 89.

Panda, B. and N. Mallick. 2007. Enhanced poly-β-hydroxybutyrate accumulation in a unicellular cyanobacterium, *Synechocystis* sp. PCC 6803. Lett Appl Microbiol. 44: 194–198.

Parikh, M.R., D.N. Greene, K.K. Woods and I. Matsumura. 2006. Directed evolution of Rubisco hypermorphs through genetic selection in engineered *E. coli*. Protein Eng Des Sel. 19: 113–119.

Park, J.H., J.J. Yoon, H.D. Park, Y.J. Kim, D.J. Lim and S.H. Kim. 2011. Feasibility of biohydrogen production from *Gelidium amansii*. Int J Hydrogen Energy. 36(21): 13997–14003.

Pearce, J., C.K. Leach and N.G. Carr. 1969. The incomplete tricarboxylic acid cycle in the blue–green alga *Anabaena variabilis*. J Gen Microbiol. 55: 371–378.

Pearce, J. and N.G. Carr. 1967. The metabolism of acetate by the blue–green algae, Anabaena variabilis and Anacystis nidulans. J Gen Microbiol. 49: 301–313.

Peers, G. 2014. Increasing algal photosynthetic productivity by integrating ecophysiology with systems biology. Trends in Biotechnol. 32(11): 551–555.

Peralta-Yahya, P.P., F. Zhang, S.B. Del Cardayre and J.D. Keasling. 2012. Microbial engineering for the production of advanced biofuels. Nature. 488: 320–328.

Pérez-Martínez, C., P. Sánchez-Castillo and M.V. Jiménez-Pérez. 2010. Utilization of immobilized benthic algal species for N and P removal. J Appl Phycol. 22(3): 277–282.

Pinzon-Gamez, N.M., S. Sundaram and L.K. Ju. 2005. Heterocyst differentiation and $H_2$ production in $N_2$-fixing cyanobacteria. Technical program.

Poolman, M.G., D.A. Fell and S. Thomas. 2000. Modeling photosynthesis and its control. J Exp Bot. 51: 319–328.

Poolman, M.G., H. Olcer, J.C. Lloyd, C.A. Raines and D.A. Fell. 2001. Computer modeling and experimental evidence for two steady states in the photosynthetic Carbin cycle. Eur J Biochem. 268: 2810–2816.

Pragya, N., K.K. Pandey and P.K. Sahoo. 2013. A review on harvesting, oil extraction and biofuels production technologies from microalgae. Renew Sustain Energy Rev. 24: 159–171.

Quintana, N., F. Van der Kooy, M.D. Van de Rhee, G.P. Voshol and R. Verpoorte. 2011. Renewable energy from cyanobacteria: energy production optimization by metabolic pathway engineering. Appl Microbiol Biotechnol. 91: 471–490.

Radakovits, R., R.E. Jinkerson, A. Darzins and M.C. Posewitz. 2010. Genetic engineering of algae for enhanced biofuel production. Eukaryotic Cell. 9(4): 486–501.

Ranjan, A., C. Patil and V.S. Moholkar. 2010. Mechanistic assessment of microalgal lipid extraction. Ind Eng Chem Res. 49(6): 2979–2985.

Rao, K.K. and D.O. Hall. 1996. Hydrogen production by cyanobacteria: potential, problems and prospects. J Marine Biotechnol. 4: 10–15.

Rasala, B.A., S.-S. Chao, M. Pier, D.J. Barrera and S.P. Mayfield. 2014. Enhanced genetic tools for engineering multigene traits into green algae. PLoS One. 9(4): e94028.

Rawat, I., R. Ranjith Kumar, T. Mutanda and F. Bux. 2011. Dual role of microalgae: phycoremediation of domestic wastewater and biomass production for sustainable biofuels production. Appl Energy. 88(10): 3411–3424.

Rawson, D.M. 1985. The effects of exogenous aminoacids on growth and nitrogenase activity in the cyanobacterium *Anabena cylindrica* PCC 7122. J Gen Microbiol. 134: 2549–2544.

Razon, L.F. 2012. Life cycle energy and greenhouse gas profile of a process for the production of ammonium sulfate from nitrogen-fixing photosynthetic cyanobacteria. Bioresour Technol. 107: 339–346.

Reijnders, M.J.M.F., vR.G.A. an HeckA, C.M.C. Lam, M.A. Scaife, V.A.P. Martins dos Santos, A.G. Smith and P.J. Schaap. 2014. Green genes: bioinformatics and systems-biology innovations drive algal biotechnology. Trends in Biotechnol. 32(12): 617–626.

Ren, Q., K. Ruth, L. Thony-Meyer and M. Zinn. 2010. Enatiometrically pure hydroxycarboxylic acids: current approaches and future perspectives. Appl Microbiol Biotechnol. 87: 41–52.

Renault, F., B. Sancey, P.M. Badot and G. Crini. 2009. Chitosan for coagulation/flocculation processes–an eco-friendly approach. European Polymer J. 45(5): 1337–1348.

Richter, N., A. Zienert and W. Hummel. 2011. A single-point mutation enables lactate dehydrogenage from *Bacillus subtilis* to utilize $NAD^+$ and $NADP^+$ as cofactor. Eng Lif Sci. 11: 26–36.

Rosenberg, J., G. Oyler, L. Wilkinson and M. Betenbaugh. 2008. A green light for engineered algae: redirecting metabolism to fuel a biotechnology revolution. Biotechnol. 19: 430–436.

Rosenberg, J.N., A. Mathias, K. Korth, M.J. Betenbaugh and G.A. Oyler. 2011. Microalgal biomass production and carbon dioxide sequestration from an integrated ethanol biorefinery in Iowa: A technical appraisal and economic feasibility evaluation. Biomass Bioenergy. 35(9): 3865–3876.

Rosenberg, J.N., G.A. Oyler, L. Wilkinson and M.J. Betenbaugh. 2008. A green light for engineered algae: redirecting metabolism to fuel a biotechnology revolution. Curr Opin Biotechnol. 19(5): 430–436.

Ruiz-Marin, A. and L. Mendoza-Espinosa. 2008. Ammonia removal and biomass characteristics of alginate-immobilized *Scenedesmus obliquus* cultures treating real wastewater. Fresenius Environ Bull. 17: 1236–1241.

Ruiz-Marin, A., L.G. Mendoza-Espinosa and T. Stephenson. 2010. Growth and nutrient removal in free and immobilized green algae in batch and semi-continuous cultures treating real wastewater. Bioresour Technol. 101(1): 58–64.

Ruzicka, L. 1953. The isoprene rule and the biogenesis of terpenic compounds. Experientia. 9: 357–67.

Sage, R.F., T.L. Sage and F. Kocacinar. 2012. Photorespiration and evolution of C4 photosynthesis. Annu Rev Plant Biol. 63: 19–47.

Sakai, M., T. Ogawa, M. Matsuoka and H. Fukuda. 1997. Photosynthetic conversion of carbon dioxide to ethylene by the recombinant cyanobacterium, *Synechococcus* sp. PCC 7942, which harbors a gene for the ethylene-forming enzyme of *Pseudomonas syringae.* J Ferment Bioeng. 84: 434–443.

Sander, K. and G.S. Murthy. 2010. Life cycle analysis of algae biodiesel. Int J Life Cycle Assess. 15(7): 704–714.

Sarkar, D. and K. Shimizu. 2015. An overview on biofuel and biochemical production by photosynthetic microorganisms with understanding of the metabolism and metabolic engineering together with cultivation method and down-stream processing. Bioresour Bioproc. 2: 17.

Savage, D.F., B. Afonso, A.H. Chen and P.A. Silver. 2010. Spacially ordered dynamics of the bacterial carbon fixation machinery. Science. 327: 1258–1261.

Savakis, P. and K.J. Hellingwerf. 2015. Engineering cyanobacteria for direct biofuel production from $CO_2$. Curr Opin Biotechnol. 33: 8–14.

Sawangkeaw, R., K. Bunyakiat and S. Ngamprasertsith. 2010. A review of laboratory-scale research on lipid conversion to biodiesel with supercritical methanol (2001–2009). The J Supercrit Fluids. 55(1): 1–13.

Saxena, R.K., P. Anand, S. Saran and J. Isar. 2009. Microbial production of 1,3-propanediol: Recent developments and emerging opportunities. Biotechnol Adv. 27: 895–913.

Schenk, P.M., S.R. Thomas-Hall, E. Stephens, U.C. Marx, J.H. Mussgnug, C. Posten, O. Kruse and B. Hankamer. 2008. Second generation biofuels: High-efficiency microalgae for biodiesel production. Bioenerg Res. 1: 20–43.

Schenk, P.M., S.R. Thomas-Hall, E. Stephens, U.C. Marx, J.H. Mussgnug, C. Posten and B. Hankamer. 2008. Second generation biofuels: high-efficiency microalgae for biodiesel production. Bioenergy Res. 1(1): 20–43.

Schlebusch, M. and K. Forchhammer. 2010. Requirement of the nitrogen starvation-induced protein Sll0783 for polyhydroxybutyrate accumulation in *Synechocystis* sp. strain PCC 6803. Appl Environ Microbiol. 76(18): 6101–6107.

Schuchardt, U., R. Sercheli and R.M. Vargas. 1998. Transesterification of vegetable oils: a review. J Brazilian Chem Soc. 9(3): 199–210.

Schwarz, D., I. Orf, J. Kopla and M. Hagemann. 2014. Effects of inorganic carbon limitation on the metabolome of the *Synechocystis* sp. PCC6803 mutant defective in *glnB* encoding the central regulator PII of cyanocaterial C/N acclimation. Metabolites. 4: 232–247.

Schwender, J. 2008. Metabolic flux analysis as a tool in metabolic engineering of plants. Curr Opin Biotechnol. 19: 131–137.

Schwender, J., Y. Shachar-Hill and J.B. Ohlrogge. 2006. Mitochondrial metabolism in developing embryos of Brassica napus. J Biol Chem. 281(45): 34040–34047.

Seo, M.Y., J.W. Seo, S.Y. Heo, J.O. Baek, D. Rairakhwada, B.R. Oh, P.S. Seo, M.H. Choi and C.H. Kim. 2009. Elimination of by-product formation during production of 1, 3-propanediol in *Klebsiella pneumonia* by inactivation of glycerol oxidative pathway. Appl Microbiol Biotechnol. 84: 527–534.

Sharkey, T.D. and S. Yeh. 2001. Isoprene emission from plants. Annu Rev Plant Physiol Plant Molec Biol. 52: 407–36.

Sharkey, T.D. 1988. Estimating the rate of photorespiration in leaves. Physiol Plant. 73: 147–152.

Shen, C.R. and J.C. Liao. 2012. Photosynthetic production of 2-methyl-1-butanol from $CO_2$ in cyanobacterium *Synechococcus elongatus* PCC 7942 and characterization of the native acetohydroxyacid synthase. Energy Environ Sci. 5: 9574.

Shi, X., K.W. Jung, D.H. Kim, Y.T. Ahn and H.S. Shin. 2011. Direct fermentation of *Laminaria japonica* for biohydrogen production by anaerobic mixed cultures. Int J Hydrogen Energy. 36(10): 5857–5864.

Siaut, M., S. Cuiné, C. Cagnon, B. Fessler, M. Nguyen, P. Carrier and G. Peltier. 2011. Oil accumulation in the model green alga Chlamydomonas reinhardtii: characterization, variability between common laboratory strains and relationship with starch reserves. BMC Biotechnol. 11(1): 7.

Singh, A., P.S. Nigam and J.D. Murphy. 2011. Renewable fuels from algae: an answer to debatable land based fuels. Bioresour Technol. 102(1): 10–16.

Sivakumar, G., J. Xu, R.W. Thompson, Y. Yang, P. Randol-Smith and P.J. Weathers. 2012. Integrated green algal technology for bioremediation and biofuel. Bioresour Technol. 107: 1–9.

Slater, S., T. Gallaher and D. Dennis. 1992. Production of poly-(3-hydroxybutyrate-co-3-hydroxyvalerate) in a recombinant *Escherichia coli* strain. Appl Environ Microbiol. 58: 1089–1094.

Solovchenko, A.E., I. Khozin-Goldberg, S. Didi-Cohen, Z. Cohen and M.N. Merzlyak. 2008. Effects of light intensity and nitrogen starvation on growth, total fatty acids and arachidonic acid in the green microalga *Parietochloris incise*. J Appl Phycol. 20(3): 245–251.

Sriram, G., D.B. Fulton, V.V. Iyer, J.M. Peterson, R. Zhou, M.E. Westgate and J.V. Shanks. 2004. Quantification of compartmented metabolic fluxes in developing soybean embryos by employing biosynthetically directed fractional $^{13}C$ labeling, two-dimensional [$^{13}C$, $^{1}H$] nuclear magnetic resonance, and comprehensive isotopomer balancing. Plant Physiol. 136(2): 3043–3057.

Sriram, G., V.V. Iyer, D. Bruce Fulton and J.V. Shanks. 2007. Identification of hexose hydrolysis products in metabolic flux analytes: A case study of levulinic acid in plant protein hydrolysate. Metab Eng. 9(5): 442–451.

Stal, L.J. and W.E. Krumbein. 1985. Oxygen protection of nitrogenase the aerobically nitrogen fixing, non-heterocystous cyanobacterium *Oscillatoria* sp. Arch Microbio. 143: 72–76.

Stal, L.J. and R. Moezelaar. 1997. Fermentation in cyanobacteria. FEMS Microbiol Rev. 21: 179–211.

Stal, L.J. and R. Moezelaar. 1997. Fermentation in cyanobacteria. FEMS Microbiol Reviews. 21: 179–211.

Stanier, R.Y. and G. Cohenbazire. 1977. Phototrophic prokaryotes—Cyanobacteria. Annu Rev Microbiol. 31: 225–274.

Steen, E.J., R. Chan, N. Prasad, S. Myers, C.J. Petzold, A. Redding, M. Ouellet and J.D. Keasling. 2008. Metabolic engineering of *Saccharomyces cerevisiae* for the production of n-butanol. Microb Cell Fact. 7: 36.

Steinhauser, D., A.R. Fernie and W.L. Araujo. 2012. Unusual cyanobacterial TCA cycles: not broken just different. Trends in Plant Sci. 17(9): 503–509.

Stephanopoulos, G. 2008. Challenges in engineering microbes for biofuels production. Science. 315: 801–4.

Stephenson, A.L., E. Kazamia, J.S. Dennis, C.J. Howe, S.A. Scott and A.G. Smith. 2010. Life-cycle assessment of potential algal biodiesel production in the United Kingdom: a comparison of raceways and air-lift tubular bioreactors. Energy & Fuels. 24(7): 4062–4077.

Stomp, M., J. Huisman, L. Voros, F.R. Pick, M. Laamanen, T. Haverkamp and L.J. Stal. 2007. Colourful coexistence of red and green picocyanobacteria in lakes and seas. Ecol Lett. 10: 290–298.

Sydney, E.B., W. Sturm, J.C. de Carvalho, V. Thomaz-Soccol, C. Larroche, A. Pandey and C.R. Soccol. 2010. Potential carbon dioxide fixation by industrially important microalgae. Bioresour Technol. 101(15): 5892–5896.

Syu, M.J. 2001. Biological production of 2, 3-butanediol. Appl Microbiol Biotechnol. 55(1): 10–18.

Takahama, K., M. Matsuoka, K.Nagahama and T. Ogawa. 2003. Construction and analysis of a recombinant cyanobacterium expressing a chromosomally inserted gene for an ethylene-forming enzyme at the psbAI locus. J. Biosci Bioeng. 95(3): 302–305.

Takahashi, H., M. Miyake, Y. Tokiwa and Y. Asada. 1998. Improved accumulation of poly-3-hydroxybutyrate by a recombinant cyanobacterium. Biotechnol Lett. 20: 183–186.

Tamagnini, P., R. Axelsson, P. Lindberg, F. Oxelfelt, R. Wunschiers and P. Lindblad. 2002. Hydrogenases and hydrogen metabolism of cyanobacteria. Microbiol Mol Biol Rev. 66: 1–20.

Tan, X., L. Yao, Q. Gao, W. Wang, F. Qi and X. Lu. 2011. Photosynthesis driven conversion of carbon dioxide to fatty alcohols and hydrocarbons in cyanobacteria. Metab Eng. 13: 169–176.

Taroncher-Oldenburg, G., K. Nishina and G. Stephanopoulos. 2000. Identification and analysis of the polyhydroxyalkanoate-specific beta-ketothiolase and acetoacetyl coenzyme A reductase genes in the cyanobacterium *Synechocystis* sp. strain PCC6803. Appl Environ Microbiol. 66: 4440–4448.

Tokiwa, Y. and C. Ugwu. 2007. Biotechnological production of (*R*)-3-hydoxybutyric acid monomer. J Biotechnol. 132: 264–272.

Tran, A.V. and R.P. Chambers. 1987. The dehydration of fermentative 2, 3-butanediol into methyl ethyl ketone. Biotechnol Bioeng. 29(3): 343–351.

Triana, J., A. Montagud, M. Siurana, D. Fuente, A. Urchueguia, D. Gamermann, J. Torres, J. Tena, P.F. de Cordoba and J.F. Urchueguia. 2014. Generation and evaluation of a genome-scale metabolic network model of *Synechococcus elongatus* PCC7942. Metabolites. 4: 680–698.

Tsang, T., R. Roberson and W.J. Vermaas. 2013. Polyhydroxybutyrate particles in *Synechocystis* sp. PCC 6803: facts and fiction. Photosynthesis Res. 118: 37–49.

Tseng, H., C. Martin, D. Nielsen and K. Prather. 2009. Metabolic engineering of *Escherichia coli* for enhanced production of (*R*)- and (*S*)-3-hydroxybutyrate. Appl Environ Microbiol. 75: 3137–3145.

Tsygankov, A.A., V.B. Borodin, K.K. Rao and D.O. Hall. 1999. $H_2$ photoproduction by batch culture of *Anabaena variabilis* ATCC 29413 and its mutant PK84 in a photobioreactor. Biotechnol Bioeng. 64: 709–715.

Tyo, K.E., Y.S. Jin, F.A. Espinoza and G. Stephanopoulos. 2009. Identification of gene disruptions for increased poly-3-hydroxybutyrate accumulation in *Synechocystis* PCC 6803. Biotechnol Prog. 25: 1236–1243.

Ungerer, J., L. Tao, M. Davis, M. Ghirardi, P.-C. Maness et al. 2012. Sustained photosynthetic conversion of $CO_2$ to ethylene in recombinant cyanobacterium *Synechocystis* 6803. Energy Environ Sci. 5: 8998–9006.

Valladares, A., M.L. Montesinos, A. Herrero and E. Flores. 2002. An ABC-type, high-affinity urea permease identified in cyanobacteria. Mol Microbiol. 43: 703–715.

van der Woude, A.D., S.A. Angermayr, V.P. Veetil, A. Osnato and K.J. Hellingwerf. 2014. Carbon sink removal: increased photosynthetic production of lactic acid by *Synechocystis* sp. PCC 6803 in a glycogen storage mutant. J Biotechnol. 184: 100–102.

van Haveren, J., E.L. Scott and J. Sanders. 2008. Bulk chemicals from biomass. Biofuels Bioprod Bioref. 2(1): 41–57.

Varman, A.M., Y. Yu, L. You and Y.J. Tan. 2013. Photoautotrophic production of D-lactic acid in an engineered cyanobacterium. Microb Cell Fact. 12: 117.

Vazquez-Bermudex, M.F., A. Herrero and E. Flores. 2000. Uptake of 2-oxoglutarate in *Synechococcus* strain transformed with the *Escherichia coli kgtP* gene. J Bacteriol. 182: 211–215.

Veyel, D., A. Erban, I. Fehrle, J. Kopka and M. Schroda. 2014. Rationals and approaches for studying metabolism in eukaryotic microalgae. Metabolites. 4: 184–217.

Wada, N., T. Sakamoto and S. Matsugo. 2013. Multiple roles of photosynthetic and sunscreen pigments in cyanobacteria focusing on the oxidative stress. Metabolites. 3: 463–483.

Wahlen, B.D., R.M. Willis and L.C. Seefeldt. 2011. Biodiesel production by simultaneous extraction and conversion of total lipids from microalgae, cyanobacteria, and wild mixed-cultures. Bioresour technol. 102(3): 2724–2730.

Wang, B., S. Pugh, D.R. Nielsen, W. Zhang and D.R. Meldrum. 2013. Engineering cyanobacteria for photosynthetic production of 3-hydroxybutyrate directly from $CO_2$. Metab Eng. 16: 68–77.

Wang, G.S., H. Grammel, K. Abou-Aisha, R. Sagesser and R. Ghosh. 2012. High-level production of the industrial product lycopene by the photosynthetic bacterium *Rhodospirillum rubrum*. Appl Environ Microbiol. 78: 7205–7215.

Wang, K.L., H. Li and J.R. Ecker. 2002. Ethylene biosynthesis and signaling networks. Plant Cell. 14 Suppl: S131–151.

Wang, L., A. Czedik-Eysenberg, R.A. Mertz, Y. Si, T. Tohge, A. Nunes-Nesi and T.P. Brutnell. 2014. Comparative analysis of C$_4$ and C$_3$ photosynthesis in developing leaves of maize and rice. Nature Biotechnol. 32(11): 1158–1165.

Wang, W., X. Liu and X. Lu. 2013. Engineering cyanobacteria to improve photosynthetic production of alka(e)nes. Biotechnol Biofuels. 6: 69.

Wang, Y., M. Shi, X. Niu, X. Zhang, L. Gao, L. Chen, J. Wang and W. Zhang. 2014. Metabolomic evolution of butanol tolerance in photosynthetic *Synechocystis* sp. PCC6803. Microb Cell Fact. 13: 151.

Wang, Z.T., N. Ullrich, S. Joo, S. Waffenschmidt and U. Goodenough. 2009. Algal lipid bodies: stress induction, purification, and biochemical characterization in wild-type and starchless *Chlamydomonas reinhardtii*. Eukaryotic Cell. 8(12): 1856–1868.

Waterbury, J.B., S.W. Watson, R.R.L. Guillard and L.E. Brand. 1979. Widespread occurrence of a unicellular, marine, planktonic, cyanobacterium. Nature. 277: 293–4.

Wiechert, W. and K. Noh. 2005. From stationary to instationary metabolic flux analysis. Adv Biochem Eng Biotechnol. 92: 145–172.

Wijffels, R.H. and M.J. Barbosa. 2010. An outlook on microalgal biofuels. Science. 329: 796–799.

Work, V.H., R. Radakovits, R.E. Jinkerson, J.E. Meuser, L.G. Elliott, D.J. Vinyard and M.C. Posewitz. 2010. Increased lipid accumulation in the *Chlamydomonas reinhardtii* sta7-10 starchless isoamylase mutant and increased carbohydrate synthesis in complemented strains. Eukaryotic Cell. 9(8): 1251–1261.

Wu, G., Q. Wu and Z. Shen. 2001. Accumulation of poly-beta-hydroxybutyrate in cyanobacterium *Synechocystis* sp. PCC6803. Bioresour Technol. 76: 85–90.

Wu, G.F., Z.Y. Shen and Q.Y. Wu. 2002. Modification of carbon partitioning to enhance PHB production in *Synechocystis* sp. PCC 6803. Enzyme Microb Technol. 30: 710–715.

Wu, Z., Y. Zhu, W. Huang, C. Zhang, T. Li, Y. Zhang and A. Li. 2012. Evaluation of flocculation induced by pH increase for harvesting microalgae and reuse of flocculated medium. Bioresour Technol. 110: 496–502.

Xie, M., W. Wang, W. Zhang, L. Chen and X. Lu. 2017. Versatility of hydrocarbon production in cyanobacteria. Appl Microbiol Biotechnol. 101: 905–919.

Xiong, W., J.A. Morgan, J. Ungerer, B. Wang, P. Maness and J. Yu. 2015. The plasticity of cyanobacterial metabolism supports direct CO$_2$ conversion to ethylene. Nature Plants. 1: 15053.

Xu, L., P.J. Weathers, X.R. Xiong and C.Z. Liu. 2009. Microalgal bioreactors: challenges and opportunities. Eng Life Sci. 9(3): 178–189.

Xu, Y., R.M. Alvey, P.O. Byrne, J.E. Graham, G. Shen and D.A. Bryant. 2011. Expression of genes in cyanobacteria: adaptation of endogenous plasmids as platforms for high-level gene expression in *Synechococcus* sp. PCC 7002. Photosynthesis Res. Protoc. 684: 273–293.

Yacoby, I., S. Pochekailov, H. Toporik, M.L. Ghirardi, P.W. King and S. Zhang. 2011. Photosynthetic electron partitioning between [FeFe]-hydrogenase and ferredoxin: NADP+-oxidoreductase (FNR) enzymes *in vitro*. PNAS USA. 108(23): 9396–9401.

Yang, C., Q. Hua and K. Shimizu. 2000. Energetics and carbon metabolism during growth of microalgal cells under photoautotrophic, mixotrophic and cyclic light-autotrophic/dark-heterotrophic conditions. Biochem Eng J. 6(2): 87–102.

Yang, C., Q. Hua and K. Shimizu. 2002a. Quantitative analysis of intracellular metabolic fluxes using GC-MS and two-dimensional NMR spectroscopy. J Biosci Bioeng. 93(1): 78–87.

Yang, C., Q. Hua and K. Shimizu. 2002b. Metabolic flux analysis in *Synechocystis* using isotope distribution from $^{13}$C-labeled glucose. Metab Eng. 4(3): 202–216.

Yang, C., Q. Hua and K. Shimizu. 2002c. Integration of the information from gene expression and metabolic fluxes for the analysis of the regulatory mechanisms in *Synechocystis*. Appl Microbiol Biotechnol. 58(6): 813–822.

Yang, Z.H., L. Luo, X. Chang, W. Zhou, G.H. Chen, Y. Zhao and Y.J. Wang. 2012. Production of chiral alcohols from prochiral ketones by microalgal photo-biocatalytic asymmetric reduction reaction. J Ind Microbiol Biotechnol. 39: 835–841.

Yao, L., F. Qi, X. Tan and X. Lu. 2014. Improved production of fatty alcohols in cyanobacteria by metabolic engineering. Biotechol Biofuels. 7: 94.

Yim, H., R. Haselbeck, W. Niu, C. Pujol-Baxley, A. Burgard, J. Boldt, J. Khandurina, J.D. Trawick, R.E. Osterhout, R. Stephen, J. Estadilla, S. Teisan, H.B. Schreyer, S. Andrae, T.H. Yang, S.Y. Lee, M.J. Burk and S. Van Dien. 2011. Metabolic engineering of *Escherichia coli* for direct production of 1, 4-butanediol. Nat Chem Biol. 7: 445–452.

Yoo, C., S.Y. Jun, J.Y. Lee, C.Y. Ahn and H.M. Oh. 2010. Selection of microalgae for lipid production under high levels carbon dioxide. Bioresour Technol. 101(1): S71–S74.

Young, G., F. Nippgen, S. Titterbraudt and M.J. Cooney. 2010. Lipid extraction from biomass using co-solvent mixtures of ionic liquids and polar covalent molecules. Separation and Purification Technol. 72: 118–121.

Young, J.D., A.A. Shastri, G. Stephanopoulos and J.A. Morgan. 2011. Mapping photoautotrophic metabolism with isotopically nonstatinary $^{13}C$ flux analysis. Metab Eng. 13: 656–665.

Yu, J. 2014. Bio-based products from solar energy and carbon dioxide. Trends in Biotechnol. 32 (1): 5–10.

Yu, J., M. Liberton, P.F. Cliften, R.D. Head, J.M. Jacobs, R.D. Smith, D.W. Koppenaal, J.J. Brand and H.B. Pakrasi. 2015. *Synechococcus elongatus* UTEX 2973, a fast growing cyanobacterial chassis for biosynthesis using light and CO2. Sci Rep. 5: 8132.

Zarzycki, J., S.D. Axen, J.N. Kinney and C.A. Kerfeld. 2013. Cyanobacterial-based approaches to improving photosynthesis in plants. J Exp Bot. 64: 787–798.

Zhang, S. and D.A. Bryant. 2011. The tricarboxylic acid cycle in cyanobacteria. Science. 334: 1551–1553.

Zhang, S., Y. Liu and D.A. Bryant. 2015. Metabolic engineering of *Synechococcus* sp. PCC 7002 to produce poly-3-hydroxybutyrate and poly-3-hydroxybutyrate-co-4-hydroxybutyrate. Metab Eng. 32: 174–183.

Zhao, B., Y. Zhang, K. Xiong, Z. Zhang, X. Hao and T. Liu. 2011. Effect of cultivation mode on microalgal growth and $CO_2$ fixation. Chem Eng Res & Des. 89(9): 1758–1762.

Zhou, J., H. Zhang, Y. Zhang, Y. Li and Y. Ma. 2012. Designing and creating a modularized synthetic pathway in cyanobacterium *Synechocystis* enables production of acetone from carbon dioxide. Metab Eng. 14: 394–400.

Zhou, J., H. Zhang, H. Meng, Y. Zhang and Y. Li. 2014. Production of optically pure D-lactate from $CO_2$ by blocking the PHB and acetate pathways and expressing D-lactate dehydrogenase in cyanobacterium *Synechocystis* sp. PCC6803. Proc Biochem. 49: 2071–2077.

Zhou, J., T. Zhu, Z. Cai and Y. Li. 2016. From cyanochemicals to cyanofactories: A review and perspective. Microb Cell Fact. 15: 2.

Zhu, L.Y., M.H. Zong and H. Wu. 2008. Efficient lipid production with *Trichosporon fermentans* and its use for biodiesel preparation. Bioresou Technol. 99(16): 7881–7885.

Zhu, T., X.M. Xie, Z.M. Li, X.M. Tan and L. XF. 2015. Enhancing photosynthetic production of ethylene in genetically engineered *Synechocystis* sp. PCC 6803. Green Chem. 17(1): 421-434.

Ziino, M., R.B. Lo Curto, F. Salvo, D. Signorino, B. Chiofalo and D. Giuffrida. 1999. Lipid composition of *Geotrichum candidum* single cell protein grown in continuous submerged culture. Bioresour Technol. 67(1): 7–11.

# Systems Biology Approach and Modeling for the Design of Microbial Cell Factories

## ABSTRACT

It becomes more and more important to develop appropriate models for the efficient design of the cell factory for microbial biofuel and biochemical production, since the appropriate model can predict the effect of culture environment and/or the specific pathway genes knockout on the growth characteristics. Among various modeling approaches, kinetic modeling is promising in the sense of realizing the essential feature of metabolic regulation. For the proper modeling, it is important to realize the systems behavior by integrating different levels of information to understand and unravel the underlying principles of the living organisms. Namely, it is important to properly understand how the environmental stimuli are detected by the cell, how those are transduced, and how the cell metabolism is regulated, and to express these into the model. The metabolism under micro-aerobic and anaerobic conditions may be made by incorporating the global regulators such as ArcA/B and Fnr. It is quite important to develop quantitative kinetic models, which incorporate enzyme level and gene level regulations from the biological science and metabolic engineering points of view.

### Keywords

Systems biology, synthetic biology, metabolic flux analysis, flux balance analysis, virtual microbe, kinetic model, metabolic regulation, metabolic engineering, dynamics, microbial cell factories

## 1. Introduction

For the development of the microbial cell factories for biofuel and biochemical production, systems biology approach and modeling with computer simulation may play important roles. It is quite important to quantitatively understand the complex and highly interrelated cellular behavior for the efficient design and construction of microbial cell factories. The molecular knowledge alone is in many cases not sufficient to understand the cell systems behavior, where the systems behavior emerges from the interactions between the characterized molecules (Kitano 2002a). This may be attained with the help of informatics and systems biology by integrating different levels of ever increasing data with deep insight into the available data by biological knowledge (Kitano 2002a, 2002b, Stelling 2004). In living organisms, metabolic network, defined as the set and topology of metabolic biochemical reactions in a cell, plays an essential role for the cell to survive, where it is under well organized control. Thousands of different biochemical reactions as well as transport processes are linked together to break down organic compounds to generate energy (catabolism) and to synthesize macromolecular compounds (anabolism) for the cell synthesis and survival. Similarly, complex signaling networks interconvert signals or stimuli that are important for the cellular function and interactions with the environment.

In order to understand the cellular systems in response to culture environments, the coupling between the recognition or sensing of the environmental condition and the adjustment of the metabolic system must be properly analyzed for the multiple levels of information such as genome, transcriptome, proteome, metabolome, and fluxome. In particular, it is important to incorporate the coupling between enzymatic reactions and the transcriptional regulation (Kotte et al. 2010). Moreover, although local regulation mechanisms are known to exist, it is not clear how those local regulation systems are coordinated on the systems level, where this may be made by "distributed sensing of the intracellular metabolic fluxes" (Kotte et al. 2010).

Among different levels of information, the metabolic fluxes are located on top of those, and it is the most important information from the phenotypic fermentation point of view (Sauer 2006, Long and Antoniewics 2014, Quek and Nielsen 2014), and it can be used for the analysis of the specific pathway gene knockout on the metabolism (Shimizu 2004, 2013). $^{13}$C-metabolic flux analysis has been shown to be useful for the metabolic regulation analysis (Sauer 2006, Shimizu 2004, 2009, 2013, Matsuoka and Shimizu 2014, Wittman 2007). However, this is essentially the analysis method for the physiological state of the organism based on mass balance together with isotopomer balance, and it does not have the predictability. It is highly desirable, and useful to be able to predict the cell growth characteristics and the metabolic changes in response to the change in culture environment and/or the specific pathway mutation.

Several approaches for the proper metabolic design of the efficient microbial cell factories for biofuel and biochemical production may be considered, where the **synthetic biology** focuses on more genomic aspects, and in some sense bottom-up approach, while **metabolic engineering** based on **metabolic flux analysis** focuses on more phenotypes towards systems level understanding of the cell metabolism, and rather top-down approach.

The ultimate goal of systems biology is to reconstruct a cell system into the computer which can predict observable phenotypes. If this could be attained, the effects of culture environment and/or the specific genetic mutation on the metabolism can be predicted without conducting many exhaustive experiments, and metabolic engineering may be made more efficiently with verification by the experiment for the selected mutants in the optimized condition based on the computer simulation. Thus, the appropriate model can contribute for the efficient design of cell factories from the practical application point of view.

A variety of models have been proposed in the past, where they are discriminated from the others depending on the underlying assumptions for the modeling, the data they require, and the accuracy of the model prediction (Selinger et al. 2003). The types of modeling formalism depend on such characteristics (Machado et al. 2011).

The model development may start with considering network structure with kinetic rate expressions, model structure, parameter identification, and model validation, which may differ depending on the purpose of using a model (Almquist et al. 2014). It must be careful that the determination of kinetic rate expression is not straightforward due to difficulty in identifying the mechanisms of enzymes and transporters (Costa et al. 2011), and therefore, some appropriate model simplification may be considered. Moreover, parameter identification, sensitivity analysis, identifiability, experimental design, and optimization are important for the modeling in practice (Almquist et al. 2014, Ashyraliyev et al. 2009, Crijovic et al. 2011, Shimizu and Matsuoka 2015).

In this chapter, various systems biology approaches, in particular modeling approaches are explained, where metabolic regulation analysis is critical for the proper modeling, and has to be made in evaluating the performance of the designed cell as well as for reengineering the cell factories (Vemuri and Aristidou 2005, Matsuoka and Shimizu 2015).

## 2. Flux Balance Analysis and its Extensions

Flux balance analysis (FBA) and its extension to genome-scale has made significant progress as it requires only basic knowledge of metabolic reaction stoichiometry, and has reasonably accurate predictability. Significant efforts have also been made to integrate gene level regulation and metabolic networks to reveal the regulation mechanism (Herrgard et al. 2006, O'Brien et al. 2013). In such approach, however, some appropriate objective functions such as the maximization of the cell growth rate, the specific substrate consumption rate, and/or the metabolite production rate must be introduced due to excess degrees of freedom. It was, however, shown that no single objective function can accurately represent the flux data for the different culture condition (Schuetz et al. 2007). Rather, a vector-valued objective function or multiple objective functions must be considered, resulting in Pareto optimal set to represent the metabolic fluxes (Schuetz et al. 2012), where the influential objective function may be the maximum ATP yield, maximum biomass yield, and minimum sum of absolute fluxes (which corresponds to minimum enzyme investment).

The formulation may be as follows:

$$\max \quad J = [j_1, j_1, ..., j_k]^T \tag{1a}$$

Subject to

$$\sum_{j=1}^{n} S_{ij} v_j = 0 \qquad (i = 1, 2, ..., m) \tag{1b}$$

where $j_i$ is the ith objective function, $S_{ij}$ is the stoichiometric coefficient, and $v_j$ is the jth flux. However, it is not easy to determine the optimal point on the Pareto optimal set in practice without further information such as costs available.

FBA approach together with MFA information may be considered for the metabolic engineering purpose such as OptKnock (Burgard et al. 2003), a bi-level programming framework for identifying gene knockout for the strain improvement. This has been extended as OptReg to consider not only knockouts but also overexpression and down regulations of various reactions in the network (Pharkya and Maranas 2006). Another extension has also been made as OptForce (Ranganathan et al. 2010), OptFlux (Rocha et al. 2010), and DFFBA (Differential Bees FBA) with OptKnock to identify optimal gene knockout strategies for maximizing the yield of the desired phenotypes while sustaining the growth rate (Choon et al. 2014). Further extension has been made as OptStrain aiming at guiding pathway modifications, through addition or deletion of reactions, of microbial networks for the overproduction of targeted compounds based on stoichiometrically balanced approach imposing maximum product yield requirements, pinpointing the optimal substrates and evaluation of different microbial hosts such as *Helicobacter pylori*, *E. coli*, *S. cerevisiae*, and other microorganisms (Pharkya et al. 2014).

Stoichiometry-based strain design algorithms are often formulated as bi-level mixed integer linear programming problems (Burgard et al. 2003,

Pharkya et al. 2006, Ranganathan et al. 2010, Yang et al. 2011, Cotton and Reed 2013), where outer level optimizes the objective function(s), while the inner level optimizes the cellular objective that counteract any externally imposed genetic or environmental perturbations (Ibarra et al. 2002, Segre et al. 2002). Different fitness functions may be considered (Park et al. 2009, Zomorrodi et al. 2012).

The linear property of stoichiometric equations underlying FBA is the computational advantage, and allows for genome-scale extension. However, it is not easy to confirm the designed cell metabolism in view of enzymatic reactions with intracellular metabolite concentrations.

The problem in FBA and its extension to genome-scale is the difficulty for the dynamic analysis as compared to kinetic modeling approach. Some extension has been made by incorporating kinetic expressions of multiple carbon sources and other nutrients into the quasi steady-state (Covert et al. 2008, Meadows et al. 2010, Feng et al. 2012). The dynamic multi-species metabolic modeling (DMMM) approach has been considered by incorporating the metabolites uptake kinetics into stoichiometric models of a microbial consocium (Zhuang et al. 2011, Salimi et al. 2011). On the other hand, the steady-state flux distributions obtained from FBA and stoichiometric information have been used to parameterize the genome-scale kinetic models applicable for small perturbations (Jamshidi et al. 2008, Jamshidi and Palsson 2010, Smallbone et al. 2007, Smallbone et al. 2010). Lin-log kinetic expression and thermodynamics may be incorporated to constrain FBA simulation (Fleming et al. 2010).

Although some attempts have been made by the hybrid type of stoichiometric/kinetics based modeling approach (Jamshidi and Palsson 2008, Antoniewicz 2013, Hoffner et al. 2013), its potential may not be fully investigated. The dynamic flux balance analysis (dFBA) may be considered for diauxic growth of *E. coli* consuming glucose and acetate by taking into account the constraints that govern the cell growth at different phases in the batch culture (Mahadevan et al. 2002). Moreover, dFBA may be used for the co-culture with multiple sugars for the cellulosic biofuels production (Hanly and Henson 2011, Hanly et al. 2012, Hanly and Henson 2013). Recently, OptForce formalism has been extended as k-OptForce by bridging the gap between stoichiometric approach and kinetics based approach, where the procedure seamlessly integrates the mechanistic detail given by kinetic models within a constraint-optimization framework tractable for genome-scale models (Chawdhury et al. 2014).

The proper formulation for the interaction between the metabolism and gene expression by applying the principle of growth optimization, enables the accurate prediction of multi-scale phenotypes (O'Brien et al. 2013), where constitutively expressed genes show growth-rate dependent expression trends (Klumpp and Hwa 2008, Klumpp et al. 2009). This implies the economic ways of the cell system that regulates in response to global change in metabolic efficiency (Valgepea et al. 2013). Moreover, such optimality model may be used for the adaptive laboratory evolution (Harcomb et al. 2013).

The construction of a virtual microbe will be an ambitious but realistic target that builds a novel resource that can provide significant benefits in the variety of practical applications. As an extension of the constraint-based genome-scale models (Edwards et al. 2002), a whole cell computational model was developed for *Mycoplasm genetarium*, a urogenital parasite adored by synthetic biologists for its reduced genome (Karr et al. 2012). This model constitutes of 28 processes of cell's operation, where these include processes that track exchanges with the extracellular medium and all the metabolic fluxes, the state of the supercoiled chromosome, transcription of all active genes, processing of all mRNAs, translation of all proteins, formation of all macromolecular complexes including RNA polymerases and ribosomes, and progresses of cytokinesis and FtsZ polymerization (Gunawardera 2012). This may be the dawn of virtual cell biology (Freddolino and Tavazoie 2012), and this might even go beyond the previous attempt of the so-called "a grand challenge of the 21st century" (Tomita 2001).

Once again, although powerful and attractive for the possible extension to the whole cell modeling or the so-called virtual microbes, the main drawback of the above approach is the difficulty in incorporating "explicitly" the metabolic regulation mechanisms.

## 3. Kinetic Modeling and Incorporation of Metabolic Regulation

Kinetic models for the metabolism require quantitative expressions that connect fluxes and metabolite concentrations. The changes in metabolite concentrations and fluxes can be obtained with respect to time by solving a set of ordinary differential equations (ODEs) such as

$$\frac{dx_i}{dt} = \sum_{j=1}^{n_i} S_{ij} v_j (v_j^{max}, x_k, p_j) \qquad (i=1,\ldots,m) \qquad (2a)$$

where the typical mechanistic expression may be the Michaelis-Menten type such as

$$v = \frac{v_m S}{K_s + S} \qquad (2b)$$

or Hill type expression may be considered, where $v_m$ and $K_s$ are the model parameters, and S is the substrate concentration for the corresponding pathway reaction. These equations require detailed enzyme reaction mechanism and characterization (Heinrich and Rapoport 1974, van Riel 2006). In order to reduce the computational burden in association with kinetic modeling, various approximate kinetic forms such as lin-log (Heijnen

2005, Wu et al. 2004, del Rosario et al. 2008), and log-lin (Hatzimanikatis et al. 1998) kinetics, power low kinetic expressions such as S-system (Wang et al. 2006), and generalized mass action (Jamshidi and Palsson 2008), and others (Pozo et al. 2011, Sorribas et al. 2007, Liebermeister and Klipp 2006) have been proposed.

As mentioned in the previous section, although the stoichiometric constraint-based genome-scale metabolic models have been developed for a variety of organisms with Cybernetic consideration (Kim et al. 2012), it is not easy to incorporate or express the effects of intracellular metabolites and enzyme activities appropriately with such approach. Although some attempts have been made for incorporating transcriptional regulation into FBA framework in the form of Boolean rules (Covert and Palsson 2002, 2003, Covert et al. 2001, Herrgard et al. 2006), the regulatory rules are not based on the metabolic regulation mechanisms, but on the basis of the available data which may be the manifestation of part or snapshot of the real regulation mechanism (Song et al. 2009).

In contrast to the stoichiometric models, the kinetic modeling approach is attractive in the sense that such mechanism can be incorporated into the model appropriately. The primary attempts of incorporating the regulation mechanisms into kinetic models have been made by cybernetic modeling approach, where the organisms are considered to utilize the available nutrient sources with maximum efficiency by the optimal strategy (Ramkrishna et al. 1987). This approach has been extended to more structure models that contain detailed pathways (Varner and Ramkrishna 1999). More recently, this approach has been considered for the potential applications to metabolic engineering (Young 2005, Young et al. 2008). In such modeling approach, an elementary mode was considered as a metabolic subunit to model cellular regulatory processes, where the elementary modes are a set of metabolic pathways by which the cellular metabolic routes can be completely described, and any feasible fluxes can be represented by their combinations at steady state (Schuster et al. 2000). The elementary modes consist of a minimal set of reactions that function at steady state, which implies that the elementary mode cannot be a functional unit if any reaction is removed (Schuster et al. 2000). The hybrid type modeling has also been developed by assuming quasi-steady-state for the intracellular metabolites (Kim and Dang 2005, Kim et al. 2008), where several applications were made for *E. coli* (Kim et al. 2008) and for yeast (Song et al. 2009).

In the kinetic modeling approach, it is critical to identify kinetic parameter values and kinetic rate laws applicable to a variety of genetic and/or environmental perturbations. Moreover, the large-scale extension may be limited by considering unambiguous kinetic model parameterization (Teusink et al. 2000). Several attempts have been made towards postulating a generalized uniform kinetic expression such as approximate enzyme kinetic equations (Heijnen 2005, Pozo et al. 2011, Chakrabarti et al. 2013, Hatzimanikatis and Bailey 1996, Smallbone et al. 2013, Stanford et al. 2013), S-system formalism (Savageau 1970, Voit 2013), or a combination *in vitro*-based lumped and approximate rate equations (Drager et al. 2009, Costa et al. 2010). However, the predictability may not be the satisfied level (Heijnen 2005, Liebermeister and Klipp 2006).

Recently, the ensemble modeling (EM) has been considered to cope with large-scale kinetic modeling by successively reducing the size of parameter space based on the available experimental data such as fluxes and/or intracellular metabolite concentrations together with thermodynamic constraints for the direction of the net fluxes (Rizk and Liao 2009). In EM approach, any type of pathway reaction mechanism can be considered as well as already known mechanism, where each reaction is decomposed into elementary reaction steps with mass action kinetics (Tan and Liao 2012) such as

$$A + E \underset{k_{-1}}{\overset{k_1}{\Leftrightarrow}} AE \underset{k_{-2}}{\overset{k_2}{\Leftrightarrow}} BE \underset{k_{-3}}{\overset{k_3}{\Leftrightarrow}} B + E \tag{3}$$

where A,B are the metabolites and E is the enzyme.

The EM procedure starts with initially assumed kinetic models that predict the experimentally observed phenotypic characteristics, and the additional data such as those of the strain under environmental and/or genetic perturbations are used to screen the models until a minimal set of kinetic models are obtained (Tan and Liao 2012). This modeling approach has been successfully applied for lysine production (Contador et al. 2009), fatty acid production (Dean et al. 2010), aromatic production (Rizk and Liao 2009), robustness analysis for engineered non-native pathways (Lee et al. 2014), and modeling cancer cells (Khazaei et al. 2012). Moreover, this approach has been applied for the modeling of *E. coli* that reasonably predicts the fluxes and intracellular metabolite concentrations of wild type and its single gene knockout mutants (Khodayari et al. 2014) based on the available multi-omics data (Ishii et al. 2007).

## 4. Modeling of the main Metabolism for Catabolite Regulation

The metabolic reactions of the central metabolism play important roles for energy generation and the production of the precursors for biosynthesis, and those form the hub on which nearly all catabolic and anabolic processes are built. Metabolic regulation of the central metabolism plays a key role in the adaptation of the organisms in response to the change in living environment. The overall structure of the central metabolic pathways is remarkably well conserved in a wide variety of living organisms. Thus the metabolic model of the central metabolism will provide a platform for further extension to peripheral metabolism.

An attempt has been made for the modeling of the main metabolic pathways to simulate the dynamic behavior of *Saccharomyces cerevisiae* in response to the pulse addition of the carbon limited growth condition and measurement by fast sampling system (Rizzi et al. 1997, Theobald et al. 1997). The kinetic model equations for the glycolysis and pentose phosphate (PP) pathway have been developed for *E. coli* to simulate the transient data obtained by the fast sampling system (Chassagnole et al. 2002). The kinetic models for the TCA cycle and anaplerotic pathways as well as glycolysis and PP pathway were also considered to simulate the typical batch and continuous cultures with some rule based approach, where the cell growth rate was estimated based on the specific ATP production rate computed from the fluxes (Kadir et al. 2010). The kinetic modeling for the main metabolism of *E. coli* has also been made based on fluxomics and metabolomics data (Peskov et al. 2012).

A wealth of information is available on genetic regulation, biochemistry, and physiology of cellular metabolism in response to culture environment, and some attempt has been made for the modeling and simulation, where it is important to make modeling based on the integrated information from gene level to flux level by incorporating the roles of transcription factors (Kotte et al. 2010, Usuda et al. 2010, Matsuoka and Shimizu 2013). The important steps are how to incorporate (i) the effect of culture environment on global regulators, (ii) the effects of global regulators on the metabolic pathway genes, (iii) the effects of metabolic pathway genes on the corresponding enzyme activities, as well as (iv) enzyme level regulations (Fig. 1) (Matsuoka and Shimizu 2015).

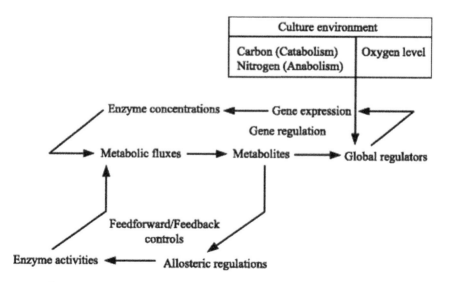

**Figure 1.** Overall metabolic regulation scheme (Matsuoka and Shimizu 2015).

## 5. Importance of the Modeling for the main Metabolic Pathways

It is by far important to model the whole main metabolic pathways such as glycolysis, TCA cycle, PP pathway, together with anaplerotic and gluconeogenic pathways. This enables us to simulate the typical batch culture, where the metabolic transition occurs.

In relation to the model development of the main metabolism, the accurate estimation of the cell growth rate is critical for the practical application point of view. In general, the cell growth rate may be expressed as a function of substrate such as Monod type model such as

$$\frac{dX}{dt} = \mu(S)X \tag{4a}$$

with

$$\mu(S) = \frac{\mu_m S}{K_s + S} \tag{4b}$$

where X and S are the cell and substrate concentrations, and $\mu$ is the specific cell growth rate. However, the saturation constant $K_s$ is small, and the dynamics depend on the maximum specific growth rate parameter $\mu_m$ which is usually constant, resulting in the difficulty in estimating the reasonable cell growth rate by Monod type model and its modification. The importance of accurate estimation of the cell growth rate is more eminent for the dynamic simulation of the specific gene knockout mutants, and for the effect of culture condition. Thus, it is desirable to be able to accurately predict the cell growth rate.

The cell growth rate is determined by the catabolic reactions such as ATP production as well as anabolic reactions under typical growth conditions. Once the main metabolic pathways could be appropriately modeled, the ATP production rate can be estimated. The model equations for the metabolic pathways are established by the mass balance with kinetic equations for the main metabolic pathways (Fig. 2). The solution to such ordinary differential equations (ODEs) can be used to compute the fluxes with respect to time. This enables us to compute the specific ATP production rate with respect to time such as

$$v_{ATP} = OP_{NADH} + OP_{FADH_2} + v_{Pgk} + v_{Pyk} + v_{Ack} + v_{SCS} - v_{Pfk} - v_{Pck} - v_{Acs} \tag{5}$$

where v is the reaction rate of the pathway (Fig. 2). $OP_{NADH}$ and $OP_{FADH_2}$ are the specific ATP production rate via oxidative phosphorylation by NADH and $FADH_2$, respectively, and those may be expressed as

$$OP_{NADH} = (v_{GAPDH} + v_{PDH} + v_{KGDH}(+v_{ICDH}) + v_{MDH}) \times (P/O) \tag{6a}$$

$$OP_{FADH_2} = v_{SDH} \times (P/O)' \qquad (6b)$$

where $(P/O)$ and $(P/O)'$ are the P/O ratios for NADH and $FADH_2$, respectively, and those are most likely to be 2.5 and 1.5, respectively, under typical aerobic condition. Those may be considered as model parameters, and can be adjusted by the experimental data (Kadir et al. 2010).

Now, $^{13}$C-metabolic flux analysis shows the correlation between the ATP production rate and the cell growth rate (Kadir et al. 2010, Yao et al. 2011, Toya et al. 2010, 2012). This indicates that the above $v_{ATP}$ can be used

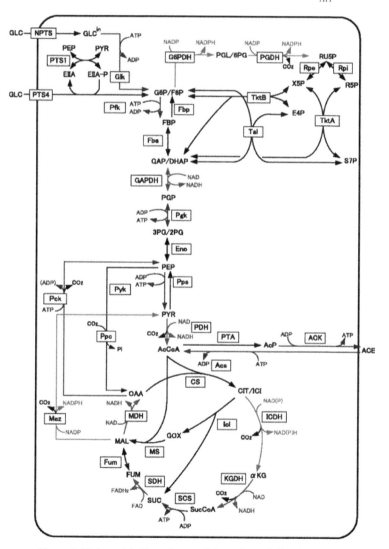

**Figure 2.** Main metabolic pathways (Matsuoka and Shimizu 2015).

to estimate the specific growth rate, and in fact it was shown that this approach allows us to estimate the cell growth rate and fluxes of the specific gene knockout mutant *E. coli* to some extent (Kadir et al. 2010, Matsuoka and Shimizu 2013).

In particular, NADH re-oxidation and substrate level phosphorylation for ATP generation are important, and ATP generation by acetate kinase (Ack) pathway is critical for survival in the case of using only xylose as a carbon source under anaerobic condition (Hasona et al. 2004). This may be simulated by the model with the cell growth rate taking into account the effect of ATP production rate as mentioned above.

Moreover, if the main metabolism was appropriately modeled, the specific $CO_2$ production rate can be also estimated by

$$v_{CO_2} = v_{PGDH} + v_{PDH} + v_{ICDH} + v_{KGDH} + v_{Mez} + v_{Pck} - v_{Ppc} \qquad (7)$$

where this can be used to estimate $CO_2$ evolution rate (CER), and thus the cell yield may be estimated together with other metabolite production rates and the cell growth rate. Since CER can be measured in practice, this may be also validated by the experimental data.

In relation to NADH production as mentioned above, the specific NADPH production rate can be also estimated as

$$v_{NADPH} = v_{G6PDH} + v_{PGDH} (+v_{ICDH}) + v_{Mez} \qquad (8)$$

where NADH is produced in many bacteria, while some bacteria such as *E. coli* produce NADPH at ICDH. It has also been observed that the specific NADPH production rate is linearly correlated with the specific growth rate from the point of view of anabolism. This means that the flux from G6P to the oxidative PP pathway can be estimated as far as the oxidative PP pathway is dominant for NADPH production, once the specific growth rate was determined from the catabolic ATP production rate (Kadir et al. 2010).

## 6. Metabolic Regulation Mechanisms to be Incorporated in the Kinetic Model

As mentioned before, several efforts have been made for the development of the appropriate kinetic models which can describe the metabolic regulation in response to genetic and/or environmental perturbations. Here, we consider the metabolic regulation mechanisms that have to be incorporated into the kinetic models, where the enzyme level regulation such as allosteric regulation may be incorporated into the kinetic rate expression, while the transcriptional regulation may be expressed in the kinetic rate expression as functions of transcription factors (TFs), where the activities of TFs may be considered to be functions of intracellular metabolites and the fluxes as will be mentioned next.

The cell system achieves the coupling between recognition and adjustment through TFs, whose activities respond to the culture environment, and regulate the expression of the associated genes. This combined recognition and adjustment forms the reaction networks that over arch the metabolic and genetic layers (Kotte et al. 2010). In general, fast action is made by the enzyme level regulation such as the feed-forward activation of pyruvate kinase (Pyk) by fructose 1,6-bisphosphate (FBP), and the feedback inhibition of phosphofruct kinase (Pfk) by phosphoenol pyruvate (PEP), a motif that enables a high level of the upstream metabolite to lower the level of the downstream metabolite (Kremling et al. 2008) (Fig. 3a). The slow action is made through the transcriptional regulation, where cAMP-Crp activates the expression of TCA cycle genes, while (FBP-inhibited) catabolite repressor/activator, Cra activates the expression of gluconeogenic pathway genes as well as some of the TCA cycle genes and the glyoxylate pathway genes, and represses the expression of the glycolysis genes in the case of *E. coli* (Fig. 3b), which will be explained later.

The levels of the flux-signaling metabolites become coupled, enabling a robust, coherent response of the TFs. The coherent behavior of the overall system is not established by a common transcriptional master regulator, but arises from the molecular interactions within the system itself (Kotte et al. 2010). It may be considered that the system of reactions of the lower glycolysis and the feed-forward activation of Pyk by FBP translate flux information into the concentration of FBP, and that this feed-forward activation affects the linearization of the glycolytic kinetics, where the glycolysis from FBP to PEP may be expressed as the reversible Michaelis-Menten (MM) equation, while Pyk may be expressed as the irreversible MM equation (Kochanowski et al. 2013) such as

$$v \to FBP \underset{k_{-1}}{\overset{k_1}{\rightleftharpoons}} PEP \xrightarrow{Pyk} PYR \tag{9}$$

where feed forward activation of FBP on Pyk may be expressed as Monod-Wyman-Changeux (MWC) kinetics (Kochanowski et al. 2013). In fact, feed-forward regulation has been known to ensure the structural robustness against perturbations (Kremling et al. 2008). This mechanism may be conserved in many organisms. For example, in *S. cerevisiae*, sugar uptake rate is well correlated with the respiratory and fermentative pathways, or the specific ethanol production rate, and the similar relationship may be seen between the glycolysis flux and FBP (Huberts et al. 2012, Christen and Sauer 2011), where Pyk is also feed-forward activated by FBP in *S. cerevisiae* (Boels and Hollenberg 1997).

In order to realize the above mechanism, the kinetic expression of Pyk (and also Ppc) must be a positive function of FBP such as

$$v_{Pyk} = v_{Pyk}([PEP],[PYR],[FBP],p_{Pyk}) \tag{10}$$

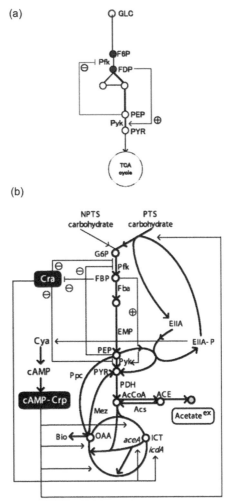

**Figure 3.** Metabolic regulation of the main metabolism: (a) Enzyme level regulation of glycolysis, (b) Transcriptional and enzyme level regulations of the main metabolic pathways.

where [·] denotes the concentration, and PEP, PYR are the substrate and product of Pfk reaction. FBP is the allosteric activator, and $\mathbf{p}_{Pfk}$ is the kinetic parameter vector for Pyk. The feed-forward activation of Pyk by FBP may be enhanced by the feedback inhibition of Pfk by PEP, where the kinetic expression for Pfk must be a negative function with respect to PEP

$$v_{Pfk} = v_{PFk}\left([F6P],[FBP],[PEP],p_{Pfk}\right) \tag{11}$$

where F6P and FBP are the substrate and product of Pfk reaction, respectively, while PEP is the allosteric inhibitor, and $p_{Pfk}$ is the parameter vector for Pyk reaction. Namely, if the sugar up-take rate or the upper glycolysis flux were increased, FBP increases, and in turn allosterically activates Pyk, which decreases PEP concentration, and the feedback inhibition of Pfk by PEP is relaxed, and causes further increase in the upper glycolysis flux (Fig. 3a). In the nominal growth condition, the feedback inhibition of Pfk by PEP may not be important, but this may cause oscillatory behavior under certain condition (Ricci 1996).

Moreover, the increased pyruvate (PYR) goes down through pyruvate dehydrogenase (PDH) to acetyl CoA (AcCoA), where AcCoA is homeostatic. Namely, the increase in AcCoA activates Ppc activity (Yang et al. 2003, Lee et al. 1999), thus reducing the upcoming Pyk-PDH fluxes and increases oxaloacetate (OAA) and activates citrate synthase (CS) reaction, thus activating the outgoing flux from AcCoA. In this way, AcCoA concentration decreases, forming the feed-back regulation against the initial increase in AcCoA (Fig. 4). This mechanism can be realized by expressing the Ppc activity as a positive function of AcCoA as well as FBP such as

$$v_{Ppc} = v_{Ppc}([PEP],[OAA],[FBP],[AcCoA],p_{Pyk}) \qquad (12)$$

where PEP and OAA are the substrate and product of Ppc reaction, and FBP and AcCoA are the allosteric activators. In another view, this phenomenon may be considered as the feed-forward regulation in the sense that the repression of TCA cycle activity is detected by the increase in AcCoA, which causes the activation of the anaplerotic Ppc pathway, and backs up the precursor metabolite such as OAA, where it is expected to be decreased due to deactivated TCA cycle.

As the glucose uptake rate increases, the TCA cycle flux tends to increase by the increased OAA and AcCoA, and then NADH is overproduced. The accumulated NADH inhibits citrate synthase (CS) and isocitrate dehydrogenase (ICDH) allosterically (Nizam et al. 2009), forming feedback regulation, and thus result in AcCoA accumulation, which in turn causes

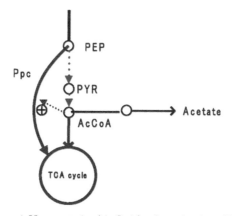

**Figure 4.** Homeostasis of AcCoA by the activation of Ppc.

acetate overflow metabolism. This enzyme level regulation by NADH in the TCA cycle can be verified by incorporating NADH oxidase (NOX) (Vemuri et al. 2006a), or nicotinic acid (Nizam et al. 2009), whereby activating TCA cycle. This effect is more enhanced under *arcA* mutant (Vemuri et al. 2006b). In the long run, the expression of TCA cycle genes is eventually repressed by the transcriptional regulation by cAMP-Crp toward steady state as will be explained later. The inhibitory effect of NADH on CS and ICDH may be expressed explicitly in the rate equation, but the problem is that the estimation of $NADH/NAD^+$ pool is difficult without detailed proper modeling of the respiratory chain, which will be mentioned later.

The typical growth condition changes from glucose-rich to acetate-rich condition in the batch culture of *E. coli*. This requires a significant reorganization of the central metabolism from glycolysis to gluconeogenesis. Although the molecular mechanism underlying the metabolic transition from glucose to acetate has been extensively investigated in *E. coli* (Wolfe 2005), its dynamics have been poorly understood. Since it is critical for the cell to efficiently and quickly reprogram the metabolism under the changing environmental condition, the cell must have the elaborate managing system.

The reaction rate for Ppc may be expressed as the function of both FBP and AcCoA as mentioned above, which then enables us to simulate the ultrasensitive regulation of anapleurosis (Xu et al. 2012). Namely, after glucose depletion, FBP concentration decreases accordingly, where Ppc and Pyk activities decrease in turn by the allosteric regulation, and PEP consumption is almost completely turned off. These make PEP concentration to be increased, and this buildup of PEP is kept during certain period (Toya et al. 2010), and this may serve to quickly uptake the glucose by PTS if it becomes available again (Xu et al. 2012). This mechanism is important for the fed-batch culture compensated by DO-stat or pH-stat, where carbon limitation often occurs periodically, and the uptake of carbon source can be made quickly and efficiently without delay. Such phenomenon can be simulated by the model as mentioned above as compared to the case without feed-forward regulation mechanism. This feed-forward regulation mechanism is also important for the modeling and simulation of lactic acid bacteria, where LDH as well as Pyk is also activated by FBP, thus producing lactate quickly and lowering the pH around the cell as soon as the glucose is available (Voit et al. 2006). Although the kinetic model for lactic acid bacteria has been developed previously (Hoefnagel et al. 2002), the above mechanism is not incorporated, and thus the simulation result does not properly reflect the real characteristics.

Moreover, after glucose depletion, FBP level drops, and thus Ppc activity decreases, while PEP carboxykinase (Pck) activity is activated by the activated Cra caused by the decreased FBP. This reveals the mechanism of avoiding the futile cycling caused by Ppc and Pck during gluconeogenic phase (Xu et al. 2012), where ATP generation becomes important. During the active glycolysis with enough sugars available, this futile cycling occurs, and loses ATP without efficient use for the compensation of the flexible metabolic fluxes and the metabolic regulation (Yang et al. 2003). This may be simulated by the appropriate models taking into account both enzymatic and transcriptional regulations.

Now, the enzyme level regulation in the glycolysis made by Pyk and Pfk as well as FBP and PEP as mentioned above keeps increasing the substrate uptake rate, where this makes the system unstable. This is counter balanced by the transcriptional regulation by cAMP-Crp, where cAMP level decreases due to the lower level of phosphorylated E II A (E II A–P), and lower activity of adenylate cyclase (Cya) at higher glucose consumption rate. Since *ptsG* which encodes a gene of PTS is under control of cAMP-Crp, the glucose uptake rate is repressed by the lower level of cAMP-Crp (Fig. 3b). Thus the incorporation of the PTS model, and the transcriptional repression of PTS by cAMP-Crp must be taken into account to realize such feedback regulation for the glucose uptake rate. The molecular mechanism for catabolite regulation has been illustrated by several researchers (Kremling et al. 2000, Kremling and Gilles 2001, Kremling et al. 2004, Sauter and Gilles 2004, Bettenbrock et al. 2006), and the PTS and catabolic regulation have been modeled by several researchers (Kotte et al. 2010, Kadir et al. 2010, Usuda et al. 2010, Nishio et al. 2008).

Moreover, the TCA cycle is transcriptionally repressed by the lower level of cAMP-Crp as the glucose consumption rate was increased. In the continuous culture (Chemostat), the effect of the cell growth rate or the dilution rate on the metabolism can be appropriately simulated for *E. coli* (Matsuoka and Shimizu 2013). Namely, as the specific growth rate was increased, FBP increases due to the increase in the fluxes of the upper glycolysis and Cra decreases. Moreover, PEP concentration decreases as the PTS flux increases, since PEP is the co-substrate of PTS, and in turn EIIA–P decreases, resulting in the decrease in Cya activity and cAMP-Crp level. The decrease in cAMP-Crp causes the repression of the TCA cycle, and causes acetate overflow metabolism as the specific growth rate increases. The trend of the simulation result by reflecting the above mechanism (Matsuoka and Shimizu 2013) is the similar as the experimental data (Ishii et al. 2007) (Fig. 5).

In *E. coli*, acetate is formed from AcCoA by Pta-Ack and also from pyruvate by pyruvate oxidase, Pox (Wolfe 2005). Acetate can be metabolized to AcCoA either by the reversed reactions of Pta-Ack or by acetyl CoA synthetase (ACS). Acetate formation has been known to be due to metabolic imbalance, where the rate of AcCoA formation via glycolysis surpasses the capacity of the TCA cycle in *E. coli* (Majewski and Domach 1990). Pox and Acs may be expressed as functions of the sigma factor ($\sigma^{38}$) RpoS, but it may be difficult to predict the behavior of RpoS, while Acs may be expressed

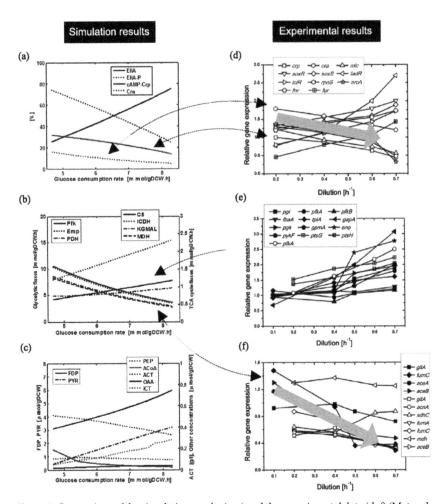

**Figure 5.** Comparison of the simulation results (a–c) and the experimental data (d–f) (Matsuoka and Shimizu 2015).

as a function of cAMP-Crp, where ACS is activated by cAMP-Crp during gluconeogenic phase (Matsuoka and Shimizu 2013).

Among intracellular metabolites, α-keto acid such as αKG turns to be a mater regulator for catabolite regulation and co-ordination of different regulations (Rabinowitz and Silhavy 2013). Namely, when favored carbon source such as glucose was depleted, αKG level fall, and cAMP increases to stimulate other carbon catabolic machinery. Namely, when preferred carbon source such as glucose is abundant, the cell growth rate becomes higher with lower cAMP level, while if it is scarce, the cell growth rate declines with higher cAMP level. In particular, under nitrogen (N)-limitation, αKG accumulates due to decreased activity of glutamate dehydrogenase

(GDH) and inhibits carbon assimilation, where there is less need for carbon-catabolic enzymes, and more demand for those involved in such nutrient assimilation. On the other hand, when anabolic nutrient such as ammonia is in excess, αKG concentration decreases due to activated GDH, producing glutamate (Glu) from αKG, cAMP level increases, and carbon catabolic enzymes increases to accelerate carbon assimilation. Namely, αKG coordinates the catabolic (C) regulation and N-regulation by inhibiting EI of PTS (Doucette et al. 2011) or cAMP via Cya (Klumpp et al. 2008, Scott et al. 2010). Moreover, the physiological function of cAMP signaling goes beyond simply enabling hierarchical utilization of carbon sources as will be mentioned later, but also controls the function of the proteome (Rabinowitz and Silhavy 2013, You et al. 2013). In order to model this phenomenon, EI of PTS has to be expressed by incorporating the inhibition by αKG, or Cya may be expressed to be inhibited by keto acids such as OAA and PYR as well as αKG, where the modeling for nitrogen regulation will be mentioned later.

In the case of biofuels production from cellulosic biomass, the hydrolyzed biomass contains multiple sugars, and those sugars are selectively assimilated with catabolite repression depending on the type of microorganism used as mentioned in Chapter 4 (Vinuselvi et al. 2012, Gorke and Stulke 2008). The metabolic regulation differs depending on the carbon sources used.

Since glycerol is the byproduct of biodiesel production (Yazdani et al. 2007, Pagliaro et al. 2007), and it is abundant and inexpensive carbon source, it is preferred to utilize such substrate for the production of biochemicals and biofuels (da Silva et al. 2009). Glycerol has a highly reduced nature as compared to other sugars such as glucose, xylose, etc., which indicates that glycerol may be more useful for the production of succinate, ethanol, lactate, and diols (Yazdani et al. 2007). Glycerol has thus been paid recent attention for the production of biofuels and biochemicals, since it is a by-product of the biodiesel production (Vasdevan and Briggs 2008, Dharmadi et al. 2006, Clomburg and Gonzalez 2013, Almeida et al. 2012, Martinez-Gomez et al. 2012). In *E. coli*, glycerol is transported, and phosphorylated to produce dihydroxy acetone phosphate (DHAP) of the central metabolism via the pathway encoded by *glpF*, *glpK*, and *glpD*, where PEP (or ATP in certain case) is used for the phosphorylation at GlpK reaction, while NADH is produced at GlpD reaction under aerobic condition (Fig. 6). These genes are under catabolic regulation by cAMP-Crp, so that glycerol is assimilated after glucose was depleted if glucose co-exists. NADH production at GlpD becomes important for the biofuels production under anaerobic condition affecting NADH/NAD$^+$ balance for dehydrogenase reactions. In the case of using glycerol as a single carbon source, cAMP-Crp increases due to the increase in the phosphorylated EIIA$^{Glc}$, where cAMP-Crp induces *glpFKD* genes. Since FBP concentration decreases in the case of using glycerol as a carbon source, Cra is activated, and this together with up-regulation of

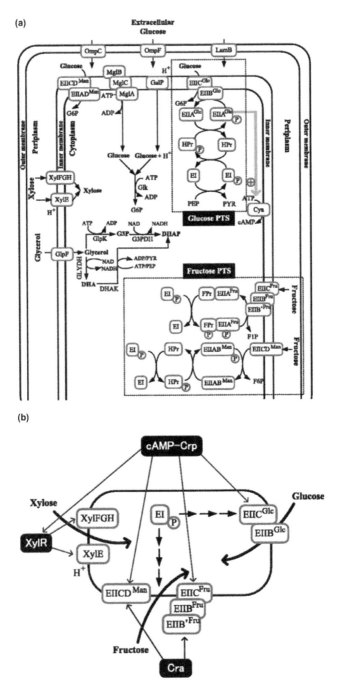

**Figure 6.** Glucose-, fructose-, and xylose-assimilating pathways (a, b) (Matsuoka and Shimizu 2015).

cAMP-Crp causes *pckA* gene as well as TCA cycle genes to be up-regulated (Oh and Liao 2000, Peng and Shimizu 2003). The kinetic expression for the glycerol uptake pathways has been proposed (Cintolesi et al. 2012). This together with the inclusion of PTS and the transcriptional regulation as mentioned above enables the simulation for the case of using multiple carbon sources such as glucose and glycerol. Moreover, the enhancement of the TCA cycle caused by the increase in cAMP-Crp can be also simulated in the case of using glycerol as a single carbon source.

In the case of using fructose, it is transported by fructose-PTS, which has its own HPr like protein domain called FPr. Namely, the phosphate of PEP is first transferred to EI (as EI-P), but then this phosphate is transferred to FPr instead of HPr, and in turn the phosphate is transferred via fructose specific EIIA$^{Frc}$ and EIIBC$^{Frc}$, and phosphorylates fructose, where phosphorylated fructose becomes fructose 1-phosphate (F1P) (Saier and Ramseier 1996). The *fruBKA* operon is under control of cAMP-Crp, and thus glucose is preferentially consumed by glucose PTS when glucose co-exists. On the other hand, this operon is repressed by Cra (Kornberg 2001). Because of this, *cra* gene knockout enables co-consumption of glucose and fructose with fructose to be consumed faster as compared to glucose (Yao and Shimizu 2013), where activated FruB in *cra* mutant competes with HPr (for glucose phosphorylation) for the phosphate of EI-P. Since phosphorylation of EIIA$^{Glc}$ via HPr becomes lower (Crasnier-Mednansky et al. 1997), the glucose up-take rate decreases as compared to the wild type strain (Yao et al. 2013). This phenomenon may be also simulated by the similar expression as glucose-PTS, but compete the phosphate of EI with glucose-PTS (Fig. 6).

In the case of using xylose as a carbon source, it is transported either by an ATP dependent high affinity ABC transporter encoded by *xylFGH*, or ATP independent low affinity proton symporter encoded by *xylE* (Griffith et al. 1992, Sumiya et al. 1995). In the case of xylose utilization, the transcription factor XylR regulates *xylAB/xylFGH* (Song and Park 1997), where *xylR* is under control of cAMP-Crp, and then catabolite repression occurs when glucose co-exists, where glucose is preferentially consumed first. The kinetic model for xylose uptake pathways as well as Entner-Doudoroff (ED) pathways has been proposed for *Zymomonas mobilis* (Altintas et al. 2006), and thus it is necessary to incorporate the activation of the transporter by cAMP-Crp, and this can be made for the catabolite repression when co-exist with glucose (Matsuoka and Shimizu 2013) (Fig. 6).

## 7. Modeling for the Metabolism under Oxygen Limitation

Most of the biofuel and biochemical production by microbes is made by the fermentation under anaerobic conditions, and therefore it is important to properly model such fermentation as well as under aerobic condition, where the latter is often employed for the enhancement of the cell growth rate before anaerobic condition to improve the productivity of the target metabolites.

Although the kinetic modeling with computer simulation has been attemped for several fermentations such as lactate fermentation (Hoefnagel et al. 2002), and acetone-butanol-ethanol fermentation (Li et al. 2011, Shinto et al. 2007, 2008), the regulatory mechanisms are rarely incorporated. Moreover, cofactor balance such as NADH balance becomes important under anaerobic condition, and thus it may be better to appropriately incorporate these in the model equations.

Microbial cells are able to adapt to different oxygen availability, where this requires reorganization of the metabolism by both enzymatic and transcriptional regulations. Many facultative anaerobes such as *E. coli* can survive at various oxgen levels (Unden et al. 1995, Green et al. 2009). The metabolism of a cell will change from fully anaerobiosis to fully aerobiosis, where fermentation pathways are active in the complete absence of oxygen (anaerobiosis), while fermentation diminishes and respiratory pathways become active upon oxygen availability. Under micro-aerobic (semi-aerobic) conditions with intermediate range of oxygen availability, both fermentative and respiratory pathways are active, where molecular oxygen is present, but at concentrations too low to allow for a complete aerobic respiration (Bettenbrock et al. 2014).

Oxygen serves as a final electron acceptor of the electron transport chain (ETC) (Ingredew and Poole 1984, Bettenbrock et al. 2014), where the function of ETC is to successively transport electrons from donors to acceptors, while translocating protons from cytoplasm through inner (cytosolic) membrane into periplasm. The resulting proton motive force (PMF) caused by the proton gradient across the membrane between cytosol and periplasm can be used to generate ATP by ATPase. In relation to ETC, dehydrogenases oxidize cytoplasmic electron donors such as NADH and $FADH_2$ by reducing membrane-associated quinones to quinoles, where the related enzymes are NADH dehydrogenase I (Nuo) (Hayashi et al. 1989) and II (Ndh) (Young and Wallace 1976), succinate dehydrogenate (SDH), and fumarate reductase (Frd) (Hirsch et al. 1963, Guest 1981), where NADH-I encoded by *nuoABCDE* operon translocates proton contributing ATP generation, while NDH-II encoded by *ndh* does not translocate proton, and not contributing to the generation of PMF ($H^+/O = 0$).

The terminal oxidases (Cytochlomes) reoxidize the quinols by giving electron to oxygen (electron acceptor). The *E. coli* ETC uses the redox pairs such as ubiquinone/ubiquinol, menaquinone/menaquinol, as well as demethylmenaquinon/demethylmenaquinol (Bettenbrock et al. 2014).

In order to properly model the metabolic transition from aerobic to anaerobic conditions, the roles of global regulators such as ArcA/B and

Fnr must be properly incorporated, where the effect of dissolved oxygen concentration on the activation of such TFs has been reported (Alexeeva et al. 2003, Levanon et al. 2005), and this may be taken into account for the simulation under microaerobic conditions. In particular, it is important to properly simulate the behavior at the branch point of PYR, where the reaction rate through PDH, $v_{PDH}$ must be the negative function of ArcA (or phosphorylated ArcA, ArcA-P), while the reaction rate through pyruvate formate lyase (Pfl), $v_{Pfl}$ is the positive function with respect to ArcA and Fnr, where PYR is converted to formate and AcCoA. Moreover, ethanol forming pathway from AcCoA, alcohol dehydrogenase (ADH) must be activated where NADH is required for this reaction. As for TCA cycle, α-ketoglutarate dehydrogenase (KGDH) may be repressed, while formate reductase (Frd) is activated by Fnr, thus the TCA cycle will be branched under anaerobic conditions. Some attempt has been made to estimate the fluxes in relation to such global regulators (Cox et al. 2005).

## 8. Concluding Remarks

Completeness of the model may not be necessary for it to improve predictions or rationalizations. The uncertainty of the model or the simulation result comes either from uncertainty of the model parameters, uncertainty of the model structure due to ambiguity in the selection of rate laws, and uncertainty caused by neglecting the regulation mechanism, and neglecting cofactor balances, etc. The importance of the above factors depends on the strains and culture conditions. It is useful in understanding cellular mechanism, and the process towards the virtual microbe could well pay off. The appropriate model will be of immense value to (Westerhoff et al. 2015):

- biotechnologists aiming to improve fermentation performances such as the yield and productivity of the target metabolite,
- microbial engineers aiming to design novel microbes able to capture available carbon, and produce bio-fuels and biochemicals,
- basic scientists aiming to understand the metabolic regulation system in the living organisms, which can be used for the synthetic biology,
- systems biologists aiming to advance the science of modeling.

The modeling approach will exceed the importance of the microbial genome sequencing projects, as it will be much closer to understanding biological function and will have widespread practical application.

In the present chapter, it is stressed the importance of incorporating the enzyme level and transcriptional regulations appropriately in the kinetic model to predict the cell's growth characteristics under environmental and/or genetic perturbations to attain the above merits. The drawback of the kinetic modeling is the increase in the kinetic model parameters as the system becomes large, and thus this may be difficult to expand to genome-scale. The reasonable idea may be to consider the kinetic modeling only for the main metabolism, and the simplified model may be considered for the peripheral metabolisms (Mannan et al. 2015).

Moreover, it is quite important to combine the catabolic regulation model with nitrogen regulation model for the coordination between C- and N-regulations, where the intracellular pool sizes of α-keto acids play important roles affecting PTS and cAMP level. Moreover, it is critical to develop a model that can express the metabolic behavior under both aerobic and micro-aerobic or anaerobic conditions for the biofuel and biochemical production by microbes.

The simulation result based on the model developed must be verified by experiments, or the simulation result may give hint for additional experimental design. In this way, modeling approach together with experimental works contribute to the innovation for the efficient design of the cell factories for biofuel and biochemical production.

## References

Alexeeva, S., K.J. Hellingwerf] and J.T. de Mattos. 2003. Requirement of ArcA for redox regulation in *Escherichia coli* under microaerobic but not anaerobic or aerobic conditions. J Bacteriol. 185: 204–209.

Almeida, J.R.M., L.C.L. Fávaro, F. Betania and B.F. Quirino. 2012. Biodiesel biorefinery: opportunities and challenges for microbial production of fuels and chemicals from glycerol waste. Biotechnol Biofuels. 5: 48.

Almquist, J., M. Cvijovic, V. Hatzimanikatis, J. Nielsen and M. Jirstrand. 2014. Kinetic models in industrial biotechnology-improving cell factory performance. Metabolic Eng. 24: 38–60.

Altintas, M.M., C.K. Eddy, M. Zhang, J.D. McMillan and D.S. Kompala. 2006. Kinetic modeling to optimize pentose fermentation in *Zymomonas mobilis*. Biotechnol Bioeng. 94: 273–295.

Antoniewicz, M.R. 2013. Dynamic metabolic flux analysis—tools for probing transient states of metabolic networks. Curr Opin Biotechnol. 24: 973–978.

Ashyraliyev, M., Y. Fomekong-Nanfack, J.A. Kaandorp and J.G. Blom. 2009. Systems biology: parameter estimation for biochemical models. FEBS J. 276: 886–902.

Bettenbrock, K., H. Bai, M. Ederer, J. Green, K.J. Hellingwerf, M. Holcombe, S. Kunz, M.D. Rolfe, G. Sanguinetti, O. Sawodny, P. Sharma, S. Steinsiek and R.K. Poole. 2014. Towards a systems level understanding of the oxygen response of *Escherichia coli*. Adv Microb Physiol. 64: 65–114.

Bettenbrock, K., S. Fischer, A. Kremling, K. Jahreis, T. Sauter and E.D. Gilles. 2006. A quantitative approach to catabolite repression in *Escherichia coli*. J Biol Chem. 281: 2578–2584.

Boels, E. and C.P. Hollenberg. 1997. The molecular genetics of hexose transport in yeasts. FEMS Microbiol Rev. 21: 85–111.

Burgard, A.P., P. Pharkya and C.D. Maranas. 2003. Optknock: a bilevel programming framework for identifying gene knockout strategies for microbial strain optimization. Biotechnol Bioeng. 84: 647–657.

Chakrabarti, A., L. Miskovic, K.C. Soh and V. Hatzimanikatis. 2013. Towards kinetic modeling of genome-scale metabolic networks without sacrificing stoichiometric, thermodynamic and physiological constraints. Biotechnol J. 8: 1043–1057.

Chassagnole, C., N. Noisommit-Rizzi, J.W. Schmid, K. Mauch and M. Reuss. 2002. Dynamic modeling of the central carbon metabolism of *Escherichia coli*. Biotechnol Bioeng. 79: 53–73.

Choon, Y.W., M.S. Mohamad, S. Deris, R.M. Illias, C.K. Chong, L.E. Chai, S. Omatu and J.M. Corchado. 2014. Differential bees flux balance analysis with Opt Knock for *in silico* microbial strains optimization. PLoS One. 9: e102744.

Chowdhury, A., A.R. Zomorrodi and C.D. Maranas. 2014. k-OptForce: integrating kinetics with flux balance analysis for strain design. PLoS Comput Biol. 10: e1003487.

Christen, S. and U. Sauer. 2011. Intracellular characterization of aerobic glucose metabolism in seven yeast species by 13C flux analysis and metabolomics. FEMS Yeast Res. 11: 263–272.

Cintolesi, A., J.M. Clomburg, V. Rigou, K. Zygourakis and R. Gonzalez. 2012. Quantitative analysis of the fermentative metabolism of glycerol in *Escherichia coli*. Biotechnol Bioeng. 109: 187–198.

Clomburg, J.M. and R. Gonzalez. 2013. Anaerobic fermentation of glycerol: a platform for renewable fuels and chemicals. Trends Biotechnol. 1: 20–28.

Contador, C.A., M.L. Rizk, J.A. Asenjo and J.C. Liao. 2009. Ensemble modeling for strain development of L-lysine-producing *Escherichia coli*. Metab Eng. 11(4-5): 221–233.

Costa, R.S., D. Machado, I. Rocha and E.C. Ferreira. 2010. Hybrid dynamic modeling of *Escherichia coli* central metabolic network combining michaelis–menten and approximate kinetic equations. Biosystems. 100: 150–157.

Costa, R.S., D. Machado, I. Rocha and E.C. Pereira. 2011. Critical perspective on the consequences of the limited availability of kinetic data in metabolic dynamic modeling. IET Syst Biol. 5: 157–163.

Cotton, C. and J.L. Reed. 2013. Constraint-based strain design using continuous modifications (CosMos) of flux bounds finds new strategies for metabolic engineering. Biotechnol J. 8: 595–604.

Covert, M.W. and B.O. Palsson. 2002. Transcriptional regulation in constraints-based metabolic models of *Escherichia coli*. J Biol Chem. 277: 28058–28064.

Covert, M.W. and B.O. Palsson. 2003. Constraints-based models: regulation of gene expression reduces the steady-state solution space. J Theor Biol. 221: 309–325.

Covert, M.W., C.H. Schilling, I. Famili, J.S. Edwards, I.I. Goryanin, E. Selkov and B.O. Palsson. 2001. Metabolic modeling of microbial strains *in silico*. Trends Biochem Sci. 26: 179–186.

Covert, M.W., N. Xiao, T.J. Chen and J.R. Karr. 2008. Integrating metabolic, transcriptional regulatory and signal transduction models in *Escherichia coli*. Bioinform. 24: 2044–2050.

Cox, S.J., S.S. Levanon, G.N. Bennett and K.Y. San. 2005. Genetically constrained metabolic flux analysis. Metab Eng. 7: 445–456.

Crasnier-Mednansky, M., M.C. Park, W.K. Studley and M.H. Saier Jr. 1997. Cra-mediated regulations of *Escherichia coli* adenylate cyclase. Microbiol. 143: 785–792.

Cvijovic, M., S. Bordel and J. Nielsen. 2011. Mathematical models of cell factories: moving towards the core of industrial biotechnology. Microb Biotechnol. 4: 572–584.

da Silva, G.P., M. Mack and J. Contiero. 2009. Glycerol: a promising and abundant carbon source for industrial microbiology. Biotechnol Adv. 27(1): 30–9.

Dean, J.T., M.L. Rizk, Y. Tan, K.M. Dipple and J.C. Liao. 2010. Ensemble modeling of hepatic fatty acid metabolism with a synthetic glyoxylate shunt. Biophys J. 98: 1385–1395.

del Rosario, R.C.H., E. Mendoza and E.O. Voit. 2008. Challenges in lin-log modelling of glycolysis in *Lactococcus lactis*. Iet Syst Biol. 2: 136–149.

Dharmadi, Y., A. Murarka and R. Gonzalez. 2006. Anaerobic fermentation of glycerol by *Escherichia coli*: A new platform for metabolic engineering. Biotechnol Bioeng. 94: 821–829.

Doucette, C.D., D.J. Schwab, N.S. Wingreen and J.D. Rabinowitz. 2011. Alpha-ketoglutarate coordinates carbon and nitrogen utilization via enzyme I inhibition. Nat Chem Biol. 7: 894–901.

Dräger, A., M. Kronfeld, M.J. Ziller, J. Supper, H. Planatscher, J.B. Magnus, M. Oldiges, O. Kohlbacher and A. Zell. 2009. Modeling metabolic networks in *C. glutamicum*: a comparison of rate laws in combination with various parameter optimization strategies. BMC Syst Biol. 3: 5.

Edwards, J.S., M.W. Covert and B.O. Palsson. 2002. Metabolic modelling of microbes: the flux-balance approach. Environ Microbiol. 4: 133–140.

Feng, X., Y. Xu, Y. Chen and Y.J. Tang. 2012. Microbes flux: a web platform for drafting metabolic models from the KEGG database. BMC Syst Biol. 6: 94.

Fleming, R.M., I. Thiele, G. Provan and H.P. Nasheuer. 2010. Integrated stoichiometric, thermodynamic and kinetic modelling of steady state metabolism. J Theor Biol. 264: 683–692.

Freddolino, P.L. and S. Tavazoie. 2012. The dawn of virtual cell biology. Cell. 150: 248–250.

Gorke, B. and J. Stulke. 2008. Carbon catabolite repression in bacteria: many ways to make the most out of nutrients. Nature Rev Microbiol. 6: 613–624.

Green, J., J.C. Crack, A.J. Thomson and N.E. LeBrun. 2009. Bacterial sensors of oxygen. Curr Opin Microbiol. 12(2): 145–51.

Griffith, J.K., M.E. Baker, D.A. Rouch, M.G. Page, R.A. Skurray, I.T. Paulsen, K.F. Chater, S.A. Baldwin and P.J. Henderson. 1992. Membrane transport proteins: implications of sequence comparisons. Curr Opin Cell Biol. 4: 684–695.

Guest, J.R. 1981. Partial replacement of succinate dehydrogenase function by phage- and plasmid-specified fumarate reductase in *Escherichia coli*. J Gen Microbiol. 122(2): 171–9.

Gunawardera, J. 2012. Silicon dreams of cells into symbols. Nature. 30: 838–840.

Hanly, T.J., M. Urello and M.A. Henson. 2012. Dynamic flux balance modeling of *S. cerevisiae* and *E. coli* co-cultures for efficient consumption of glucose/xylose mixtures. Appl Microbiol Biotechnol. 93: 2529–2541.

Hanly, T.J. and M.A. Henson. 2011. Dynamic flux balance modeling of microbial co-cultures for efficient batch fermentation of glucose and xylose mixtures. Biotechnol Bioeng. 108: 376–385.

Hanly, T.J. and M.A. Henson. 2013. Dynamic metabolic modeling of a microaerobic yeast co-culture: predicting and optimizing ethanol production from glucose/xylose mixtures. Biotechnol Biofuels. 6: 44.

Harcomb, W.R., N.F. Delaney, N. Leiby, N. Klitgord and C.J. Marx. 2013. The ability of flux balance analysis to predict evolution of central metabolism scales with the initial distance to the optimum. PLoS Comput Biol. 9: e1003091.

Hasona, A., Y. Kim, F.G. Healy, L.O. Ingram and K.T. Shanmugam. 2004. Pyruvate formate lyase and acetate kinase are essential for anaerobic growth of *Escherichia coli* on xylose. J Bacteriol. 186: 7593–7600.

Hatzimanikatis, V. and J.E. Bailey. 1996. MCA has more to say. J Theor Biol. 182: 233–242.

Hatzimanikatis, V., M. Emmerling, U. Sauer and J.E. Bailey. 1998. Application of mathematical tools for metabolic design of microbial ethanol production. Biotechnol Bioeng. 58: 154–161.

Hayashi, M., T. Miyoshi, S. Takashina and T. Unemoto. 1989. Purification of NADH-ferricyanide dehydrogenase and NADH-quinone reductase from *Escherichia coli* membranes and their roles in the respiratory chain. Biochim Biophys Acta. 977: 62–69.

Heijnen, J.J. 2005. Approximative kinetic formats used in metabolic network modeling. Biotechnol Bioeng. 91: 534–545.

Heinrich, R. and T.A. Rapoport. 1974. A linear steady-state treatment of enzymatic chains. General properties, control and effector strength. Eur J Biochem. 42: 89–95.

Herrgard, M.J., B.-S. Lee, V. Portnoy and B.O. Palsson. 2006. Integrated analysis of regulatory and metabolic networks reveals novel regulatory mechanisms in *Saccharomyces cerevisiae*. Genome Research. 16: 627–635.

Herrgård, M.J., S.S. Fong and B.O. Palsson. 2006. Identification of genome-scale metabolic network models using experimentally measured flux profiles. Plos Comp Biol. 2: 676–686.

Hirsch, C.A., M. Rasminsky, B.D. Davis and E.C. Lin. 1963. A fumarate reductase in *Escherichia coli* distinct from succinate dehydrogenase. J Biol Chem. 238: 3770–3774.

Hoefnagel, M.H.N., M.J.C. Starrenburg, D.E. Martens, J. Hugenholtz, M. Kleerebezem, I.I. Van Swam, R. Bongers, H.V. Westerhoff and J.L. Snoep. 2002. Metabolic engineering of lactic acid bacteria, the combined approach: kinetic modeling, metabolic control and experimental analysis. Microbiol. 148: 1003–1013.

Hoffner, K., S.M. Harwood and P.I. Barton. 2013. A reliable simulator for dynamic flux balance analysis. Biotechnol Bioeng. 110: 792–802.

Huberts, D.H., B. Niebel and M. Heinemann. 2012. A flux-sensing mechanism could regulate the switch between respiration and fermentation. FEMS Yeast Res. 12: 118–128.

Ibarra, R.U., J.S. Edwards and B.O. Palsson. 2002. *Escherichia coli* K-12 undergoes adaptive evolution to achieve *in silico* predicted optimal growth. Nature. 420: 186–189.

Ingledew, W.J. and R.K. Poole. 1984. The respiratory chains of *Escherichia coli*. Microbiol Rev. 48: 222–271.

Ishii, N., K. Nakahigashi, T. Baba, M. Robert, T. Soga, A. Kanai, T. Hirasawa, M. Naba, K. Hirai, A. Hoque, P.Y. Ho, Y. Kakazu, K. Sugawara, S. Igarashi, S. Harada, T. Masuda, N. Sugiyama, T. Togashi, M. Hasegawa, Y. Takai, K. Yugi, K. Arakawa, N. Iwata, Y. Toya, Y. Nakayama, T. Nishioka, K. Shimizu, H. Mori and M. Tomita. 2007. Multiple high-throughput analyses monitor the response of *E. coli* to perturbations. Science. 316: 593–597.

Jamshidi, N. and B.O. Palsson. 2008. Formulating genome-scale kinetic models in the post-genome era. Mol Syst Biol. 4: 171.

Jamshidi, N. and B.O. Palsson. 2010. Mass action stoichiometric simulation models: incorporating kinetics and regulation into stoichiometric models. Biophysical J. 98: 175–185.

Kadir, T.A., A.A. Mannan, A.M. Kierzek, J. McFadden and K. Shimizu. 2010. Modeling and simulation of the main metabolism in *Escherichia coli* and its several single-gene knockout mutants with experimental verification. Microb Cell Fact. 9: 88.

Karr, J.R., J.C. Sanghvi, D.N. Macklin, M.W. Gutschow, J.M. Jacobs, B. Bolival Jr, N. Assad-Garcia, J.I. Glass and M.W. Covert. 2012. A whole-cell computational model predicts phenotype from genotype. Cell. 150: 389–401.

Khazaei, T., A. McGuigan and R. Mahadevan. 2012. Ensemble modeling of cancer metabolism. Front Physiol. 3: 135.

Khodayari, A., A.R. Zomorrodi, J.C. Liao and C.D. Maranas. 2014. A kinetic model of *Escherichia coli* core metabolism satisfying multiple sets of mutant flux data. Metab Eng. 25: 50–62.

Kim, J.I., H.S. Song, S.R. Sunkara, A. Lali and D. Ramkrishna. 2012. Exacting predictions by cybernetic model confirmed experimentally: steady state multiplicity in the chemostat. Biotechnol Prog. 28: 1160–1166.

Kim, J.I., J.D. Varner and D. Ramkrishna. 2008. A hybrid model of anaerobic *E. coli* GJT001: combination of elementary flux modes and cybernetic variables. Biotechnol Prog. 24: 993–1006.

Kim, J.W. and C.V. Dang. 2005. Multifaceted roles of glycolytic enzymes. Trends Biochem Sci. 30: 142–150.

Kitano, H. 2002a. Systems biology: a brief overview. Science. 295: 1662–1664.

Kitano, H. 2002b. Computational systems biology. Nature. 420: 206–210.

Klumpp, S. and T. Hwa. 2008. Growth-rate dependent partitioning of RNA polymerases in bacteria. PNAS USA. 105: 20245–20250.

Klumpp, S., Z. Zhang and T. Hwa. 2009. Growth-rate dependent global effects on gene expression in bacteria. Cell. 139: 1366–1375.

Kochanowski, K., B. Volkmer, L. Gerosa, B.R. Haverkorn van Rijsewijk, A. Schmidt and M. Heinemann. 2013. Functioning of a metabolic flux sensor in *Escherichia coli*. PNAS USA. 110: 1130–1135.

Kornberg, H.L. 2001. Routes for fructose utilization by *Escherichia coli*. J Mol Microbiol Biotechnol. 3: 355–359.

Kotte, O., J.B. Zaugg and M. Heinemann. 2010. Bacterial adaptation through distributed sensing of metabolic fluxes. Mol Sys Biol. 6: 355.

Kremling, A. and E.D. Gilles. 2001. The organization of metabolic reaction networks. II. Signal processing in hierarchical structured functional units. Metab Eng. 3: 138–150.

Kremling, A., K. Bettenbrock and E.D. Gilles. 2008. A feed-forward loop guarantees robust behavior in *Escherichia coli* carbohydrate uptake. Bioinformatics. 24: 704–710.

Kremling, A., K. Jahreis, J.W. Lengeler and E.D. Gilles. 2000. The organization of metabolic reaction networks: A signal-oriented approach to cellular models. Metab Eng. 2: 190–200.

Kremlng, A., S. Fischer, T. Sauter, K. Bettenbrock and E.D. Gilles. 2004. Time hierarchies in the *Escherichia coli* carbohydrate uptake and metabolism. BioSystems. 73: 57–71.

Lee, B., J. Yen, L. Yang and J.C. Liao. 1999. Incorporating qualitative knowledge in enzyme kinetic models using fuzzy logic. Biotechnol Bioeng. 63: 722–729.

Lee, Y., J.G. Lafontaine Rivera and J.C. Liao. 2014. Ensemble modeling for robustness analysis in engineering non-native metabolic pathways. Metab Eng. 25: 63–71.

Li, R.-D., Y.-Y. Li, L.-Y. Lu, C. Ren, Y.-X. Li and L. Liu. 2011. An improved kinetic model for the acetone-butanol-etahnol pathway of *Clostridium acetobutyricum* and model-based perturbation analysis. BMC Sys Biol. 5: S12.

Liebermeister, W. and E. Klipp. 2006. Bringing metabolic networks to life: convenience rate law and thermodynamic constraints. Theor Biol Med Model. 3: 41.

Long, C.P. and M.R. Antoniewicz. 2014. Metabolic flux analysis of *Escherichia coli* knockouts: lessons from the Keio collection and future outlook. Curr Opin Biotechnol. 28: 127–133.

Machado, D., R. Costa, M. Rocha, E. Ferreira, B. Tidor and I. Rocha. 2011. Modeling formalisms in systems biology. AMP Expre. 1: 1–34.

Mahadevan, R., J.S. Edwards and F.J. Doyle. 2002. Dynamic flux balance analysis of diauxic growth in *Escherichia coli*. Biophys J. 83: 1331–1340.

Majewski, R.A. and M.M. Domach. 1990. Simple constrained-optimization view of acetate overflow in *Escherichia coli*. Biotech Bioeng. 35: 732–738.

Mannan, A.A., Y. Toya, K. Shimizu, J. McFadden, A.M. Kierzek and A. Rocco. 2015. Integrating kinetic model of *E. coli* with genome scale metabolic fluxes overcomes its open system problem and reveals bistability in central metabolism. PLoS One. 10(10): e0139507.

Martínez-Gómez, K., N. Flores, H.M. Castañeda, G. Martínez-Batallar, G. Hernández-Chávez, O.T. Ramírez, G. Gosset, S. Encarnación and F. Bolivar. 2012. New insights into *Escherichia coli* metabolism: carbon scavenging, acetate metabolism and carbon recycling responses during growth on glycerol. Micob Cell Fact. 11: 46.

Matsuoka, Y. and K. Shimizu. 2013. Catabolite regulation analysis of *Escherichia coli* for acetate overflow mechanism and co-consumption of multiple sugars based on systems biology approach using computer simulation. J Biotechnol. 168: 155–173.

Matsuoka, Y. and K. Shimizu. 2014. Metabolic flux analysis. *In*: J.O. Kromer, L.K. Nielsen and L.M. Blank (eds.). Human press.

Matsuoka, Y. and K. Shimizu. 2015. Current status and future perspectives of kinetic modeling for the cell metabolism with incorporation of the metabolic regulation mechanism. Bioresour Bioprocess. 2: 4.

Meadows, A.L., R. Karnik, H. Lam, S. Forestell and B. Snedecor. 2010. Application of dynamic flux balance analysis to an industrial *Escherichia coli* fermentation. Metab Eng. 12: 150–160.

Nishio, Y., Y. Usuda, K. Matsui and H. Kurata. 2008. Computer-aided rational design of the phosphotransferase system for enhanced glucose uptake in *Escherichia coli*. Mol Syst Biol. 4: 160.

Nizam, S.A., J. Zhu, P.Y. Ho and K. Shimizu. 2009. Effects of *arcA* and *arcB* genes knockout on the metabolism in *Escherichia coli* under aerobic condition. Biochem Eng J. 44: 240–250.

O'Brien, E.J., J.A. Lerman, R.L. Chang, D.R. Hyduke and B.O. Palsson. 2013. Genome-scale models of metabolism and gene expression extend and refine growth phenotype prediction. Mol Sys Biol. 9: 693.

Oh, M.K. and J.C. Liao. 2000. Gene expression profiling by DNA microarrays and metabolic fluxes in *Escherichia coli*. Biotechnol Prog. 16: 278–286.

Pagliaro, M., R. Ciriminna, H. Kimura, M. Rossi and C. Della Pina. 2007. From glycerol to value-added products. Angew Chem Int Ed Engl. 46(24): 4434–4440.

Peng, L. and K. Shimizu. 2003. Global metabolic regulation analysis for *Escherichia coli* K12 based on protein expression by 2-dimensional electrophoresis and enzyme activity measurement. Appl Microbiol Biotechnol. 61: 163–178.

Peskov, K., E. Mogilevskaya and O. Demin. 2012. Kinetic modelling of central carbon metabolism in *Escherichia coli*. FEBS J. 279: 3374–3385.

Pharkya, P., A.P. Burgard and C.D. Maranas. 2014. OptStrain: a computational framework for redesign of microbial production systems. Genom Res. 14: 2367–2376.

Pharkya, P. and C.D. Maranas. 2006. An optimization framework for identifying reaction activation/inhibition or elimination candidates for overproduction in microbial systems. Metab Eng. 8: 1–13.

Pozo, C., A. Marín-Sanguino, R. Alves, G. Guillén-Gosálbez, L. Jiménez and A. Sorribas. 2011. Steady-state global optimization of metabolic non-linear dynamic models through recasting into power-law canonical models. BMC Syst Biol. 5: 137.

Quek, L.-E. and L.K. Nielsen. 2014. A depth-first search algorithm to compute elementary flux modes by linear programming. BMC Syst Biol. 8: 94.

Rabinowitz, J. and T.J. Silhavy. 2013. Metabolite turns master regulator. Nature. 500: 283–284.

Ramkrishna, D., D.S. Kompala and G.T. Tsao. 1987. Are microbes optimal strategists. Biotechnol Prog. 3: 121–126.

Ranganathan, S., P.F. Suthers and C.D. Maranas. 2010. OptForce: an optimization procedure for identifying all genetic manipulations leading to targeted overproductions. Plos Comp Biol. 6: e1000744.

Rark, J.M., T.Y. Kim and S.Y. Lee. 2009. Constraints-based genome-scale metabolic simulation for systems metabolic engineering. Biotechnol Adv. 27: 979–988.

Ricci, J.C.D. 1996. Influence of phosphenol pyruvate on the dynamic behavior of phosphofructokinase of *Escherichia coli*. J Theor Biol. 178: 145–150.

Rizk, M.L. and J.C. Liao. 2009. Ensemble modeling for aromatic production in *Escherichia coli*. PLoS One. 4: e6903.

Rizzi, M., M. Baltes, U. Theobald and M. Reuss. 1997. *In vivo* analysis of metabolic dynamic in *Saccharomyces cerevisiae*: II. Mathematical model. Biotechnol Bioeng. 55: 592–608.

Rocha, I., P. Maia, P. Evangelista, P. Vilaca, S. Soares, J.P. Pinto, J. Nielsen, K.R. Patil, E.C. Ferreira and M. Rocha. 2010. OptFlux: an open-source software platform for *in silico* metabolic engineering. BMC Syst Biol. 4: 45.

Saier, M.H. and T.M. Ramseier. 1996. The catabolite repressor/activator (Cra) protein of enteric bacteria. J Bacteriol. 178: 3411–3417.

Salimi, F., K. Zhuang and R. Mahadevan. 2010. Genome-scale metabolic modeling of a clostridial co-culture for consolidated bioprocessing. Biotechnol J. 5: 726–738.

Sauer, U. 2006. Metabolic networks in motion: $^{13}$C-based flux analysis. Mol Systems Biol. 2: 62.

Sauter, T. and E.D. Gilles. 2004. Modeling and experimental validation of the signal transduction via the *Escherichia coli* sucrose phosphotransferase system. J Biotechnol. 110: 181–199.

Savageau, M.A. 1970. Biochemical systems analysis. 3. Dynamic solutions using a power-law approximation. J Theor Biol. 26: 215–226.

Schuetz, R., L. Kuepfer and U. Sauer. 2007. Systematic evaluation of objective functions for predicting intracellular fluxes in *Escherichia coli*. Mol Syst Biol. 3: 119.

Schuetz, R., N. Zamboni, M. Zampieri, M. Heinemann and U. Sauer. 2012. Multidimensional optimality of microbial metabolism. Science. 336: 601–604.

Schuster, S., D.A. Fell and T. Dandekar. 2000. A general definition of metabolic pathways useful for systematic organization and analysis of complex metabolic networks. Nature Biotechnol. 18: 326–332.

Scott, M., C.W. Gunderson, E.M. Mateescu, Z. Zhang and T. Hwa. 2010. Interdependence of cell growth and gene expression: origins and consequences. Science. 330: 1099–1102.

Segrè, D., D. Vitkup and G.M. Church. 2002. Analysis of optimality in natural and perturbed metabolic networks. PNAS USA. 99: 15112–15117.

Selinger, D.W., M.A. Wright and G.M. Church. 2003. On the complete determination of biological systems. Trends Biotechnol. 21: 251–254.

Shalel-Levanon, S., K.-Y. San and G.N. Bennett. 2005. Effect of oxygen, and ArcA and FNR regulators on the expression of genes related to the electron transfer chain and the TCA cycle in *Escherichia coli*. Metab Eng. 7: 364–374.

Shimizu, K. 2004. Metabolic flux analysis based on $^{13}$C-labeling experiments and integration of the information with gene and protein expression patterns. Adv Biochem Eng Biotechnol. 91: 1–49.

Shimizu, K. 2009. Toward systematic metabolic engineering based on the analysis of metabolic regulation by the integration of different levels of information. Biochem Eng J. 46: 235–251.

Shimizu, K. 2013. Bacterial Cellular Metabolic Systems. Woodhead Publ Ltd., Oxford.

Shimizu, K. and Y. Matsuoka. 2015. Fundamentals of systems analysis and modeling of biosystems and metabolism. e-Book. Bentham Sci Publ Co. Ltd.

Shinto, H., Y. Tashiro, G. Kobayashi, T. Sekiguchi, T. Hanai, Y. Kuriya, M. Okamoto and K. Sonomoto. 2007. Kinetic modeling and sensitivity analysis of acetone-butanol-ethanol production. J Biotechnol. 131: 45–56.

Shinto, H., Y. Tashiro, G. Kobayashi, T. Sekiguchi, T. Hanai, Y. Kuriya, M. Okamoto and K. Sonomoto. 2008. Kinetic study of substrate dependency for higher butanol production in acetone-butanol-ethanol fermentation. Proc Biochem. 43: 1452–1461.

Smallbone, K., E. Simeonidis, D.S. Broomhead and D.B. Kell. 2007. Something from nothing-bridging the gap between constraint-based and kinetic modelling. FEBS J. 274: 5576–5585.

Smallbone, K., E. Simeonidis, N. Swainston and P. Mendes. 2010. Towards a genome-scale kinetic model of cellular metabolism. BMC Syst Biol. 4: 6.

Smallbone, K., H.L. Messiha, K.M. Carroll, C.L. Winder, N. Malys, W.B. Dunn, E. Murabito, N. Swainston, J.O. Dada, F. Khan, P. Pir, E. Simeonidis, I. Spasić, J. Wishart, D. Weichart, N.W. Hayes, D. Jameson, D.S. Broomhead, S.G. Oliver, S.J. Gaskell, J.E. McCarthy, N.W. Paton, H.V. Westerhoff, D.B. Kell and P. Mendes. 2013. A model of yeast glycolysis based on a consistent kinetic characterisation of all its enzymes. FEBS Lett. 587: 2832–2841.

Song, H.S., J.A. Morgan and D. Ramkrishna. 2009. Systematic development of hybrid cybernetic models: application to recombinant yeast co-consuming glucose and xylose. Biotechnol Bioeng. 103: 984–1002.

Song, S. and C. Park. 1997. Organization and regulation of the d-xylose operons in *Escherichia coli* K-12: XylR acts as a transcriptional activator. J Bacteriol. 179: 7025–7032.

Sorribas, A., B. Hernandez-Bermejo, E. Vilaprinyo and R. Alves. 2007. Cooperativity and saturation in biochemical networks: A saturable formalism using Taylor series approximations. Biotechnol Bioeng. 97: 1259–1277.

Stanford, N.J., T. Lubitz, K. Smallbone, E. Klipp, P. Mendes and W. Liebermeister. 2013. Systematic construction of kinetic models from genome-scale metabolic networks. PLoS One. 8: e79195.

Stelling, J. 2004. Mathematical models in microbial systems biology. Curr Opin Microbiol. 7: 513–518.

Sumiya, M., E.O. Davis, L.C. Packman, T.P. McDonald and P.J. Henderson. 1995. Molecular genetics of a receptor protein for d-xylose, encoded by the gene *xylF*, in *Escherichia coli*. Receptors Channels. 3: 117–128.

Tan, Y.K. and J.C. Liao. 2012. Metabolic ensemble modeling for strain engineers. Biotechnol J. 7: 343–353.

Teusink, B., J. Passarge, C.A. Reijenga, E. Esgalhado, C.C. van der Weijden, M. Schepper, M.C. Walsh, B.M. Bakker, K. van Dam, H.V. Westerhoff and J.L. Snoep. 2000. Can yeast glycolysis be understood in terms of *in vitro* kinetics of the constituent enzymes? Testing biochemistry. Eur J Biochem. 267: 5313–5329.

Theobald, U., W. Mailinger, M. Baltes, M. Rizzi and M. Reuss. 1997. *In vivo* analysis of metabolic dynamic in *Saccharomyces cerevisiae*: I. Experimental observations. Biotechnol Bioeng. 55: 305–316.

Tomita, M. 2001. Whole-cell simulation: a grand challenge of the 21st century. Trends in Biotech. 19: 205–210.

Toya, Y., K. Nakahigashi, M. Tomita and K. Shimizu. 2012. Metabolic regulation analysis of wild-type and *arcA* mutant *Escherichia coli* under nitrate conditions using different levels of omics data. Mol Biosyst. 8: 2593–2604.

Toya, Y., N. Ishii, K. Nakahigashi, M. Tomita and K. Shimizu. 2010. $^{13}$C-metabolic flux analysis for batch culture of *Escherichia coli* and its Pyk and Pgi gene knockout mutants based on mass isotopomer distribution of intracellular metabolites. Biotechnol Prog. 26: 975–992.

Unden, G., S. Becker, J. Bongaerts, G. Holighaus, J. Schirawski and S. Six. 1995. O$_2$-sensing and O$_2$-dependent gene regulation in facultatively anaerobic bacteria. Arch Microbiol. 164(2): 81–90.

Usuda, Y., Y. Nishio, S. Iwatani, S.J. Van Dien, A. Imaizumi, K. Shimbo, N. Kageyama, D. Iwahata, H. Miyano and K. Matsui. 2010. Dynamic modeling of *Escherichia coli* metabolic and regulatory systems for amino-acid production. J Biotechnol. 147: 17–30.

Valgepea, K., K. Adamberg, A. Seiman and R. Vilu. 2013. *Escherichia coli* achieves faster growth by increasing catalytic and translational rates of proteins. Mol Biosyst. 9: 2344–2358.

van Riel, N.A. 2006. Dynamic modelling and analysis of biochemical networks: mechanism-based models and model-based experiments. Brief Bioinform. 7: 364–374.

Varner, J. and D. Ramkrishna. 1999. Metabolic engineering from a cybernetic perspective. 1. Theoretical preliminaries. Biotechnol Prog. 15: 407–425.

Vasudevan, P. and M. Briggs. 2008. Biodiesel production–current state of the art and challenges. J Ind Microbiol Biotechnol. 35: 421–430.

Vemuri, G.N., E. Altman, D.P. Sangurdekar, A.B. Khodursky and M.A. Eiteman. 2006b. Overflow metabolism in *Escherichia coli* during steady-state growth: transcriptional regulation and effect of the redox ratio. Appl Environ Microbiol. 72: 3653–3661.

Vemuri, G.N., M.A. Eiteman and E. Altman. 2006a. Increased recombinant protein production in *Escherichia coli* strains with overexpressed water-forming NADH oxidase and a deleted ArcA regulatory protein. Biotechnol Bioeng. 94: 538–542.

Vinuselvi, P., M.K. Kim, S.K. Lee and C.-M. Ghim. 2012. Rewiring carbon catabolite repression for microbial cell factory. BMB Reports. 45(2): 59–70.

Voit, E., A.R. Neves and H. Santos. 2006. The intricate side of systems biology. PNAS USA. 103: 9452–9457.

Voit, E.O. 2013. Biochemical systems theory: A review. ISRN Biomathematics. 2013: 897658.

Wang, L. and V. Hatzimanikatis. 2006. Metabolic engineering under uncertainty-II: analysis of yeast metabolism. Metab Eng. 8: 142–159.

Westerhoff, H., J. McFadden and K.Shimizu. 2015. Private communication on UK-Japan collaboration project.

Wittman, C. 2007. Fluxome analysis using GC-MS. Microb Cell Fact. 6: 6.

Wolfe, A.J. 2005. The acetate switch. Microbiol Mol Biol Rev. 69: 12–50.

Wu, L., W.M. Wang, W.A. van Winden, W.M. van Gulik and J.J. Heijnen. 2004. A new framework for the estimation of control parameters in metabolic pathways using lin-log kinetics. Eur J Biochem. 271: 3348–3359.

Xu, Y.F., D. Amador-Noguez, M.L. Reaves, X.J. Feng and J.D. Rabinowitz. 2012. Ultrasensitive regulation of anapleurosis via allosteric activation of PEP carboxylase. Nat Chem Biol. 8: 562–568.

Yang, C., Q. Hua, T. Baba, H. Mori and K. Shimizu. 2003. Analysis of *Escherichia coli* an aprelotic metabolism and its regulation mechanisms from the metabolic responses to altered dilution rates and phosphoenol pyruvate carboxykinase knockout. Biotechnol Bioeng. 84: 129–144.

Yang, L., W.R. Cluett and R. Mahadevan. 2011. EMILiO: a fast algorithm for genome-scale strain design. Metab Eng. 13: 272–281.

Yao, R., H. Kurata and K. Shimizu. 2013. Effect of *cra* gene mutation on the metabolism of *Escherichia coli* for a mixture of multiple carbon sources. Adv Biosci Biotechnol. 4: 477–486.

Yao, R. and K. Shimizu. 2013. Recent progress in metabolic engineering for the production of biofuels and biochemicals from renewable sources with particular emphasis on catabolite regulation and its modulation. Proc Biochem. 48: 1409–1417.

Yao, R., Y. Hirose, D. Sarkar, K. Nakahigashi, Q. Ye and K. Shimizu. 2011. Catabolic regulation analysis of *Escherichia coli* and its *crp*, *mlc*, *mgsA*, *pgi* and *ptsG* mutants. Microb Cell Fact. 10: 67.

Yazdani, S.S. and R. Gonzalez. 2007. Anaerobic fermentation of glycerol: a path to economic viability for the biofuels industry. Curr Opin Biotechnol. 18(3): 213–9.

You, C., H. Okano, S. Hui, Z. Zhang, M. Kim, C.W. Gunderson, Y.-P. Wang, P. Lenz, D. Yan and T. Hwa. 2013. Coordination of bacterial proteome with metabolism by cyclic AMP signaling. Nature. 500: 301–306.

Young, I.G. and B.J. Wallace. 1976. Mutations affecting the reduced nicotinamide adenine dinucleotide dehydrogenase complex of *Escherichia coli*. Biochim. Biophys. Acta 449(3): 376–85.

Young, J.D. 2005. A system-level mathematical description of metabolic regulation combining aspects of elementary mode analysis with cybernetic control laws. PhD thesis, Purdue University.

Young, J.D., K.L. Henne, J.A. Morgan, A.E. Konopka and D. Ramkrishna. 2008. Integrating cybernetic modeling with pathway analysis provides a dynamic, systems-level description of metabolic control. Biotechnol Bioeng. 100: 542–559.

Zhuang, K., M. Izallalen, P. Mouser, H. Richter, C. Risso, R. Mahadevan and D.R. Lovley. 2011. Genome-scale dynamic modeling of the competition between Rhodoferax and Geobacter in anoxic subsurface environments. Isme Journal. 5: 305–316.

Zomorrodi, A.R., P.F. Suthers, S. Ranganathan and C.D. Maranas. 2012. Mathematical optimization applications in metabolic networks. Metab Eng. 14: 672–686.

# Appendix A

Appendix A Effect of global regulators on the metabolic pathway gene expression

| Global regulator | Metabolic pathway genes |
|---|---|
| Cra | + : *aceBAK, acn , cydB, fbp, icdA, pckA, pgk, ppsA,*<br>− : *acnB, adhE, cyoA, eda, edd, fbaB, fruBKA, glgCAP, glk, gpmM, maeB, marR, mpl, ndh, pdhR-aceEF-lpd, pfkA, ppc, prpB, ptsH, pykFA, sdhCDAB-sucABCD, setB, sgrST, tpiA, zwf* |
| Crp (cAMP-Crp) | + : *aceEF,acnAB,acs, focA, fumA, gltA, lpdA, malT, manXYZ, mdh, mlc, ppsA, pckA, pdhR, pflB, pgk, ptsG,* ptsH, *sdhCDAB, sucABCD, ugpABCEQ, fruBKA, mtlAD, tktAB, talAB, ndh, cyoABCD, nuoA~N, atpIBFHAGDC*<br>− : *cyaA, lpdA, rpoS* |
| ArcA/B | + : *cydAB, focA, pflB, ackA*<br>− : *aceBAK, aceEF, acnAB, cyoABCDE, fumAC, gltA, icdA, lpdA, mdh, nuoABCDEFGHIJKLMN, pdhR, sdhCDAB, sodA, sucABCD* |
| Mlc | − : *crr, manXYZ, malT, ptsG, ptsHI* |
| PdhR | − : *aceEF, lpdA (ndh,cyoABCDE)* |
| CsrA | + : *eno,pfkA,pgi,pykF,tpiA*<br>− : *fbp,glgC, glgA, glgB,pgm, ppsA,pckA,* |
| Fur | − : *entABCDEF, talB,sodA* |
| RpoS | + : *acnA, acs, ada, appAR, appB, argH, aroM, dps, bolA, fbaB, fumC, gabP, gadA, gadB, katE, katG, ldcC, narY, nuv, pfkB, osmE, osmY, poxB, sodC,talA, tktB,ugpE,C, xthA, yhgY,*<br>− : *ompF* |
| SoxR/S | + : *acnA, cat, fumC, fur, sodA, sox, zwf* |
| OxyR | + : *ahpC, ahpF, katG* |
| PhoR/B | + : *phoBR, phoA-psiF, asr, pstSCAB-phoU*<br>− : *phoH, phnCHN, ugpA, argP* |
| Fnr | + : *acs, focA, frdABCD, pflB, yfiD*<br>− : *acnA, cyoABCDE, cydAB, fumA, fnr, icdA, ndh, nuoABCEFGHIJKLMN, sdhCDAB, sucABCD* |
| Fis | + : *adhE, hyaA-F, nuoA-N, ndh, lpdA*<br>− : *fis, crp, cya, acnB, acs, fumB, aldB, pflB* |
| Ihf | + : *fis, fnr, aceBAK, glcDEFG, hycA-I, gltA, focA, pflB*<br>− : *pta, ackA, sucABCD, ndh, nuoA* |

# Index

# About the Author

Prof. **Kazuyuki Shimizu** is based at the Kyushu Institute of Technology (KIT) and the Institute of Advanced Biosciences (IAB), Keio University, Japan. Following a BS and an MS in Chemical Engineering at Nagoya University, and a PhD in Chemical Engineering at Northwestern University, USA, he started his career in 1981 as a research associate at Nagoya University, Japan, and was promoted to associate professor in 1990. He then moved to the KIT as a professor in 1991. In 2000, he became an adjunct professor at IAB, Keio University, Japan. He has long been involved in research into $^{13}$C-metabolic flux analysis ($^{13}$C-MFA), and studies on modeling and systems biology with the aim of gaining insight into the basic principles governing living cell systems. He recognizes the importance of uncovering the metabolic regulation mechanism of a cell system, based on both experimental (wet) and computational (dry) approaches. He has also organized a UK–Japan collaboration project on microbial systems biology toward developing virtual microbes by integration of the different levels of hierarchical omics information, such as transcriptomics, proteomics, metabolomics, and fluxomics data.

Printed and bound by CPI Group (UK) Ltd, Croydon, CR0 4YY

01/11/2024

01782624-0015